Nonisotopic
Immunoassay

Nonisotopic Immunoassay

Edited by

T. T. Ngo
University of California, Irvine
Irvine, California

Plenum Press • New York and London

Library of Congress Cataloging in Publication Data

Nonisotopic immunoassay / edited by T. T. Ngo.
    p.   cm.
  Bibliography: p.
  Includes index.

    ISBN-13: 978-1-4684-5468-0   e-ISBN-13: 978-1-4684-5466-6
    DOI: 10.1007/978-1-4684-5466-6

    1. Immunoassay. I. Ngo, T. T. (That Tjien), 1944–
  QP519.9.I42N66   1988
  616.07′56—dc19   87-36105
                                                                        CIP

Cover illustration: Immune complex formed by
enzyme-antigen conjugate and antibodies.

© 1988 Plenum Press, New York
A Division of Plenum Publishing Corporation
233 Spring Street, New York, N.Y. 10013

Softcover reprint of the hardcover 1st edition 1988

Dedicated to the memory of my parents

Mr. Ngo Nam Seng 伍南生 (1913–1984)

and

Mrs. Thjin Hie Foeng 陳喜鳳 (1913–1986)

PREFACE

The basis of all immunoassays is the interaction of antibodies with antigens. The most widely used immunoassay technique is radioimmunoassay (RIA) which was first developed by Yalow and Berson in 1959. The principle of RIA is elegantly simple. It utilizes a competitve binding reaction between analytes and a radio-labeled analog of the analytes (the tracer) for anti-analyte antibodies. In addition to its exquisite specificity, extraordinary sensitivity, good accuracy and precision, ease and rapidity of assay and simplicity of assay development, the applicability of RIA to a wide variety of substances has made it one of the most powerful and versatile analytical methods of the 20th century and beyond. Millions of RIA's are being performed annually on clinical, biological and environmental samples in licensed laboratories. In order to expand the use of RIA beyond the confines of these laboratories to areas like physician's offices, patients' homes, economically less developed countries, agricultural fields, large scale and continuing screening tests for infectious diseases, it has become necessary to develop non-isotopic labels. Indeed the last fifteen years have seen the development of a great number of ingenious non-isotopic labels in immunoassay so that a whole new industry capitalizing on the potential market for non-isotopic immunoassays has appeared.

It is the purpose of this volume to present in depth, state-of-the-art reviews on techniques used in non-isotopic immunoassays. Topics covered include: (1) Enzyme-labeled immunoassay; (2) Luminescene immunoassay; (3) Immunoassay at liquid-solid interface; (4) Membrane immunoassay and (5) "Particle"-mediated immunoassay.

This volume complements two recently published books on immunoassay: I. "Enzyme-Mediated Immunoassay", edited by T.T. Ngo and H.M. Lenhoff, 1985 and II. "Electrochemical Sensors in Immunological Analysis", edited by T.T. Ngo, 1987. Both volumes were published by Plenum Press.

I am very grateful to my wife, <u>Ping Ying</u> (梓影), for her understanding and patience. The support of my brothers (達銳，達澐，達敏), my sisters (碧友，碧玉) and my uncle and aunt, Dr. and Mrs. Richard Chen-Tan (陳順華) during my early university years are thankfully acknowledged. I thank Professor Howard Lenhoff for encouragement and useful discussions.

October, 1987
T.T. Ngo

伍達穎

CONTENTS

SECTION I

ENZYME-LABELED IMMUNOASSAY

RECENT DEVELOPMENTS IN ENZYME IMMUNOASSAYS

Barry J. Gould and Vincent Marks

Department of Biochemistry
University of Surrey
Guildford, Surrey GU2 5XH, U.K.

INTRODUCTION

Since the initial studies by Yalow and Berson in 1959 radioimmunoassay (RIA) has developed into an extremely versatile analytical technique. It has been used particularly in clinical laboratories to quantitate a wide variety of compounds. The specificity is dependent upon the antibodies and the sensitivity on both the antibodies and the radiolabel.

However, radioisotopes do have their drawbacks. The preparation of the radiolabelled antigen involves real risks, which are cumulative. Even when these are prepared commercially the product shows batch-to-batch variation and, generally has a half-life limited to two months. Their toxic nature necessitates the application of strict regulatory control and their measurement requires the use of specialised, sophisticated and hence expensive equipment. The necessity for a separation step has prevented the development of simple automation.

These disadvantages of radiolabels encouraged the search for alternatives. Engvall and Perlmann (1971) and Van Weemen and Schuurs (1971) independently described the use of enzyme-labelled reagents. These are now well-established and have allowed the development of a diverse range of assay protocols. The most general advantage of enzyme-labels compared to radiolabels is their improved shelf-life. Enzyme labels may often be stored under sterile conditions for more than a year at 4°C or at room temperature when freeze-dried. Radiation hazards are obviously avoided.

PRINCIPLES OF ENZYME IMMUNOASSAY

We use the term enzyme immunoassay (EIA) to describe all immunoassays where the activity of an enzyme is measured to determine the analyte. EIA is generally subdivided into heterogeneous (separation required) assays and homogeneous (separation free) assays. In heterogeneous assays the activity of the enzyme is not affected by the presence of analyte, whereas in homogeneous assays it is because of alteration in enzyme activity by it that a separation step is not required.   There are two basic types of heterogeneous EIA, the classical "competitive" EIA, and the immunoenzymometric or sandwich assay. The principles of these various systems are considered below together with their advantages and disadvantages.

Competitive EIA

This type of assay is analogous to "traditional" RIA. The assay contains three components namely: limited and constant quantities of the enzyme-labelled antigen and its antibody together with either (i) a variable amount of antigen for calibration purposes or (ii) an unknown amount of antigen in the test sample (Fig. 1).

Since the number of antibody binding sites available is less than the total number of unlabelled and labelled antigen molecules present there is competition between the two molecular species. The greater the quantity of unlabelled antigen present, the lower will be the amount of enzyme-labelled antigen combining with antibody. After separation of the unbound and antibody-bound enzyme-labelled antigen the enzyme activity in the bound fraction is measured. The use of varying amounts of antigen is necessary for the production of a calibration curve from which the unknown amounts of antigen can be determined.

This type of EIA requires a separation step. The simplest way of doing this is to have the antibody adsorbed on to a solid phase which allows for easy washing of the antibody-bound enzyme-labelled antigen complex prior to determination of enzyme activity. This particular type of EIA has no inherent advantage over RIA unless substitution of the radiolabel is desired either on grounds of safety or

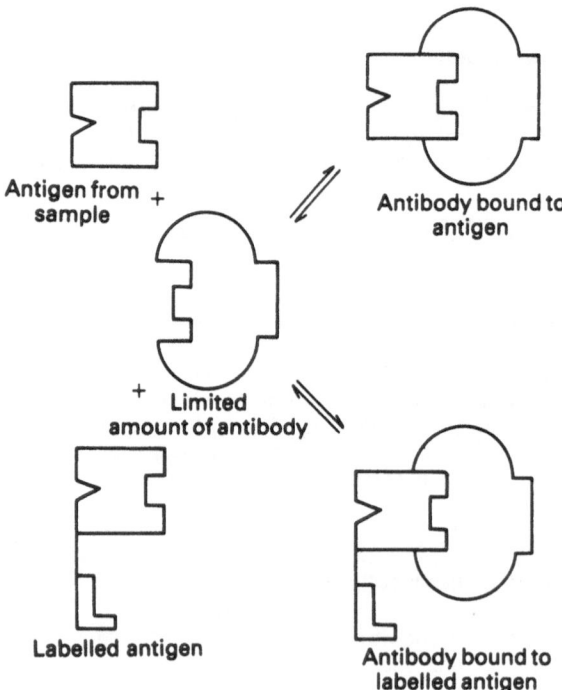

Antigen from sample +

Antibody bound to antigen

+ Limited amount of antibody

Labelled antigen

Antibody bound to labelled antigen

Fig. 1. Classical competitive EIA for antigen. All components are mixed and incubated. Isolation of antibody-bound enzyme-labelled-antigen is necessary before measurement of enzyme activity. (Reprinted by permission, from Blake and Gould, 1984).

availability of appropriate equipment. In this case it is the method of choice for small haptens if a sensitive assay (e.g. picogram quantities of hormones) is required. Such assays have a sensitivity comparable to RIA. It is, however, important to use a different chemical link between the hapten-enzyme conjugate and the hapten-protein conjugate used to raise the antisera. Otherwise the phenomenon referred to as "bridge recognition" can occur resulting in considerable loss of sensitivity (Van Weemen and Schuurs, 1975).

An alternative assay for antigens depends on competition between a fixed amount of solid-phase antigen and antigen in the sample for a limited amount of specific enzyme-labelled antibody (Fig. 2). These assays are also called inhibition assays because the amount of enzyme that will be bound to the solid phase decreases with increasing concentration of the antigen in the sample to be measured. This type of assay can also be used to measure haptens although the attachment of hapten to the solid phase is not always easy. The sensitivity is comparable with that achieved with RIA or better.

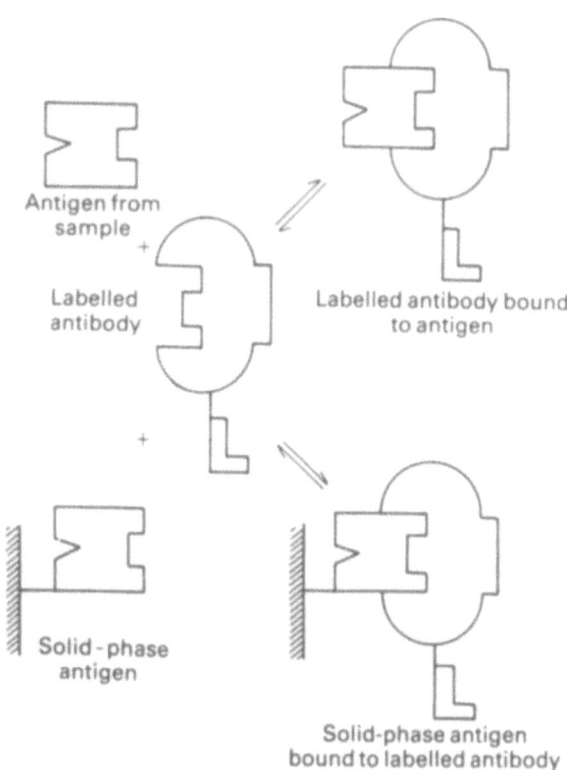

Antigen from
sample

+

Labelled
antibody

Labelled antibody bound
to antigen

+

Solid-phase
antigen

Solid-phase antigen
bound to labelled antibody

Fig. 2.   Competitive EIA for antigen using solid-phase antigen. Enzyme-labelled antibody reacts specifically with antigen in the sample and is then added to excess of solid-phase antigen. After washing, the enzyme label still attached to the solid phase is measured. (Reprinted by permission, from Blake and Gould, 1984).

Competitive heterogeneous assays are generally simpler to develop than the immunoenzymometric assays described below. However they are also potentially less sensitive and more likely to be affected by the presence of sample interferants during the antigenic reaction. Matrix effects are reduced by the use of excess reagents (Fritz and Bunker, 1982).

Immunoenzymometric Assays

Although competitive assays were used in early EIA there has subsequently been a progressive increase in the use of non-competitive or immunoenzymometric assays. These assays, which are also referred to as sandwich assays, rely on the presence of excess reagent. In consequence competition is no longer involved and the final amount of bound enzyme label that is measured is proportional to the concentration of analyte. These assays all use solid-phase reagent, either an antibody or antigen, which allows for easy separation and washing between stages.

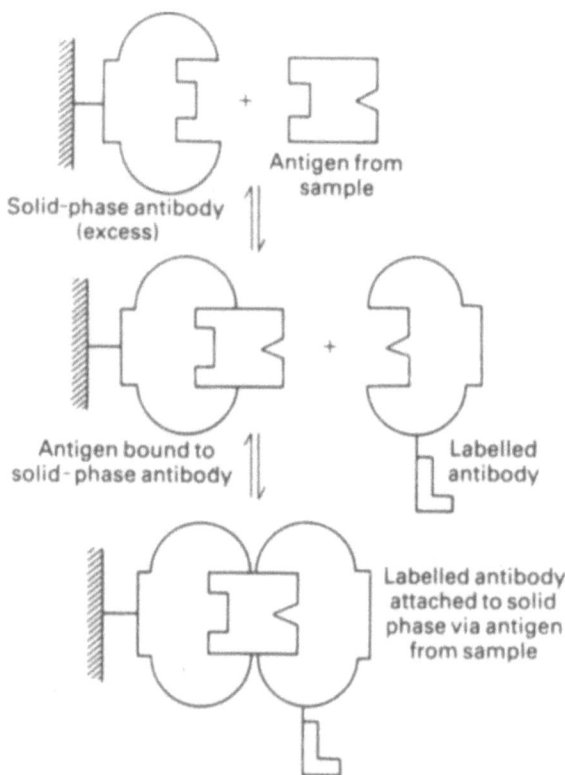

Fig. 3. Immunoenzymometric assay for antigen. Antigen in the sample is mixed with excess solid-phase antibody. After washing of the solid phase, enzyme-labelled antibody which is specific for another site on the antigen is added. The enzyme label which remains bound after washing is measured. (Reprinted with permission, from Blake and Gould, 1984).

Because of this, assays of this kind are frequently called "Enzyme-linked immunosorbent assays" (ELISA) but this term can also include the competitive EIA for which the principles are quite different.

A simple immunoenzymometric assay for antigen is illustrated in Fig. 3. The excess of solid-phase antibody is first incubated with the sample containing the antigen. After washing, the enzyme-labelled antibody is added to the mixture and after a further washing stage the enzyme activity retained is measured.

These methods have been widely used for detecting infectious diseases. The use of stable reagents, often combined with simple visual assessment of test results of as many as 100 samples at a time, have made them very suitable for tests on blood bank specimens and for epidemiological surveys. The tests have frequently been used in the field without the need for complex equipment.

In the sandwich type of assay the analyte must be large enough to have at least two reactive antigenic sites and this normally excludes its use for haptens. With polyclonal antibodies the same antiserum can generally be used to coat the solid phase surface and to produce the enzyme-labelled antibody. An alternative method is to use two different monoclonal antibodies. This eliminates one incubation

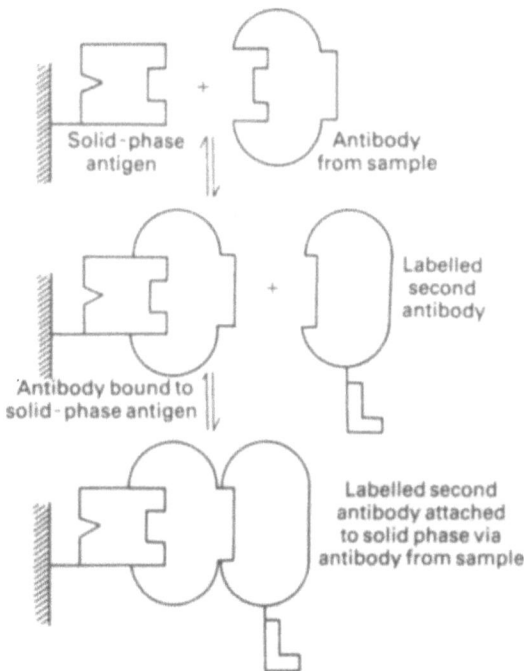

Fig. 4. Immunoenzymometric assay for antibody. Antibody in the sample is mixed with excess solid-phase antigen. After washing of the solid phase, enzyme-labelled second antibody is added. Bound enzyme activity is measured after washing. (Reprinted by permission, from Blake and Gould, 1984).

step since both the sample and enzyme-labelled monoclonal antibody can be added to the solid-phase (Sevier et al., 1981) which has already been coated with the second monoclonal antibody. These "one-step" solid-phase assays have high specificity, since two antigenic sites are involved, and can be both quick and convenient. By using a large excess of the monoclonal antibody reagents it is possible to minimise the saturation effect called the "high dose hook effect" (Ryall et al., 1982). In some cases linear calibration curves have been obtained which may enable a single calibration sample to be used (Shimizu et al., 1985).

Single, "one-step" assays for human choriogonadotrophin in urine, with a sensitivity of 4μg/L have been described which can be completed in 5 min. The method uses an immunoconcentration device which reduces the diffusion distances and makes rapid assay possible (Valkirs and Barton, 1985).

Another important application of immunoenzymometric assays is in the detection and measurement of antibodies (Fig. 4). Here, an immobilised or solid-phase antigen is required to bind initially with antibody in the sample. After washing, the bound antibody is detected and quantified by the addition of an enzyme-labelled species-specific anti-immunoglobin and measuring the residual enzyme activity after washing away surplus reagent.

False positives may occur in the detection of IgM class-specific antibodies in the presence of IgG antibodies and IgM anti-IgG rheumatoid factor; false negatives may be found when competition occurs between IgM and high levels of IgG antibody. These problems can be overcome if a solid-phase anti-IgM is used to capture IgM. Under these circumstances the specific IgM can be detected by addition of the appropriate antigen which is then visualised by means of an enzyme-labelled antibody

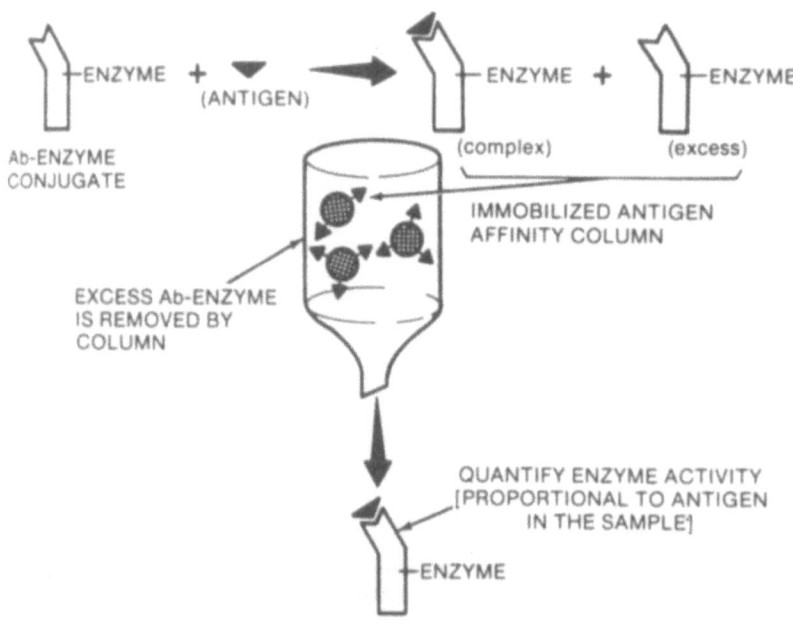

Fig. 5.   Schematic representation of assay procedure of affinity-column-mediated immunoenzymometric assay (ACMIA). (Reprinted by permission, from Freytag, 1985).

specific for the antigen. This type of assay requires an additional step to the simple immunoenzymometric technique described earlier. However it does achieve improved assay specificity.

Several variants of the immunoenzymometric method include additional steps for convenience. These allow the use of general, commercially produced, enzyme-labelled reagents in the final step and thus make it unnecessary for the user-laboratory to make their own enzyme-labelled reagents. These reagents include enzyme-labelled species specific antisera (Engvall and Perlmann, 1972; Voller et al., 1978), and enzyme-labelled protein A (Schuurs and Weeman, 1980). It is, however, essential to avoid non-specific binding of these labels if reliable results are to be obtained.

A novel immunometric technique, termed affinity column mediated immunometric assay (ACMIA), is illustrated in Fig.5. It combines the advantages of the non-competitive type of sandwich assay with the requirement for only one antigenic site (Freytag et al., 1984). It can therefore be used for haptens as well as larger antigens. For maximum sensitivity, purified monovalent antibodies are conjugated to enzyme on a one to one basis using procedures described by Ishikawa and co-workers (Imagawa et al., 1984). Excess enzyme labelled monovalent antibody is incubated with sample containing antigen and the mixture passed through an affinity column containing immobilised antigen in large excess. The column retains excess antibody but permits the enzyme-labelled antibody-antigen complex to pass through and be collected. Its enzyme activity is measured and is directly proportional to the original concentration of antigen in the sample. An assay for human IgG using this principle has been described and enabled 0.1fmol of the protein to be measured (Freytag, 1985).

The rapid growth of immunometric assays has a sound theoretical basis (Ekins, 1985). Among the potential advantages of this type of assay design are that,

(i)   the affinity of the antibody is less important than in competitive assays, since more is used, and the laws of mass action favour association of the antigen-antibody complex

(ii)   the precision of antibody pipetting is now generally of minor consequence,

(iii) a shorter incubation time can be used.

These advantages should lead to more rapid, rugged and reliable immunoassays being introduced with improved sensitivity and specificity. The main disadvantage is one of increased costs, due to the larger amounts of antibodies used but since these ordinarily constitute only a small proportion of the total cost of the assay it should not prove insuperable.

To take advantage of the full potential of immunometric assays, low levels of non-specific binding must be attained. It is also necessary to use labels with higher specific activities than that of commonly used isotopes. Utilisation of enzymes with high turnover numbers is one approach but for maximum sensitivity they will be enhanced by the use of improved detection methods. These might employ fluorescent, chemi- or bioluminescent products in combination with appropriate measurement instrumentation.

The Solid Phase

The use of a solid phase permits easy washing. It is usually plastic in the form of a tube, bead, peg or, most commonly, a microtitre plate with 96 wells. Many instruments are now available that allow all 96 wells of a microtitre plate to be washed and read automatically in about a minute. Attention to detail, such as mixing of the well contents before measurement of their absorbance improves duplication (Kemp et al., 1985).

Normally the surface of the plastic is coated with the antigen or antibody by passive absorption. Although this method has been criticised, methods using covalent binding have not, so far, given improved results. It is probably more important to use plates from a reliable source than to spend effort improving attachment. For reproducible results a consistent coating technique is essential. It is customary to try to reduce subsequent non-specific binding by using blocking agents such as Tween and/or bovine serum albumin in the reaction mixture.

Those interested in further information about such practical matters are referred to recent reviews (Voller and Bidwell, 1980; Blake and Gould, 1984; Sauer et al., 1985; Standefer, 1985).

Homogeneous EIA

The so-called enzyme multiplied immunoassay technique (EMIT; Syva, Maidenhead, U.K.) is probably still the most widely used homogeneous enzyme immunoassay system. The principle upon which it is based is shown in Fig. 6. It is a competitive type assay and was developed by Rubenstein et al. (1972). It depends on a change in specific enzyme activity when enzyme-labelled antigen binds to antibody. The assay mixture contains limited amounts of specific antibody and hapten-

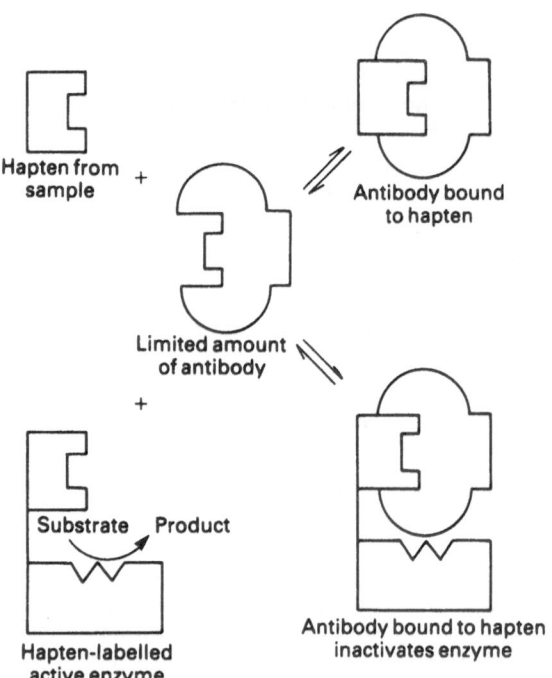

Fig. 6. Principle of the EMIT system of homogeneous EIA. Hapten-labelled enzyme is enzymically active but combination with hapten-specific antibody causes marked inhibition of enzyme activity. The relative amounts of free hapten and hapten-labelled enzyme determine the final measured enzyme activity. (Reprinted by permission, from Blake and Gould, 1984).

10

labelled enzyme. In the absence of unlabelled hapten, enzyme activity of the label is virtually abolished by the binding of antibody. However in the presence of increasing amounts of free hapten, such as required to produce the standard curve or in the test samples, enzyme activity is increased.

This type of assay has been employed for the measurement of haptens being commonly used for the assay of therapeutic drugs and simple hormones as well as for the detection of drugs of abuse. The two most commonly used enzymes are glucose-6-phosphate dehydrogenase and malate dehydrogenase. Hapten conjugates of these enzymes are usually inhibited by up to about 80% by an excess of antibody directed against the hapten probably as a result of conformational changes. Paradoxically, when thyroxine is conjugated to malate dehydrogenase, there is a substantial inhibition of enzyme activity which can be partially reversed by binding to thyroxine antibodies. This phenomenon has been used to develop a homogeneous EIA for thyroxine (Ullman et al., 1979).

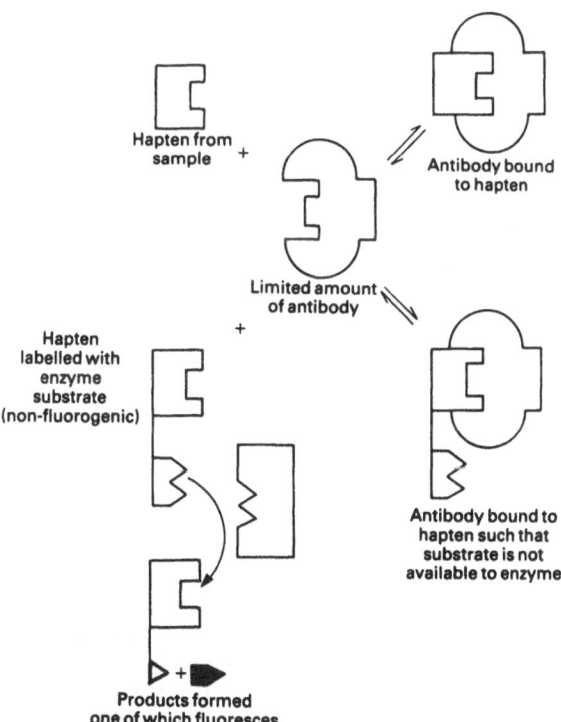

Fig. 7. Principle of "Substrate-Labelled Fluorescent Immunoassay" (SLFIA). Only a limited amount of substrate-labelled hapten can be used and therefore only a small amount of substrate is available for reaction, which necessitates the formation of a fluorescent product to obtain adequate sensitivity. Either the rate of formation of fluorescent product or the total amount formed can be measured. (Reprinted by permission, from Blake and Gould, 1984).

EMIT was originally applied only to small molecules but has been extended to the measurement of proteins (Gibbons et al., 1980). The enzyme is a protein-β-galactosidase conjugate and the normal substrate, o-nitrophenylgalactoside is converted into a macromolecular form by attachment to a dextran carrier. On addition of protein-specific IgG enzyme activity of the label is inhibited by up to 95% due to steric exclusion of the substrate. Using this chromogenic substrate the sensitivity limit for serum proteins is about 10 μg/ml but greater sensitivity can be achieved by changing to an umbelliferone galactoside substrate and measuring the fluorescent product formed within 30 sec (Armenta et al., 1985). Individual sera sometimes cause gross interference with enzymic activity. This takes the form of inhibition of enzymic action and can be almost completely eliminated by the addition of excess inactivated enzyme. It is possible that the interference is caused by the presence of antibodies to β-galactosidase in the serum sample.

The major advantages of the EMIT system are that it is simple to perform, easily automated and, if required, can produce results rapidly at nanomole concentrations. The commercially available "kit" has been adapted for use with continuous-flow systems (Nolan et al., 1981), automated reaction-rate analysers (Galen and Forman, 1977) and centrifugal analysers (Lasky et al., 1977). More recently a very cheap automated method for IgG using flow injection analysis has been reported. The assay involves peroxidase and is monitored using a fluorometric end point (Kelly and Christian, 1982).

Other Homogeneous Enzyme-Based Immunoassays

A potential disadvantage of the EMIT system is that binding of the hapten-enzyme to antibody complex does not necessarily produce a reversible effect on its enzymic activity. This requires the ligand to be bound close to the active site of the enzyme or to amino acids involved in essential conformational changes. At present the likelihood of such reactions occurring cannot be predicted for virtually any enzyme. However, other molecules such as substrates, enzyme cofactors and inhibitors are known to interact directly with the active sites of enzymes. If these small molecules are conjugated to hapten, their binding to antibody will more certainly interfere with their normal combination with enzyme and hence its enzyme activity. Since the enzyme itself is not involved in the conjugation reactions a wider range of reaction conditions can be used than might otherwise be possible.

Enzyme substrates have been used as immunoassay labels. However since the hapten and hapten-labelled substrate are involved in competition, only a small amount of hapten-labelled substrate can be used, thereby preventing the normal amplification of the signal usually associated with enzymic catalysis. In order to achieve adequate sensitivity at micromolar concentrations it is customary to use a fluorescent product (Fig. 7). This method has been given the name "substrate-labelled fluorescent immunoassay" (SLFIA) and the enzyme normally used is β-galactosidase. The common substrate employed is an umbelliferyl-galactoside derivative of the hapten (Burd et al., 1977) which does not fluoresce until hydrolysed by the β-galactosidase. The rate of umbelliferone production and its final amount are directly proportional to the amount of free hapten in the sample and either can be used for quantitation (Davis and Marks, 1983). This method has been adapted to the measurement of proteins (Ngo et al., 1981).

The principles of prosthetic group-labelled immunoassay (PGLIA) or apoenzyme reactivation immunoassay system (ARIS) are illustrated in Fig. 8. Here the flavin N^6-(6-aminohexyl) adenine dinucleotide (aminohexyl-FAD) is conjugated to the substance to be measured. This labelled ligand, when not combined with antibody, is still able to act as the prosthetic group for the glucose oxidase apoenzyme thereby restoring its enzymic activity. In this competitive assay the amount of active holoenzyme formed is directly proportional to the amount of drug-prosthetic group complex displaced from antibody binding sites by drug in the test sample. The glucose-oxidase reaction is monitored by coupling to peroxidase and measurement of a coloured product formed by oxidation of a colourless chromogen. An assay for theophylline at concentrations as low as 2 μg/ml has been developed (Morris et al., 1981). It has also been applied to the measurement of protein molecules.

Another type of homogeneous assay using combination of a prosthetic group with its apoenzyme to give an active enzyme was described by Ngo and Lenhoff (1983). These authors conjugated the holoenzyme glucose-oxidase to the ligand to be measured - dinitrophenyl groups in this instance - and then removed its prosthetic group, FAD, by acid denaturation. The DNP-labelled apoglucose oxidase can still recombine with FAD to give an active enzyme. However, if the apoenzyme is first allowed to combine with antibodies to DNP, active holoenzyme is not produced on incubation with FAD, presumably because the antibody-apoenzyme complex has restricted mobility.

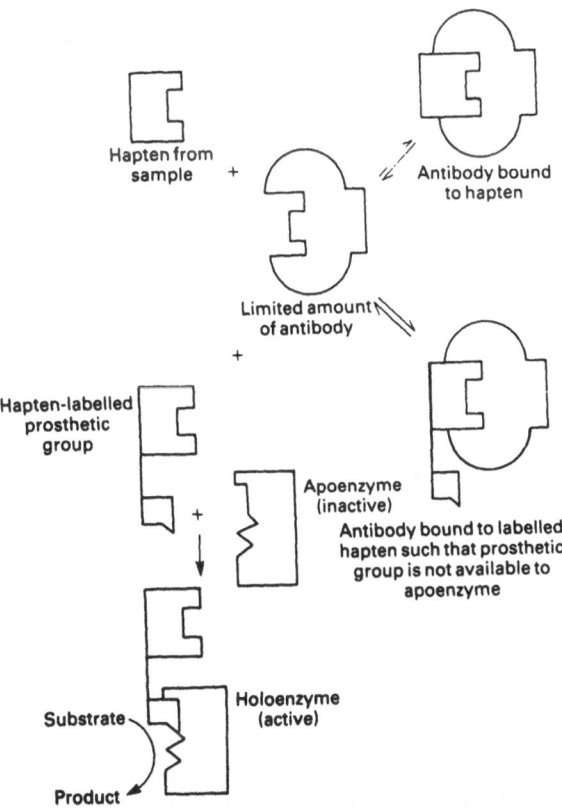

Fig. 8.   Principle of "Prosthetic Group-Labelled-Immunoassay" (PGLIA). The apo-enzyme only becomes active when it has been converted to holoenzyme by binding to the prosthetic group, which forms part of the active site of the enzyme. (Reprinted by permission, from Blake and Gould, 1984).

13

In the system described by Ngo and Lenhoff (1983) incubation of labelled apoenzyme with increasing amounts of DNP and a fixed amount of anti-DNP produce increasing amounts of active enzyme on incubation with excess FAD. The authors described this type of assay as an antibody-induced conformational restriction enzyme immunoassay (AICREIA) the principles of which are illustrated in Fig 9

Gonneli et al. (1981) described a somewhat similar assay for thyroxine. They made ribonuclease S-peptide conjugates with thyroxine which readily combined with S-protein to give active ribonuclease. The assay could be used to determine thyroxine at plasma concentrations of 0.1μM.

The use of enzyme inhibitors in homogeneous enzyme immunoassay systems is only feasible if the inhibitor has a high affinity for the enzyme ($< 10^{-8}$ M) whose activity serves as the label. However inhibitors do have the advantage over substrate based assays in that the intrinsic amplifying effect of the enzyme is still used.

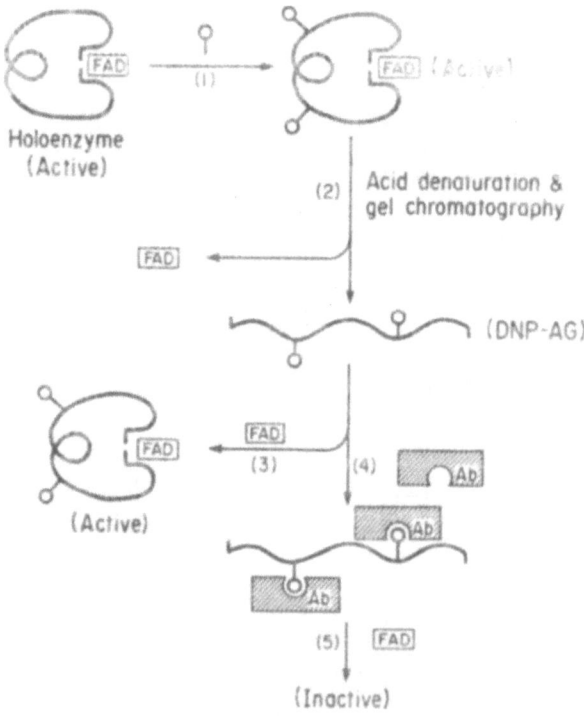

Fig. 9. Principle of an "Antibody-Induced Conformational Restriction EIA" (AICREIA). The holoenzyme used is glucose oxidase and ♀ represents DNP, 2,4-dinitrophenyl groups; DNP-AG is denatured apo-glucose oxidase labelled with DNP groups; Ab is antibody to DNP-groups. (Reprinted by permission, from Ngo and Lenhoff, 1983).

The basic principles of inhibitor based assays are shown in Fig. 10. In the homogeneous assay illustrated enzyme activity decreases with increasing concentration of hapten in the test sample. The first example of this type of immunoassay used anti-enzyme antibodies as the inhibitor (Ngo and Lenhoff, 1980). Subsequent examples have used phosphonate-based inhibitors of acetylcholinesterase (Finley et al., 1980), and methotrexate derivatives of dihydrofolate reductase (Place et al., 1983). Bacquet and Twumasi (1984) made use of the very tight binding between the protein avidin and the biotin prosthetic group of the enzyme in their assay system; an avidin-diphenylhydantoin conjugate was employed as the inhibitor of pyruvate carboxylase in the presence of a fixed amount of anti-diphenylhydantoin antibodies. Enzyme activity decreased as the concentration of free diphenylhydantoin in the test sample increased.

An entirely different approach to the use of enzymes as labels in immunoassay systems is in an "enzyme channelling immunoassay". This requires the presence of

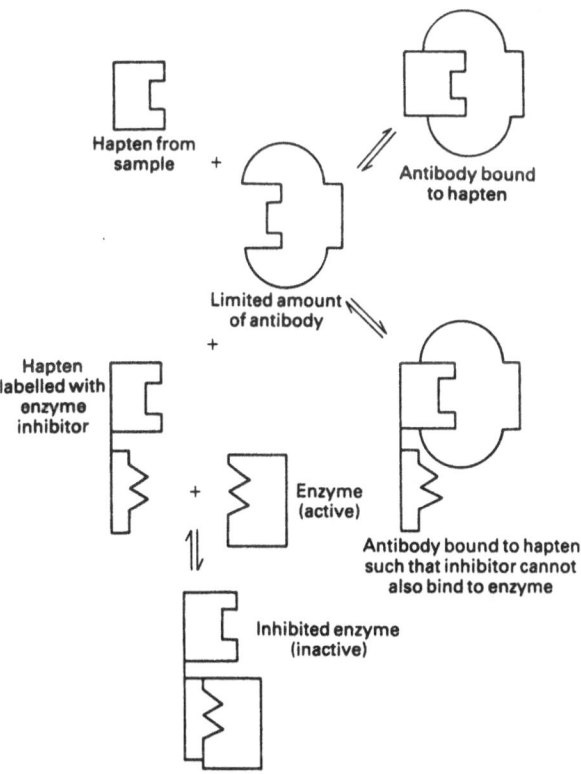

Fig. 10. Principle of an inhibitor-based homogeneous EIA. The inhibitor-labelled hapten combines with enzyme which loses activity. In principle the reaction between enzyme and inhibitor could be reversible (as shown) or irreversible. (Reprinted by permission, from Blake and Gould, 1984).

two enzymes that catalyse consecutive reactions. In the original version an antigen-glucose-6-phosphate dehydrogenase conjugate competed with free antigen for a limited amount of antibody which was itself coimmobilised on agarose beads with the other enzyme, hexokinase. Consequently, the amount of glucose-6-phosphate dehydrogenase, that became attached to the solid-phase decreased as the concentration of free antigen increased. The product of the first enzyme hexokinase i.e. glucose-6-phosphate was the substrate for the second enzyme i.e. glucose-6-phosphate dehydrogenase and the close proximity of the two enzymes - as a result of the immune reaction - resulted in an accelerated overall rate of formation of the final product. To minimise background reaction a scavenger enzyme is required free in solution, to remove product of the first enzyme that has escaped reaction with the bound antibody. Using this system Litman et al. (1980) measured proteins at concentrations as low as 5 ng/ml.

A more efficient model assay system uses glucose oxidase and peroxidase, both conjugated to antibodies. On addition of antigen the two enzymes are brought close together and subsequently precipitated with anti-glucose oxidase. Using this system, with catalase as the scavenger enzyme, an assay for polyribose phosphate with detection limit of 5 pg was developed by Gibbons et al. (1984). The method can also be used for the detection of microorganisms.

Yet a further approach to homogeneous enzyme immunoassay uses stable, unilamellar liposomes containing entrapped enzyme (Litchfield et al., 1984). The liposomes are readily lysed by hapten-cytolytic-agent conjugates which are, themselves, inactivated on combination with hapten specific antibody. Thus there is an increasing amount of liposome lysis, in the presence of increasing concentrations of test hapten, which is accompanied by release of enzyme. Assays based on this principle are sensitive to nanomolar concentrations of analyte. A similar liposome preparation may also be useful in heterogeneous immunoassays.

One of the methods of amplifying the physical signal produced by the labels employed in homogeneous enzyme immunoassays is by use of naturally occurring cascade systems. Blake et al. (1984) demonstrated such an application when they established a model assay for biotin which used part of the blood coagulation system. Here successive zymogen activation stages occur with final formation of thrombin which is itself assayed colorimetrically.

CHOICE OF ENZYME LABEL

Enzymes are used as labels in immunoassay systems because of their amplifying effect. A single molecule of enzyme typically converts $10^3 - 10^4$ molecules of substrate into product per minute, although with some enzymes values as high as $10^6 - 10^7$ are obtained. Ideally enzymes should have the properties listed in Table 1.

The initial velocity (v) of an enzyme-catalysed reaction,

$$E + S \rightleftharpoons ES \longrightarrow E + P$$

enzyme   substrate   enzyme-substrate   product
intermediate

can be described by the Michaelis-Menten equation,

$$v = \frac{V.S}{K_m + S}$$

where S is the substrate concentration, and V (maximal velocity) and K_m (Michaelis constant) are two constants of the enzyme. From this equation it follows that enzymes with high V and low K_m will provide the most sensitive labels. The other parameters listed in Table 1 are self explanatory.

Table 1. Properties of an ideal enzyme label

(i) High enzyme activity (V) at low substrate concentrations (low K_m).

(ii) Enzyme stable at the pH required for good antibody - antigen binding (generally neutral pH) and active at this pH, particularly for homogeneous EIA.

(iii) Cheap, accurate and sensitive assay method, preferably with a spectrophotometric end-point.

(iv) Presence of reactive groups through which enzymes can be covalently linked to antibody, antigen or hapten with minimum loss of enzyme or immune activities.

(v) Enzyme-labelled conjugates which are stable under routine storage and assay conditions.

(vi) Availability of soluble, purified enzyme at low cost.

(vii) Absence of health hazards attributable to enzyme, substrates and cofactors.

(viii) Absence of enzyme activity and factors affecting enzyme activity from the test fluid (particularly for homogeneous EIA).

From Blake and Gould, 1984.

The enzymes which are most commonly used as labels in immunoassay systems are listed in Table 2 together with some of their properties. The enzymes most commonly used in heterogeneous EIA are peroxidase, alkaline phosphatase and β-galactosidase. Peroxidase is currently the cheapest and has 10-15% of its mass as carbohydrate which is useful in some conjugation reactions.

A wide range of chromogens, yielding colours ranging from blue with o-toluidine and green with 2,2'-azino-di(3-ethylbenzothiazoline-6-sulphonate) (ABTS) to red orange with o-phenylenediamine (OPD) and black of 5-aminosalicylic acid (5-AS), are available. Alkaline phosphatase obtained from calf intestine is relatively expensive, but generally gives very stable conjugates. It is, moreover, easily detected colorimetrically or by one of the more sensitive methods such as fluorescence, which may be necessary for ultrasensitive methods.

β-Galactosidase is absent from human plasma and therefore suitable for use in both homogeneous and heterogeneous EIA. The availability of multiple (12-20) free thiol groups is useful in certain conjugation reactions. By using colorimetric or fluorogenic substrates rapid and very sensitive assays can be developed.

Other enzymes commonly used in homogenous EIA are glucose-6-phosphate dehydrogenase and malate dehydrogenase; lysozyme is now only of historical interest having been the first enzyme used for this purpose. Glucose-6-phosphate dehydrogenase used in homogenous EIA derives from <u>Leuconostoc mesenteroides</u> and has high activity with NAD^+. This avoids the possibility of interference by glucose-6-phosphate dehydrogenase from red cells which is specific for $NADP^+$. Malate dehydrogenase has a greater specific activity than glucose-6-phosphate dehydrogenase but its use has been limited to assays on urine since it can be present in plasma. Both dehydrogenases are measured by the production of NADH using absorbance at 340 nm or fluorescence.

(a) Heterogeneous EIA

Table 2. Enzymes commonly used as labels for EIA

Enzyme	Source	pH optimum	Specific activity at 37°C/units mg^{-1}	K_m	Relative molecular mass	End-point
Acetylcholinesterase	Electrophorus electricus	7-8	1,400	90 μM	54,000	pH electrode or spectrophotometric
Adenosine deaminase	Calf intestine	7.5-9	200	60 μM		Ammonia gas-sensing electrode
Alkaline phosphatase	Calf intestine	8-10	1,000	0.2 mM for PNPP	100,000	Spectrophotometric or fluorimetric
Catalase	Calf liver	6-8	40,000	*	250,000	UV absorbance or thermometry
β-Galactosidase	Escherichia coli	6-8	600	1 mM	540,000	Spectrophotometric or fluorimetric
Glucose oxidase	Aspergillus niger	4-7	200	$K_m^{Glu}=33$ mM	186,000	H_2O_2 combines with chromogen
Peroxidase	Horseradish	5-7	4,500	$K_m^{H_2O_2}=0.2$ mM	40,000	H_2O_2 combines with chromogen
Urease	Jack beans	6.5-7.5	10,000	10 mM	483,000	Ammonia reacts with chromogen or gas-sensing electrode

(continued)

Table 2. (continued)

(b) Homogeneous EIA

Enzyme	Source	pH optimum	Specific activity at 37°C/units mg^{-1}	K_m	Relative molecular mass	End-point
Acetylcholinesterase	Electrophorus electricus	7-8	1,400	90 μM	54,000	pH electrode or spectrophotometric
β-Galactosidase	Escherichia coli	6-8	600	1 mM	540,000	Spectrophotometric or fluorimetric
Glucose 6-phosphate dehydrogenase	Leuconostoc mesenteroides	7.8	400	K_m^{G6P} =0.1 mM K_m^{NAD}=0.15 mM	104,000	Formation of NADH by UV absorbance or fluorimetric
Lysozyme	Chicken egg white	4.5-5.5	-		14,500	Formation of cell wall fragments (ΔA at 450 nm)
Malate dehydrogenase	Pig heart	8.5-9.5	1,000	K_m^{Mal}=0.3 mM K_m^{NAD}=0.1 mM	70,000	Formation of NADH

* K_m depends on substrate.

19

SIMULTANEOUS ASSAY OF TWO ANALYTES

Dual assays have been developed using RIA, generally based on the use of the two iodine isotopes ^{125}I and ^{131}I, but the usefulness of these assays is reduced by the short half-life of ^{131}I and the non-ideal distinction between the gamma emissions of the two isotopes.

A potential advantage of using enzymic endpoints for the simultaneous assay of two or even more analytes is the comparative ease of measuring two or more chromogens (or fluorophores) in the same reaction mixture. This approach (Blake et al., 1982) has yet to be thoroughly exploited however. The advantages of simultaneous multiple-analyte measurement compared to normal single-analyte methods include savings in reagent costs and sample volumes and a reduction in the overall assay time. Such assays would, however, only be useful where two analytes were normally measured in the same specimen, when it might be more accurate to have the two values in an identical specimen. Examples of such situations are fairly common e.g. the thyroid hormones; cortisol and ACTH; drugs and their metabolite; analytes that are expressed as a ratio of each other.

In the first use of enzymes for simultaneous assay of two analytes, alkaline phosphatase was conjugated to thyroxine and detected at 540 nm by the formation of phenolphthalein from its monophosphate. β-Galactosidase was similarly conjugated to triiodothyronine and detected at 420 nm by the formation of o-nitrophenol from the β-galactoside substrate. Competitive heterogeneous assays were performed for each analyte using double antibody precipitation. The use of these two enzymes allowed each reaction to be measured in the presence of the other without cross-detection. The results obtained for the two thyroid hormones determined in this way in the same tube agreed well with the results of individual RIA measurements (Blake et al., 1982).

Another example used a dual homogenous assay, based on the SLFIA system, to measure the plasma concentration of two drugs, phenytoin and phenobarbital, which are frequently prescribed in combination. The two enzymes used were β-galactosidase and phosphodiesterase I and the substrate labels a β-galactosylcoumarin derivative of phenobarbital and a 4-methylcoumarin phosphodiester derivative of phenytoin respectively. It was found necessary in this case to stagger the readings, however, because both fluorophore products gave maximum fluorescent emissions at the same wavelength. Nevertheless, the assays were of comparable sensitivity to single drug immunoassays (Dean et al., 1983).

INVOLVEMENT OF ENZYMES IN ULTRASENSITIVE TECHNIQUES

The full potential of immunometric techniques can only be exploited if large signal amplifications are achieved. Many such amplification systems involve the use of enzyme-catalysed reactions, although there are other techniques which are discussed in the following chapters. It is, however, necessary to ensure that the sensitivity and specificity of the assay is not limited by other factors such as the purity, binding efficiency, and the amount of enzyme-labelled antibodies used (Ishikawa et al., 1982).

The simplest approach to signal amplification is to use enzyme reactions which involve the formation of fluorescent products. The most promising enzymes are β-galactosidase (Labrousse et al., 1982), glucose-6-phosphate dehydrogenase (Shah et al., 1984), and alkaline phosphatase (Yolken and Stopa, 1982). Using the latter enzyme, for example, a 100-fold increase in sensitivity can be achieved. This can be enhanced still further by increasing the incubation time and using a greater surface area for binding antigen. Simple steps such as these have produced assays capable of detecting 6 ag/ml of mouse IgG (Shalev et al., 1980).

A different approach is to produce amplification by enzymic cycling reactions. A simple system based on this principle uses alkaline phosphatase as the primary enzyme in an immunoenzymometric method to convert $NADP^+$ to NAD^+. This is then used in a secondary amplification system involving alcohol dehydrogenase and ethanol. The NADH produced is transformed back into NAD^+ utilising the enzyme diaphorase with simultaneous formation of a formazan dye. This dye accumulates rapidly since each molecule of NAD^+ formed in the "primary" reaction is recycled many times. Typically a 100-fold increase in absorbance can be obtained for the same amount of enzyme activity in the primary system. The primary and secondary systems may take place at the same time in the original vessel or separately if this is more convenient (Johannsson et al., 1985; Self, 1985).

An alternative to the use of enzymes alone as labels in immunoassay systems is to use them in combination with another physico-chemical method of detection. One such is referred to as "ultrasensitive enzymatic radioimmunoassay" (USERIA). The radioisotope used in this procedure is tritium which allows the development of sensitive assays avoiding the use of the potentially more hazardous radioactive iodine. In one example, immobilised alkaline phosphatase was used as the enzyme label and the substrate was tritiated adenosine monophosphate. DEAE-Sephadex columns were used to separate the product, tritiated adenosine, from the substrate. When this assay was applied to the measurement of cholera toxin it became 100-1000 times more sensitive than the regular RIA or ELISA techniques. It was estimated that as few as 600 molecules of the toxin could be detected by the USERIA method (Harris et al., 1979).

Another sensitive method of detection is to use enzymes as labels in conjunction with chemiluminescent substrates. Horseradish peroxidase catalyses the oxidation of luminol and, if this reaction takes place in the presence of an enhancer, such as luciferin, an intense light is produced which can be measured in a luminometer (Whitehead et al., 1983). If qualitative assays alone are satisfactory these can be produced by using high speed photographic film (Thorpe et al., 1984). A similar system using glucose-6-phosphate dehydrogenase and bacterial bioluminescence reagents has been used to monitor urinary human chorionic gonadotrophin (Kohen et al., 1984) and may be suitable as a dipstick.

EVALUATION OF THE CURRENT STATUS OF ENZYME LABELS

The main benefits of the use of enzymes as labels in immunoassay have been the development of a wide range of homogenous immunoassays and the enthusiastic search for improved immunoenzymometric assays. Enzyme immunoassays have been used in many biological disciplines. These include their use clinically for measuring serum proteins, hormones and drugs, screening for antibodies and the diagnosis of infectious diseases and, in the food and veterinary industries for the investigation of viral antigens food contaminants and the detection of antibiotics, herbicides and other noxious compounds. Some areas where there will undoubtedly be an increased use is in the measurement of blood clotting factors (Amiral et al., 1984), and enzymes, particularly specific isoenzymes, as proteins rather than as enzyme activities (Kato et al., 1981; Jackson et al., 1984). They will also find application for estimating changes in various glycoproteins in disease (Kohn et al., 1983), where lectins replace antibodies as the specific binders for the various carbohydrate moieties.

Homogeneous assays were originally restricted by their relative lack of sensitivity to the assay of haptens in high i.e. micromolar concentrations. However progress has been made towards measuring larger antigens in lower concentrations. Many of the newer assays can be automated to give results within a few minutes of sample collection. There has always been the worry that homogeneous assays might be adversely affected by interfering substances present in biological fluids. In general this has not been a major problem and is minimised by increasing assay sensitivity so that smaller samples are analysed. Extraneous interference is not a problem with immunometric assays because of the many washing stages that are required for their proper execution.

Enzyme-based immunoassays have been used to develop simple dipstick tests suitable for use at the bedside or even by the patient themselves at home. A recent example is the application of a heterogeneous assay for the early detection of pregnancy based on the detection of human chorionic gonadotrophin (HCG) in blood or urine. The test involves two seperate stages each lasting 15 minutes separated by a rinse. The dipstick incorporates its own blank for detection of non-specific binding (Norman et al., 1985).

Dipstick tests have also been developed based on the SLFIA (Greenquist et al., 1981; Waller et al., 1983), PGLIA (Tyhach et al., 1981), and the "enzyme-channelling immunoassay" (Litman et al., 1983). These allow what would previously have been technically complicated tests to be performed with good precision and limited equipment simply and quickly. A simple reflectance photometer is generally required to measure the endpoint though for semiquantitation visual inspection may be adequate. Ultrasound, which enhances mass transport and hence antigen-antibody binding, can accelerate the rate of development and hence completion of dipstick tests (Chen et al., 1984).

A variant of the "enzyme-channelling immunoassay" has been developed by Zuk et al. (1985) who utilised a dry paper test strip containing excess immobilized antibody. The strip is placed in an aliquot of liquid enzyme reagent containing glucose oxidase and a peroxidase-antigen conjugate to which sample containing the antigen to be measured, e.g. theophylline, has been added. Immunocapillary migration occurs: the distance travelled by the peroxidase-antigen conjugate increases with increasing amounts of sample antigen which competes with it for binding to the immobilised antibody. The glucose oxidase is however, evenly distributed throughout the strip. After "development" the test strip is immersed in reagent containing glucose and a chromogen. The function of the glucose oxidase is to generate hydrogen peroxide which serves as substrate for the peroxidase which is

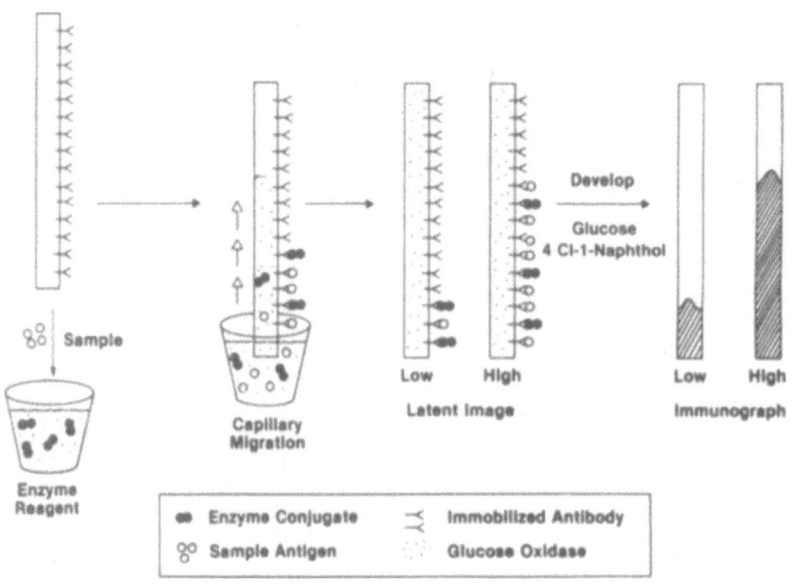

Fig. 11. Schematic representation of the enzyme immunochromatographic test strip assay. (Reprinted by permission, from Zuk et al., 1985).

22

visualised by formation of an insoluble blue product. The total test time is about 15 minutes and results in a colour zone or rocket with a sharp front. The length of the colour strip is measured and is directly proportional to the sample antigen concentration. This technique has been termed enzyme immunochromatography and is illustrated in Fig. 11.

One of the problems associated with the use of enzyme labels is that the measurement of enzyme activity itself may be imprecise compared to that of radioactivity. This difficulty is minimised by using reliable automated equipment for performing homogeneous assays where rate measurements are required and by including the relevant standards. The problem is partly overcome for immunoenzymometric assays by use of the batchwise approach to assays. In this way variations in temperature, in particular, will affect all samples in the batch, including the standards (O'Sullivan, 1984).

The preparation of enzyme labels can lead to production of a range of polymers, only some of which contain both enzyme and antibody unless special precautions are adopted. This means that for the most sensitive assays it is necessary to purify the enzyme labels, preferably by affinity methods. It is often possible, to employ commercially prepared general enzyme labels to set up an immunoenzymometric assay. This involves an extra step which though taking extra time does mean that no special chemical synthesis is needed. However for the most sensitive assays commercially available general purpose reagents usually require further purification. It is worth noting that in general, homogeneous assays nearly always use commercially prepared reagents; immunoenzymometric assays on the other hand are simpler for individuals to set up themselves, although they may require preparation of specific reagents.

This individual approach to immunoenzymometric assay has encouraged the development of three types of assay. Simple qualitative assays using microtitre plates where development of a coloured product has been of great benefit in large screening surveys. Second, quantitative assays replacing RIA when the need to avoid radiation hazards is paramount or the equipment necessary for radioactive counting is not available. Thirdly, there are the very sensitive assays which are only possible using immunometric techniques. These require extremely reliable precise methodology coupled with large signal amplifications.

REFERENCES

Amiral, J., Adalbert, B., and Adam, M., 1984, Application of enzyme immunoassay to coagulation testing, Clin. Chem., 30: 1512-1516.

Armenta, R., Tarnowski, T., Gibbons, I., and Ullman, E.F., 1985, Improved sensitivity in homogeneous enzyme immunoassays using a fluorgenic macromolecular substrate: an assay for serum ferritin, Anal. Biochem., 146: 211-219.

Bacquet, C., and Twumasi, D.Y., 1984, A homogeneous enzyme immunoassay with avidin-ligand conjugate as the enzyme-modulator, Anal. Biochem., 136: 487-490.

Blake, C., Al-Bassam, M.N., Gould, B.J., Marks, V., Bridges, J.W., and Riley, C., 1982, Simultaneous enzyme immunoassay of two thyroid hormones, Clin. Chem., 28: 1469-1473.

Blake, C., and Gould, B.J., 1984, Use of enzymes in immunoassay techniques: A review, Analyst, 109: 533-547.

Blake, D.A., Skarstedt, M.T., Shultz, J.L., and Wilson, D.P., 1984, Zymogen activation: a new system for homogeneous ligand-binding assay, Clin. Chem., 30: 1452-1456.

Burd, J.F., Wong, R.C., Feeney, J.E., Carrico, R.J., and Boguslaski, R.C., 1977, Homogeneous reactant-labeled immunoassay for therapeutic drugs exemplified by genitamicin determinations in human serum, Clin. Chem., 23: 1402-1408.

Chen, R., Weng, L., Sizto, N.C., Osorio, B., Hsu, C.-J., Rodgers, R., and Litman, D.J., 1984, Ultrasound-accelerated immunoassay, as exemplified by enzyme immunoassay of choriogonadotropin, Clin. Chem., 30: 1446-1451.

Davis, S.J., and Marks, V., 1983, Measurement of serum theophylline concentrations using a modified Ames TDA^R system, Therap. Drug Mon., 5:479-484.

Dean, K.J., Thompson, S.G., Burd, J.F., and Buckler, R.T., 1983, Simultaneous determination of phenytoin and phenobarbital in serum or plasma by substrate-labelled fluorescent immunoassay, Clin. Chem., 29: 1051-1056.

Ekins, R.P., 1985, Current concepts and future developments, in: "Alternative immunoassays", W.P. Collins, ed., John Wiley and Sons, Chichester, New York, Brisbane, Toronto, Singapore.

Engvall, E., and Perlmann, P., 1971, Enzyme-linked immunosorbent assay (ELISA). Quantitative assay for immunoglobulin G, Immunochem., 8: 871-874.

Engvall, E., and Perlmann, P., 1971, Enzyme-linked immunosorbent assay (ELISA). III. Quantitation of specific antibodies by enzyme-labeled antimmunoglobulin in antigen-coated tubes, J. Immunol. Meth., 10: 161-170.

Finley, P.R., Williams, R.J., and Lichti, D.A., 1980, Evaluation of a new homogeneous enzyme inhibitor immunoassay of serum thyroxine with use of a bichromatic analyzer, Clin. Chem., 26: 1723-1726.

Freytag, J.W., 1985, Affinity column mediated immunoenzymometric assays, in: "Enzyme-mediated immunoassay", T.T. Ngo and H.M. Lenhoff, ed., Plenum Press, New York and London.

Freytag, J.W., Dickinson, J.C., and Tseng, S.Y., 1984, A high sensitivity affinity-column-mediated immunometric assay as exemplified by digoxin, Clin. Chem., 30: 417-420.

Fritz, T.J. and Bunker, D.M., 1982, Hyperlipidemia interference in radioimmunoassays, Clin. Chem., 28: 2325.

Galen, R.S. and Forman, D., 1977, Enzyme immunoassay of serum thyroxine with the 'autochemist' multichannel analyser, Clin. Chem., 23: 119-121.

Gibbons, I., Skold, C., Rowley, G.L., and Ullman, E.F., 1980, Homogeneous enzyme immunoassay for proteins employing β-galactosidase, Anal. Biochem., 102: 167-170.

Gonnelli, M., Gabellieri, G., Montagnoli, G. and Felicioli, R., 1981, Complementing S-peptide as modulator in enzyme immunoassay, Biochem. Biophys. Res. Commun., 102: 917-923.

Greenquist, A.C., Walter, B., and Li, T.M., 1981, Homogeneous fluorescent immunoasdsay with dry reagents, Clin. Chem., 27: 1614-1617.

Harris, C.C., Yolken, R.H., Krokan, H., and Hsu, I.C., 1979, Ultrasensitive enzymatic radioimmunoassay: application to detection of cholera toxin and rotovirus, Proc. Natl. Acad. Sci. USA, 76: 5336-5339.

Imagawa, M., Hashida, S., Ishikawa, E., and Freytag, J.W., 1984, Preparation of a monomeric 2,4-dinitrophenyl Fab'-β-galactosidase conjugate for immunoenzymometric assays, J. Biochem., 96: 1727-1735.

Ishikawa, E., Imagawa, M., Yoshitake, S., Niitsu, Y., Urushizaki, I., Inada, M., Imura, H., Kanazawa, R., Tachibana, S., Nakazawa, N., and Ogawa, H., 1982, Major factors limiting sensitivity of sandwich enzyme immunoassay for ferritin, immunoglobulin E, and thyroid-stimulating hormone, Ann. Clin. Biochem., 19: 379-384.

Jackson, A.P., Siddle, K., and Thompson, R.J., 1984, Two site monoclonal antibody assays for human heart and brain-type creatine kinase, Clin. Chem., 30: 1157-1162.

Johannsson, A., Stanley, C.J., and Self, C.H., 1985, A fast highly sensitive colorimetric enzyme immunoassay system demonstrating benefits of enzyme amplification in clinical chemistry, Clin. Chim. Acta, 148: 119-124.

Kato, K., Suzuki, F., and Umeda, Y., 1981, Highly sensitive immunoassays for three forms of rat brain enolase, J. Neurochem., 36: 793-797.

Kelly, T.A., and Christian, G.D., 1982, Capillary flow injection analysis for enzyme assay with fluorometric detection, Anal. Chem., 54: 1444-1445.

Kemp, M., Husky, S., and Jensenius, J.C., 1985, Enzyme immunoassay plates should be shaken before absorbance is measured, Clin. Chem., 31: 1090-1091.

Kohen, F., Bayer, E.A., Wilchek, M., Barnard, G., Kim, J.B., Collins, W.P., Beheshti, I., Richardson, A., and McCapra, F., 1984, The development of luminescence-based immunoassays for haptens and peptide hormones, in: "Analytical applications of bioluminescence and chemiluminescence", L.J. Kricka, ed., Academic Press, New York.

Kohn, J., Raymond, J., Voller, A., and Turp, P., 1983, Rapid methods for the demonstration of sugar residues in tissue extracts, fluids and lectins in plant extracts, Lectins, 3: 405-414.

Labrousse, H., Guesdon, J.-L., Ragimbeau, J., and Avremeas, S., 1982, Miniaturization of β-galactosidase immunoassays using chromogenic and fluorogenic substrates, J. Immunol. Meth., 48: 133-147.

Lasky, F.D., Ahuja, K.K., and Karmen, A., 1977, Enzyme immunoassays with the miniature centrifugal fast analyser, Clin. Chem., 23: 1444-1448.

Litchfield, W.J., Freytag, J.W., and Ademich, M., 1984, Highly sensitive immunoassays based on use of liposomes without complement, Clin. Chem., 30: 1441-1445.

Litman, D.J., Hanlon, T.M., and Ullman, E.F., 1980, Enzyme channeling immunoassay. A new homogeneous enzyme immunoassay technique, Anal. Biochem., 136: 223-229.

Litman, D.J., Lee, R.H., Jeong, H.J., Tom, H.K., Stiso, S.N., Sizto, N.C., and Ullman, E.F., 1983, An internally referenced test strip immunoassay for morphine, Clin. Chem., 29: 1598-1603.

Morris, D.L., Ellis, P.B., Carrico, R.J., Yeager, F.M., Schroeder, H.R., Albarella, J.P., and Boguslaski, R.C., 1981, Flavin adenine dinucleotide as a label in homogeneous colorimetric immunoassays, Anal. Chem., 53: 658-665.

Ngo, T.T., Carrico, R.J., Boguslaski, R.C., and Burd, J.F., 1981, Homogeneous substrate-labeled fluorescent immunoassay for IgG in human serum, J. Immunol. Meth., 42: 93-103.

Ngo, T.T., and Lenhoff, H.M., 1980, Enzyme modulators as tools for the development of homogeneous enzyme immunoassays, FEBS Lett., 116: 285-288.

Ngo, T.T., and Lenhoff, H.M., 1983, Antibody-induced conformational restriction as basis for new separation-free enzyme immunoassay, Biochem. Biophys. Res. Commun., 114: 1097-1103.

Nolan, J.P., DiBenedetto, G. and Tarsa, N.J., 1981, Continuous-flow enzyme immunoassay for thyroxine in serum, Clin. Chem., 27: 738-741.

Norman, R.J., Lowings, C., and Chard, T., 1985, Dipstick method for human chorionic gonadotropin suitable for emergency use on whole blood and other fluids, Lancet, (i): 19-20.

O'Sullivan, M.J., 1984, Enzyme Immunoassay, in: "Practical Immunoassay: the state of the art", W.R. Butt, ed., Marcel Dekker Inc., New York and Basel.

Place, M.A., Carrico, R.J., Yeager, F.M., Albarella, J.P., and Boguslaski, R.C., 1983, A colorimetric immunoassay based on enzyme inhibitor method, J. Immunol. Method, 61: 209-216.

Rowley, G.L., Rubenstein, K.E., Huisjen, J., and Ullman, E.F., 1975, Mechanism by which antibodies inhibit hapten-malate dehydrogenase conjugates, J. Biol. Chem., 250: 3759-3766.

Rubenstein, K.E., Schneider, R.S., and Ullman, E.F., 1972, "Homogeneous" enzyme immunoassay. New immunochemical technique, Biochem. Biophys. Res. Commun., 47: 846-851.

Ryall, R.G., Story, C.J., and Turner, D.R., 1982, Reappraisal of the causes of the "hook effect" in two-site immunoradiometric assays, Anal. Biochem., 127: 308-315.

Sauer, M.J., Foulkes, J.A., and Morris, B.A., 1985, Principles of Immunoassay, in: "Immunoassays in food analysis", B.A. Morris and M.N. Clifford ed., Elsevier Applied Science Publishers, London and New York.

Schuurs, A.H.W. and Van Weemen, B.K., 1980, Enzyme-immunoassay: a powerful analytical tool, J. Immunoassay, 1: 229-249.

Self, C.H., 1985, Enzyme amplification - a general method applied to provide an immunoassisted assay for placental alkaline phosphatase, J. Immun. Meth., 76: 389-393.

Sevier, E.D., David, G.S., Martinis, J., Desmond, W.J., Bartholomew, R.M., and Wang, R., 1981, Monoclonal antibodies in clinical immunology, Clin. Chem., 27: 1797-1806.

Shah, H., Saranko, A.-M., Härkönen, M., and Adlercreutz, H., 1984, Direct solid-phase fluoroenzymeimmunoassay of 5β-pregnane-3α, 20α-diol-3α-glucuronide in urine, Clin. Chem., 30: 185-187.

Shalev, A., Greenberg, A.H., and McAlpine, P.J., 1980, Detection of attograms of antigen by a high senstivity enzyme-linked immunosorbent assay (HS-ELISA) using a fluorogenic substrate, J. Immunol. Meth., 38: 125-139.

Shimizu, S.Y., Kabakoff, D.S., and Sevier, E.D., 1985, Monoclonal antibodies in immunoenzymetric assays, in: "Enzyme-mediated immunoassay", T.T. Ngo and H.M. Lenhoff, ed., Plenum Press, New York and London.

Standefer, J.C., 1985, Separation-required (heterogeneous) enzyme immunoassay for haptens and antigens, in: "Enzyme-mediated immunoassay", T.T. Ngo and H.M. Lenhoff ed., Plenum Press, New York and London.

Thorpe, G.H.G., Whitehead, T.P., Penn, R., and Kricka, L.J., 1984, Photographic monitoring of enhanced luminescent immunoassays, Clin. Chem., 30: 806-807.

Tyhach, R.J., Rupchock, P.A., Pendergrass, J.H., Skjold, A.C., Smith, P.J., Johnson, R.D., Albarella, J.P., and Profitt, J.A., 1981, Adaptation of prosthetic-group-label homogeneous immunoassay to reagent-strip format, Clin. Chem., 27: 1499-1504.

Ullman, E.F., Yoshida, R.A., Blakemore, J.I., Maggio, E., and Leute, R., 1979, Mechanism of inhibition of malate dehydrogenase by thyroxine derivatives and reactivation by antibodies, Biochem. Biophys. Acta, 567: 66-74.

Van Weemen, B.K., and Schuurs, A.H.W.M., 1971, Immunoassay using antigen enzyme conjugates, FEBS Letters, 15: 232-236.

Van Weemen, B.K., and Schuurs, A.H.W.M., 1975, The influence of heterologous combinations of antiserum and enzyme-labelled estrogen on the characteristics of estrogen enzyme-immunoassays, Immunochem., 12: 667-670.

Valkirs, G.E. and Barton, R., 1985, Immunoconcentration - a new format for solid-phase immunoassays, Clin. Chem., 31: 1427-1431.

Voller, A., Bartlett, A., and Bidwell, D.E., 1978, Enzyme immunoassay with special reference to ELISA techniques, J. Clin. Path., 31: 507-520.

Voller, A., and Bidwell, D.E., 1980, "The enzyme linked immunosorbent assay, Vol.2. A review of recent developments", Microsystems, Guernsey.

Walter, B., Greenquist, A.C., and Howard, W.E., 1983, Solid-phase reagent strips for the detection of therapeutic drugs in serum by substrate-labelled fluorescent immunoassay, Anal. Chem., 55: 873-878.

Whitehead, T.P., Thorpe, G.H.G., Carter, T.J.N., Groncutt, C., and Kricka, L.J., 1983, Enhanced luminescence procedure for sensitive determination of peroxidase-labelled conjugates in immunoassay, Nature, 305: 158-159.

Yalow, R.S., and Berson, S.A., 1959, Assay of plasma insulin in human subjects by immunological methods, Nature, 184: 1648-1649.

Yolken, R.H., and Stopa, P.J., 1979, Enzyme-linked fluorescence assay: ultrasensitive solid-phase assay for detection of human rotavirus, J. Clin. Microbiology, 10: 317-321.

Zuk, R.F., Ginsberg, V.K., Houts, T., Rabbie, J., Merrick, H., Ullman, E.F., Fisher, M.M., Sizto, C.C., Stiso, S.N., and Litman, D.J., 1985, Enzyme immunochromatography - a quantitative immunoassay requiring no instrumentation, Clin. Chem., 31: 1144-1150.

METHODS FOR ENZYME-LABELING OF ANTIGENS

ANTIBODIES AND THEIR FRAGMENTS

Eiji Ishikawa, Seiichi Hashida, Takeyuki Kohno and
Koichiro Tanaka

Department of Biochemistry, Medical College of Miyazaki
Kiyotake, Miyazaki 889-16, Japan

INTRODUCTION

A number of reagents have been used for enzyme-labeling of antigens and antibodies during the past two decades. However, most of them suffer from serious disadvantages, and only a few reagents are in current use, including glutaraldehyde, periodate, maleimide compounds and pyridyl disulfide compounds (Ishikawa et al., 1983a).

Glutaraldehyde readily reacts with amino groups of enzymes, antigens and antibodies under mild conditions to form stable cross-links. Horseradish peroxidase and antibodies can be converted to monomeric conjugates by a two-step procedure with glutaraldehyde, which is simple and reproducible (Avrameas and Ternynck, 1971). However, other enzymes, antigens and antibodies are polymerized to various extents by random coupling, causing a considerable loss of antibody activity and limiting the labeling efficiency. Only a small proportion of horseradish peroxidase is recovered in antibody conjugates (Ishikawa et al., 1983a).

Oxidation of carbohydrate moieties attached to horseradish peroxidase molecules by periodate generates aldehyde groups to react with amino groups of antibodies providing stable cross-links after reduction. This method offers high yields of conjugates in contrast to the glutaraldehyde method (Wilson and Nakane, 1978). However, the increase in the yield is accompanied by the formation of polymerized conjugates, and antibody activity is not fully retained (Ishikawa et al., 1983a).

Maleimide groups introduced into enzyme molecules readily react with thiol groups introduced into antibody molecules to form antibody-enzyme conjugates. Fab' can be conjugated to enzymes through thiol groups in the hinge by the reaction with maleimide groups introduced into enzymes (the hinge method). No polymerization takes place, and the antigen-binding activity of Fab' is fully retained (Ishikawa et al., 1983a). The hinge method reproducibly provides monomeric Fab'-horseradish peroxidase conjugates in high yields, although complex methods are required to prepare monomeric conjugates with other enzymes (Imagawa et al., 1984a; Inoue et al., 1985). Pyridyl disulfide groups can be used in place of maleimide groups, but conjugation efficiency is lower (Ishikawa et al., 1983a).

Antibodies and enzymes can be indirectly and non-covalently cross-linked using strong binding ability of biotin and avidin instead of direct and covalent cross-links described above (Guesdon et al., 1979).

A feature of this method is amplification of signals, which can be achieved by introduction of many biotin residues into antibody molecules and subsequent binding of avidin molecules linked to enzyme molecules. However, the non-specific binding or the background is also amplified.

PREPARATION OF IgG AND ITS FRAGMENTS

IgG is purified from serum and ascites by fractionation with Na_2SO_4 followed by passage through a column of DEAE cellulose or by affinity chromatography on a column of protein A-Sepharose. The yield of IgG varies among different species of animal due to difference in the concentration of IgG in serum (Lindmark et al., 1983). Mouse monoclonal IgG_{2a}, IgG_{2b} and IgG_3 are readily adsorbed to protein A-Sepharose at pH 8.0 and eluted at pH 3.5-5.0 (Ey et al., 1978; Lindmark et al., 1983; Seppälä et al., 1981). However, elution at different pH results in only a partial separation of these subclasses. Mouse monoclonal IgG_1 is only partially adsorbed to protein A-Sepharose and is purified by successive passages through a column of DEAE cellulose to remove serum proteins other than IgG and a column of protein A-Sepharose to remove IgG_{2a}, IgG_{2b} and IgG_3.

IgG is digested with pepsin to yield $F(ab')_2$, which is split to Fab' by reduction. Susceptibility of IgG to pepsin digestion varies among different species of animal and different subclasses, and different conditions of pepsin digestion are required for different subclasses. $F(ab')_2$ and Fab' are obtained in good yields from IgG of rabbit, goat, sheep, guinea pig and capybara and also from IgG_1, IgG_{2a} and IgG_3 of mouse but not from mouse IgG_{2b} (Dissanayake and Hay, 1975; Parham, 1983; Lamoyi and Nisonoff, 1983) (Fig. 1).

Purification of IgG from Most of Animals and IgG_1 from Mouse

1. Add slowly 0.18 g of Na_2SO_4 to 1 ml of serum or ascites supernatant with stirring, and continue stirring at 22-25°C for 30 min after a complete dissolution. 2. Centrifuge the mixture at 12,000 x g for 10 min at 22-25°C. 3. Dissolve the precipitates with 1 ml of sodium phosphate buffer, 17.5 mmol/l, pH 6.3 for IgG from most of animals such as rabbit, goat, sheep, guinea pig and capybara and sodium phosphate buffer, 40 mmol/l, pH 8.0, containing NaCl 30 mmol/l for mouse IgG and dialyze the solution against the same buffer. 4. Apply the supernatant to a column (0.1 ml/mg protein) of DEAE cellulose using the same buffer to remove most of other proteins. 5. Concentrate the effluent to 1 ml in a microconcentrator (CENTRICON-30, Amicon Corp., Scientific Systems Division, Danvers, Massachusetts) by centrifugation at 2,000 x g at 4°C. 6. Apply the concentrated effluent to a column (1.5 x 45 cm) of Ultrogel AcA 44 (LKB, Stockholm, Sweden) at a flow rate of 20-30 ml/h using sodium phosphate buffer, 0.1 mol/l, pH 7.0 to remove hemoglobin, if necessary. The fraction volume of 1.0 ml. (To obtain mouse monoclonal IgG_1, dialyze the IgG fractions against sodium phosphate buffer, 0.1 mol/l, pH 8.0, and apply the dialyzed IgG solution to a column (1.6 x 2.0 cm) of protein A-Sepharose CL-4B (Pharmacia Fine Chemicals AB, Uppsala, Sweden) using sodium phosphate buffer, 0.1 mol/l, pH 8.0 to remove IgG_{2a}, IgG_{2b} and IgG_3 (Ey et al., 1978; Lindmark et al., 1983; Seppälä et al., 1981). 7. Concentrate the effluent in a microconcentrator as described above. 8. Calculate the amount of IgG using the extinction coefficient at 280 nm and the molecular weight. $E_{280} = 1.5$ $g^{-1}.l.cm^{-1}$ for rabbit IgG (Palmer and Nisonoff, 1964). $E_{280} = 1.4$ $g^{-1}.l.cm^{-1}$ for mouse IgG (Ey et al., 1978). Mr = 150,000 (Dorrington and Tanford, 1970; Gorini et al., 1969).

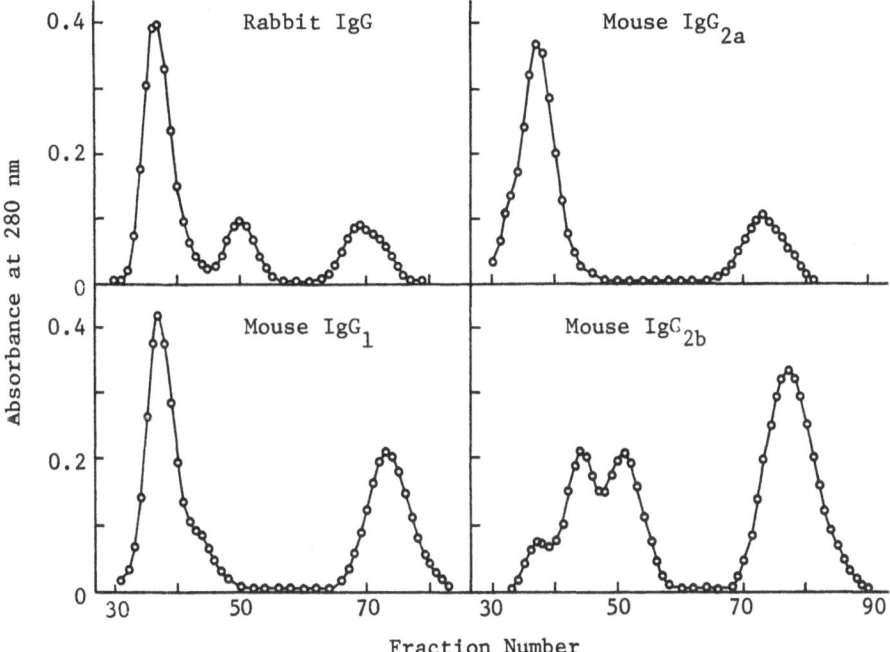

Fig. 1.   Elution profiles from a column of Ultrogel AcA 44 of F(ab')$_2$
prepared by pepsin digestion of rabbit IgG and mouse
monoclonal IgG$_1$, IgG$_{2a}$ and IgG$_{2b}$.

Purification of IgG$_{2a}$, IgG$_{2b}$ and IgG$_3$ from Mouse

1. Dialyze ascites supernatant against sodium phosphate buffer, 0.1
mol/1, pH 8.0.   2. Apply the dialyzed ascites supernatant to a column
(more than 0.2 ml/mg IgG$_{2a}$, IgG$_{2b}$ or IgG$_3$) of protein A-Sepharose CL-4B
(Pharmacia) using the same buffer to remove IgG$_1$ (Ey et al., 1978;
Lindmark et al., 1983; Seppälä et al., 1981).   3. Elute IgG$_{2a}$, IgG$_{2b}$ or
IgG$_3$ with sodium citrate buffer, 0.1 mol/1, pH 3.5-5.0.   4. Adjust the
pH of the eluate to 7 using Tris-HCl buffer, 2 mol/1, pH 8.0, and dialyze
against sodium phosphate buffer, 0.1 mol/1, pH 7.0.   5. Calculate the
amount of IgG using the extinction coefficient at 280 nm and the molecular
weight. E_{280} = 1.4 g^{-1}.1.cm^{-1} (Ey et al., 1978).   Mr = 150,000
(Gorini et al., 1969).

Pepsin Digestion of IgG to F(ab')$_2$

1. Dialyze 3-5 mg of IgG in 0.5 ml against sodium acetate buffer, 0.1
mol/1, pH 4.2-4.9, containing NaCl 0.1 mol/1 at 4°C.   pH of the buffer
is 4.5 for IgG from most of animals such as rabbit, goat, sheep, guinea
pig and capybara and mouse monoclonal IgG$_{2a}$ and IgG$_3$ (Lamoyi and Nisonoff,
1983).   pH of the buffer is 4.2 for mouse monoclonal IgG$_1$ and 4.7-4.9
for mouse monoclonal IgG$_{2b}$.   2. Dissolve 0.06-0.2 mg of pepsin from
porcine gastric mucosa in the dialyzed IgG solution.   3. Incubate the
mixture at 37°C for 6-24 h.   4. Adjust the pH of the digested IgG
solution to 7 using Tris-HCl buffer, 2 mol/1, pH 8.0.   5. Apply the
digested IgG solution at pH 7.0 to a column (1.5 x 45 cm) of Ultrogel AcA
44 (LKB) at a flow rate of 20-30 ml/h using sodium phosphate buffer, 0.1
mol/1, pH 7.0 (Fig. 1).   The fraction volume is 1.0 ml.   6.
Concentrate the F(ab')$_2$ fractions to 0.45 ml in a microconcentrator

(CENTRICON-30, Amicon Corp., Scientific Systems Division, Danvers, Massachusetts) by centrifugation at 2,000 x g at 4°C for 30 min.   7. Calculate the amount of F(ab')$_2$ using the extinction coefficient at 280 nm and the molecular weight. E_{280} = 1.48 g^{-1}.1.cm^{-1} for rabbit F(ab')$_2$ (Mandy and Nisonoff, 1963). E_{280} = 1.4 g^{-1}.1.cm^{-1} for mouse F(ab')$_2$ (Ey et al., 1978).   Mr = 92,000 for rabbit F(ab')$_2$ (Jaquet and Cebra, 1965; Utsumi and Karush, 1965).   Mr = 95,000-110,000 for mouse F(ab')$_2$ (Gorini et al., 1969; Svasti and Milstein, 1972; Lamoyi and Nisonoff, 1983).

Reduction of F(ab')$_2$ to Fab'

1. Add to 0.45 ml of F(ab')$_2$ (2-5 mg) 0.05 ml of 2-mercaptoethyl-amine 0.1 mol/l in sodium phosphate buffer, 0.1 mol/l, pH 6.0, containing EDTA 5 mmol/l.   2. Incubate the mixture at 37°C for 1.5 h.   3. Apply the reaction mixture to a column (1.5 x 45 cm) of Ultrogel AcA 44 (LKB) at a flow rate of 20-30 ml/h using sodium phosphate buffer, 0.1 mol/l, pH 6.0, containing EDTA 5 mmol/l.   The fraction volume is 1.0 ml.   (This confirms a complete split of F(ab')$_2$ to Fab' or separates Fab' from other proteins which are not split by reduction.   If this is unnecessary, apply the reaction mixture to a column (1.5 x 45 cm) of Sephadex G-25 (Pharmacia) at a flow rate of 30-40 ml/h using the same buffer.   Or a rapid gel filtration is possible with little dilution by a centrifuged column procedure (Penefsky, 1979).   Apply the reaction mixture to a column (1.0 x 6.4 cm, 5 ml) of Sephadex G-50 (fine, Pharmacia) with a fine mesh filter at the bottom, which has been equilibrated with the same buffer and centrifuged in a test tube at 100 x g for 2 min, and centrifuge the column in the same way.)   4. Concentrate the Fab' fractions in a microconcentrator (CENTRICON-30, Amicon Corp., Scientific Systems Division, Danvers, Massachusetts) by centrifugation at 2,000 x g at 4°C for 30 min.   5. Calculate the amount of Fab' using the extinction coefficient at 280 nm and the molecular weight. E_{280} = 1.48 g^{-1}.1.cm^{-1} for rabbit Fab' (Mandy and Nisonoff, 1963). E_{280} = 1.4 g^{-1}.1.cm^{-1} for mouse Fab' (Ey et al., 1978).   Mr = 46,000 for rabbit Fab' (Jaquet and Cebra, 1965; Utsumi and Karush, 1965).   Mr = 47,000-58,000 for mouse Fab' (Gorini et al., 1969; Svasti and Milstein, 1972; Lamoyi and Nisonoff, 1983).

Measurement of Thiol Groups in Fab'

1. Prepare a sample in a total volume of 0.5 ml of sodium phosphate buffer, 0.1 mol/l, pH 6.0, with an absorbance at 280 nm of 0.2-1.0 (0.14-0.71 mg/ml, 2.9-14 nmol/ml).   2. Add 0.02 ml of 4,4'-dithiodipyridine (m.w. 220.3) 5 mmol/l (1.1 g/l) to 0.5 ml of the sample solution above. 3. Incubate the mixture at room temperature for 20 min.   4. After the incubation, read absorbance at 324 nm.   5. Calculate the average number of thiol groups per Fab' molecule using the molar extinction coefficient at 324 nm of pyridine-4-thione which is 19,800 mol^{-1}.1.cm^{-1} (Grassetti and Murray, 1967).   The average number of thiol groups per Fab' molecule is approximately 1 for rabbit Fab' and 1-3 for Fab' from other animals.

Blocking Thiol Groups of Fab'

1. Incubate F(ab')$_2$ with 2-mercaptoethylamine 10 mmol/l as described above.   2. Add 0.1 ml of N-ethylmaleimide 0.1 mol/l to 0.5 ml of the reaction mixture.   N-Ethylmaleimide can be replaced by sodium monoiodo-acetate.   3. Incubate the mixture at 30°C for 60 min.   4. Apply the reaction mixture to a column (1.0 x 45 cm) of Ultrogel AcA 44 (LKB) using NaCl 0.15 mol/l.

In general, fluorimetry offers more sensitive assays of enzymes than colorimetry.   The detection limits of ß-D-galactosidase from <u>Escherichia coli</u> and alkaline phosphatase from calf intestine by fluorimetry using 4-methylumbelliferyl derivatives are 1,000 to 5,000-fold less than those by colorimetry using nitrophenyl derivatives (Ishikawa and Kato, 1978).   However, the detection limit of horseradish peroxidase by fluorimetry using 3-(4-hydroxyphenyl)propionic acid is only 5 to 20-fold less than that by colorimetry using o-phenylenediamine or 3,3',5,5'-tetramethylbenzidine (Ishikawa et al., 1983b).   A colorimetric assay of alkaline phosphatase from calf intestine by enzymatic cycling is very sensitive (Johannsson et al., 1985), and the detection limit of the enzyme by a 100 min assay is 0.3 amol in our laboratory.   A bioluminescent assay of dehydrogenases is highly sensitive (Tanaka and Ishikawa, 1984).

Colorimetric Assay of Peroxidase with o-Phenylenediamine

1. Dissolve immediately before use 0.11 g of o-phenylenediamine (m.w. 108) in 100 ml of sodium acetate buffer, 50 mmol/l, pH 5.0, containing bovine serum albumin 0.25 g/l.   2. Incubate 0.01 ml of enzyme samples plus 0.15 ml of the substrate solution with 0.05 ml of H_2O_2 1 g/l at 30°C for 10 min.   3. Stop the enzyme reaction by addition of 0.8 ml of H_2SO_4 0.5 mol/l containing Na_2SO_3 1 g/l.   4. Read absorbance at 491 nm. The detection limit of horseradish peroxidase is 25 amol (1 pg) in a 10 min assay.

Colorimetric Assay of Peroxidase with 3,3',5,5'-Tetramethylbenzidine

1. Dissolve 13.4 mg of 3,3',5,5'-tetramethylbenzidine (m.w. 240) in 1 ml of N,N-dimethylformamide followed by mixing with 100 ml of sodium acetate buffer, 0.1 mol/l, pH 5.5.   2. Incubate 0.01 ml of enzyme samples plus 0.6 ml of the substrate solution with 0.2 ml of H_2O_2 0.1 g/l at 30°C for 10-100 min.   3. Stop the enzyme reaction by addition of 0.2 ml of H_2SO_4 2 mol/l.   4. Read absorbance at 450 nm.   The detection limits of horseradish peroxidase in 10 and 100 min assays are 50 and 10 amol (2 and 0.4 pg), respectively.

Fluorimetric Assay of Peroxidase with 3-(4-Hydroxyphenyl)propionic Acid

1. Dissolve 0.3 g of 3-(4-hydroxyphenyl)propionic acid in 50 ml of sodium phosphate buffer, 0.1 mol/l, pH 8.0.   pH is lowered to 7.   2. Incubate enzyme samples in 0.01 ml of sodium phosphate buffer, 10 mmol/l, pH 7.0, containing NaCl 0.1 mol/l and bovine serum albumin 0.5 g/l plus 0.1 ml of the substrate solution with 0.05 ml of H_2O_2 0.15 g/l at 30°C for 10-100 min.   3. Stop the enzyme reaction by addition of 2.5 ml of glycine-NaOH buffer, 0.1 mol/l, pH 10.3.   4. Read fluorescence intensity using 320 nm for excitation and 405 nm for emission analysis. Use 0.2-1.0 mg/l of quinine in H_2SO_4 0.05 mol/l as a standard.   The detection limits of horseradish peroxidase in 10 and 100 min assays are 5 and 0.5 amol (200 and 20 fg), respectively.

Colorimetric Assay of ß-D-Galactosidase with o-Nitrophenyl-ß-D-Galactoside

1. Dissolve 150 mg (0.5 mmol) of o-nitrophenyl-ß-D-galactoside (m.w. 301.3) in 10 ml of deionized water.   2. Incubate enzyme samples in 0.4 ml of sodium phosphate buffer, 10 mmol/l, pH 7.0, containing NaCl 0.1 mol/l, $MgCl_2$ 1 mmol/l, bovine serum albumin 0.05-1.0 g/l and NaN_3 1 g/l with 0.1 ml of the substrate solution at 30°C for 10-100 min.   3. Stop the enzyme reaction by addition of 2 ml of Na_2CO_3 0.1 mol/l.   4. Read absorbance at 420 nm.   The detection limits of ß-D-galactosidase from

Escherichia *coli* in 10 and 100 min assays are 1 and 0.1 fmol (540 and 54 pg), respectively.

Fluorimetric Assay of ß-D-Galactosidase with 4-Methylumbelliferyl-ß-D-Galactoside

1. Dissolve 10 mg (0.03 mmol) of 4-methylumbelliferyl-ß-D-galactoside (m.w. 338.3) in 2.0 ml of N,N-dimethylformamide and add 98 ml of deionized water.   2. Incubate enzyme samples in 0.1 ml of sodium phosphate buffer, 10 mmol/l, pH 7.0, containing NaCl 0.1 mol/l, $MgCl_2$ 1 mmol/l, bovine serum albumin 0.05-1.0 g/l and NaN_3 1 g/l with 0.05 ml of the substrate solution at 30°C for 10-1,000 min.   3. Stop the enzyme reaction by addition of 2.5 ml of glycine-NaOH buffer, 0.1 mol/l, pH 10.3.   4. Read fluorescence intensity using 4-methylumbelliferone (m.w. 176.2) 10-1,000 nmol/l in glycine-NaOH buffer, 0.1 mol/l, pH 10.3, as a standard. Use 360 nm for excitation and 450 nm for emission analysis.   The detection limits of ß-D-galactosidase from *Escherichia* *coli* in 10, 100 and 1000 min assays are 0.2, 0.02 and 0.002 amol (110, 11 and 1.1 fg), respectively.

Colorimetric Assay of Alkaline Phosphatase with p-Nitrophenylphosphate

1. Dissolve 20 mg (0.054 mmol) of p-nitrophenylphosphate disodium salt (m.w. 371.2) in 10 ml of glycine-NaOH buffer, 0.1 mol/l, pH 10.3, containing $MgCl_2$ 1 mmol/l, $ZnCl_2$ 0.1 mmol/l, NaN_3 0.5 g/l and egg albumin 0.25 g/l.   2. Incubate enzyme samples in 0.5 ml of the same buffer with 0.5 ml of the substrate solution at 30°C for 10 min.   3. Stop the enzyme reaction by addition of 0.5 ml of NaOH 1 mol/l.   4. Read absorbance at 405 nm.   The detection limits of alkaline phosphatase from calf intestine in 10 min assays is 10 fmol (1 ng).

Colorimetric Assay of Alkaline Phosphatase by Enzymatic Cycling

1. Dissolve 3.0 mg (4,000 nmol) of NADP (m.w. 743.4) in 10 ml of deionized water (the substrate solution).   2. Dissolve 4 mg of alcohol dehydrogenase, 4 mg of diaphorase and 2.8 mg (5,500 nmol) of p-iodonitrotetrazolium violet (m.w. 505.7) in 10 ml of sodium phosphate buffer, 25 mmol/l, pH 7.2, containing ethanol 40 g/l (the cycling solution).   3. Incubate enzyme samples in 0.08 ml of diethanolamine buffer, 0.1 mol/l, pH 9.5, containing $MgCl_2$ 2 mmol/l, $ZnCl_2$ 0.2 mmol/l, bovine serum albumin 0.2 g/l and NaN_3 2 g/l with 0.08 ml of the substrate solution at 30°C for 10-100 min.   4. Add 0.3 ml of the cycling solution to the reaction mixture.   5. Continue the incubation at 30°C for 20 min.   6. Stop the cycling reaction by addition of 0.075 ml of HCl 0.4 mol/l.   7. Read absorbance at 495 nm.   The detection limit of alkaline phosphatase from calf intestine by a 100 min assay is 0.3 amol (30 fg).

Fluorimetric Assay of Alkaline Phosphatase with 4-Methylumbelliferyl Phosphate

1. Dissolve 2.6 mg (0.01 mmol) of 4-methylumbelliferyl phosphate (m.w. 256.2) in 33.3 ml of glycine-NaOH buffer, 0.1 mol/l, pH 9.5.   2. Incubate enzyme samples in 0.1 ml of glycine-NaOH buffer, 0.1 mol/l, pH 9.5, containing $MgCl_2$ 1 mmol/l, $ZnCl_2$ 0.1 mmol/l, NaN_3 0.5 g/l and egg albumin 0.25 g/l with 0.05 ml of the substrate solution at 30°C for 10-100 min.   3. Stop the enzyme reaction by addition of 2.5 ml of K_2HPO_4-KOH buffer, 0.5 mol/l, pH 10.4, containing EDTA 10 mmol/l.   4. Read fluorescence intensity using 4-methylumbelliferone 100 nmol/l in the same buffer as a standard.   Use 360 nm for excitation and 450 nm for emission analysis.   The detection limits of alkaline phosphatase from calf intestine in 10 and 100 min assays are 10 and 1.0 amol (1.0 and 0.1 pg), respectively.

Bioluminescent Assay of Glucose-6-Phosphate Dehydrogenase

1. Dissolve 10 mg (0.033 mmol) of glucose-6-phosphate disodium salt (m.w. 304.2) and 7.3 mg (0.011 mmol) of NAD$^+$ (m.w. 663.4) in 10 ml of Tris-HCl buffer, 95 mmol/l, pH 7.8, containing MgCl$_2$ 3.5 mmol/l.   2. Dissolve 1 U of lyophilized NAD(P)H:FMN oxidoreductase from Photobacterium fischeri in 1.0 ml of sodium phosphate buffer, 50 mmol/l, pH 7.0, containing 40 % (v/v) glycerol, EDTA 1 mmol/l and dithiothreitol 0.1 mmol/l.   3. Dissolve 1 mg of lyophilized luciferase from Photobacterium fischeri in 1.0 ml of deionized water.   4. Dissolve 0.25 mg of myristic aldehyde in a mixture of bovine serum albumin 10 g/l (1.25 ml) and Triton X-100 2 g/l (1.25 ml).   5. Dissolve 2.5 mg of FMN in 5 ml of sodium phosphate buffer, 10 mmol/l, pH 7.0, and adjust the concentration to 0.1 mmol/l by taking its extinction coefficient at 450 nm to be 1.22×10^4 mol^{-1}.l.cm^{-1}.   6. Incubate 0.05 ml of the substrate solution with enzyme samples in a total volume of 0.005 ml of Tris-HCl buffer, 0.1 mol/l, pH 7.8, containing MgCl$_2$ 4 mmol/l and egg albumin 1 g/l at 30°C for 10-100 min.   7. Stop the enzyme reaction by incubation at 100°C for 30 sec.   8. Add to a polypropylene vial 0.355 ml of sodium phosphate buffer, 0.1 mol/l, pH 6.5, incubated at 28°C for 5 min, 0.01 ml of FMN, 0.005 ml of luciferase, 0.02 ml of NAD(P)H:FMN oxidoreductase and 0.01 ml of myristic aldehyde.   9. Add 0.05 ml of the heated mixture incubated at 28°C for 3 min to the polypropylene vial, and measure luminescence between 15 and 45 sec after the addition of samples.   The detection limits of glucose-6-phosphate dehydrogenase from Leuconostoc mesenteroides in 10 and 100 min assays are 0.055 amol and 0.0055 amol, respectively.

ENZYME-LABELING OF ANTIBODIES AND THEIR FRAGMENTS

Hinge Method for Enzyme-Labeling of Fab' Using Maleimide Compounds

In the hinge method using maleimide compounds, thiol groups in the hinge of Fab' are reacted with maleimide groups introduced into enzyme molecules (Figs. 2 and 3) (Ishikawa et al., 1983a and c).   No polymerization takes place and the antigen-binding activity is fully retained.

Introduction of maleimide groups is performed using N,N'-o-phenylenedimaleimide or N,N'-oxydimethylenedimaleimide for ß-D-galactosidase from Escherichia coli which has thiol groups in the native form (Ishikawa et al., 1983a).   For other enzymes including horseradish peroxidase, alkaline phosphatase from calf intestine and glucose-6-phosphate dehydrogenase from Leuconostoc mesenteroides is used N-succinimidyl-4-(N-maleimidomethyl) cyclohexane-1-carboxylate, N-succinimidyl-6-maleimidohexanoate or sulfosuccinimidyl-4-(N-maleimidomethyl)cyclohexane-1-carboxylate (Hashida et al., 1984).   The highest conjugation efficiency has been obtained using N-succinimidyl-6-maleimidohexanoate.   There is no loss of peroxidase activity by the introduction of maleimide groups. However, 40-50 % of alkaline phosphatase and 50-70 % of glucose-6-phosphate dehydrogenase are lost.

This method provides monomeric Fab'-horseradish peroxidase conjugates in high yields, and the conjugates can be easily separated from unconjugated Fab' and peroxidase by gel filtration (Fig. 4) (Ishikawa et al., 1983a and c).   Other enzymes such as ß-D-galactosidase from Escherichia coli and alkaline phosphatase from calf intestine have much larger molecular weights than Fab' and can not be separated from the Fab'-enzyme conjugates by gel filtration.   Therefore, other enzymes after introduction of maleimide groups are reacted with 3 to 10-fold molar excess of Fab' to be completely converted to conjugates (Figs. 5 and 6). Therefore, the number of Fab' molecules conjugated per enzyme molecule is

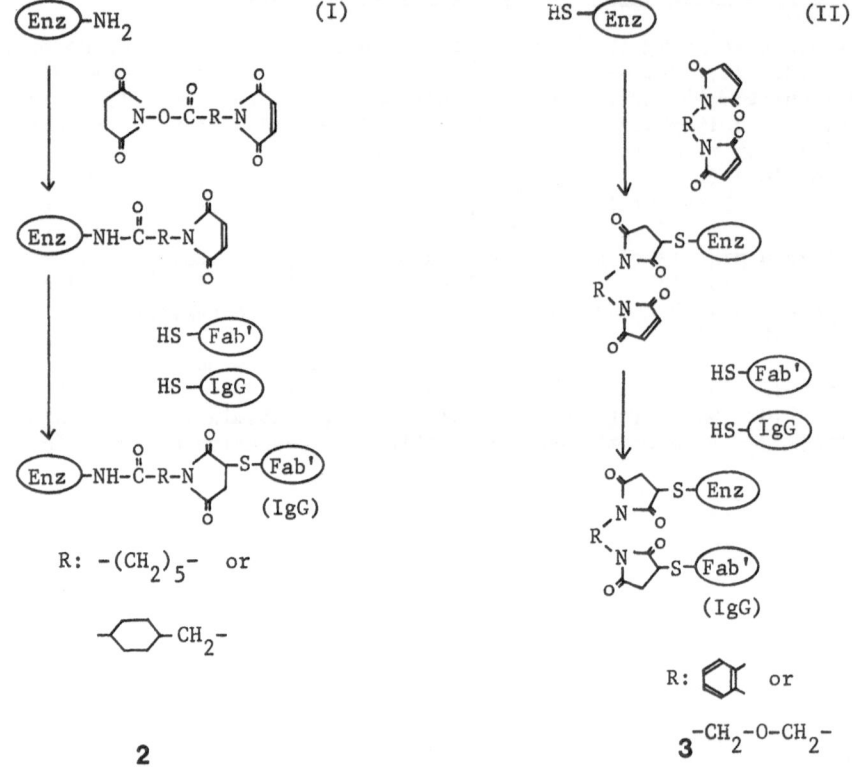

2

Fig. 2.   Hinge method using maleimide compounds (I) (Ishikawa et al., 1983a and c).   Thiol groups in the hinge of Fab' or thiol groups generated by reduction in the hinge of IgG are reacted with maleimide groups introduced into enzyme molecules using N-succinimidyl-6-maleimidohexanoate or N-succinimidyl-4-(N-maleimidomethyl)cyclohexane-1-carboxylate.

Fig. 3.   Hinge method using maleimide compounds (II) (Ishikawa et al., 1983a; Ruan et al., 1984).   Thiol groups in the hinge of Fab' or thiol groups generated by reduction in the hinge of IgG are reacted with maleimide groups introduced into enzyme molecules using N,N'-o-phenylenedimaleimide or N,N'-oxydimethylenedimaleimide.   This is applied to the conjugation with enzymes such as ß-D-galactosidase from <u>Escherichia coli</u>, which has many thiol groups in the native form.

more than one.   For the preparation of monomeric conjugates with these enzymes, more complex methods are required (Imagawa et al., 1984a; Inoue et al., 1985).

Fab' from animals such as rabbit, goat, sheep, guinea pig, capybara and mouse (monoclonal IgG_1 and IgG_{2a}) has been efficiently conjugated to horseradish peroxidase in our laboratory, and conjugation of Fab' from mouse monoclonal IgG_3 may also be possible.   However, conjugation of fragments obtained by pepsin digestion of mouse monoclonal IgG_{2b} has been much less efficient.

In the hinge method using maleimide compounds, there is an alternative conjugation method, in which maleimide groups are introduced into the hinge of Fab' or the reduced IgG by treatment with N,N'-o-phenylenedimaleimide or N,N'-oxydimethylenedimaleimide and reacted with thiol groups

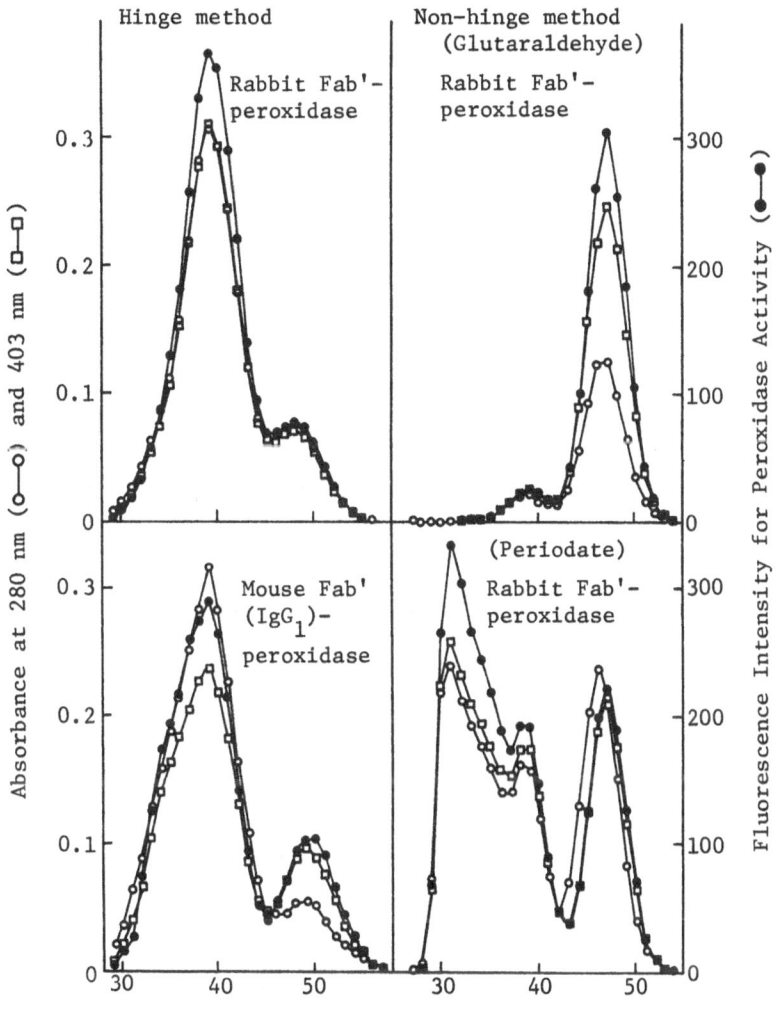

Fig. 4.   Elution profiles from a column of Ultrogel AcA 44 of Fab'-
horseradish peroxidase conjugates prepared by the hinge method
using N-succinimidyl-6-maleimidohexanoate and by the non-hinge
methods using glutaraldehyde and periodate (Ishikawa et al.,
1983a and c).   The conjugation efficiency by the hinge
method was 85 % for rabbit Fab' and 86 % for mouse Fab' (IgG_1).
Fab' from IgG_{2a} can be conjugated with a similar efficiency.

in the native form of ß-D-galactosidase.   However, the recovery of Fab'
or the reduced IgG in the conjugate is lower than that in the hinge method
described above (Fig. 5) (Ruan et al., 1984).

   Stability of maleimide groups and thiol groups   Maleimide groups
are stable at low pH but labile at high pH, while N-hydroxysuccinimide
ester of maleimide compounds efficiently reacts with amino groups of
proteins at neutral or higher pH.   Maleimide groups of N-succinimidyl-
4-(N-maleimidomethyl)cyclohexane-1-carboxylate and N-succinimidyl-6-malei-
midohexanoate are fairly stable at neutral pH.   However, maleimide
groups attached to benzene ring are labile at neutral pH.   Therefore,

Table 1.   Stability of Maleimide Groups

Maleimide compound	Maleimide groups (%) remaining after incubation at 30°C at pH 7.0 for			
	0.5 h	1 h	3 h	6 h
Succinimidylmaleimidoacetate	99	95	90	83
Succinimidyl-4-maleimidobutyrate	96	95	86	79
Succinimidyl-6-maleimidohexanoate	98	99	92	88
Succinimidyl-4-(N-maleimidomethyl)cyclohexane-1-carboxylate	96	97	94	90
Sulfosuccinimidyl-4-(N-maleimidomethyl)cyclohexane-1-carboxylate	102	100	98	100
Succinimidyl-m-maleimidobenzoate	62	43	15	7
Succinimidyl-4-(p-maleimidophenyl)butyrate	78	65	32	13
Sulfosuccinimidyl-m-maleimidobenzoate	87	79	40	17
Sulfosuccinimidyl-4-(p-maleimidophenyl)butyrate	77	71	39	25
N,N'-o-Phenylenedimaleimide	–	75	29	8
N,N'-Oxydimethylenedimaleimide	–	82	48	–

Each maleimide compound (0.7 mmol/1) was incubated in sodium phosphate buffer, 0.1 mol/1, pH 7.0.

N-succinimidyl-m-maleimidobenzoate and related compounds are not suitable for enzyme-labeling (Table 1) (Hashida et al., 1984).   Maleimide groups of N,N'-o-phenylenedimaleimide and N,N'-oxydimethylenedimaleimide are also labile at neutral pH.   However, introduction of maleimide groups into ß-D-galactosidase from Escherichia coli with these dimaleimides is efficiently performed at pH 6, and, under this condition, maleimide groups of these dimaleimides are sufficiently stable (Yoshitake et al., 1979).
    Thiol groups are not very stable especially at a low concentration. However, thiol groups are fairly stable in the presence of EDTA (Yoshitake et al., 1979).

    Introduction of maleimide groups into horseradish peroxidase
1. Incubate 2 mg (50 nmol) of horseradish peroxidase (E_{403} = 2.275 g^{-1}.1.cm^{-1} and Mr = 40,000 (Keilin and Hartree, 1951)) in 0.3 ml of sodium phosphate buffer, 0.1 mol/1, pH 7.0, with 0.3-0.7 mg (900-2,100 nmol) of N-succinimidyl-4-(N-maleimidomethyl)cyclohexane-1-carboxylate (m.w. 334.3) in 0.03 ml of N,N-dimethylformamide at 30°C for 0.5-1.0 h with continuous stirring.   The maleimide compound can be replaced by N-succinimidyl-6-maleimidohexanoate which is more soluble in the reaction mixture (Hashida et al., 1984).   2. Centrifuge briefly the reaction mixture to remove the precipitated reagent.   3. Apply the clear supernatant to a

column (1.0 x 45 cm) of Sephadex G-25 (Pharmacia) at a flow rate of 30-40 ml/h using sodium phosphate buffer, 0.1 mol/l, pH 6.0.   The fraction volume is 0.5-1.0 ml.   A more rapid gel filtration is possible with little dilution by a centrifuged column procedure (Penefsky, 1979).   A column (1.0 x 6.4 cm, 5 ml) of Sephadex G-50 (fine, Pharmacia) with a fine mesh filter at the bottom is equilibrated with the same buffer and centrifuged in a test tube at 100 x g for 2 min.   The reaction mixture (0.5 ml) is placed on the column, followed by centrifugation in the same way.   4. Concentrate pooled fractions in a microconcentrator (CENTRICON-30, Amicon Corp., Scientific Systems Division, Danvers, Massachusetts) by centrifugation at 2,000 x g at 4°C.   Do not use NaN_3 as a preservative, since it inactivates peroxidase and accelerates the decomposition of maleimide groups.

Measurement of maleimide groups   1. Prepare a sample in 0.45 ml of sodium phosphate buffer, 0.1 mol/l, pH 6.0, with an absorbance at 403 nm of 0.7-1.0 (0.31-0.44 mg/ml, 7.7-11 nmol/ml).   Use 0.45 ml of the same buffer as a control.   2. Mix 0.01 ml of 2-mercaptoethylamine-HCl (m.w. 113.6) 0.1 mol/l freshly prepared and 2 ml of EDTA, 50 mmol/l, pH 6.0. 3. Add 0.05 ml of the 2-mercaptoethylamine-EDTA mixture to 0.45 ml of the sample.   4. Incubate the reaction mixture at 30°C for 20 min. 5. Add 0.02 ml of 4,4'-dithiodipyridine (m.w. 220.3) 5 mmol/l (1.1 g/l), and incubate the mixture at 30°C for 10 min.   6. Read absorbance at 324 nm.   Extinction coefficient at 324 nm of pyridine-4-thione = 19,800 $mol^{-1}.l.cm^{-1}$ (Grassetti and Murray, 1967).   The average number of maleimide groups introduced per peroxidase molecule is 1-2, and the enzyme activity is fully retained.

Conjugation of Fab' to the maleimide-peroxidase (Figs. 2 and 4) 1. Incubate 1.8 mg (45 nmol) of the maleimide-peroxidase with 2.0 mg (43 nmol) of Fab' in 1 ml of sodium phosphate buffer, 0.1 mol/l, pH 6.0, containing EDTA 2.5 mmol/l at 4°C for 20 h or at 30°C for 1 h.   The final concentrations of the maleimide-peroxidase and Fab' in the reaction mixture for conjugation should be 0.01-0.05 mmol/l.   It may be better to block the remaining thiol groups with N-ethylmaleimide.   2. Apply the reaction mixture to a column (1.5 x 45-100 cm) of Ultrogel AcA 44 (LKB) at a flow rate of 10-30 ml/h using sodium phosphate buffer, 0.1 mol/l, pH 6.5.   The fraction volume is 1.0 ml.   3. Read absorbance at 280 and 403 nm, and measure peroxidase activity in each fraction (Fig. 4).   4. Store the conjugate at 4°C after addition of 1/98 volume of thimerosal 2-5 g/l and 1/98 volume of bovine serum albumin 100 g/l.   Do not use NaN_3 as a preservative, since it inactivates peroxidase.

Introduction of maleimide groups into ß-D-galactosidase   1. Incubate 2.5 mg (4.6 nmol) of ß-D-galactosidase (E_{280} = 2.09 $g^{-1}.l.cm^{-1}$ and Mr = 540,000 (Craven et al., 1965)) in 0.5 ml of sodium phosphate buffer, 0.1 mol/l, pH 6.0, with 0.25 mg (930 nmol) of N,N'-o-phenylenedimaleimide (m.w. 268) in 0.005 ml of N,N-dimethylformamide at 30°C for 20 min. N,N'-o-Phenylenedimaleimide can be replaced by N,N'-oxydimethylenedimaleimide.   2. Apply the reaction mixture to a column (1.0 x 30-45 cm) of Sephadex G-25 or G-50 (Pharmacia) at a flow rate of 30-40 ml/h using sodium phosphate buffer, 0.1 mol/l, pH 6.0.   The fraction volume is 0.5-1.0 ml.   Or a rapid gel filtration is possible with little dilution by a centrifuged column procedure (Penefsky, 1979).   A column (1.0 x 6.4 cm, 5 ml) of Sephadex G-50 (fine, Pharmacia) with a fine mesh filter at the bottom is equilibrated with the same buffer and centrifuged in a test tube at 100 x g for 2 min.   The reaction mixture (0.5 ml) is placed on the column, followed by centrifugation in the same way.   3. Measure the content of maleimide groups as described for the maleimide-peroxidase.   The average number of maleimide groups introduced per ß-D-galactosidase molecule is 13-16, and the introduction of maleimide

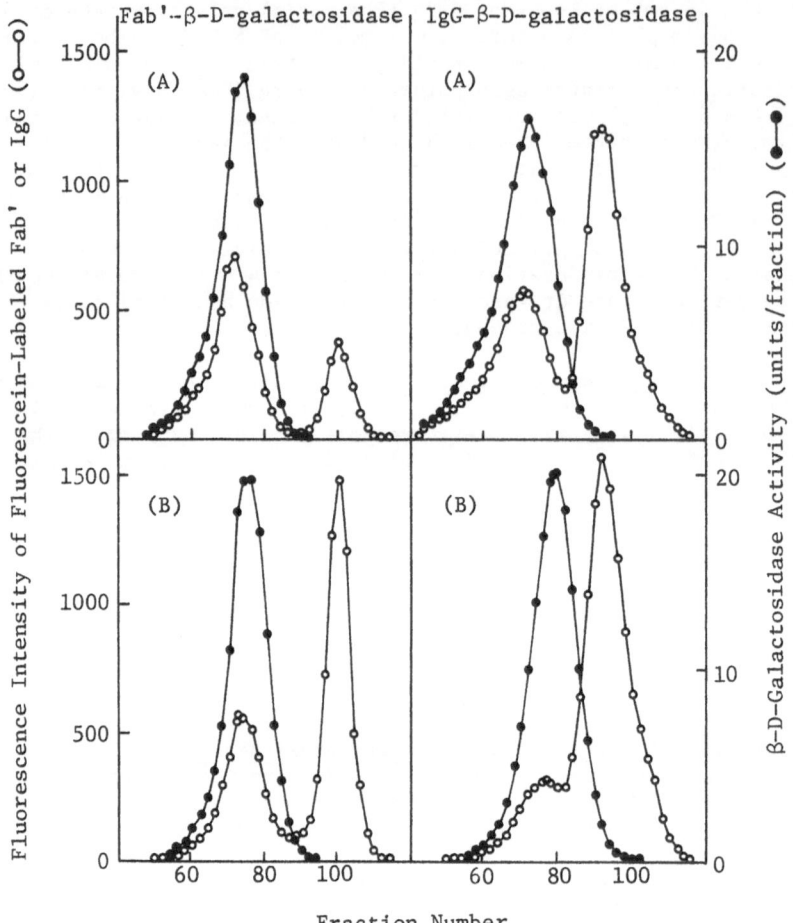

Fig. 5.   Elution profiles from a column of Sepharose 6B of rabbit Fab'-
           and rabbit IgG-ß-D-galactosidase conjugates prepared by the
           hinge method using N,N'-o-phenylenedimaleimide (Ruan et al.,
           1984).   Maleimide groups introduced into ß-D-galactosidase
           molecules from Escherichia coli are reacted with thiol groups
           of Fab' or the reduced IgG (A).   Or thiol groups in the
           native form of ß-D-galactosidase are reacted with maleimide
           groups introduced into the hinge of Fab' or the reduced IgG
           (B).

groups causes less than 10 % loss of the enzyme activity.

Conjugation of Fab' to the maleimide-ß-D-galactosidase (Figs. 3 and
5)   1. Incubate 2 mg (3.7 nmol) of the maleimide-ß-D-galactosidase with
0.51-0.85 mg (11-18 nmol) of Fab' in 2 ml of sodium phosphate buffer, 0.1
mol/l, pH 6.0, containing EDTA 2.5 mmol/l at 4°C for 15-24 h.   Addition
of a small amount of fluorescein-labeled normal Fab' (2.5 %) helps
monitoring the conjugation efficiency of Fab'.   2. Apply the reaction
mixture to a column (1.6 x 65 cm) of Sepharose 6B (Pharmacia) at a flow
rate of 20 ml/h using sodium phosphate buffer, 10 mmol/l, pH 6.5, containing NaCl 0.1 mol/l, MgCl$_2$ 1 mmol/l, NaN$_3$ 1 g/l and bovine serum albumin 1
g/l (buffer A).   The fraction volume is 1.0 ml.   3. Read fluorescence intensity of fluorescein-labeled Fab' in each fraction using 490 nm

for excitation and 510 nm for emission with a standard of fluorescein 1 nmol/l to monitor the conjugation efficiency of Fab', and measure ß-D-galactosidase activity in each fraction (Fig. 5).   4. Store the conjugate in buffer A at 4°C.

Introduction of maleimide groups into alkaline phosphatase

1. Dialyze 2 mg (20 nmol) of alkaline phosphatase from calf intestine (E_{280} = 0.99 g^{-1}.l.cm^{-1} (Morton, 1955); Mr = 100,000 (Engström, 1961)) in 0.5 ml against sodium borate buffer, 50 mmol/l, pH 7.6, containing MgCl$_2$ 1 mmol/l and ZnCl$_2$ 0.1 mmol/l.   2. Incubate 2 mg of the dialyzed enzyme in 0.5 ml with 0.17 mg (500 nmol) of N-succinimidyl-4-(N-maleimidomethyl) cyclohexane-1-carboxylate (m.w. 334.3) in 0.01 ml of N,N-dimethylformamide at 30°C for 0.5 h.   The maleimide compound can be replaced by N-succinimidyl-6-maleimidohexanoate.   3. Apply the reaction mixture to a column (1.0 x 45 cm) of Sephadex G-25 or G-50 (Pharmacia) at a flow rate of 30-40 ml/h using Tris-HCl buffer, 0.1 mol/l, pH 7.0, containing MgCl$_2$ 1 mmol/l and ZnCl$_2$ 0.1 mmol/l (buffer T).   The fraction volume is 0.5-1.0 ml.   Or a rapid gel filtration is possible with little dilution by a centrifuged column procedure (Penefsky, 1979).   A column (1.0 x 6.4 cm, 5 ml) of Sephadex G-50 (fine, Pharmacia) with a fine mesh filter at the bottom is equilibrated with the same buffer and centrifuged in a test tube at 100 x g for 2 min.   The reaction mixture (0.5 ml) is placed on the column, followed by centrifugation in the same way.   4. Measure the content of maleimide groups as described for the maleimide-peroxidase. The average number of maleimide groups introduced per enzyme molecule is 4-6, and 40-50 % of the enzyme activity is lost by the introduction of maleimide groups.

Conjugation of Fab' to the maleimide-alkaline phosphatase (Figs. 2 and 6)

1. Incubate 1 mg (10 nmol) of the maleimide-alkaline phospha-tase in 0.25 ml of buffer T with 2.3 mg (50 nmol) of Fab' plus 0.03 mg of fluorescein-labeled Fab' in 0.25 ml of sodium phosphate buffer, 0.1 mol/l, pH 6.0, containing EDTA 5 mmol/l at 4°C for 20 h.   Addition of a small amount of fluorescein-labeled normal Fab' (2.5 %) helps monitoring the conjugation efficiency of Fab'.   2. Incubate the reaction mixture with 0.01 ml of 2-mercaptoethylamine 10 mmol/l at room temperature for 20 min to block remaining maleimide groups.   3. Apply the reaction mixture to a column (1.5 x 45 cm) of Ultrogel AcA 34 (LKB) at a flow rate of 20 ml/h using Tris-HCl buffer, 10 mmol/l, pH 6.8, containing NaCl 0.1 mol/l, MgCl$_2$ 1 mmol/l, ZnCl$_2$ 0.1 mmol/l, and NaN$_3$ 0.5 g/l.   The fraction volume is 1.0 ml.   4. Read fluorescence intensity of fluorescein-labeled Fab' in each fraction using 490 nm for excitation and 510 nm for emission with a standard of 1 nmol/l fluorescein to monitor the conjugation efficiency of Fab', and measure alkaline phosphatase activity in each fraction (Fig. 6). 5. Store the conjugate at 4°C after addition of 1/99 volume of bovine serum albumin 100 g/l.

Introduction of maleimide groups into glucose-6-phosphate dehydro-genase

1. Dialyze 0.5 mg (4.9 nmol) of glucose-6-phosphate dehydro-genase from Leuconostoc mesenteroides (E_{280} = 1.15 g^{-1}.l.cm^{-1} and Mr = 103,700 (Olive and Levy, 1971)) suspended in 0.5 ml of ammonium sulfate 3.2 mol/l against sodium phosphate buffer, 0.2 mol/l, pH 7.0.   2. Dissolve 0.76 mg (2500 nmol) of glucose-6-phosphate disodium salt (m.w. 304.2) in 0.005 ml of deionized water and 0.66 mg (1000 nmol) of NAD$^+$ (m.w. 663.4) in 0.005 ml of deionized water.   3. Dissolve 0.26-0.70 mg (780-21000 nmol) of N-succinimidyl-4-(N-maleimidomethyl)cyclohexane-1-carboxylate (m.w. 334.3) in 0.02 ml of N,N-dimethylformamide.   4. Incubate 0.49 ml of the enzyme solution with 0.005 ml of the glucose-6-phosphate solution and 0.005 ml of the NAD$^+$ solution at 30°C for 5 min. 5. Add 0.02 ml of the maleimide reagent solution to the mixture and continue the incubation at 30°C for 30 min with continuous shaking.   6.

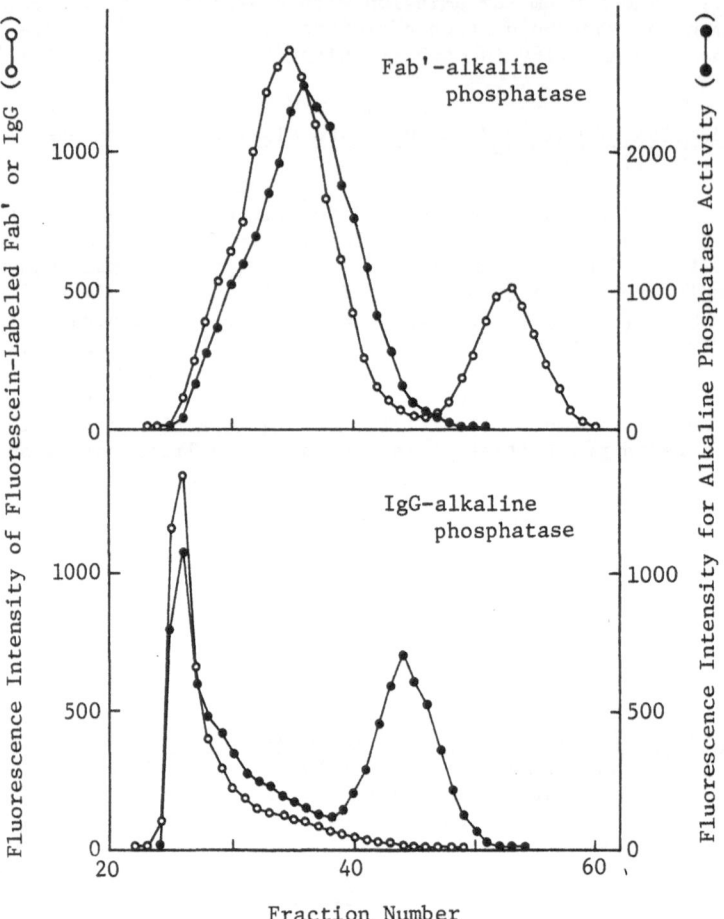

Fig. 6. Elution profiles from a column of Ultrogel AcA 34 of rabbit
Fab'- and rabbit IgG-alkaline phosphatase conjugate prepared by
the hinge method and the non-hinge method, respectively, using
N-succinimidyl-6-maleimidohexanoate (Ishikawa et al., 1983a).

Apply the reaction mixture to a Sephadex G-25 column (1.0 x 45 cm) at a
flow rate of 30-40 ml/h using sodium phosphate buffer, 0.1 mol/l, pH 6.0.
7. Measure the content of maleimide groups as described for the maleimide-
peroxidase. The average number of maleimide groups introduced per
enzyme molecule is 4.6-6.4, and 50-70 % of the enzyme activity was lost by
the introduction of maleimide groups.

Conjugation of Fab' to the maleimide-glucose-6-phosphate dehydrogenase 1. Incubate 0.5 mg (4.9 nmol) of the maleimide-glucose-6-phos-
phate dehydrogenase in 0.6 ml of sodium phosphate buffer, 0.1 mol/l, pH
6.0, with 2.2 mg (48 nmol) of Fab' plus 0.06 mg of fluorescein-labelled
normal Fab' in 1 ml of sodium phosphate buffer, 0.1 mol/l, pH 6.0,
containing EDTA 5 mmol/l at 4°C for 20 h. 2. Add to the reaction
mixture 0.01 ml of 2-mercaptoethylamine 100 mmol/l, and incubate the
reaction mixture at room temperature for 20 min to block remaining
maleimide groups. 3. Apply the reaction mixture to a column (1.5 x 45
cm) of Ultrogel AcA 34 at a flow rate of 20 ml/h using sodium phosphate
buffer, 0.1 mol/l, pH 6.5. 4. Store the fraction containing the

conjugate at 4°C after addition of 1/98 volume of bovine serum albumin 100 g/1 and 1/98 volume of NaN$_3$ 100 g/1.

Hinge Method for Enzyme-Labeling of Fab' Using Pyridyl Disulfide Compounds

Fab' can be conjugated to enzymes by the hinge method using pyridyl disulfide compounds.   Pyridyl disulfide groups in place of maleimide groups are introduced into enzymes using N-succinimidyl-3-(2-pyridyl-dithio)propionate and reacted with thiol groups in the hinge of Fab' (Fig. 7) (Imagawa et al., 1982b).   A monomeric Fab'-horseradish peroxidase conjugate is formed, which has the same properties as that prepared by the hinge method using maleimide compounds. No polymerization takes place, and the antigen-binding activity is fully retained after conjugation. However, the pyridyl disulfide method is less efficient than the maleimide method.

Introduction of pyridyl disulfide groups into peroxidase   1. Incubate 2 mg (50 nmol) of horseradish peroxidase (E$_{403}$ = 2.275 g^{-1}.1.cm^{-1} and Mr = 40,000 (Keilin and Hartree, 1951)) in 0.3 ml of sodium phosphate buffer, 0.1 mol/1, pH 7.5, with 0.78 mg (2,500 nmol) of N-succinimidyl 3-(2-pyridyldithio)propionate (m.w. 312.5) in 0.06 ml of ethanol at 25°C for 30 min.   2. Apply the reaction mixture to a column (1.0 x 45 cm) of Sephadex G-25 (Pharmacia) at a flow rate of 30-40 ml/h using sodium phosphate buffer, 0.1 mol/1, pH 6.0.   The fraction volume is 0.5-1.0 ml.   Or a rapid gel filtration is possible with little dilution by a centrifuged column procedure (Penefsky, 1979).   A column (1.0 x 6.4 cm, 5 ml) of Sephadex G-50 (fine, Pharmacia) with a fine mesh filter at the bottom is equilibrated with the same buffer and centrifuged in a test tube at 100 x g for 2 min.   The reaction mixture (0.5 ml) is placed on the column, followed by centrifugation in the same way.   4. Concentrate pooled fractions in a microconcentrator (CENTRICON-30, Amicon Corp., Scientific Systems Division, Danvers, Massachusetts) by centrifugation at 2,000 x g at 4°C.   Do not use NaN$_3$ as a preservative, since it inactivates peroxidase.

Measurement of pyridyl disulfide groups   1. Prepare a sample in 0.5 ml of sodium phosphate buffer, 0.1 mol/1, pH 6.0, with an absorbance at 403 nm of 0.35-1.0 (0.15-0.44 mg/ml, 3.8-11 nmol/ml).   2. Incubate the sample with 0.02 ml of dithiothreitol (m.w. 154.3), 0.1 mol/1 (15.4 mg/ml) at 30°C for 20 min.   3. Read absorbance at 343 nm. Extinction coefficient at 343 nm of pyridine-2-thione = 8,080 mol^{-1}.1.cm^{-1} (Stuchbury et al., 1975).   The average number of pyridyl disulfide groups introduced per peroxidase molecule is 2.5-2.7.

Conjugation of Fab' to the pyridyl disulfide-peroxidase (Fig. 7) 1. Incubate 1.8 mg (45 nmol) of the pyridyl disulfide-peroxidase with 2 mg (43 nmol) of Fab' in 0.44 ml of sodium phosphate buffer, 0.1 mol/1, pH 6.0, containing EDTA 2.5 mmol/1 at 30°C for 2.5 h.   The final concentrations of the pyridyl disulfide-peroxidase and Fab' should be 100 nmol/ml or higher.   2. Apply the reaction mixture to a column (1.5 x 45-100 cm) of Ultrogel AcA 44 (LKB) at a flow rate of 10-30 ml/h using sodium phosphate buffer, 0.1 mol/1, pH 6.5.   The fraction volume is 1.0 ml.   3. Read absorbance at 280 and 403 nm, and measure peroxidase activity in each fraction (Fig. 4).   4. Store the conjugate at 4°C after addition of 1/98 volume of thimerosal 2-5 g/1 and 1/98 volume of bovine serum albumin 100 g/1.   Do not use NaN$_3$ as a preservative, since it inactivates peroxidase.

Non-hinge Method for Enzyme-Labeling of Fab' and Fab

In the non-hinge method, Fab' after the treatment with N-ethylmalei-mide or monoiodoacetate to block thiol groups or Fab is conjugated to

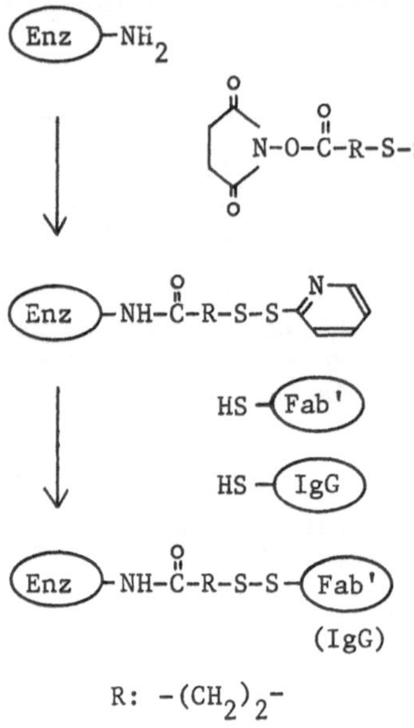

Fig. 7.   Hinge method using pyridyl disulfide compounds (Imagawa et al., 1982b; Ishikawa et al., 1983a and c).

enzymes through amino groups using various cross-linking reagents including glutaraldehyde, periodate, maleimide compounds and pyridyl disulfide compounds (Ishikawa et al., 1983a and c).

     Conjugation of Fab' to the glutaraldehyde-treated peroxidase   1. Incubate 10 mg of horseradish peroxidase in 0.19 ml of sodium phosphate buffer, 0.1 mol/l, pH 6.8, with 0.01 ml of glutaraldehyde 250 g/l at room temperature for 18 h.   2. Apply the reaction mixture to a column (1 x 45 cm) of Ultrogel AcA 44 (LKB) using NaCl 0.15 mol/l to remove excess of glutaraldehyde and dimer and/or polymers of the peroxidase.   3. Incubate a mixture of 4.8 mg (120 nmol) of the glutaraldehyde-treated peroxidase in 0.5 ml of NaCl 0.15 mol/l, 1.2 mg (25 nmol) of the SH-blocked Fab' in 0.5 ml of NaCl 0.15 mol/l, and 0.1 ml of sodium carbonate buffer, 1 mol/l, pH 9.5, at 4°C for 24 h and then with 0.05 ml of L-lysine-HCl 0.2 mol/l in sodium phosphate buffer, 0.25 mol/l, pH 8.0, at 4°C for 2 h.   4. Apply the reaction mixture to a column (1.5 x 45 cm) of Ultrogel AcA 44 (LKB) (Fig. 4) and store the conjugate as described in the hinge method.

     Conjugation of Fab' to the periodate-oxidized peroxidase   1. Incubate 2 mg of horseradish peroxidase in 0.5 ml of deionized water with 0.1 ml of NaIO$_4$ 0.1 mol/l at room temperature for 10 min.   2. Add 0.05 ml of ethylene glycol to the reaction mixture and incubate at room temperature for 5 min.   3. Apply the reaction mixture to a column of Sephadex G-25 or G-50 (Pharmacia) using sodium acetate buffer, 1 mmol/l, pH 4.4.   4. Incubate a mixture of 1.6 mg (40 nmol) of the oxidized peroxidase in 0.4 ml of sodium acetate buffer, 1 mmol/l, pH 4.4, 18 mg (40 nmol) of the SH-blocked Fab' in 0.4 ml of NaCl 0.15 mol/l, and 0.02 ml of sodium carbonate buffer, 1 mol/l, pH 9.5, at 25°C for 2 h.   5. Add 0.04 ml of sodium borohydride 4 g/l and incubate the reaction mixture at 4°C for 2 h.   6. Apply the reaction mixture to a column (1.5 x 45 cm) of Ultrogel AcA 44 (LKB) (Fig. 4) and store the conjugate as described in the hinge method.

42

Maleimide Method for Enzyme-Labeling of IgG and F(ab')$_2$

In the hinge method using maleimide compounds, thiol groups are generated by reduction of disulfide bonds in the hinge of IgG and reacted with maleimide groups introduced into enzymes (Figs. 2 and 3).

The hinge method using N-succinimidyl-6-maleimidohexanoate or N-succinimidyl-4-(N-maleimidomethyl)cyclohexane-1-carboxylate provides a monomeric IgG-horseradish peroxidase conjugate, since thiol groups are generated only in a small number (2-4) and the number of maleimide groups introduced into peroxidase molecules is limited. However, IgG remains unconjugated in a significant proportion and is not easily separated from the conjugate by gel filtration. This is also the case in the conjugation of IgG with alkaline phosphatase from calf intestine.

The hinge method using N,N'-o-phenylenedimaleimide or N,N'-oxydimethylenedimaleimide is useful for labeling of IgG with ß-D-galacto-sidase from _Escherichia coli_. Maleimide groups are introduced into ß-D-galactosidase molecules by treatment with one of the dimaleimides and reacted with excess of thiol groups generated in the hinge of IgG to completely convert ß-D-galactosidase to the conjugate (Figs. 3 and 5). This complete conversion is important, since unconjugated ß-D-galacto-sidase with a molecular weight of 540,000 can not easily separated from the conjugate by gel filtration. No polymerization takes place under the condition described below, since the number of thiol groups generated in the hinge of IgG is limited.

In the non-hinge method, thiol groups introduced into IgG or F(ab')$_2$ are reacted with maleimide groups introduced into enzymes. This method is suitable for the conjugation of IgG to horseradish peroxidase and alkaline phosphatase from calf intestine. Thiol groups are introduced into IgG and F(ab')$_2$ using S-acetylmercaptosuccinic anhydride and reacted with excess of maleimide groups introduced into the enzyme molecules to completely convert the mercaptosuccinylated IgG to the conjugate. This complete conversion is important, since unconjugated IgG is not easily separated from the conjugates with these enzymes by gel filtration.

Introduction of thiol groups into IgG and F(ab')$_2$ 1. Incubate 5 mg of IgG (33 nmol) or F(ab')$_2$ (54 nmol) in 0.5 ml of sodium phosphate buffer, 0.1 mol/l, pH 6.5, with 0.6 mg (3,450 nmol) of S-acetylmercapto-succinic anhydride (m.w. 174.2) in 0.01 ml of N,N-dimethylformamide at room temperature for 30 min.   2. Add 0.02 ml of EDTA 0.1 mol/l, 0.1 ml of Tris-HCl buffer, 0.1 mol/l, pH 7.0 and 0.1 ml of hydroxylamine-HCl, 1 mol/l, pH 7.0.   3. Incubate the reaction mixture at 30°C for 4 min. 4. Apply the reaction mixture to a column of Sephadex G-25 or G-50 (Pharmacia) at a flow rate of 30-40 ml/h using sodium phosphate buffer, 0.1 mol/l, pH 6.0.   The fraction volume is 1.0 ml.   Or a rapid gel filtration is possible with little dilution (Penefsky, 1979).   Apply the reaction mixture to a column (1.0 x 6.4 cm, 5 ml) of Sephadex G-50 (fine, Pharmacia) with a fine mesh filter at the bottom, which has been equilibrated with the same buffer and centrifuged in a test tube at 100 x g for 2 min, and centrifuge the column in the same way.   5. Measure the content of thiol groups as described for Fab'.   The average number of thiol groups introduced per IgG or F(ab')$_2$ molecule is 4-7.

Conjugation of the mercaptosuccinylated IgG and F(ab')$_2$ to the maleimide-peroxidase (Fig. 8)   1. Incubate the maleimide-peroxidase (3 mg, 75 nmol) (see the hinge method for labeling of Fab') with the mercaptosuccinylated IgG (2.3 mg, 15 nmol) or F(ab')$_2$ (1.4 mg, 15 nmol) in 0.5 ml of sodium phosphate buffer, 0.1 mol/l, pH 6.0, containing EDTA, 2-3 mmol/l at 4°C for 20 h.   2. Apply the reaction mixture to a column (1.5 x 45 cm) of Ultrogel AcA 34 (LKB) at a flow rate of 20-30 ml/h using sodium phosphate buffer, 0.1 mol/l, pH 6.5.   The fraction volume is 1.0 ml.   3. Read absorbance at 280 and 403 nm, and measure peroxidase

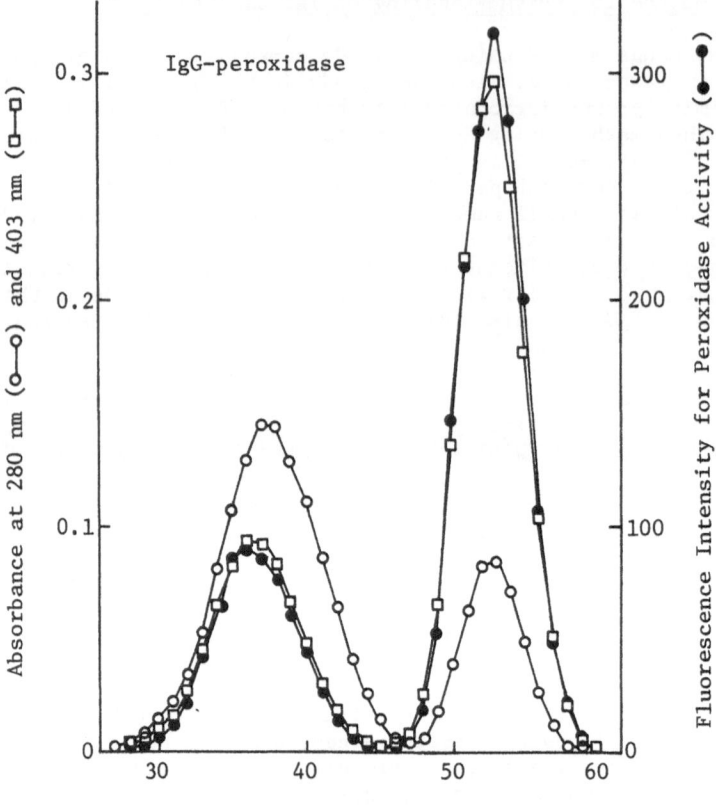

Fig. 8.   Elution profile from a column of Ultrogel AcA 34 of mouse
monoclonal IgG_1-horseradish peroxidase conjugate prepared by
the non-hinge method using N-succinimidyl-4-(N-maleimidomethyl)
cyclohexane-1-carboxylate.   The number of peroxidase
molecules conjugated per IgG molecule was 1.0-2.7.

activity in each fraction (Fig. 8).   4. Store the conjugate at 4°C
after addition of 1/98 volume of thimerosal 2-5 g/l and 1/98 volume of
bovine serum albumin 100 g/l.   Do not use NaN_3 as a preservative, since
it inactivates peroxidase.

Conjugation of the mercaptosuccinylated IgG and $F(ab')_2$ to the
maleimide-alkaline phosphatase (Fig. 6)   1. Incubate the
maleimide-alkaline phosphatase (7.5 mg, 75 nmol) in 0.25 ml of buffer T
(see the hinge method for labeling of Fab') with the mercaptosuccinylated
IgG (2.3 mg, 15 nmol) plus fluorescein-labeled IgG (0.058 mg) or the
mercaptosuccinylated $F(ab')_2$ (1.4 mg, 15 nmol) plus fluorescein-labeled
$F(ab')_2$ (0.035 mg) in 0.25 ml of sodium phosphate buffer, 0.1 mol/l, pH
6.0, containing EDTA, 5 mmol/l at 4°C for 20 h.   2. Apply the reaction
mixture to a column (1.5 x 45 cm) of Ultrogel AcA 34 (LKB) at a flow rate
of 20 ml/h using Tris-HCl buffer, 10 mmol/l, pH 6.8, containing NaCl 0.1
mol/l, $MgCl_2$ 1 mmol/l, $ZnCl_2$ 0.1 mmol/l, and NaN_3 0.5 g/l.   The
fraction volume is 1.0 ml.   3. Read fluorescence intensity of fluorescein-labeled IgG or $F(ab')_2$ in each fraction using 490 nm for excitation
and 510 nm for emission with a standard of 1 nmol/l fluorescein to monitor
the conjugation efficiency of IgG or $F(ab')_2$ and measure alkaline phos-

phatase activity in each fraction (Fig. 6).   4. Store the conjugate at 4°C after addition of 1/99 volume of bovine serum albumin 100 g/l.

Reduction of IgG   1. Add to 0.45 ml of IgG (2-5 mg) 0.05 ml of 2-mercaptoethylamine 0.1 mol/1 in sodium phosphate buffer, 0.1 mol/1, pH 6.0, containing EDTA 5 mmol/1.   2. Incubate the mixture at 37°C for 1.5 h.   3. Apply the reaction mixture to a column (1.5 x 45 cm) of Sephadex G-25 (Pharmacia) at a flow rate of 30-40 ml/h using sodium phosphate buffer, 0.1 mol/1, pH 6.0, containing EDTA 5 mmol/1.   Or a rapid gel filtration is possible with little dilution (Penefsky, 1979). Apply the reaction mixture to a column (1.0 x 6.4 cm, 5 ml) of Sephadex G-50 (fine, Pharmacia) with a fine mesh filter at the bottom, which has been equilibrated with the same buffer and centrifuged in a test tube at 100 x g for 2 min, and centrifuge the column in the same way.   4. Concentrate the reduced IgG fractions in a microconcentrator (CENTRICON-30, Amicon Corp., Scientific Systems Division, Danvers, Massachusetts) by centrifugation at 2,000 x g at 4°C.   5. Calculate the amount of the reduced IgG using the extinction coefficient at 280 nm and the molecular weight. $E_{280} = 1.5\ g^{-1}.l.cm^{-1}$ for rabbit IgG (Palmer and Nisonoff, 1964). $E_{280} = 1.4\ g^{-1}.l.cm^{-1}$ for mouse IgG (Ey et al., 1978). $Mr = 150,000$ (Dorrington and Tanford, 1970; Corini et al., 1969).

Conjugation of the reduced IgG to the maleimide-ß-D-galactosidase (Figs. 3 and 5)   1. Incubate 2 mg (3.7 nmol) of the maleimide-ß-D-galactosidase (see the hinge method for labeling of Fab') with 1.8 mg (12 nmol) of the reduced IgG in 2 ml of sodium phosphate buffer, 0.1 mol/1, pH 6.0, containing EDTA 2.5 mmol/1 at 4°C for 15-24 h.   Addition of a small amount of fluorescein-labeled normal IgG (2.5 %) helps monitoring the conjugation efficiency of IgG.   2. Apply the reaction mixture to a column (1.6 x 65 cm) of Sepharose 6B (Pharmacia) at a flow rate of 20 ml/h using sodium phosphate buffer, 10 mmol/1, pH 6.5, containing NaCl 0.1 mol/1, $MgCl_2$ 1 mmol/1, NaN_3 1 g/l and bovine serum albumin 1 g/l (buffer A).   The fraction volume is 1.0 ml.   3. Read fluorescence intensity of fluorescein-labeled IgG in each fraction using 490 nm for excitation and 510 nm for emission with a standard of 1 nmol/1 fluorescein to monitor the conjugation efficiency of IgG, and measure ß-D-galactosidase activity in each fraction (Fig. 5).   4. Store the conjugate in buffer A at 4°C.

Biotin-Avidin Method

Biotinylation of IgG   1. Incubate 1 mg (6.7 nmol) of IgG in 0.1 ml of $NaHCO_3$, 0.1 mol/1, pH 8.2-8.6, with 0.047-0.75 mg (138-2,200 nmol) of biotin-N-hydroxysuccinimide (m.w. 341.4) in 0.025 ml of N,N-dimethylform-amide at room temperature for 1 h.   2. Apply the reaction mixture to a column of Sephadex G-25 or G-50 using sodium phosphate buffer, 0.1 mol/1, pH 7.0.   The average number of biotin residues introduced into IgG is calculated by measuring the number of amino groups.

Determination of free amino groups in IgG   1. Prepare IgG (1-100 nmol as amino groups) in 2.5 ml of sodium borate buffer, 0.2 mol/1, pH 9.0.   2. Add 0.8 ml of fluorescamine in acetone, 0.1 g/l, to the IgG solution.   3. Measure fluorescence intensity using 390 nm for excitation and 475 nm for emission.   Use L-leucine as a standard (Udenfriend et al., 1972).

Conjugation of the mercaptosuccinylated avidin to the maleimide-peroxidase (Fig. 9)   1. Incubate 2 mg (30 nmol) of avidin (Vector Laboratories Inc., California) in 0.3 ml of sodium phosphate buffer, 0.2 mol/1, pH 7.0, with 0.105 mg (600 nmol) of S-acetylmercaptosuccinic anhydride in 0.005 ml of N,N-dimethylformamide at 30°C for 30 min.   2.

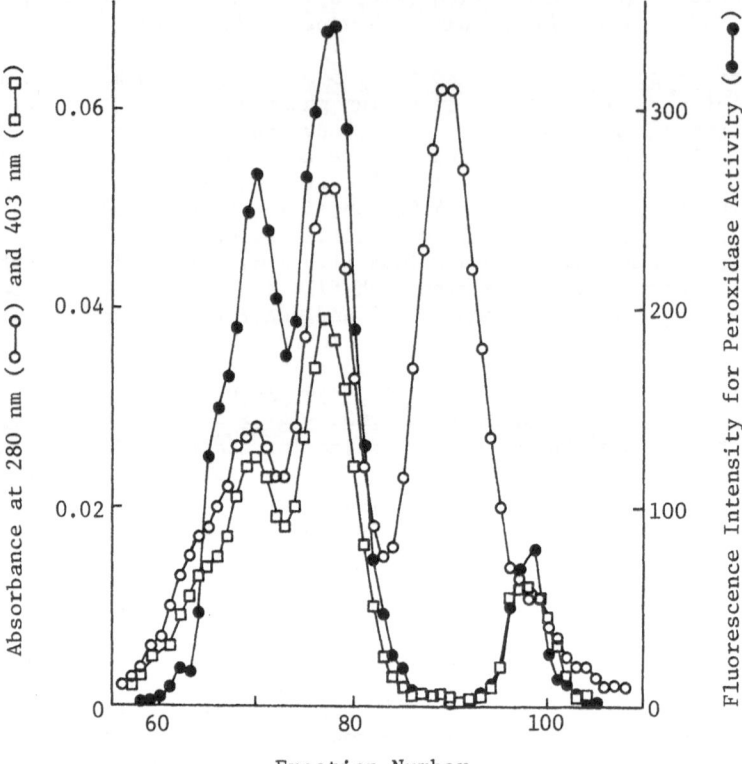

Fig. 9.   Elution profile from a column of Ultrogel AcA 44 of avidin-
           horseradish peroxidase conjugate prepared by the maleimide
           method.

Add 0.01 ml of EDTA 0.1 mol/l, pH 7.0, 0.06 ml of Tris-HCl buffer, 0.1
mol/l, pH 7.0, and 0.06 ml of hydroxylamine-HCl, 1.0 mol/l, pH 7.0.
3. Incubate the reaction mixture at 30°C for 5 min.   4. Apply the
reaction mixture to a column of Sephadex G-25 or G-50 (Pharmacia) at a
flow rate of 30-40 ml/h using sodium phosphate buffer, 0.1 mol/l, pH 6.0,
containing EDTA 5 mmol/l.   The fraction volume is 1.0 ml.   5.
Concentrate pooled fractions in a microconcentrator (CENTRICON-30, Amicon
Corp.) by centrifugation at 2,000 x g at 4°C for 30 min.   6. Measure
the content of thiol groups as described for Fab'.   The average number
of thiol groups is 1.6.   7. Incubate 0.32 mg (8 nmol) of the
maleimide-peroxidase (see the hinge method for labeling of Fab') in 0.03
ml of sodium phosphate buffer, 0.1 mol/l, pH 6.0, with 0.82 mg (12 nmol)
of the mercaptosuccinylated avidin in 0.13 ml of sodium phosphate buffer,
0.1 mol/l, pH 6.0, containing EDTA 5 mmol/l at 4°C for 20 h.   8. Apply
the reaction mixture to a column (1.5 x 100 cm) of Ultrogel AcA 44 (LKB)
at a flow rate of 20-30 ml/h using sodium phosphate buffer, 0.1 mol/l, pH
6.5.   The fraction volume is 1.0 ml.   9. Read absorbance at 280 nm
and measure peroxidase activity in each fraction (Fig. 9).   10. Store
the conjugate at 4°C after addition of 1/98 volume of thimerosal 2-5 g/l
and 1/98 volume of bovine serum albumin 100 g/l.   Do not use NaN$_3$ as a
preservative, since it inactivates peroxidase.

CHARACTERIZATION AND EVALUATION OF ANTIBODY-ENZYME CONJUGATES PREPARED BY VARIOUS METHODS

Molecular Size

Fab'-horseradish peroxidase conjugates prepared by the hinge method, whether maleimide compounds or pyridyl disulfide compounds are used, are largely monomeric as being evident in the elution profile by gel filtration (Fig. 4) (Ishikawa et al., 1983a and c).   This is based on the facts that Fab' has a limited number (1-3) of thiol groups in the hinge and that the number of maleimide groups introduced per horseradish peroxidase molecule is very small (1-2) due to a limited number of amino groups.   The two-step glutaraldehyde method also provides a monomeric Fab'(or Fab)-horseradish peroxidase conjugate, while in the periodate method, polymers are formed besides a monomeric conjugate (Fig. 4) (Table 2) (Ishikawa et al., 1983a and c).

The conjugates of Fab' and other enzymes such as ß-D-galactosidase from Escherichia coli and alkaline phosphatase from calf intestine are heterogeneous.   In the Fab' conjugates with these enzymes prepared by the hinge method described above, enzyme molecules are not polymerized but associated with different numbers of Fab' molecules, while these enzymes are polymerized in the glutaraldehyde method (Table 2) (Ishikawa et al., 1983a).   Monomeric conjugates with these enzymes are obtained only by other methods (Imagawa et al., 1984a; Inoue et al., 1985).

The conjugates of IgG and enzymes prepared by the maleimide method and the pyridyl disulfide method are also heterogeneous (Imagawa et al., 1982a; Ishikawa et al., 1983a).   IgG molecules are not polymerized but associated with different numbers of peroxidase or alkaline phosphatase molecules.   In IgG-ß-D-galactosidase conjugates, ß-D-galactosidase molecules, which are not polymerized, are associated with different numbers of IgG molecules (Table 2).

Purity

Monomeric Fab'-horseradish peroxidase conjugates prepared by the hinge method are well separated from unconjugated peroxidase, Fab' and IgG, if any, by gel filtration on a column of Ultrogel AcA 44, provided that the column is sufficiently long (Fig. 4).   The purity of Fab'-peroxidase conjugate preparation was examined in two ways (Ishikawa et al., 1983a and b).   1) The conjugate preparation was applied to a column of (anti-IgG) IgG-Sepharose 4B to compare peroxidase activity in the effluent with the applied activity.   Peroxidase activity in the conjugate preparation obtained by the hinge method was almost completely (98-99 %) adsorbed to the column, indicating that there was little unconjugated peroxidase in the conjugate preparation.   2) Fluorescein-labeled Fab' was conjugated to peroxidase by the hinge method, and the conjugate preparation was applied to a column of concanavalin A-Sepharose 4B.   The adsorption of fluorescence intensity was 90-95 %, indicating that unconjugated $F(ab')_2$ was present only in a small proportion.   The conjugate with high purity can be readily obtained by elution from the column with 2-methyl-D-mannoside (Ishikawa et al., 1983a).

The purity of Fab'-peroxidase conjugates prepared by the glutaraldehyde method and the periodate method tends to be lower (Ishikawa et al., 1983a).   In the glutaraldehyde method, separation of the conjugate from unconjugated peroxidase is often not complete, since peroxidase remains unconjugated in a large proportion.

Fab' conjugates with ß-D-galactosidase and alkaline phosphatase prepared by the hinge method are well separated from unconjugated Fab', $F(ab')_2$ and IgG, if any, by gel filtration (Figs. 5 and 6).   However, the conjugates are not separated from unconjugated enzymes by gel filtration, and the proportion of unconjugated enzymes in the conjugate

Table 2.   Characteristics of Antibody-Enzyme Conjugates Prepared by Various Methods

Conjugation method	Conjugate	Polymeri-zation of enzyme	Molar ratio (Fab' or IgG : enzyme)	Recovery in the conjugate	
				Fab' or IgG	Enzyme
				%	
Hinge method					
with maleimide	Fab'-HRP	no	1 : 1	75-85	75-85
with pyridyl disulfide	Fab'-HRP	no	1 : 1	55	55
with maleimide	Fab'-GAL	no	1-3 : 1	70-80	100
with maleimide	Fab'-ALP	no	1-3 : 1	78	100
Non-hinge method					
with glutar-aldehyde	Fab'-HRP	no	1 : 1	25	5
	Fab'-GAL	yes	polymers	-	-
	Fab'-ALP	yes	polymers	-	-
with periodate	Fab'-HRP	yes and	polymers and	55	55
		no	monomer	20	20
Maleimide method	IgG-HRP	no	1 : 1-3	100	45
(Pyridyl disulfide	IgG-GAL	no	1-3 : 1	40	100
method)	IgG-ALP	no	1 : 1-3	100	55

HRP: horseradish peroxidase.   GAL: ß-D-galactosidase from Escherichia coli.   ALP: alkaline phosphatase from calf intestine.

preparations can be examined by application to a column of (anti-IgG) IgG-Sepharose 4B and comparison of the enzyme activity in the effluent with the applied activity.

Yield

In the hinge method for labeling of Fab' with horseradish peroxidase using N-succinimidyl-4-(N-maleimidemethyl)-cyclohexane-1-carboxylate and N-succinimidyl-6-maleimidohexanoate, the recovery of both the enzyme and Fab' in the conjugate is very high (75-85 %), while lower (55 %) in the hinge method using N-succinimidyl-3-(2-dithiopyridyl)propionate.   In the glutaraldehyde method, the recovery of horseradish peroxidase is very low (Table 2).

In the hinge method for labeling of Fab' with ß-D-galactosidase from Escherichia coli and alkaline phosphatase from calf intestine, the recovery of the enzymes in the conjugates is almost complete, and the recovery of Fab' is 70-80 % (Table 2).

Stability

The activity of horseradish peroxidase and ß-D-galactosidase from Escherichia coli in the conjugates prepared by the hinge method using maleimide compounds is stable for at least 4 years, when stored under the conditions described above.   Alkaline phosphatase from calf intestine may be similarly stable.

The antigen-binding activity of the conjugates prepared by the hinge method using maleimide compounds appears to be stable for years under the conditions described above, although the stability may varies among different kinds of antibodies.   In our laboratory, no significant loss of the antigen-binding activity is observed for 3-4 years of rabbit anti-human growth hormone Fab'-peroxidase conjugate, goat anti-human IgE Fab'-ß-D-galactosidase conjugate, goat anti-human alpha-fetoprotein Fab'-peroxidase conjugate, goat anti-human alpha-fetoprotein Fab'-ß-D-galactosidase conjugate, rabbit anti-human chorionic gonadotropin Fab'-peroxidase conjugate, rabbit anti-human ferritin Fab'-peroxidase and rabbit anti-human ferritin Fab'-ß-D-galactosidase conjugate.

The cross-link formed by the hinge method using maleimide compounds is also stable for years under the conditions described above.   This is confirmed in the following way.   After the storage under the conditions described above for 3-4 years, Fab'-peroxidase conjugate is subjected to gel filtration on a column of Ultrogel AcA 44, and fluorescein-labeled Fab'-ß-D-galactosidase conjugate is subjected to gel filtration on a column of Sepharose 6B.   There is little free peroxidase activity or little fluorescence intensity corresponding to free Fab'.

Usefulness

Comparison of Fab'- and IgG-enzyme conjugates   Fab'-enzyme conjugate can be readily prepared from IgG of most animals such as rabbit, goat, sheep, guinea pig and capybara and from mouse IgG_1, IgG_{2a} and IgG_3 (Figs. 1 and 4) and is more useful than IgG-enzyme conjugate in both immunohistochemical staining of tissue sections and enzyme immunoassay (Imagawa et al., 1982a and b; Ishikawa et al., 1983a, b and c). Fab'-enzyme conjugate gives a lower nonspecific binding and a higher sensitivity in the detection of antigen on the solid phase by enzyme immunoassay technique (Fig. 10).   Fab'-horseradish peroxidase conjugate more readily penetrates into tissue sections and provides a lower background staining in immunoenzymatic staining.   However, pepsin digestion of mouse IgG_{2b} gives fragments different from $F(ab')_2$ (Fig. 1), which are much less efficiently conjugated to horseradish peroxidase.   In some cases, the antigen-binding ability may be partially or totally lost by pepsin digestion and/or by reduction with 2-mercaptoethylamine (Hashida et al., 1985c).

Comparison of the hinge method and the non-hinge method   In the hinge method, thiol groups in the hinge which is located remote from the antigen-binding site are used for the conjugation to enzymes with full retention of the antigen-binding activity, while amino groups of Fab' or Fab are used in the non-hinge method, being accompanied by loss of the antigen-binding activity (Imagawa et al., 1982a and b; Ishikawa et al., 1983a, b and c) (Fig. 10).   In addition, Fab'-enzyme conjugate prepared by the hinge method gives a lower non-specific binding than Fab'- and Fab-enzyme conjugates prepared by the non-hinge method (Imagawa et al., 1982a and b; Ishikawa et al., 1983a, b and c) (Fig. 10).   Therefore, Fab'-enzyme conjugates prepared by the hinge method provide more sensitive enzyme immunoassays and more efficient immunohistochemical stainings of tissue sections than those prepared by the non-hinge method.

Comparison of the hinge method and the biotin-avidin method   In the biotin-avidin method, there is no need to conjugate antibodies to enzymes, and enzyme activity bound to antibodies through the biotin-avidin link increases with increasing numbers of biotin residues introduced into antibody molecules.   However, the non-specific binding of biotinylated antibodies to solid phase increases with increasing numbers of biotin residues introduced, and avidin-enzyme conjugate also binds to non-specifically bound biotinylated antibody molecules, enhancing the background signal.

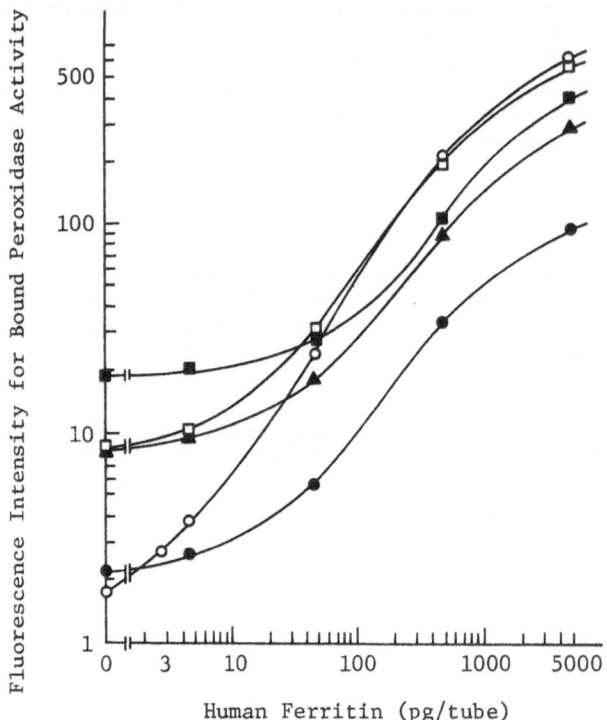

Fig. 10.   Dose response curves of human ferritin by sandwich enzyme
            immunoassay technique using horseradish peroxidase conjugates
            with rabbit IgG and its fragments (Imagawa et al., 1982a).
            Open circles: Fab'-peroxidase conjugate prepared by the hinge
            method using N-succinimidyl-4-(N-maleimidomethyl)cyclohexane-
            1-carboxylate.   Closed circles: Fab'-peroxidase conjugate
            by the glutaraldehyde method.   Closed triangles: monomeric
            Fab'-peroxidase conjugate by the periodate method.   Closed
            squares: polymeric Fab'-peroxidase conjugate prepared by the
            periodate method.   Open squares: IgG-horseradish peroxidase
            conjugate by the maleimide method.

     Application   Using affinity-purified Fab'-peroxidase and ß-D-
galactosidase conjugates prepared by the hinge method, the detection
limits of macromolecular antigens by sandwich enzyme immunoassay technique
are as little as less than 1 attomole (Table 3).   We have described a
highly sensitive sandwich enzyme immunoassay for human chorionic gonadotropin (hCG) using affinity-purified rabbit anti-hCG Fab'-ß-D-galacto-
sidase conjugate prepared by the hinge method (Kondo et al., 1984).
This has made it possible to measure the basal serum level of hCG in
normal subjects using 0.05 ml of serum (Imagawa et al., 1984c), which has
not been possible by radioimmunoassay (Borkowski et al., 1984).   We
have described a highly sensitive sandwich enzyme immunoassay for human
growth hormone (hGH) using affinity-purified rabbit anti-hGH Fab'-peroxi-
dase conjugate prepared by the hinge method using a maleimide compound
(Hashida et al., 1983).   The detection limit of hGH is 60 fg/tube or 3
ng/1 of serum.   This sensitive assay has made it possible to measure
the basal serum level (Hashida et al., 1985a) and the urine level (Hashida

Table 3.   Detection Limit of Macromolecular Antigens by Sandwich Enzyme
Immunoassay Technique Using Affinity-Purified Fab'-Enzyme
Conjugate Prepared by the Hinge Method

Human antigen	Label enzyme	Detection limit	Reference
		amol/tube	
Ferritin	GAL	0.05	Imagawa et al., 1982c
	HRP	0.1	Imagawa et al., 1984b
IgE	GAL	0.1	Hashida and Ishikawa, 1985d
Alpha-fetoprotein	GAL	0.3	ibid.
	HRP	0.8	ibid.
Chorionic gonadotropin	GAL	0.7	ibid.
	HRP	0.7	ibid.
Thyroid-stimulating hormone	HRP	0.1	Inoue et al., 1986
Growth hormone	HRP	3	Hashida et al., 1983
Insulin	HRP	4	Ruan et al., 1985a and b

An antibody IgG-coated polystyrene ball of 3.2 mm in diameter was
incubated with antigen in 0.15 ml and subsequently with affinity-purified
Fab'-enzyme conjugate in 0.15 ml.
GAL: ß-D-galactosidase from _Escherichia coli_
HRP: horseradish peroxidase

et al., 1985b) of hGH in normal subjects using 0.02 ml of serum and 0.05
ml of urine, which has not been possible by radioimmunoassay (Drobny et
al., 1983; Baumann and Abramson, 1983).   We have also described a
highly sensitive sandwich enzyme immunoassay for human thyroid-stimulating
hormone (hTSH) using affinity-purified rabbit anti-hCG Fab'-peroxidase
conjugate and affinity-purified rabbit anti-hTSH Fab'-peroxidase conjugate
prepared by the hinge method using a maleimide compound (Inoue et al.,
1986).   Using the former conjugate, the detection limit of hTSH is 0.05
nU (0.1 amol)/tube and 1 nU/ml of serum.   Using the latter conjugate,
the detection limit of hTSH is 0.3 nU/tube or 6 nU/ml, and it has been
possible to measure extremely low serum levels even in pregnant and
postmenopausal women, which have not been measured by radioimmunoassay.

ENZYME-LABELING OF ANTIGENS

    Antigens can be conjugated to enzymes in the same way as IgG.
Thiol groups or maleimide groups are introduced into antigen molecules and
reacted with maleimide groups or thiol groups introduced into enzyme
molecules.
    In general, however, antigen-enzyme conjugates are heterogeneous.
Antigen molecules are not polymerized but associated with different
numbers of enzyme molecules.   Or enzyme molecules are not polymerized
but associated with different numbers of antigen molecules.   Monomeric

antigen-enzyme conjugates can be prepared only by complex methods (Imagawa et al., 1984a; Inoue et al., 1985).   And it is difficult to conjugate enzyme molecules to a specified site of antigen molecules except for very particular ones.   As a result, the binding of monoclonal antibody to antigen-enzyme conjugate may happen to be prevented by steric hindrance.

REFERENCES

Avrameas, S., and Ternynck, T., 1971, Peroxidase labelled antibody and Fab conjugates with enhanced intracellular penetration, Immunochemistry, 8: 1175.

Baumann, G., and Abramson, E. C., 1983, Urinary growth hormone in man: Evidence for multiple molecular forms, J. Clin. Endocrinol. Metab., 56: 305.

Beaucamp, K., Bergmeyer, H. U., and Beutler, H-O., 1974, Coenzymes, metabolites, and other biochemical reagents, in: "Methods of Enzymatic Analysis" Bergmeyer, H. U., ed., Academic Press, New York, p. 545.

Borkowski, A., Puttaert, V., Gyling, M., Muquardt, C., and Body, J. J., 1984, Human chorionic gonadotropin-like substance in plasma of normal nonpregnant subjects and women with breast cancer, J. Clin. Endocrinol. Metab., 58: 1171.

Craven, G. R., Steers, Jr., E., and Anfinsen, C. B., 1965, Purification, composition, and molecular weight of the ß-galactosidase of Escherichia coli K12, J. Biol. Chem., 240: 2468.

Dissanayake, S., and Hay, F. C., 1975, Pepsin digestion of mouse IgG immunoglobulins subfragments of the Fc region, Immunochemistry, 12: 373.

Dorrington, K. J., and Tanford, C., 1970, Molecular size and conformation of immunoglobulins, Adv. Immunol., 12: 333.

Drobny, E. C., Amburn, K., and Baumann, G., 1983, Circadian variation of basal plasma growth hormone in man, J. Clin. Endocrinol. Metab., 57: 524.

Engström, L., 1961, Studies on calf-intestinal alkaline phosphatase. I. Chromatographic purification, microheterogeneity and some other properties of the purified enzyme, Biochim. Biophys. Acta, 52: 36.

Ey, P. L., Prowse, S. J., and Jenkin, C. R., 1978, Isolation of pure IgG_1, IgG_{2a} and IgG_{2b} immunoglobulins from mouse serum using protein A-Sepharose, Immunochemistry, 15: 429.

Gorini, G., Medgyesi, G. A., and Doria, G., 1969, Heterogeneity of mouse myeloma γG globulins as revealed by enzymatic proteolysis, J. Immunol., 103: 1132.

Grassetti, D. R., and Murray, Jr., J. F., 1967, Determination of sulfhydryl groups with 2,2'- or 4,4'-dithiodipyridine, Arch. Biochem. Biophys., 119: 41.

Guesdon, J.-L., Ternynck, T., and Avrameas, S., 1979, The use of avidin-biotin interaction in immunoenzymatic techniques, J. Histochem. Cytochem., 27: 1131.

Hashida, S., Nakagawa, K., Imagawa, M., Inoue, S., Yoshitake, S., Ishikawa, E., Endo, Y., Ohtaki, S., Ichioka, Y., and Nakajima, K., 1983, Use of inorganic salts to minimize serum interference in a sandwich enzyme immunoassay for human growth hormone using Fab'-horseradish peroxidase conjugate, Clin. Chim. Acta, 135: 263.

Hashida, S., Imagawa, M., Inoue, S., Ruan, K-h., and Ishikawa, E., 1984, More useful maleimide compounds for the conjugation of Fab' to horseradish peroxidase through thiol groups in the hinge, J. Appl. Biochem., 6: 56.

Hashida, S., Nakagawa, K., Ishikawa, E., and Ohtaki, S., 1985a, Basal level of human growth hormone (hGH) in normal serum, Clin. Chim. Acta, 151: 185.

Hashida, S., Ishikawa, E., Nakagawa, K., Ohtaki, S., Ichioka, T., and Nakajima, K., 1985b, Demonstration of human growth hormone in normal urine by a highly specific and sensitive sandwich enzyme immunoassay, Anal. Lett., 18(B13): 1623.

Hashida, S., Imagawa, M., Ishikawa, E., and Freytag, J. W., 1985c, A simple method for the conjugation of affinity-purified Fab' to horseradish peroxidase and ß-D-galactosidase from Escherichia coli, J. Immunoassay, 6: 111.

Hashida, S., and Ishikawa, E., 1985d, Use of normal IgG and its fragments to lower the non-specific binding of Fab'-enzyme conjugates in sandwich enzyme immunoassay, Anal. Lett., 18(B9): 1143.

Imagawa, M., Yoshitake, S., Hamaguchi, Y., Ishikawa, E., Niitsu, Y., Urushizaki, I., Kanazawa, R., Tachibana, S., Nakazawa, N., and Ogawa, H., 1982a, Characteristics and evaluation of antibody-horseradish peroxidase conjugates prepared by using a maleimide compound, glutaraldehyde, and periodate, J. Appl. Biochem. 4: 41.

Imagawa, M., Hashida, S., Ishikawa, E., and Sumiyoshi, A., 1982b, Evaluation of Fab'-horseradish peroxidase conjugates prepared using pyridyl disulfide compounds, J. Appl. Biochem., 4: 400.

Imagawa, M., Yoshitake, S., Ishikawa, E., Niitsu, Y., Urushizaki, I., Kanazawa, R., Tachibana, S., Nakazawa, N., and Ogawa, H., 1982c, Development of a highly sensitive sandwich enzyme immunoassay for human ferritin using affinity-purified anti-ferritin labelled with ß-D-galactosidase from Escherichia coli, Clin. Chim. Acta, 121: 277.

Imagawa, M., Hashida, S., Ishikawa, E., and Freytag, J. W., 1984a, Preparation of a monomeric 2,4-dinitrophenyl Fab'-ß-D-galactosidase conjugate for immunoenzymometric assay, J. Biochem., 96: 1727.

Imagawa, M., Hashida, S., Ishikawa, E., Niitsu, Y., Urushizaki, I., Kanazawa, R., Tachibana, S., Nakazawa, N., and Ogawa, H., 1984b, Comparison of ß-D-galactosidase from Escherichia coli and horseradish peroxidase as labels of anti-human ferritin Fab' by sandwich enzyme immunoassay technique, J. Biochem., 96: 659.

Imagawa, M., Yoshitake, S., Ishikawa, E., Kanoh, E., Tsunetoshi, Y., Iwasa, S., Konishi, E., Kondo, K., Ichioka, Y., and Nakajima, K., 1984c, Level of human chorionic gonadotropin-like substance in serum of normal subjects, Anal. Lett., 17(B7): 575.

Inoue, S., Hashida, S., Tanaka, K., Imagawa, M., and Ishikawa, E., 1985, Preparation of monomeric affinity-purified Fab'-ß-D-galactosidase conjugate for immunoenzymometric assay, Anal. Lett., 18(B11): 1331.

Inoue, S., Hashida, S., Ishikawa, E., Mori, T., Imura, H., Ogawa, H., Ichioka, T., and Nakajima, K., 1986, Highly sensitive sandwich enzyme immunoassay for human thyroid-stimulating hormone (hTSH) in serum using monoclonal anti-hTSH ß-subunit IgG_1-coated polystyrene balls and polyclonal anti-human chorionic gonadotropin Fab'-horseradish peroxidase conjugate, Anal. Lett., in press.

Ishikawa, E., and Kato, K., Ultrasensitive enzyme immunoassay, 1978, Scand. J. Immunol. 8 (Suppl. 7): 43.

Ishikawa, E., Imagawa, M., Hashida, S., Yoshitake, S., Hamaguchi, Y., and Ueno, T., 1983a, Enzyme-labeling of antibodies and their fragments for enzyme immunoassay and immunohistochemical staining, J. Immunoassay, 4: 209.

Ishikawa, E., Imagawa, M., and Hashida, S., 1983b, Ultrasensitive enzyme immunoassay using fluorogenic, luminogenic, radioactive and related substrates and factors to limit the sensitivity, Develop. Immunol., 18: 219.

Ishikawa, E., Yoshitake, S., Imagawa, M., and Sumiyoshi, A., 1983c, Preparation of monomeric Fab'-horseradish peroxidase conjugate using thiol groups in the hinge and its evaluation in enzyme immunoassay and immunohistochemical staining, Ann. New York Acad. Sci., 420: 74.

Jaquet, H., and Cebra, J. J., 1965, Comparison of two precipitaing derivatives of rabbit antibody : Fragment I dimer and the product of pepsin digestion, Biochemistry, 4: 954.

Johannsson, A., Stanley, C. J., and Self, C. H., 1985, A fast highly sensitive colorimetric enzyme immunoassay system demonstrating benefits of enzyme amplification in clinical chemistry, Clin. Chim. Acta, 148, 119.

Keilin, D., and Hartree, E. F., 1951, Purification of horse-radish peroxidase and comparison of its properties with those of catalase and methaemoglobin, Biochem. J., 49: 88.

Kondo, K., Imagawa, M., Iwasa, S., Kitada, C., Konishi, E., Suzuki, N., Yamoto, M., Nakano, R., Yoshitake, S., and Ishikawa, E., 1984, A specific and sensitive sandwich enzyme immunoassay for human chorionic gonadotropin using antibodies against the carboxy-terminal portion of the ß-subunit, Clin. Chim. Acta, 138: 229.

Lamoyi, E., and Nisonoff, A., 1983, Preparation of F(ab')$_2$ fragments from mouse IgG of various subclasses, J. Immunol. Methods, 56: 235.

Lindmark, R., Thorén-Tolling, K., and Sjöquist, J., 1983, Binding of immunoglobulins to protein A and immunoglobulin levels in mammalian sera, J. Immunol. Methods, 62: 1.

Mandy, W. J., and Nisonoff, A., 1963, Effect of reduction of several disulfide bonds on the properties and recombination of univalent fragments of rabbit antibody, J. Biol. Chem., 238: 206.

Morton, R. K., 1955, Some properties of alkaline phosphatase of cow's milk and calf intestinal mucosa, Biochem. J., 60: 573.

Olive, C., and Levy, H. R., 1971, Glucose 6-phosphate dehydrogenase from Leuconostoc mesenteroides, J. Biol. Chem., 246; 2043.

Palmer, J. L., and Nisonoff, A., 1964, Dissociation of rabbit γ-globulin into half molecules after reduction of one labile disulfide bond, Biochemistry, 3: 863.

Parham, P., 1983, On the fragmentation of monoclonal IgG_1, IgG_{2a}, and IgG_{2b} from BALB/c mice, J. Immunol., 131: 2895.

Penefsky, H. S., 1979, A centrifuged-column procedure for the measurement of ligand binding by beef heart F_1. in: "Methods in Enzymology" Colowick, S. P., and Kaplan, N. O., eds., Vol. LVI, Academic Press, New York, p. 527.

Ruan, K-h., Imagawa, M., Hashida, S., and Ishikawa, E., 1984, An improved method for the conjugation of Fab' to ß-D-galactosidase from Escherichia coli, Anal. Lett., 17(B7): 539.

Ruan, K-h., Hashida, S., Yoshitake, S., Ishikawa, E., Wakisaka, O., Yamamoto, Y., Ichioka, T., and Nakajima, K., 1985a, A micro-scale affinity-purification of Fab'-horseradish peroxidase conjugates and its use for sandwich enzyme immunoassay of insulin in human serum, Clin. Chim. Acta, 147: 167.

Ruan, K-h., Hashida, S., Yoshitake, S., Ishikawa, E., Wakisaka, O., Yamamoto, Y., Ichioka, T., and Nakajima, K, 1985b, A more sensitive and less time-consuming sandwich enzyme immunoassay for insulin in human serum with less serum interference, Ann. Clin. Biochem., in press.

Seppälä, I., Sarvas, H., Péterfy, F., and Mäkelä, O., 1981, The four subclasses of IgG can be isolated from mouse serum by using protein A-Sepharose, Scand. J. Immunol., 14: 335.

Stuchbury, T., Shipton, M., Norris, R., Malthouse, J. P. G., and Brocklehurst, K., 1975, A reporter group delivery system with both absolute and selective specificity for thiol groups and an improved fluorescent probe containing the 7-nitrobenzo-2-oxa-1,3-diazole moiety, Biochem. J., 151: 417.

Svasti, J., and Milstein, C., 1972, The disulphide bridges of a mouse immunoglobulin G1 protein, Biochem. J., 126: 837.

Tanaka, K., and Ishikawa, E., 1984, Highly sensitive bioluminescent assay of dehydrogenases using NAD(P)H:FMN oxidoreductase and luciferase from Photobacterium fischeri, Anal. Lett., 17(B18): 2025.

Udenfriend, S., Stein, S., Böhlen, P., Dairman, W., Leimgruber, W., and
    Weigele, M., 1972, Fluorescamine: A reagent for assay of amino
    acids, peptides, proteins, and primary amines in the picomole range,
    Science, 178: 871.
Utsumi, S., and Karush, F., 1965, Peptic fragmentation rabbit γG-immuno-
    globulin, Biochemistry, 4: 1766.
Wilson, M. B., and Nakane, P. K., 1978, Recent developments in the
    periodate method of conjugating horseradish peroxidase (HRPO) to
    antibodies, in: "Immunofluorescence and Related Staining Techniques"
    Knapp, W., Holubar, K., and G. Wick, eds., Elsevier/North-Holland
    Biomedical Press, Amsterdam, p. 215.
Yoshitake, S., Hamaguchi, Y., and Ishikawa, E., 1979, Efficient conju-
    gation of rabbit Fab' with ß-D-galactosidase from Escherichia coli,
    Scand. J. Immunol., 10: 81.

CHROMOGENIC SUBSTRATES FOR ENZYME IMMUNOASSAY

Bärbel Porstmann and Tomas Porstmann

Institute of Pathological and Clinical Biochemistry and
Institute of Medical Immunology, Faculty of Medicine (Charite)
Humboldt University, Berlin, GDR

INTRODUCTION

General Remarks

Out of a variety of enzymes which have been described according to their kinetics and structure only a few have turned out to be marker enzymes in the enzyme immunoassay (EIA). The ones that are mostly used are horseradish peroxidase (HRP), calf intestinal alkaline phosphatase (AP) and ß-galactosidase from E. coli (ßGal).The suitability of enzymes as markers is not only determined by their great molar activity that is to be largely maintained even after binding to immunoreactants, their stability and commercial availability but especially by their simple, practicable and sensitive detectability.

Enzyme reactions are quantifiable by various measuring techniques that are determined by the respective substrate and the features of the product resulting from this, such as luminescence technology, fluori-, potentio-, ampero- or colorimetry. The most frequently applied technique however is measuring a coloured product of a reaction which is formed from a colourless substrate due to the catalytic activity of the marker enzyme. In this the chromogen can be a direct product of the reaction of the marker enzyme or be formed as a consequence of a reaction coupled with an indicator enzyme or an auxiliary / indicator enzyme system. Such combined reactions as recycling or amplifying systems result in a multiplication of the chromogen formation and through this in a particularly sensitive detection of enzymes.

Colorimetry offers many advantages: great sensitivity, easy handling, simple and cheap measuring equipment and visual way of evaluating the material. Haptens and antigens can be detected and quantified with same sensitivity compared with radio immunoassay (pico- to femtomol range). Chromogen concentrations can be differentiated both by simple photometers within the visible range (mostly 400 nm to 600 nm) and semi-quantitatively or qualitatively by naked eye. Hence the EIA makes visual yes-or-no decisions possible, which opened up a new era: for highly sensitive, specific and practicable screening tests under field conditions.

Principles of Measuring Enzymes in EIA

In enzymes activity determination with chromogens, too, the enzyme activity is proportional in a linear way to the chromogenic product concentration within defined enzyme concentration levels. Depending on either the EIA or EMIT technique the chromogen formation is directly (e.g. sandwich technique) or indirectly (e.g. competitive technique) proportional to the antigen concentration. To determine the enzyme activity in EIA various measuring techniques have been employed and described: kinetic measuring, two-point and end-point measuring.In order to realize a high throughput of samples in kinetic measuring techniques it is necessary to achieve short times of reaction. As the extinction changes during this time are essentially smaller than during the long substrate reaction times of two- and end-point measuring the EIA's employing kinetic measuring techniques are usually less sensitive. On top of this stopping reagents are often applied in two- and end-point measuring which change the pH-value (strong acids or leaches) dramatically thus producing a bathochromic or hypsochromic shift in the absorption maximums due to the feature of many chromogens to act as pH-indicator. Resulting from this the absorption coefficient is partly rising tremendously. The lower sensitivity resulting from kinetic measuring techniques makes them unsuitable for proteins or hormons whose levels in biological fluids are placed within the ng/l range, e.g. prostatic acid phosphatase, somatotropic hormone and alpha-1-fetoprotein (AFP), or when a low detection limit is required, e.g.for hepatitis B-surface antigen (HBsAg). These are suitable for the quantification of proteins in higher concentrations ,e.g. C-reactive protein in newborns or IgM in cerebrospinal fluid. When using a high quality photometer the kinetic activity determination will produce more precise results than two-point measuring. When applying the latter the use of stopping reagents can cause differences between the substrate reaction times of the samples in comparison with the standards in particular with a large number of tests and manual handling. Time differences influence precision and accuracy of the determinations the more dramatically the greater the enzyme activity and the shorter the enzyme reaction times. When using photometers with horizontal light path the results are even more incorrect due to unprecise dosage of the stopping reagent.

Two-point measuring is most widely used in EIAs. The reaction time of the marker enzyme is chosen according to which the measuring range will be required for the respective antigen. Mostly it is between 30 and 60 min but can also last hours with a continuous substrate turnover. However the reaction temperature is decisive. As a rule marker enzymes effect a higher substrate turnover at 37°C than at room temperature but also become inactive sooner. Partly temperature-dependent enzyme kinetics differ widely when using various chromogens. For reasons of practicability and due to a lack of measuring equipment which would ensure exactly the same temperature in all wells of microtitre plates enzyme reactions applying two-point measuring in EIA are mostly carried out at room temperature (RT). The great variation in RT is a major cause of differing extinctions of the standard curves in several series. Starting the enzyme reaction it is often neglected that a newly prepared substrate solution, whose buffer is stored at 4°C, must be warmed up to reach RT. The different standard curves influence the measuring range which, if not taken into account, can lead to less precision and accuracy of antigen determination in EIA. This fact must be taken into consideration for the design of future measuring gauges. A mathematical evaluation and a definition of the measuring range by computers is to be demanded for all EIAs with inherent variations in temperature for each series.

What May be Checked by Experiment Prior to Use of Chromogens?

(a) The optimum reaction conditions for the marker enzyme must be determined, e.g. kind of buffer, ionic strenght, pH-value. The optimum chromogen concentration must be found. It should be at least ten times the Km-value of the enzyme with a good solubility of the chromogen and a lack of substrate inhibition. With two substrates (e.g. for HRP), that influence each other in their binding the optimum concentrations of both must be determined depending on the pH-value. It must be checked if the optimum substrate concentration depends on reaction time and temperature.
(b) For the respective chromogen temperature-dependent kinetics of the marker enzyme must be determined for a given period of time chosen for the substrate reaction in EIA.
(c) Two-point measuring requires a stopping reagent which fully inhibits the enzyme activity. It must be tested as well if the stopping reagent influences the extinction and stability of the blank value. The stability of the chromogenic product after reaction stop must be evaluated under different light conditions, a precipitation must be excluded.
(d) The absorption spectrum of the chromogen must be defined, a concentration- dependent batho- or hypsochromic shift of the absorption maximum by the stopping reagent must be evaluated. Absorption values in EIA should be measured at the absorption maximum determined before.

Optimum Characteristics for Using a Chromogen in EIA

The substrate should be odourless and colourless, neither be mutagenic, nor carcinogenic or poisonous nor cause allergic reactions.It should not be volatile or aggressive to vessels or photometers. It must be soluble in water and stable both as substrate and product. The chromogen must be inexpensive and commercially available. The substrate's Km-value should be very low, the turnover rate of the enzyme very high. The formation of chromogenic products must be in proportion to the enzyme concentration (or activity) within an extinction range to be defined. The colour-intensity and enzyme concentration must be in linear proportion over a wide range. The chromogenic product should have a great molar extinction coefficient preferably with a broad absorption maximum. Chromogenic products of green, red and blue colours are particularly suitable for visual evaluations. There must be a stopping reagent for the chromogen that terminates the enzyme reaction and keeps the substrate colourless,thus ensuring a stable blank value, and also keeps the colour of a chromogenic product stable for several hours. The extinction coefficient should not decline but rather rise owing to the stopping solution. Stability of chromogenic substrate and product should not be influenced by light.

There is hardly any chromogenic substrate for all marker enzymes which meets all requirement in full. But a great number of chromogens have quite a few advantages so that they turn out to be often used as substrates in EIA. Following this we want to present and discuss experimental data regarding chromogenic substrate reactions of the marker enzymes most frequently used. Listed in Material and Methods are the optimum substrate and stopping solutions found out by various working groups for different chromogens and enzymes.

MATERIAL AND METHODS

The enzyme immunoassays described in the text for the quantification of HBsAg, human lysozyme, AFP and antibodies against tetanus toxin are all based on the principle of immunoenzymometric (sandwich) assays.The wells of polystyrene plates were coated with 0.1 ml of antigen-specific IgG or tetanus toxin respectively.The concentrations ranged between 2 and 10 mg/ l according to the test applied. After washing three times with PBS containing 0.1 % Tween 20 (v/v) 0.1 ml of the standard samples, diluted in PBS containing 0.1 % Tween 20 (v/v) and 5 % horse serum (v/v) are incubated. Depending on the parameter the incubation was performed at room temperature or 37°C for 60 to 120 min. After having been washed the wells thrice again IgG-enzyme-conjugate (0.1 to 1 mg/ l) is incubated under reaction conditions as described for standard samples. Three times washing is followed then by incubation with 0.1 ml of the enzyme substrate solution (30 min at room temperature if not stated otherwise). The reaction is stopped by the same volume (0.1 ml) of stopping reagent and the absorbances are measured against the EIA blank values (analyte free serum). The detection limit in EIA is defined as that concentration of the analyte per litre of sample material which gives an absorbance more than the mean absorbance of an analyte free serum + 3 S. The detection limits of enzymes in our experiments were determined at 25°C, 30 min reaction time and after termination the reaction with equal volumes of stopping solution and defined as concentration per litre of stopped substrate solution (Porstmann et al.,1985 a).

Subsequently we put together the optimum composition of substrate and stopping solutions for activity measuring of various marker enzymes:

Chromogens for Horseradish Peroxidase

O-phenylenediamine (oPD),solution A and B: o-phenylenediamine (Merck, Darmstadt, FRG), (20 mmol/ l for solution A , 10 mmol/ l for solution B) and hydrogen peroxide (12 mmol/ l for solution A , 5.5 mmol/ l for solution B) dissolved in citrate buffer (0.1 mol/ l) pH 5.0.
Stopping solution: sulphuric acid (2 mol/1) containing sodium sulphite (0.1 mol/ l).
Absorbance measurement at λ = 492 nm.

2,2'-Azino-di(3-ethylbenzthiazoline sulphonic acid-6) (ABTS): ABTS (Boehringer, Mannheim, FRG) (2 mmol/ l) and hydrogen peroxide (2.5 mmol/1) dissolved in acetate buffer (0.1 mol/ l) pH 4.2.
Stopping solution: sodium azide (10 mmol/ l).
Absorbance measurement at λ = 414 nm or λ = 405 nm.

O-dianisidine (oDia): o-dianisidine (Serva,Heidelberg,FRG) (0.35 mmol/ l, predissolved in methanol , 1% w/v) and hydrogen peroxide (0.7 mmol/ l) dissolved in acetate buffer (0.01 mol/ l) pH 5.0 containing Tween 20 (0.1 % v/v) and chelaplex III (0.5 mmol/ l).
Stopping solution: hydrochloric acid (5 mol/ l).
Absorbance measurement at λ = 530 nm.

5-Aminosalicylic acid (5AS): 5-aminosalicylic acid (Ferrak,Berlin,FRG) (1 g/ l) and hydrogen peroxide (1.5 mmol/ l)dissolved in phosphate buffer (0.01 mol/ l) pH 6.8 containing chelaplex III (0.5 mmol/ l). The pH-value of the substrate solution is adjusted to 6.0 before adding the peroxide.
Stopping solution: not described.
Absorbance measurement at λ = 474 nm.

γ,γ'-4,-4'-Diamino-3,3'-(biphenylylenedioxy)dibutyric acid (dicarboxidine):dicarboxidine (Kabi Diagnostica, Stockholm, Sweden) (1 mmol/ l)and hydrogen peroxide (0.2 mmol/ l) dissolved in phosphate buffer (0.05 mol/ l) pH 6.9.
Stopping solution: not described.
Absorbance measurement at λ = 440 nm.

3,3',5,5'-tetramethylbenzidine (TMB): tetramethylbenzidine (Fluka,Buchs,Switzerland) (1mmol/ 1) , predissolved in dimethylsulfoxide (0.1 mol/ 1) and hydrogen peroxide (3.0 mmol/ 1) dissolved in sodium acetate/citric acid buffer (0.2 mol/ 1) pH 4.0 .
Stopping solution: sulphuric acid (1 mol/ 1).
Absorbance measurement at λ = 450 nm.

Phenol and 4-aminoantipyrine (Trinder-Emerson-reagent): aminoantipyrine (Riedel-de Haen, Hannover, FRG) (2 mmol/ 1), phenol (25 mmol/ 1) and hydrogen peroxide (0.8 mmol/ 1) dissolved in phosphate buffer (0.1 mol/ 1) pH 7.2.
Stopping solution: not described.
Absorbance measurement at λ = 492 nm.

3-Methyl-2-benzothiazolinone hydrazone (MBTH) and 3-(dimethylamino) benzoic acid (DMAB) (Ngo-Lenhoff-reagent):stock solution A - MBTH (0.6 mmol/ 1) and stock solution B - DMAB (30 mmol/ 1) (Aldrich Chemical Company, Inc.,Milwaukee, Wiskonsin), both dissolved in citric acid / phosphate buffer (0.1 mol/ 1) pH 7.0.
Immediately prior to use, the components A and B are combined as follows: 22 ml of stock A, 22 ml of stock DMAB, 0.24 ml hydrogen peroxide (30 % w/v) and 156 ml citric acid / phosphate buffer (0.1 mol/ 1) pH 7.0.
Stopping solution: sulphuric acid (0.2 mol/ 1).
Absorbance measurement at λ = 590 nm.

Chromogen for Alkaline Phosphatase

P-nitrophenyl phosphate (pNPP) :p-nitrophenylphosphate (Boehringer, Mannheim,FRG) (10 mmol/ 1) dissolved in diethanolamine buffer (0.1 mol/ 1) pH 9.8 containing magnesium chloride (1 mmol/ 1).
Stopping solution: sodium carbonate (0.5 mol/ 1).
Absorbance measurement at λ = 405 nm.

Chromogens for ß-Galactosidase

O-nitrophenyl-ß-D-galactopyranoside (oNPG): o-nitrophenyl-galacto-pyranoside (Boehringer,Mannheim,FRG) (2.7 mmol/ 1), magnesium titriplex (2 mmol/ 1), mangane sulphate (0.2 mmol/ 1), magnesium sulphate (1 mmol/1) and ß-mercaptoethanol (0.6 mmol/ 1) dissolved in sodium phosphate buffer (0.1 mol/ 1) pH 7.0.
Stopping solution: sodium carbonate (0.5 mol/ 1)
Absorbance measurement at λ = 420 nm.

Chlorophenolic red -ß-D-galactopyranoside (CPRG): chlorophenolic red -galactopyranoside (Boehringer,Mannheim,FRG) (2.7 mmol/ 1), buffer as for the oNPG reaction.
Absorbance measurement at λ = 574 nm.

Resorufin-ß-D-galactopyranoside (RG): resorufin-galactopyranoside (Boehringer,Mannheim,FRG) (0.4 mmol/ 1), buffer as for the oNPG reaction.
Absorbance measurement at λ = 570 nm.

Chromogen for Urease

Bromcresol purple (BCP): 80 mg bromcresol purple (Gurr,London,UK) are dissolved in 14.8 ml NaOH (0.01 mol/ 1) and the volume is made to 1 l with deionized water containing urea (1 g/ 1) and EDTA (0.2 mmol/ 1). The pH is adjusted to 4.8.
Stopping solution: thiomersal,optimum concentration is to be tested.
Absorbance measurement at λ = 588 nm.

Chromogen for Acetylcholine Esterase

Ellman's reagent: acetylthiocholine iodide (0.75 mmol/ 1) and 5,5'-dithiobis (2-nitrobenzoic acid) (Merck,Darmstadt,FRG) (0.5 mmol/ 1) dissolved in phosphate buffer (0.1 mol/ 1) pH 7.4.
Stopping solution: not described.
Absorbance measurement at λ = 412 nm.

Coupled Enzyme Reaction with Glucose 6-Phosphate Dehydrogenase as Marker Enzyme (homogeneous assay)

Solution A: citric acid (18 mmol/ 1) / disodium hydrogen phosphate (32 mmol/ 1) buffer, glucose 6-phosphate (52 mmol/ 1), nicotinamide adenine dinucleotide (NAD) (15 mmol/ 1), nitro blue tetrazolium salt (NBT) (1.5 mmol/ 1), Triton X-100 (7% v/v) and sodium azide (1.5 mmol/ 1) dissolved in citric acid (18 mmol/ 1) / disodium hydrogen phosphate (32 mmol/ 1) buffer pH 5.7.

Solution B: diaphorase (DI) (2960 IU/ 1) , oxamic acid (0.05 mol/ 1),magnesium chloride (4 mmol/ 1), sodium azide (1.5 mmol/ 1), bovine serum albumin (10 g/ 1), glucose 6-phosphate dehydrogenase (covalently linked to an immunoreactant, concentration dependent on the test) dissolved in tris buffer (0.1 mol/ 1) pH 8.1.
The volumes of solution A and B are to be mixed in proportions of 1 : 2.
Stopping solution: hydrochloric acid (1 mol/ 1) containing Triton X-100 (1 % v/v).
Absorbance measurement at λ = 580 nm.

Amplified System with Alkaline Phosphatase

First step: Incubation of solution A for 20 min at 25°C.
Solution A consists on diethanolamine buffer (50 mmol/ 1) pH 9.5 containing nicotinamide adenine dinucleotide phosphate (NADP) (0.1 mmol/ 1) and magnesium chloride (1.0 mmol/ 1).

Second step: Incubation of solution B (addition to solution A) for a further 10 min. Solution B consists on sodium phosphate buffer (25 mmol/1) pH 7.2 containing alcohol dehydrogenase (ADH) (100 mg/ 1), diaphorase (50 mg/ 1) and p-iodonitrotetrazolium violet (INT) (0.55 mmol/ 1).
Stopping solution: sulphuric acid (0.2 mol/ 1).
The volumes of solution A, B and C are to be added in proportions of 1 : 2 : 0.5 .
Absorbance measurement at λ = 492 nm.

RESULTS AND DISCUSSION

Chromogens for Horseradish Peroxidase (HRP)

HRP is, by far, the most frequently used marker enzyme for enzyme immunoassays. In the meantime many chromogenic substrates have been described for HRP. Peroxidase acts as catalyst for dehydrogenation of many organic compounds (phenols, aromatic amines, hydroquinones, benzidine derivatives) according to the following principle:

$$2 \text{ SH} + H_2O_2 \xrightarrow{\text{HRP}} 2 \text{ S} + 2 H_2O \ .$$

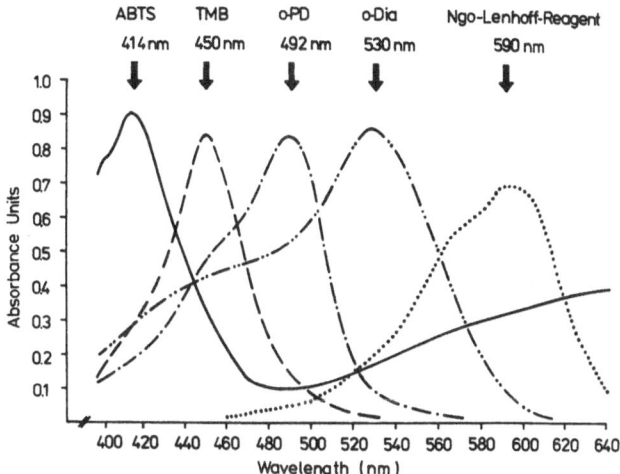

Fig.1. Absorbance spectra of different chromogenic products
after conversion by HRP and termination the reaction
as described in Material and Methods. Maximum absorbance is indicated by an arrow.
——— ABTS, — — — TMB, —·— oPD, —··— oDia,
••••••• Ngo-Lenhoff reagent.

It has little specificity refering to hydrogen donators, a big one,
however, refering to peroxide. Apart from hydrogen peroxide only O-OH
groups are used (also acetyl, methyl and ethylhydroperoxides). Two main
types of reaction have been distinguished : (a) oxidation of redox
indicators (such as ABTS or o-dianisidine) and (b) oxidative coupling of an
amino aromatic with another aryl compound, e.g. 4-aminoantipyrine with
phenol or substituted phenol and 3-methyl-2-benzothiazolinone hydrazone
(MBTH) with 3-(dimethylamino) benzoic acid (DMAB), N,N-dimethylaniline or
with 2-hydroxy-3,5-dichloro-benzene-sulfonate. The absorption spectra with
maximum absorbances for the most commonly used chromogens for HRP are
evident from fig. 1.

(a) Oxidation of redox indicators

The extent of HRP action during reaction depends on the structure and redox
potential of the chromogen.

O-phenylenediamine (oPD) is currently the most sensitive chromogen for
HRP (fig. 2). It is oxidized by HRP and condenses to 2,2-diamino-azobenzol
(2,2'DAB).

2,2'DAB shows a wide absorption maximum at λ = 450 nm with a pH-value ranging around 5. It is shifted to λ = 492 nm when the pH-value is lowered to 1.0 with the extinction coefficient growing, depending on the final proton concentration.The extinction coefficient at λ = 492 nm in 1 mol/ 1 acid was determined as 19.5 cm^2/ μmol (Hildebrandt, 1986).Differences in HRP activity have been recorded due to use of different preparations of oPD. The use of o-phenylenediamine-dihydrochloride generally results in extinctions twice or thrice lower than achieved when using the free base. Under otherwise identical reaction conditions up to 40 per cent increase in activity was recorded in citrate buffer compared to acetate, tartrate or phthalate buffer (Gallati and Brodbeck, 1982 a). Like all the other chromogens of HRP oPD has an optimum hydrogen peroxide concentration depending on the reaction time (fig. 3). For the reaction times usual in EIA (10 to 30 min) a peroxide concentration of 5.5 mmol/1 can be considered the optimum. For very fast EIAs with extremely short substrate reaction times the peroxide concentration should be raised to 15 mmol/ 1 of substrate solution. The optimum peroxide concentration differs widely with various chromogens being used. Depending on the peroxide concentration an accelerated formation of Compound III, i.e. an inactive enzyme-hydrogen peroxide complex, takes place. The rate of Compound III formation depends on the chromogen used.

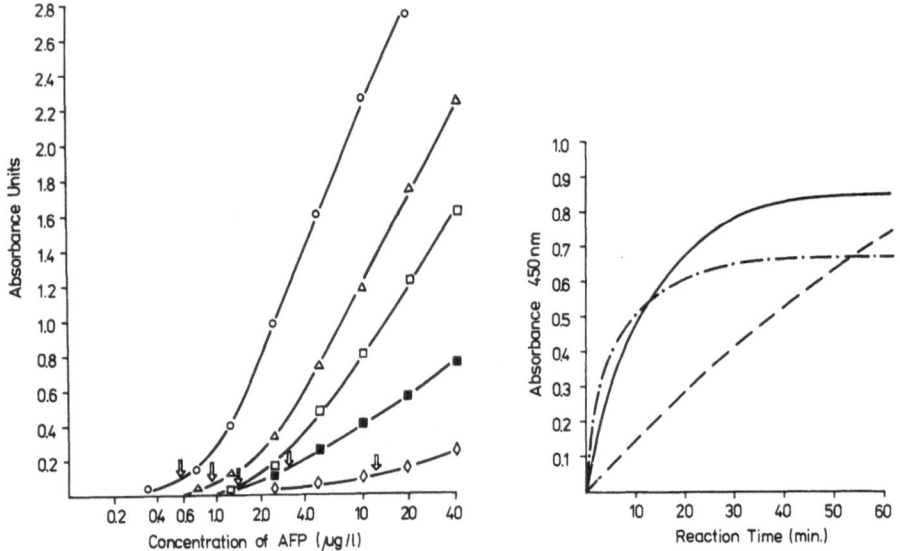

Fig. 2 (left). Standard curves for AFP in the EIA applying HRP labelled
      antibodies and different chromogens for enzyme quantification.
      O — O oPD (solution A), Δ — Δ ABTS, □ — □ oDia (conditions as
      described in Material and Methods),■ — ■ oDia (substrate solu-
      tion without detergents, yellow colour, measurement at 403 nm),
      ◊ — ◊ Trinder reagent. Arrows indicate the detection limits.

Fig. 3 (right). Influence of hydrogen peroxide concentration on the
      time course of product formation by HRP using oPD (15 mmol/ 1)
      as substrate. Hydrogen peroxide concentrations (mmol/ 1):
      — — 1.25, ——— 5, — · —10.

Many worker described that oPD-substrate reactions are sensitive to light. Porstmann et al. (1985 b) showed that the instability of the oPD substrate is caused by spontaneous decay of hydrogen peroxide, which is much less pronounced by exposure to light as compared to the influence of the pH-value of the solution. The decrease of the pH-value as a result of the termination of the enzyme reaction is the main reason for the decay of remaining peroxide responsible for the non-enzymatic conversion of oPD to 2,2'DAB. Adding supra pure hydrochloric and sulphuric acid to the substrate solution in absence of HRP caused a 6.5 fold or a 3.5 fold increase in absorption after 60 min irrespectively of storing samples in the dark or not (fig. 4). The non-enzymatic formation of 2,2'DAB follows zero-order reaction kinetics if sufficient quantities of unconverted peroxide are available (low basic absorption values, which in the EIA represent small quantities of antigen) but changes to first-order reaction kinetics in the high basic absorptions (representing high concentrations of antigen in EIA) due to a lack of available peroxide. This, with a general increase in extinction results in a flattening of the standard curves (fig. 5). The addition of sulphite ions to the stopping solution stabilizes the reaction product of HRP for many hours even without light protection and ensures both a visual and a photometric evaluation on the day following the termination of the reaction (Porstmann, 1985b).

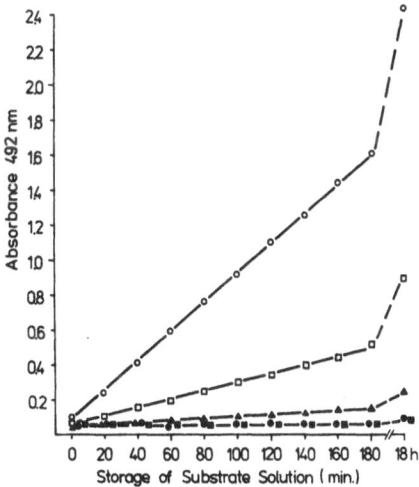

Fig. 4. Time dependent increase of absorbances of the oPD-substrate solution in the absence of HRP, mixed with equal volumes of 4 val/ 1 hydrochloric acid suprapure in the absence(O — O) and presence (● — ●) of 0.1 mol/ 1 sodium sulphite or 4 val/ 1 sulphuric acid suprapure in the absence (□ — □) and presence (■ — ■) of 0.1 mol/ 1 sodium sulphite, and 0,1 mol/ 1 citrate buffer, pH 5.0 (▲ — ▲).

2,2'-Azino-di(3-ethyl-benzthiazoline-sulphonic acid-6) (ABTS), a heterocyclic azine, is available as another substrate for a highly sensitive HRP detection.

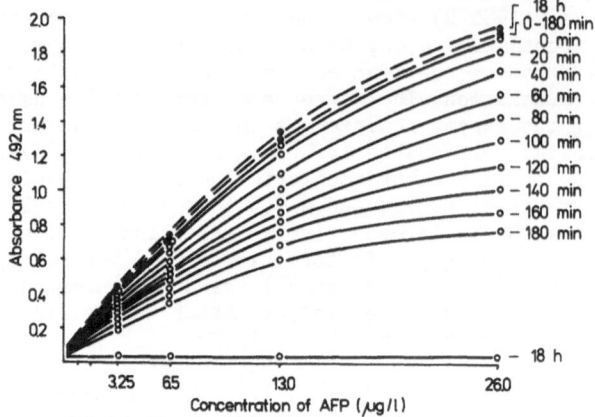

Fig. 5. Standard curves of an AFP-EIA at different times after termination the oPD-substrate reaction by O — O 4 val/ 1 hydrochloric
acid suprapure and by ● — —● 4 val/ 1 hydrochloric acid suprapure containing 0.1 mol/ 1 sodium sulphite.

ABTS (reduced form)

In the presence of peroxide the ammonium salt of ABTS is converted into a
radical cation which tends to disproportionate.

azine
λ_{max} = 340 nm

radical cation
λ_{max} = 414 nm

disproportionation

2 radical cation

+ ABTS

azodication

One can largely diminish the disproportionation when using a 100 fold azine excess. The absorption maximum of the radical cation was determined at $\lambda = 414$ nm and the extinction coefficients as 21.6 $cm^2/\mu mol$ at $\lambda = 420$ nm and 18.4 $cm^2/\mu mol$ at $\lambda = 405$ nm (Werner et al.,1970). Due to its green colour the radical cation is particularly suitable for visual evaluations compared to the yellow oxidation products of 5-aminosalicylic acid and oPD, but especially nitrophenols, the reaction products of AP and ßGal. Same as with oPD the affinity of HRP to ABTS and hydrogen peroxide depends on the pH-value. Furthermore a high ABTS concentration reduces the affinity to peroxide and vice versa. In acetate buffer the activities increase up to 44 per cent as opposed to phosphate, citrate, tartrate or phthalate buffer (Gallati, 1979).

The termination of enzyme reactions when using ABTS still entials the problem as colour cannot be stabilized for hours. Persijn and Jonker (1978) suggested sodium azide in a final concentration of 0.5 mmol/ 1 which would ensure stability at least for 15 min. Greater azide concentrations push the disproportionate decrease in colour. According to the stopping solution suggested by Persijn we were able to record 48 hours of stability (Porstmann et al., 1981 b). We propose a final azide concentration of at least 5 mmol/ 1 to ensure a sufficient inhibition of HRP. Klapper and Hackett (1963) found that an azide concentration of 4 mmol/ 1 only effects an inhibition of 57 per cent. The extent seems depending on the isoenzyme composition. Due to its explosive properties in connection with heavy metals sodium azide must be handled carefully. Although NaCN inhibits HRP nearly completely already at a concentration of 0.8 mmol/ 1 it must be refused as stopping reagent because of its extremely toxic features. EDTA/fluoride mixtures accelerate the disproportionation of the ABTS radical cation and fluorides cause the formation of fluoric acid which corrodes the reaction vessels. Gallati and Brodbeck (1982 b) suggested alkylsulfates (Teepol-610) and sodium dodecyl- hydrogensulfates as stopping reagents. Alkylphosphate as of a final concentration of 0.4 per cent (w/v) leads to a total inactivation of HRP with the radical cation being stabilized for at least 80 minutes. The same effect can be achieved by dodecylhydrogensulfate in concentrations ranging from 0.1 to 0.5 per cent (w/v).

O-dianisidine (oDia) was often used as chromogen in EIA, but only effected insensitive peroxidase measuring with the substrate and stopping solutions applied.
A quinone-diimine dye is formed by peroxidatic oxidation.

3,3'-dimethoxybenzidine   quinone-diimine dye
  (o-dianisidine)

We created optimum reaction conditions so that, using oDia an increase in sensitivity by the 9-fold was achieved and the chromogen is only a little less suitable than oPD and ABTS (Porstmann et al., 1980, Porstmann et al., 1981b).We have shown that non-ionic detergents such as polyoxyethylene-octyl-phenol or -sorbitolester (e.g. Triton X-100 or Tween-20) increase the activity of HRP due to delay of inactivation in the course of substrate reaction. This rise in activity was investigated using different chromogens and was highest with oDia (fig. 6). The factor of activity increase depends

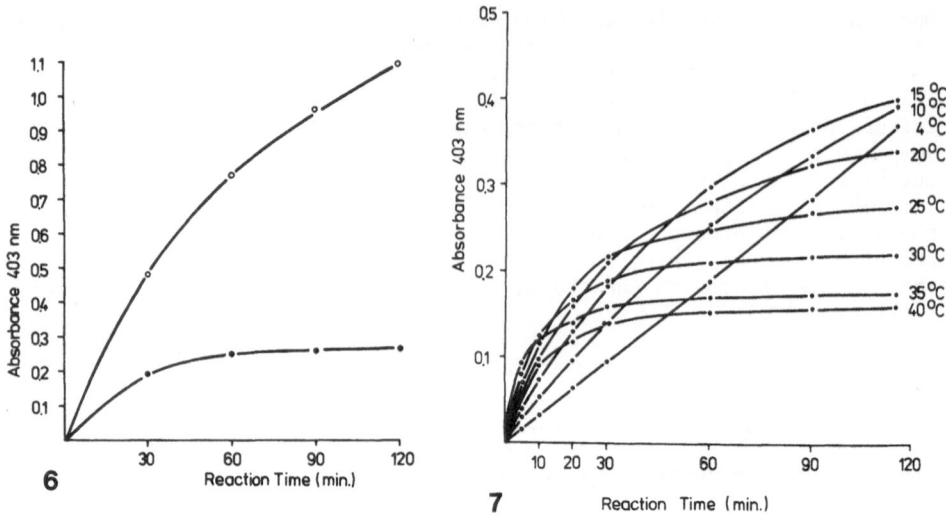

Fig. 6 (left). Time course of oDia-oxidation by HRP-IgG-conjugates
        in the presence of 1 ml/ 1 Tween-20 (O——O) and in the
        absence of detergent (●——●).

Fig. 7 (right). Temperature-dependent time course of product
        accumulation using oDia as chromogen without addition
        of detergent to the substrate.

on the enzyme reaction time and is 4.0 after 120 min. It is characteristic
of the HRP kinetics that it becomes the more inactive the higher the
reaction temperature although the initial rate is steadily increasing
(tested up to 40°C) (fig. 7). HRP is very thermostable (heating above 60°C
for several minutes does not cause inactivation). Fast inactivation during
enzyme reaction is caused by temperature-dependent formation of Compound
III which differs according to the use of different chromogens. (oDia
causes a fast formation). When detergents were added to the substrate
mixture the point of enzyme inactivation was delayed for a given reaction
temperature. This effect caused an increase in analytical sensitivity of
the test system for quantification of HBsAg and a decrease of at least one
titre step of the detection limit of HBsAg (Porstmann et al., 1981 a). The
chromogen-specific detergent effect has not yet been fully clarified, but
we suspect a decreased formation of Compound III. Chromogen differences in
protection might be attributable to an inherent protective potential of the
dyes. Hofmann et al.(1983) in spectrophotometry gave evidence of a
detergent binding to HRP. Formation of polyglycoletherperoxide with
hydrogen peroxide is suspected acting as a hydrogen acceptor. Oxidized
oDia, like many chromogens, shows features of a pH-indicator. Depending on
the acid concentration its point of maximum absorption shifts from 403 nm
(up to 0.75 mol/ 1 hydrochlorid acid) to 530 nm (above 2.5 mol/ 1 acid)
with absorption increasing 2.2 fold. The red colour is perfectly suitable
for visual evaluations. The stability of the product is kept 20 hours after
termination. Hence we recommend hydrochloric acid as stopping reagent
(final concentration at least 2.5 mol/1). Amongst the chromogens tested by
us in EIA oDia ranks third after oPD and ABTS according to the sensitivity
achieved in the optimum method described (fig. 2).

oDia is carcinogenic, however, and oPD and ABTS have mutagenic properties (Ames et al., 1975). Although careful handling of the EIA excludes skin contact with the chromogens, inhaling powder cannot be ruled out. That is why despite great sensitivity of these chromogens other non-mutagenic or -carcinogenic substances have been searched for. Mention must be made in this field of 5-aminosalicylic acid (5AS), tetramethylbenzidine (TMB) and dicarboxidine.

5AS as commercially available preparation is often more or less yellow coloured. Ellens and Gielkens (1980) described a relatively simple purification technique which results in a colourless substrate solution and remains colourless after adding chelaplex. In an EIA for rotavirus antigen the 5AS purified in this way, was compared to oPD as chromogen. The extinction (oPD as acid-stopped chromogen, 5AS unstopped) produced in one hour of reaction was more than three times as high using oPD in comparison with 5AS, the PN-ratios, however, were identical for all concentration ranges since the blank values for oPD were nearly four times higher than those for 5AS. This stresses the importance of colourless substrate solutions for an increase in sensitivity in EIA which shows effect particularly in evaluation as P/N ratio. But it must be pointed out that also oPD is almost colourless after recrystallization. Al-Kaissi and Mostratos (1983) achieved that HRP was more sensitive with 5AS than with ABTS or oPD. It should be noted, however, that the latter substrate solutions had not been optimized. The stopping solution they used (1mol/ 1 NaOH) causes an additional extinction decrease. Such chromogen comparisons under unoptimized reaction conditions are to be refused.

Dicarboxidine as chromogenic substrate for HRP was described by Paul et al. (1982):

$$H_2N - \text{[benzene ring]} - \text{[benzene ring]} - NH_2$$

$$\begin{array}{cc} O & O \\ | & | \\ HOOC(CH_2)_3 & (CH_2)_3COOH \end{array}$$

dicarboxidine

The oxidation product of dicarboxidine is brown at pH 6.9 and has a broad extinction maximum at λ = 440 nm with a millimolar extinction coefficient of 13.4 (Paul, 1982). No stopping reagent was tested. Adding acid leads to precipitation of the oxidized chromogen.

Benzidine, still rather often used in histochemistry cannot be used in EIA because of its precipitation. Benzidine is carcinogenic due to its in vivo hydroxylation in ortho-position.

3,3', 5,5'-Tetramethylbenzidine (TMB) is a derivative in which o-hydroxylation becomes impossible. It was used in EIA by Bos et al. (1981).The oxidation product of TMB is a yellow diphenoquinone dye with an extinction maximum at λ = 450 nm. Gallati and Pracht (1985) described the formation of a blue radical cation if there are small quantities of HRP and an excess of hydrogen peroxide and TMB.

$$\text{TMB} \xrightarrow[-e^-]{\text{HRP}} \text{TMB-radical cation}$$

TMB
(colourless)

TMB-radical cation (blue)
λ_{max} = 370nm and 650 nm.

$$\xrightarrow[\text{HRP}]{-e^-}$$

3,3',5,5'-tetramethyl-1,1'-diphenoquinone-
4,4'-diimoniumion, λ_{max} = 450 nm

By lowering the pH-value the blue radical cation (absorption maximum at
λ =370 nm) can be converted to yellow diphenoquinone (total conversion at
pH = 1.0) with a higher extinction coefficient at λ = 450 nm. The product
remains stable at least 90 min when using sulphuric acid as stopping
reagent. As in all the other chromogen reactions of HRP the initial rate of
the HRP increases when using TMB with peroxide concentration and reaction
temperature increasing. But inactivation is accelerated as well. The
hydrogen peroxide concentration of 3 mmol/ 1 listed in Material and Methods
is suitable for a reaction time of 30 min at room temperature.TMB was used
in EIA for HCG, testosterone and HBsAg and compared with oPD, the latter
not always under optimum conditions . Under optimum conditions both
chromogens show nearly identical EIA sensitivity both with respect to the
detection limit and the stepness of standard curves (Bos et al., 1981).
This seems to make TMB a favourite chromogen for HRP. It is limited
soluble, but due to its great sensitivity and lack of mutagenic properties
it should be widely applied. Apart from the chromogenic substrates
mentioned there is a multitude of other ones from which chromogenic
products can be formed through peroxidatic oxidation. They are prefered in
clinical chemistry as substrates in cases of coupled reactions with
peroxidase as indicator enzyme for enzyme or analyt detection (most
frequently for evaluation of blood sugar levels). Their sensitivity is
sufficient for these matters but is insufficient for EIA. Among them are
pyrogallol, o-methoxyphenol (guaiacol), o-tolidine and
3-amino-9-ethylcarbazole to mention only a few. There are indications that
different HRP isoenzymes prefer different substrates (chromogens) and have
varying pH - or concentration - optima. Thus the use of eugenol
(fluorescent product) or benzidine on electrophoretically separated HRP
produced different pattern. Such possible differences have not been
investigated for most of HRP's chromogens, which is of importance however,
since different preparations are commercially available (alkali, acid or
neutral isoenzymes or mixtures) and these enter differently into bonds with

immunoreactants according to the coupling method as well. Optimization studies for chromogens should be made using enzyme-conjugates, with an exactly stated HRP - preparation and coupling method.

(b) Oxidative coupling of two different reactants by HRP

Trinder reagent (Trinder, 1969) consists on amino-antipyrine and phenol which are condensed by HRP in the presence of peroxide.

$$2\ H_2O_2 + \quad \text{[4-amino-antipyrine]} + \text{[phenol, OH]} \xrightarrow{\text{HRP}} \text{[quinone imine]} + 4\ H_2O$$

4-amino-antipyrine   phenol   quinone imine

The chromophoric compound is red with an absorption maximum at λ =492 nm and an extinction coefficient of 9 cm^2/μmol. The reaction was optimized by Gallati (1977) and the chromogen was compared in EIA (Porstmann et al., 1981b). We found that the use of Trinder reagent produces utterly insensitive test results (fig. 2). Additional disadvantages arise from the light-dependent instability and polymerization of phenol and the lack of a stopping reagent. The HRP reaction cannot even be stopped by alkylsulphate (Gallati and Brodbeck, 1982 b). Meiattini et al. (1978) replaced phenol by 4-hydroxybenzoic acid. Barham and Trinder (1972) suggested to replace phenol by 2-hydroxy-3,5- dichlorobenzenesulfonate (HDCBS), which would achieve a raise in sensitivity by 4 to 5 fold. This reagent was used successfully for quantification of glucose, triglycerides, creatine kinase and choline esterase in coupled enzyme reactions (Artiss et al., 1982, McGowan et al., 1983, Wimmer et al., 1985).
So far there is no experience concerning application in EIA. The emergent chromogenic condensate is red and has an extinction maximum of between λ = 505 nm and 510 nm.

Ngo-Lenhoff reagent is another oxidative coupling reagent (Gochman and Schmitz, 1971, Ngo and Lenhoff, 1980).

$$\text{[MBTH]} + \text{[DMA, N(CH}_3)_2\text{]} \xrightarrow{\text{HRP}} \text{[indamine dye]}$$

2-methyl-2-benzo-   N,N'-dimethyl-   indamine dye
thiazolinone   aniline λ_{max} = 590 nm
hydrazone (MBTH)   (DMA)

The molar extinction coefficient at λ = 590 nm was determined as 47,000 (Ngo and Lenhoff,1980).
According to our investigations there is a great sensitivity to light of the substrate solution in spite of the system optimized by Geoghegan et al. (1983). An increase in the blank values of the substrate occurs, however,

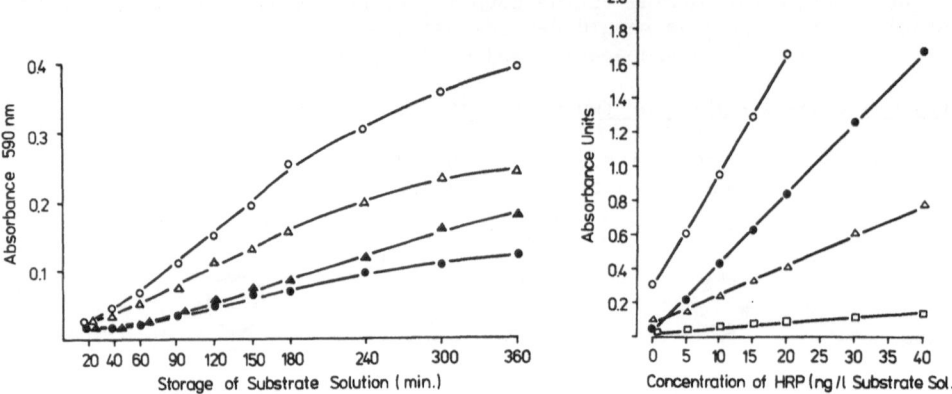

Fig. 8 (left). Time course of absorbances of the Ngo–Lenhoff substrate
in the absence of HRP with hydrogen peroxide (O —— O) or without
(△ —— △), stored in the dark (closed symbols) or exposed to
diffuse light (opened symbols).

Fig. 9 (right). Standard curves for HRP (purity number 3.0) quantified
by two –point measurement of activity (reaction time 30 min at
25°C) comparing different chromogens. O —— O oPD (solution A),
● —— ● oPD (solution B), △ —— △ ABTS, □ —— □ Ngo–Lenhoff reagent.

also in total darkness (fig. 8). After termination of the reaction by means
of hydrochloric or sulphuric acid the pH–value should be about 3.0.
The decrease in pH resulted in a colour change from a purple–blue to a

Table 1. Enzyme activities of native and IgG–coupled HRP, AP
and ßGal using the chromogenic substrates ABTS,
pNPP and oNPG (if not otherwise stated).

determined parameter	HRP	AP	ßGal
molar activity of en-zyme (mol/s x 1 x mol)	2600	850	354
specific activity (mmol/s x 1 x g)	65	8.5	0.68
specific activity of conjugates (mmol/s x 1 x g)	9.4	2.9	0.41
detection limit of enzyme (ng/ 1)	ABTS: 4 oPD : 0.5 Ngo–L.:10	20	oNPG:1000 RG: 600 CPRG: 150
detection limit of conjugates (ng/ 1)	28	70	1500

definite clear blue. This caused a shift in the absorption maximum from 590 nm to 595 nm (Geoghegan et al., 1983). A further decrease in pH will lead to extinction losses. The detection limit for HRP (purity number 3.0) was determined by us with 10 μg/ l of reaction solution using Ngo-Lenhoff reagent at a reaction time of 30 min. With various oPD substrate solutions it comes to between 0.4 and 2 ng/l and 5ng/l with ABTS (fig. 9, table 1).In comparison with oPD solution A, Ngo-Lenhoff reagent produces an analytical sensitivity 20 times lower in HRP concentration determinations. The low analytical sensitivity using Ngo-Lenhoff reagent results in very flat standard curves in EIA as can be seen from fig. 10. The detection limits for lysozyme only differ by 1 geometric step of dilution using oPD (solution B), ABTS and Ngo-Lenhoff reagent. It must be pointed out that Ngo-Lenhoff reagent warrants a very good visual differentiation. MBTH was also used as cosubstrate for 2-hydroxy-3,5-dichlorobenzenesulfonate and 3-(dimethylamino)benzoic acid. Capaldi and Tayler (1983) showed that hydrogen peroxide in the presence of peroxidase was capable of oxidatively coupling MBTH with its azine to give a cationic tetraazapentamethine dye with a maximum absorbance at 630 nm. The mixture of MBTH and its azine was formed by incubating formaldehyde with excess MBTH.

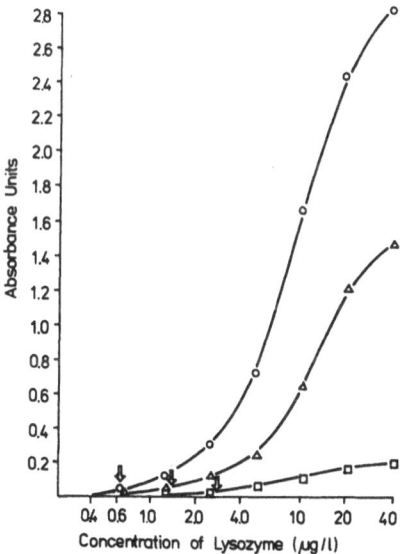

Fig.10. Standard curves for human lysozyme in an EIA with HRP labelled antibodies using different chromogens. O — O oPD (solution B), △ — △ ABTS, □ — □ Ngo-Lenhoff reagent. Arrows indicate detection limit.

Chromogens for Alkaline Phosphatase (AP)

AP is the marker enzyme that is used second most. It is prepared from calf intestine and shows a great molar activity with phenylphosphate and p-nitrophenylphosphate as substrates. Due to simple product measuring the latter is almost exclusively used as substrate in colorimetric EIA with AP as marker enzyme.

p-nitrophenyl-   p-nitrophenol
phosphate

The absorbance maximum was determined at λ = 405 nm with an extinction
coefficient in 0.25 mol/ 1 sodium carbonate of 17.5 cm^2/ μmol (Porstmann et
al., 1985 a). Sodium carbonate is a stopping reagent which, due to
alkalinization enhances the extinction coefficient of pNP and turns the
enzyme inactive. The diethanolamine buffer should be free from monoethanol-
amine as there is a temperature-dependent AP inhibition. Anorganic
phosphate in higher concentrations can inhibit AP competitively e.g. when
using strong yellow pNPP that has been stored too long (emergence of
nitrophenol by non-enzymatic hydrolysis). For highly sensitive EIA test
systems exclusively colourless pNPP should be used. The extinction
coefficient of pNP is dependent on pH, temperature, ionic strength and
protein content of the solution. An increase in temperature from 15 to $45^{\circ}C$
raises extinction by 5-7 % at λ = 405 nm. That is why in the determination
of α-amylase the substrate 4-nitrophenyl- maltopentasoide, which is
converted to pNP after enzymatic degradation is replaced by ß-2-chloro-
4-nitrophenyl-maltopentasoide. The product 2-chloro-4-nitrophenol
has a high extinction coefficient (16.6 cm^2/μmol at λ = 400 nm) and whose
absorption is not dependent on pH or temperature (Teshima et al., 1985). It
remains to be checked if AP would accept 2-chloro-4-nitrophenyl- phosphate
as substrate and yield a sufficient turnover rate.

Chromogens for ß-galactosidase

O-nitrophenyl-ß-D-galactopyranoside (oNPG) was nearly exclusively used
as chromogenic substrate for ßGal in EIA.

o-nitrophenyl-ß-   o-nitrophenol   ß-D-galacto-
D-galactopyrano-   (oNP)   pyranoside
side (oNPG)

The absorption spectra with maximum absorbances for the most important
chromogens for ßGal are evident from fig. 11. The extinction coefficient of
oNP at λ = 420 nm in 0.25 mol/ 1 sodium carbonate was determined as 4.7
cm^2/μmol (Porstmann et al., 1985 a). It is nearly 4 times lower than that
of pNP at 405 nm. Nevertheless oNPG as substrate for ßGal produces a better
sensitivity than pNPG due to a seven fold higher turnover. As can be seen
from table 2 the oNPG substrate solution shows the smallest extinction
(blank value) and the greatest stability (no increase in extinction over a
test period of 25 hours). In contrast to this a non-enzymatic temperature
-dependent hydrolysis is found in pNPP- and a radical cation formation in

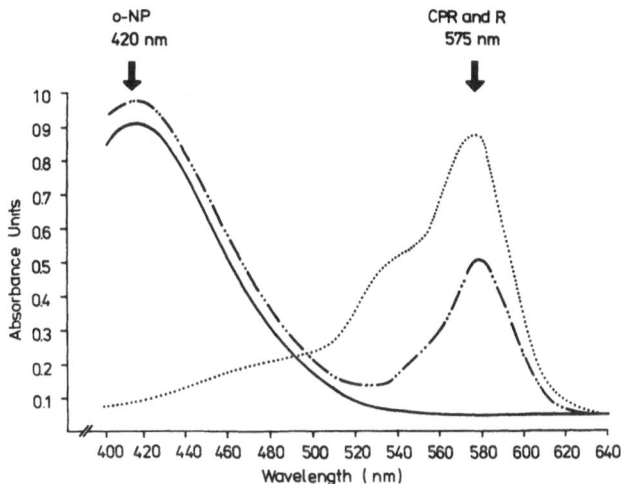

Fig.11. Absorbance spectra of different chromogenic products after conversion by ßGal and termination the reaction described in Material and Methods. Maximum absorbance is indicated by an arrow. ——— o-nitrophenol, —··—chlorophenolic red, ······· resorufin.

ABTS-solutions. This is of importance for long-term incubations of the marker enzymes with their substrates. New chromogenic substrates for ßGal will facilitate a considerably more sensitive determination of activity.

Chlorophenolic red-ß-D-galactopyranoside (CPRG) is a new very sensitive substrate.

$$\text{CPRG} \qquad\qquad \text{RG}$$

It is well soluble in water. The substrate solution is yellow, in higher concentrations orange. The degradation product CPR is purple. The extinction coefficient at λ = 574 nm is declared as 75 cm^2/ μmol (Boehringer, Mannheim,FRG).

Resorufine-ß-D-galactopyranoside (RG) is another chromogenic substrate of ßGal, that gives a great sensitivity. So far it was used for fluorimetric ßGal-analysis (Hofmann and Sernetz, 1984). The substrate solution is yellow-orange, the product (R) is purple with an extinction coefficient at λ= 570 nm of 66 cm^2/ μmol. Unfortunately it is of limited solubility in buffer so that concentrations above 0.4 mmol/ 1 (Km-value 0.38 mmol/ 1) cause flocculations. The substrate solutions (pH 7.0-8.0) with CPRG and RG are stable for several hours just like with oNPG (no increase in extinction). Alkaline stopping reagents like sodium carbonate cause a non-enzymatic substrate hydrolysis of all substrates (Fig. 12 b).

Table 2. Initial blank values of substrate solutions
for different marker enzymes and temperature
-dependent increase

substrate	initial blank (absorbance)	percentual increase of absorbance at different reaction temperatures		
		10°C	25°C	37°C
pNPP	0.175	0.5	0.7	2.0
ABTS	0.085	0.5	0.5	0.5
oNPG	0.015	0	0	0

It is smallest with oNPG and strongest with RG. When using RG a suitable
stopping reagent must be found or a kinetic activity evaluation must be
made. We compared the sensitivity in ßGal activity determination applying
the mentioned substrates. CPRG in its highest concentration tested (2.7
mmol/ 1) warrants extinction values ten times higher and RG at the limit of
solubility (0.4 mmol/ 1) six times higher than that of oNPG (fig. 12 a).
This is also reflected in the lower detection limit for ßGal (table 1). In
an enzyme immunoassay for quantification of antibodies against tetanus
toxin we attained a tremendous increase in sensitivity by using chromogens
CPRG and RG (fig. 13).

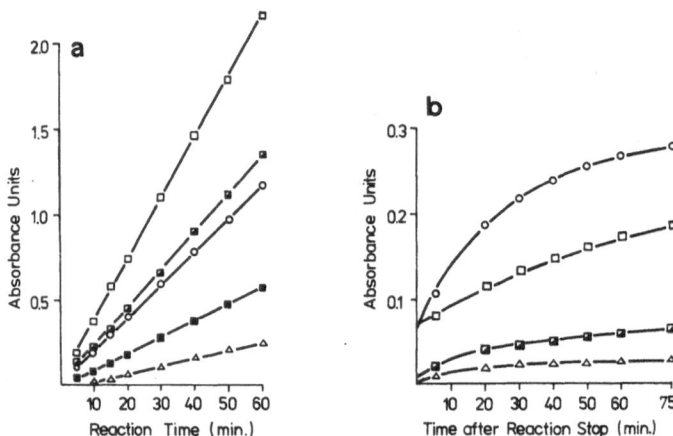

Fig.12.a (left). Chromogen comparison for activity measurement of ßGal
(3 µg/ 1 of stopped substrate solution).□—□ CPRG (2.7 mmol/
1),▧ — ▧ CPRG (1.5 mmol/ 1),■—■ CPRG (0.4 mmol/ 1),
○—○ RG (0.4 mmol/ 1),△—△ oNPG (2.7 mmol/ 1).

Fig.12.b (right). Time course of ßGal substrate stability in absence of
ßGal after addition of equal volumes of sodium carbonate.
○—○ RG (0.4 mmol/ 1),□—□ CPRG (2.7 mmol/ 1),▧ — ▧ CPRG
(1.5 mmol/ 1),△—△ oNPG (2.7 mmol/ 1).

We compared the three mostly applied marker enzymes HRP, AP and ßGal using the chromogenic substrates ABTS, pNPP and oNPG (Porstmann et al., 1985 a). HRP has the greatest molar and specific activity and the lowest detection limit. This also applies to the HRP-IgG conjugates (table 1). But ßGal and AP can be determined considerably more sensitive by long-term substrate incubation at $37^{\circ}C$ since it has linear reaction kinetics at this temperature for many hours (fig.14). Long-term measuring of HRP should be carried out only at a temperature below $10^{\circ}C$ when it comes to measuring over a period of several hours. Applying ABTS as a chromogen disproportionation of the radical cation occurs after prolonged reaction times due to which the extinction will decline again. When IgG-enzyme -conjugates of HRP, AP and ßGal are used in colorimetric EIA for quantifying AFP the greatest sensitivity was achieved in the shortest enzyme reaction time using HRP-IgG-conjugates (Porstmann et al., 1985 a).

Chromogens for Urease

Urease as marker enzyme was tested successfully by Chandler et al. (1982) and Chandler and Hurell (1982).

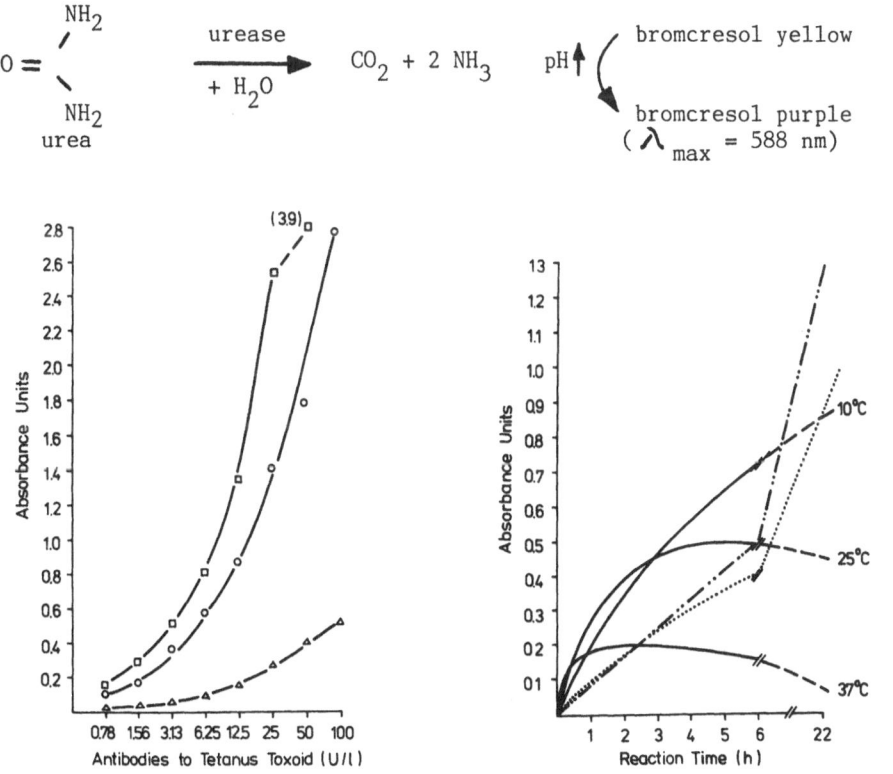

Fig.13 (left). Standard curves for the detection of human IgG-antibodies specific for tetanus toxin recorded from an EIA applying ßGal labelled antibodies and different chromogens. ☐ — ☐ CPRG (2.7 mmol/ 1), O — O RG (0.4 mmol/ 1), △ — △ oNPG (2.7 mmol/ 1).

Fig.14 (right). Time course of temperature-dependent chromogen production using ABTS (——), oNPG (—··—) and pNPP (······) for HRP, ßGal and AP. Reaction temperature for ßGal and AP: $37^{\circ}C$.

It was used in EIAs for quantifying or detecting monoclonal antibodies in cell supernatants, antibodies against varizella zoster and for identification of snake venom. Urea is a cheap and stable substrate. Thiomersal acts as an effective stopping solution. Ammonia, the product of the enzyme reaction, produces a colour change in the bromcresol purple indicator until the pH of the substrate is raised from its initial value of 4.8 to 5.2. This change is clearly visible. The colour change is linear.During a total test time of only 40 min Chandler and Hurell were able to detect 15 µg/ l of snake venom. The main advantage of the urease system is the possibility of an excellent visual differentiation.

Chromogens for Acetylcholine Esterase

Ellman's reagent is frequently used as chromogenic substrate. Acetylcholine esterase (AchE) from electrophorus electricus is a stable marker enzyme which was used in an EIA for the quantification of eicosanoids by Pradelles et al. (1985) among others.

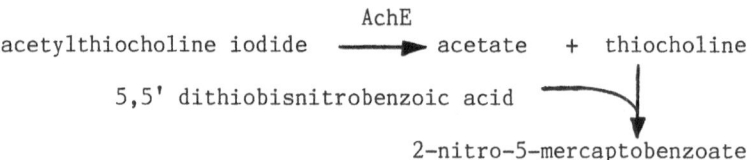

The yellow product has an extinction maximum at λ = 412 nm. The sensitivity achieved in EIA (competitive assay technique) was within the range of a RIA carried out at the same time for comparison (detection limit 15–40 ng/l).

Chromogenic Substrates in Coupled Enzyme Reactions and Enzymatic Cycling Systems

Coupled enzyme reactions are applied to all marker enzymes whose direct reaction product cannot be quantified by photometry or fluorimetry. The enzyme reaction of glucose oxidase (GOD) from aspergillus niger, as a rule, is made visible by peroxidase as indicator enzyme with the chromogens for HRP being used that have been mentioned before.

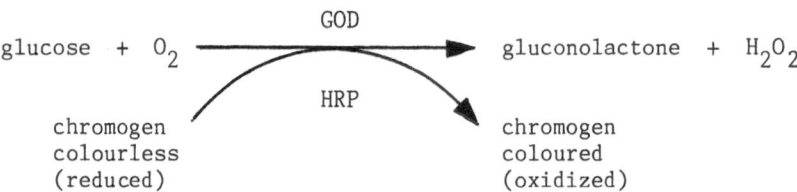

GOD/POD systems are much more insensitive in short-time measuring than HRP as marker enzyme. However a continuous chromogen production is detectable for hours as HRP is becoming inactive much slower due to low hydrogen peroxide concentration.

Nitro blue tetrazolium has proved useful when taking GOD as marker enzyme in immunohistochemistry. It was also used in EIA. All enzyme reactions being involved in formation or consumption of NAD or NADH can be made visible through enzymes like lipoamide dehydrogenase or diaphorase (DI) or through redox mediators like phenazine methosulfate (PMS) and meldola blue resulting in coloured tetrazolium salts (Orsonneau et al., 1982).

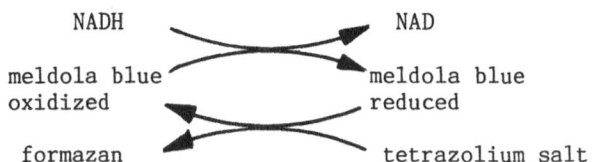

Compared to PMS meldola blue has the advantage that it is practically insensitive to light and that it effects the transfer slightly more rapidly. The oxidized form of meldola blue absorbs light at 565 nm whereas its reduced form is colourless. The reduced form is stabilized for at least 5 min working in an acid medium (pH 2.2) with Triton X-100 (2.5 % v/v) to prevent spontaneous reoxidation (Orsonneau et al., 1982) which can be brought about by two-step enzyme measuring technique (first step: NADH formation by lactic dehydrogenase or urease/glutamate dehydrogenase for example; second step: meldola blue reaction). In homogeneous EIAs for quantifying cortisol, biotin and dinitrophenylysine, bacterial glucose-6-phosphate dehydrogenase (G6PDH) was used as marker enzyme and the reaction was made visible by means of diaphorase as auxiliary enzyme with nitro blue tetrazolium (Dona, 1985). The composition of the substrate solution is stated in Material and Methods.

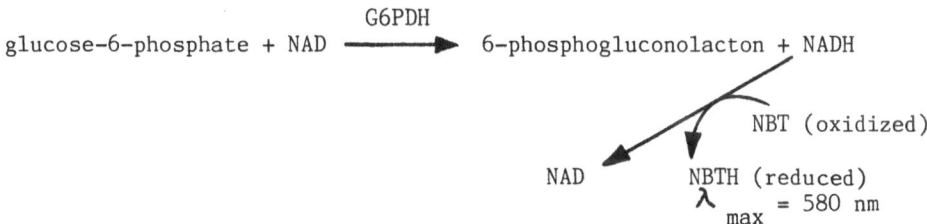

The coloured NBTH has an absorption three times higher at 580 nm than NADH at 340 nm. The detection limits for haptens with G6PDH as marker enzyme in colorimetric EMITs were determined in the μg/ l range.

Enzymatic cycling systems hold at least two enzymes which serve as catalysts for coupled reactions:

$$A + S_1 \xrightarrow{\text{enzyme 1}} B + P_1 \qquad \text{primary system}$$

$$B + S_2 \xrightarrow{\text{enzyme 2}} A + P_2 \qquad \text{secondary system (amplifier)}$$

Small quantities of A lead to the formation of large quantities of P_1 and P_2. They are either directly measurable or through an indicator reaction by receiving measurable signals in the form of luminous or fluorescent features of P_3:

$$P_1 \text{ (or } P_2) + C \xrightarrow{\text{enzyme 3}} P_3 + X \qquad \text{secondary system (indicator)}$$

Both A (e.g. a coenzyme) and enzyme 1 can be the marker substances for immunoreactants in EIA. Many enzymes can be coupled by this way. The amplification varies from 3.000 per hour to 100.000 per hour. The usefulness of these systems depends on the turnover of enzymes, on possible product inhibitions, on the Km-values of the enzymes for the substrates and particularly on whether measuring the final product is practicable. The enzymes of the secondary system should have a great substrate specificity since the turnover of the substrate in the primary system increases the blank values. But this can be minimized by separating the primary and secondary system (two-step amplification system). The accumulation of the triggering substance from the primary system is allowed to continue for a fixed period of time before the addition of amplifier. Yet, two-step amplifiers are less sensitive than one-step amplifiers (Stanley et al., 1985). For the quantification of human thyroid stimulating hormone (TSH) and prostatic acid phosphatase (PAP) a recycling system was used with AP as marker enzyme (Stanley et al., 1985, Moss et al., 1985) (stated in Material and Methods). The gain for this redox amplifier is in the order of 3.000 per hour.

P-iodonitrotetrazolium violet (INT) and thiazolyl blue (MTT) are very sensitive chromogenic substrates. The marker enzyme AP dephosphorylates NADP. The redox cycle is driven by the enzymes alcohol dehydrogenase (ADH), lipoamide dehydrogenase or diaphorase and an excess of ethanol. The formation of formazan dyes as products can be taken as a general detection system if redox cycles are used:

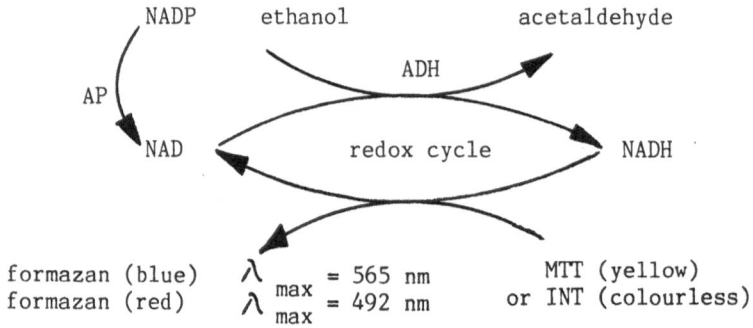

The detection limit for TSH in a sandwich assay was $1-2 \times 10^{-5}$ IU/ 1 with an amplifier system being applied. INT shows somewhat more sensitive reactions than MTT. When using p-nitrophenol as substrate (non-amplified system) the detection limit was 7×10^{-4} IU/1 (Stanley et al., 1985). The detection limit for PAP was 0.05 µg/ 1 (Johannsson et al., 1985). The amplified system makes an increase by 170 to 250-fold in measured signals possible compared to non-amplified systems. Another example for enzymatic cycling systems in immunoassay are NAD labelled immunoassays. As a rule substrate or coenzyme labelled assays mostly functioning as homogeneous assays are less sensitive than enzyme labelled assays as the latter have an amplifying factor in the formation of the product due to the catalytic activity of the enzyme. Cycling systems can also make substrate or coenzyme labelled immunoassays very sensitive. Chromogenic substrates were used in NAD-labelled homogeneous immunoassays with enzymatic cycling systems for quantifying biotin and insulin (review by Cox, 1983).

P-nitroso-N,N-dimethylaniline (NDA) and thiazolyl blue (NBT) were the chromogenic substrates used in these systems:

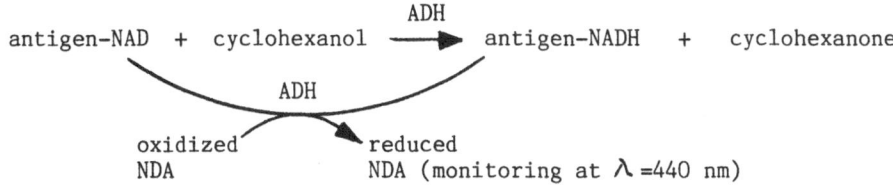

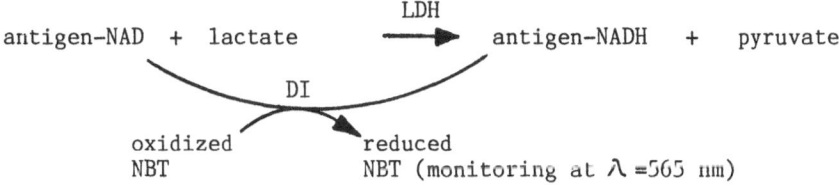

Other cycling systems in NAD-labelled immunoassays were made to form NADPH which was measured by fluorimetry or photometry, such as systems with

- ADH/malic dehydrogenase/malic enzyme   or
- G6PDH/NAD   -   peroxidase/6-phosphogluconate dehydrogenase (Cox, 1983).

Other systems use cyclic phosphorylation and dephosphorylation of fructose-6-phosphate. In this case amyloglucosidase, hexokinase, phosphoglucoisomerase or aldolase being used as marker enzymes (Harper and Orengo, 1981, Stanley et al., 1985). Here, too, a NADPH-dependent indicator reaction follows being pursued at 340 nm. However, chromogenic substrates could also be used for this in future.

SUMMARY

Chromogenic substrates for marker enzymes in enzyme immunoassays have the advantage of simple evaluation (visually or by photometers) and of high sensitivity. They are used preferably for EIA-test kits. Main requirements to chromogens are high extinction coefficients and turnover rates effected by marker enzymes, stability after termination of the reaction by the stopping reagent and no health hazarding properties. There is almost no chromogen that meets all requirements optimally. Carcinogenicity, poor solubility or unsatisfactory stability after termination of reaction limit the usefulness of one or the other chromogen. Particularly sensitive as chromogens for peroxidase measuring are o-phenylenediamine, ABTS, tetramethylbenzidine, aminosalicylic acid and o-dianisidine. For alkaline phosphatase currently p-nitrophenyl-phosphate is chosen as chromogen. Activity determination of ß-galactosidase is mostly carried out using o-nitrophenyl-galactopyranoside which, however, warrants relatively insensitive results in EIA. Extinctions up to ten times higher with the same enzyme concentrations are achieved using the chromogens chlorophenolic red- and resorufin-galactopyranoside. For these no suitable stopping reagents are available for the time being thus making them suitable only for kinetic activity determination. For urease detection pH indicators like bromcresol purple are suitable. The determination of acetylcholine esterase activity is recommendable with Ellman's reagent.

Coupled enzyme reactions or enzyme cycling systems causing the formation of NADH can be visualized by redox mediators like phenazine methosulphate or meldola blue or by oxidoreductases like diaphorase by means of reduction of tetrazolium salts to formazan dyes.

REFERENCES

Al-Kaissi, E. and Mostratos, A., 1983, Assessment of substrates for horse-
   radish peroxidase in enzyme immunoassay,
   J. Immunol. Methods, 58: 127- 132.
Ames, B.N., Kammen, H.O., and Yamasaki, E., 1975, Hairdyes are mutagenic:
   Identification of a variety of mutagenic ingredients,
   Proc. Natl. Acad. Sci., 72: 2423-2427.
Artiss, J.D., McGowan, M.W., Strandbergh, D.R., and Zak, B., 1982,
   A procedure for the kinetic colorimetric determination of serum
   cholinesterase activity,Clin.Chim.Acta, 124: 141-148.
Artiss, J.D., Strandbergh, D.R., and Zak, B., 1983, On the use of a sensi-
   tive indicator reaction for the automated glucose oxidase
   -peroxidase coupled reaction,Clin.Biochem., 16: 334-337.
Barham, D., and Trinder, P., 1972, An improved color reagent for the
   determination of blood glucose by the oxidase system,
   Analyst,97: 142-145.
Bos, E.S., van der Doelen, A.A., van Rooy, N., and Schuurs, A.H.W.M.,1981,
   3,3 ,5,5 -tetramethylbenzidine as an Ames test negative
   chromogen for horse-radish peroxidase in enzyme-immunoassay,
   J. of Immunoassay, 2: 187-204.
Capaldi, D.J., and Taylor, K.E., 1983, A new peroxidase color reaction:
   Oxidative coupling of 3-methyl-2-benzothiazoline hydrazone
   (MBTH) with its formaldehyde azine. Application to glucose
   and choline oxidases,Anal. Biochem., 129: 329-336.
Chandler, H.M., Cox, J.C., Healey, K., MacGregor, A., Premier, R.R.,
   and Hurrell, J.G.R., 1982, An investigation of the use of
   urease-antibody conjugates in enzyme immunoassays,
   J. Immunol. Methods, 53: 187-194.
Chandler, H.M., and Hurell, G.R., 1982, A new enzyme immunoassay system
   suitable for field use and its application in a snake venom
   detection kit,Clin.Chim. Acta, 121: 225-230.
Cox, C., 1983, NAD labeled immunoassays,Trends in Anal. Chem., 2: 129-131.
Dona, V., 1985, Homogeneous colorimetric enzyme inhibition immunoassay for
   cortisol in human serum with Fab anti-glucose-6-phosphate
   dehydrogenase as a label modulator,
   J. Immunol. Methods, 82 : 65-75.
Ellens, D.J., and Gielkens, A.L.J., 1980, A simple method for the puri-
   fication of 5-aminosalicylic acid. Application of the product
   as substrate in enzyme-linked immunosorbent assay (ELISA),
   J. Immunol. Methods, 37: 325-332.
Gallati, H., 1977, Aktivitätsbestimmung von Peroxidase mit Hilfe des
   "Trinder-Reagens",J., Clin.Chem.Clin.Biochem., 15: 699-703.
Gallati, H., 1979, Peroxidase aus Meerrettich: Kinetische Studien sowie
   Optimierung der Aktivitätsbestimmung mit den Substraten H_2O_2
   und ABTS,J. Clin.Chem.Clin.Biochem., 17: 1-7.
Gallati, H., and Brodbeck, H., 1982 a, Peroxidase aus Meerrettich:
   Kinetische Studien und Optimierung der Aktivitätsbestimmung
   mit den Substraten H_2O_2 und o-Phenylendiamin,
   J.Clin.Chem.Clin.Biochem., 20: 221-225.
Gallati, H., and Brodbeck, H., 1982 b, Peroxidase aus Meerrettich:
   Reagens zum Abstoppen der katalytischen Umsetzung der
   Substrate H_2O_2 und 2,2'-Azino-di(3-ethyl-benzthiazolinsulfon-
   säure-(6) (ABTS),J.Clin.Chem.Clin.Biochem., 20: 757-760.
Gallati, H., and Pracht, J., 1985, Peroxidase aus Meerrettich: Kine-
   tische Studien und Optimierung der Peroxidase-Aktivitäts-
   bestimmung mit den Substraten H_2O_2 und 3,3',5,5'-Tetra-
   methylbenzidin,J.Clin.Chem.Clin.Biochem., 23: 435-460.

Geoghegan, W.D., Struve, M.F., and Jordon, R.E., 1983, Adaption of the Ngo-Lenhoff Peroxidase assay for solid phase ELISA, J. Immunol. Methods, 60: 61-68.

Gochman, N., and Schmitz, J.M., 1971, Automated determination of uric acid with use of a uricase-peroxidase system, Clin. Chem., 17: 1154-1159.

Harper, J.R., and Orengo, A., 1981, The preparation of an immuno-globulin-amyloglucosidase conjugate and its quantitation by an enzyme-cycling assay,Anal. Biochem., 113: 51-57.

Hofmann, F., Hubl, W., and Schütting, R., 1983, Eine mechanisierte Bestimmungsmethode für Meerrettich-Peroxidase am Reaktions-geschwindigkeitsanalysator mit den Chromogenen o-Dianisidin und o-Phenylendiamin zur Anwendung beim ELISA, Z. med. Labor.-Diagn., 24: 155-160.

Hofmann, J., and Sernetz, M., 1984, Immobilized enzyme kinetics analyzed by flow-through microfluorimetry. Resorufin-ß-D-galactopyrano-side as a new fluorogenic substrate for ß-galactosidase, Anal. Chim. Acta, 163: 67-72.

Hildebrandt, A., 1986, Verfahren zur Bestimmung der Enzymaktivität gelöster und trägergebundener Peroxidase-markierter Anti-körper,Z. med. Labor.-Diagn. ,6: 149-153.

Johannsson, A., Stanley, Ch.J., and Self, H.C., 1985, A fast highly sensitive colorimetric enzyme immunoassay system demon-strating benefits of enzyme amplification in clinical chemistry,Clin. Chim. Acta., 148: 119-124.

Klapper, M.H., and Hackett, D.P., 1963, The oxidatic activity of horseradish peroxidase. I. Oxidation of hydro- and naphthohydroquinones,J. Biol. Chem., 238: 3736-3742.

Lin, E.H., and Gibson, D.M., 1977, Visualization of peroxidase iso-enzymes with eugenol, a noncarcinogenic substrate, Anal. Biochem., 79: 597-601.

McGowan, M.W., Artiss, J.D., Strandbergh, D.R., and Zak, B., 1983, A peroxidase-coupled method for the colorimetric deter-mination of serum triglycerides,Clin. Chem., 29: 538-542.

Meiattini, F., Prenzipe, L., Bardelli, F., Giannini, G., and Tarli, P., 1978, The 4-hydroxybenzoate/4-aminophenazone chromo-genic system used in the enzymic determination of serum cholesterol,Clin. Chem., 24: 2161-2165.

Moss, D.W., Self, C.H., Whitaker, K.B., Bailyes, E., Siddle, K., Johannson, A., Stanley, C.J., and Cooper, E.H., 1985, An enzyme-amplified monoclonal immuno-enzymometric assay for prostatic acid phosphatase,Clin. Chim. Acta., 152:85-94.

Ngo, T.T., and Lenhoff, H.M., 1980, A sensitive and versatile chromo-genic assay for peroxidase and peroxidase-coupled reactions, Anal. Biochem. ,105: 389-397.

Orsonneau, J.L., Meflah, K., Lustenberger, P., Cornu, G., and Bernard, S., 1982, Sensitation and visualisation of bio-chemical measurements using the NAD/NADH system by means of Meldola blue.I. Principle and application to the continous flow measurement of LDH and HBDH activities in serum, Clin. Chim. Acta, 125: 177-184.

Paul, K.G., Ohlsson, P.J., and Jönsson, N.A., 1982, The assay of peroxidases by means of dicarboxidine on enzyme-linked immunosorbent assay level,Anal. Biochem. ,124: 102-107.

Persijn, J.P., and Jonker, K.M., 1978, A terminating reagent for the peroxidase-labelled enzyme immunoassay, J. Clin.Chem.Clin.Biochem., 16: 531-532.

Porstmann, B., Porstmann, T., and Gaede, D., 1980, Optimierung der Aktivitätsbestimmung von Meerrettichperoxidase, Z.med.Labor.-Diagn., 21: 201-209.

Porstmann, B., Porstmann, T., Gaede, D., Nugel, E., and Egger, E.,
        1981 a, Temperature dependent rise in activity of horse-
        radish peroxidase caused by non-ionic detergents and its
        use in enzyme-immunoassay,Clin. Chim. Acta, 109: 175-181.
Porstmann, B., Porstmann, T., and Nugel, E., 1981 b, Comparison of
        chromogens for the determination of horseradish peroxidase
        as a marker in enzyme immunoassay,
        J. Clin.Chem.Clin.Biochem., 19: 435-439.
Porstmann, B., Porstmann, T., Nugel, E., and Evers, U., 1985 a, Which
        of the commonly used marker enzymes gives the best results
        in colorimetric and fluorimetric enzyme immunoassays:
        horseradish peroxidase, alkaline phosphatase or ß-galac-
        tosidase ,J. Immunol. Methods 79: 27-37.
Porstmann, T., Porstmann, B., Wietschke, R., von Baehr, R., and Egger,
        E., 1985 b, Stabilization of the substrate reaction of
        horseradish peroxidase with o-phenylenediamine in the
        enzyme immunoassay,J. Clin.Chem.Clin.Biochem., 23: 41-44.
Pradelles, Ph., Grassi, J., and Maclouf, J., 1985, Enzyme immunoassay
        of eicosanoids using acetylcholine esterase as label: an
        alternative to radioimmunoassay,Anal. Chem., 57: 1170-1173.
Rathlev, T., and Franks, G.F., 1982, New procedure for detecting anti-
        nuclear antibodies using glucose oxidase immunoenzyme
        technique,Am. J. Clin.Pathol., 677: 705-709.
Stanley, C.J., Paris, F., Plumb, A., Webb, A., and Johannson, A., 1985,
        Enzyme amplification: A new technique for enhancing the speed
        and sensitivity of enzyme immunoassays,
        Int. Clin.Prod.Rev., 7/8: 44-51.
Teshima, Sh., Mitsuhida, N., and Ando, M., 1985, Determination of
        -amylase in biological fluids using a new substrate
        (ß-2-chloro-4-nitrophenyl-maltopentaoside),
        Clin.Chim.Acta, 150: 165-174.
Trinder, P., 1969, Determination of glucose in blood using glucose
        oxidase with an alternative oxygen acceptor,
        Ann.Clin.Biochem. ,6: 24-27.
Werner, W., Rey, H.-G., and Wielinger, H., 1970, Über die Eigen-
        schaften eines neuen Chromogens für die Blutzuckerbe-
        stimmung nach der GOD/POD-Methode,Z.Anal. Chemie, 252: 224-228.
Wimmer, M.C., Artiss, J.D., and Zak, B., 1985, Peroxidase coupled
        method for kinetic colorimetry of total creatine kinase
        activity in serum,Clin.Chem., 31: 1616-1620.

AMPLIFICATION SYSTEMS FOR ENZYME IMMUNOASSAY

Jean-Luc Guesdon

Laboratoire des Sondes Froides
Institut Pasteur
Paris, France

INTRODUCTION

Since its development (Avrameas and Guilbert, 1971a, 1971b ; Engvall and Perlmann, 1971 ; Van Weemen and Schuurs, 1971) the enzyme immunoassay has been applied extensively for diagnosing bacterial, viral and parasitic diseases. The enzyme immunoassay method has proved to be a powerful tool in many areas and has been the focus of several reviews, symposia and workshops (Feldmann et al., 1976 ; Engvall and Pesce, 1978 ; Malvano, 1980 ; Guesdon and Avrameas, 1981 ; Ishikawa et al., 1981 ; Avrameas et al., 1983 ; Talwar, 1983). In order to broaden the application of the enzyme immunoassay, it is important to improve existing technology ; indeed, highly sensitive assays are required in fundamental research as well as in medicine. The enzyme immunoassay, involving antigen, hapten or antibody labelled with an enzyme, combines the specific recognition properties of antibodies with the high sensitivity characteristic of enzyme based analytical techniques. The physicochemical properties and the concentration of the antibodies and the enzyme directly affect the assay's sensitivity. More precisely, the best sensitivity is obtained with antibodies having a high affinity constant. Whereas the affinity constant cannot be increased beyond its upper limit, the sensitivity of the competitive immunoassay can be improved by decreasing the concentration of antibody used. Indeed, the effect of decreasing the antibody concentration is to shift the usable range to a lower concentration of antigen. As the antibody concentration decreases, however, the dose-response curve flattens and higher sensitivity is achieved at the expense of discrimination : consequently, the assay will be less accurate. Thus, using an enzyme-antibody conjugate with high immunoreactivity and enzyme activity, and amplifying the specific signal so that a low antibody concentration may be used will improve both the sensitivity and the accuracy of the enzyme immunoassay.

The enzyme used as label, which plays a major role in this kind of assay, should fulfill the following criteria :

1) The enzyme must have high specific activity and turnover rate. This activity must function satisfactorily under the conditions of temperature, ionic strengh and pH in which the immunoassay will be performed.

2) The label must be stable at room temperature.

3) The number of enzyme molecules must be measurable. Enzyme activity is usually determined by spectrophotometry. If the product absorbs light in the near ultra-violet or visible region, it is possible to measure the absorbance change during the progress of the enzyme reaction. If the product does not absorb light, it can be transformed into a colored product by adding an appropriate chemical reagent to the substrate solution. Fluorometric, calorimetric or electrode methods are also used to measure enzyme activity in the enzyme immunoassay.

Generally, in the enzyme immunoassay, enzymes are cross-linked with antibodies through covalent bonds using various organic compounds (Avrameas et al., 1978 ; Ishikawa et al., 1983a). It should always be assumed that chemical modification of an enzyme affects its activity to a certain extent. For example, the covalent linkage between an enzyme and an antibody can decrease the enzyme's activity due to steric hindrance and modification of the net charge or the charge distribution of the enzyme protein. Moreover, severe diminution of antibody activity has been observed during conjugate preparation, and the product of a labelling reaction may often form a heterogeneous population.

In order to overcome the drawbacks of techniques employing covalent coupling, several methods involving biospecific, ligand/counter-ligand interaction have been proposed : antigen-antibody (Avrameas, 1969 ; Mason et al., 1969 ; Sternberger et al., 1970) avidin-biotin (Guesdon et al., 1979) and lectin-sugar (Guesdon and Avrameas, 1980) systems.

The avidin-biotin amplified enzyme immunoassay comprises three main steps : 1) incubation of immobilized antigen with biotin-labelled antibody ; 2) incubation with avidin ; 3) incubation with biotin-labelled enzyme. The possibility with this procedure of adding the reagents sequentially could be an advantage. Since the antibody and the enzyme label are not covalently linked, they can be added at different concentrations depending on the purpose of the assay. If necessary, the assay can be shortened by using avidin covalently coupled to the enzyme or preformed biotinylated enzyme/avidin complexes instead of adding avidin and biotinylated enzyme sequentially.

The avidin-biotin system, based on the principle that avidin possesses 4 binding sites and that the avidin molecule can act as a bridge between two different biotinylated proteins, is used with increasing frequency to improve the sensitivity of staining (Heitzmann and Richards, 1974 ; Bayer et al., 1976 ; Heggeness and Ash, 1977 ; Berman and Bash, 1980 ; Hsu et al., 1981 ; Childs Moriarty and Unabia, 1982 ; Ogata et al., 1983) and immunoassay techniques (Guesdon et al., 1979 ; Costello et al., 1979 ; Kendall et al., 1983 ; Subba Rao et al., 1983 ; Yolken et al., 1983 ; Leipold and Remy, 1984 ; Bacquet and Twumasi, 1984 and Adler-Storhz et al., 1985).

Lectins, proteins found in relatively large amounts in the seeds of various plants, are able to interact specifically with carbohydrate moities. As such moities are present in certain enzyme molecules, we have tested the possibility of using lectin-antibody conjugates to develop an immunoassay (Guesdon and Avrameas, 1980). The principle of this assay, the lectin-immunotest, is that the specific sites of the lectin can operate as acceptors for glyco-enzymes secondarily added to the system.

The test is carried out as follows :
In order to quantitate an antigen, an adequate dilution of the sample to be tested is added to the solid phase coated with the corresponding antibody. After incubation, the solid phase is washed and incubated with a solution containing both antigen-specific antibody coupled to a lectin and an excess of mono- or di-saccharide specific for the lectin used. The antibody combining site of the lectin-antibody conjugate is thus allowed to react with the antigen epitopes ; because of the excess saccharide present, the lectin combining sites will not react with any polysaccharides or glycoproteins present in the reaction mixture. The solid phase is then washed again to eliminate excess saccharide and lectin-antibody conjugate, and an enzyme marker (generally peroxidase or glucose oxidase), which interacts specifcally with the lectin used, is added. After further incubation and washing, the enzyme activity associated with the solid phase is determined by adding the appropriate substrate. This enzyme activity is proportional to the concentration of the antigen being quantified.

Morh and Franz (1981) alternatively have proposed that when Concanavalin A (Con A) serves as the lectin, it can be inactivated by elimination of the bivalent metal cations Ca^{++} and Mn^{++} so that it will not react with carbohydrate components of the antigen. In this case, the reactivation of Con A takes place only after the immune reaction and the removal of unbound conjugate.

Since lectins possess two or more active sites and the enzyme marker is employed without chemical modification, molecular amplification will be optimal. However, this assay can be applied only with enzyme bearing carbohydrate moieties.

Assays have been described which use immunological binding of marker enzymes by means of bridge antibodies. The procedure is based on the principle that an anti-immunoglobulin antibody is bound by one of its combining sites to the antibody directed against the antigen to be tested ; the second combining site reacts with the anti-enzyme antibody which then binds the enzyme added during the last step. Such a method requires 4 or 5 incubation steps (Yorde et al., 1976 ; Butler et al., 1978 ; Metzger et al., 1981). Immune complexes prepared with monoclonal anti-peroxidase or anti-galactosidase antibodies and the corresponding enzymes has been used (Ternynck et al., 1983) instead of immune complexes formed with polyclonal antibodies. With this approach, it is not necessary to determine the equivalence point of the antibody preparation ; furthermore, a higher concentration of enzyme can be used because precipitation never occurs with a monoclonal antibody. However, this system at present is limited to mouse antibody detection.

A more general method, which incorporates antigen-specific antibody and anti-enzyme antibody belonging to different animal species, has been described and employed for the titration of human anti-Echinococcus granulosus antibody and for the quantitation of human IgE (Guesdon et al., 1983a) and human alpha-1-fetoprotein (Porstmann et al., 1984). This method makes use of chimera antibodies ; such antibodies are prepared by covalently coupling the antibody specific for the antigen to be quantified with the polyclonal or monoclonal antibody specific for an enzyme. This assay comprises of three main steps :

1) Incubation of the antigen-containing material on antibody-coated solid phase.

2) Incubation with the chimera antibodies.

3) Incubation with the enzyme.

Each step is separated from the other by a washing step. After the third step a substrate solution is added.

Another way to obtain a higher molecular amplification in the heterogeneous enzyme immunoassay is to increase the number of enzyme molecules specifically bound to the solid phase. This can be achieved by coupling bovine serum albumin (BSA) with the antibody directed against the antigen to be quantified. In this case, the BSA-antibody conjugate is revealed by adding anti-BSA antibodies labelled with an enzyme (Guesdon et al., 1983b).

The detection limit in an enzyme immunoassay depends ultimately on the lowest concentration at which the label can be detected via an activity assay. Indeed, the sensitivity of such a determination depends upon the turnover number of the enzyme molecule used and the method employed to detect the product formed during the enzymatic reaction. High signal amplification has been observed with hydrolases, such as alkaline phosphatase and β-galactosidase in combination with either a tritiated substrate (Harris et al., 1979) or a substrate which produces a fluorescent compound (Ishikawa and Kato, 1978 ; Neurath and Strick, 1981 ; Labrousse et al., 1982 and Ali and Ali, 1983).

More recently, Self (1985) has described an enzyme amplification method for alkaline phosphatase. With this method, amplification is achieved by the use of an indicator enzyme (alkaline phosphatase) to produce NAD, which in turn triggers the formation of a measurable formazan dye. This is achieved by the use of two enzymes, alcohol dehydrogenase and diaphorase, which interconvert NAD with NADH utilising ethanol and INT violet as substrates.

This short review will be restricted to an examination of several methods to obtain a signal amplification in enzyme immunoassays. The scope and limitations of these methods will be discussed.

MATERIALS AND METHODS

Reagents

Avidin, bovine serum albumin (BSA), horseradish peroxidase (POD) with a purity number of 3, alkaline phosphatase (calf intestinal mucosa), glucose oxidase 4-methylumbelliferyl-β-D-galactopyranoside (MUG), 4-methylumbelliferyl phosphate (MUP) and D-biotin-N-hydroxy-succinimide ester (BNHS) were purchased from Boehringer Mannheim (FRG). Glycine, D-glucose o-nitrophenyl-β-D-galactopyranoside (ONPG) were obtained from Janssen Chimica (Belgium), Mg Titriplex, Tween 20 and Perhydrol 30% H_2O_2 from Merck (FRG), o-phenylenediamine, peroxidase type II and 3-(p-hydroxyphenyl propionic) acid from Sigma Chemical (USA), 25% water solution glutaraldehyde research grade from Serva (FRG), N-ethoxy-carbonyl-2-ethoxy-1,2-dihydroquinoline (EEDQ) from Aldrich Chemical Company (France), gelatin from Prolabo (France). Escherichia coli β-galactosidase was kindly donated by Dr. Agnès Ullmann (Pasteur Institute, Paris). Agglutinins of Canavalia ensiformis, Triticum vulgare, Lens culinaris, Ricinus communis, methyl-α-D-mannoside, galactose and N-acetyl glucosamine were obtained from IBF (Villeneuve-la-Garenne, France). Flat-bottomed polystyrene microtiter plates (M29 LSE) were purchased from CML (Nemours, France).

Antisera and antibodies

Rabbit and sheep antisera against the various proteins used in the amplified enzyme immunoassays were obtained by hyperimmunizing animals using complete Freund adjuvant. Sheep anti-mouse IgG or anti-rabbit IgG antibodies and rabbit anti-BSA antibodies were isolated from whole serum by affinity chromatography using corresponding IgG or BSA coupled to Ultrogel beads (Guesdon and Avrameas, 1976).

Antiserum to human IgE and immunoadsorbent isolated anti-IgE antibody were prepared as described previously (Guesdon et al., 1978). The preparation of anti-peroxidase monoclonal antibodies has been described in detail elsewhere (Ternynck et al., 1983).

Enzyme-protein conjugates

Avidin and antibodies of different specificities were coupled with alkaline phosphatase, glucose oxidase, peroxidase or β-galactosidase using the one step glutaraldehyde coupling procedure (Avrameas et al., 1978). The resulting conjugates were dialyzed overnight at 4°C against 0.15 M NaCl containing 0.010 M phosphate buffer pH 7.4 (PBS), centrifuged, mixed with an equal volume of bidistilled glycerol and stored at -20°C.

Antibody biotinylation

Biotinyl-N-hydroxysuccinimide (BNHS) was used for introducing biotin moieties into sheep anti-mouse IgG antibodies (Guesdon et al., 1979). Ten mg of antibodies were dissolved in 1 ml 0.1 M $NaHCO_3$ solution and mixed with 0.11 ml of 0.2 M BNHS solution in distilled dimethylformamide. The reaction mixture was incubated at room temperature for 1 hour and then dialyzed for 24 hours at 4°C against several changes of PBS. After dialysis, the percentage of free amino groups in biotin substituted antibodies was determined by using the 2-4-6 trinitrobenzene sulfonic acid procedure (Habeeb, 1966). Under these conditions, the percentage of biotin substituted amino-groups was 90-95%. The preparation was stored at -20°C in 50% glycerol.

Coupling of antibody with biotinyl-glucose oxidase

Sheep anti-rabbit IgG antibodies were double labelled with biotin and glucose oxidase by following the procedure described below. First, glucose oxidase was biotinylated by mixing 3.3 mg glucose oxidase dissolved in 0.3 ml of 0.15 M NaCl containing 0.1 M $NaHCO_3$ with 2.4 mg BNHS dissolved in 0.03 ml dimethylformamide. After being left for one hour at room temperature, the mixture was dialyzed overnight at 4°C against 0.15 M NaCl. More than 90% of the amino groups were substituted. After dialysis, 0.05 ml 1 M KH_2PO_4 and 0.05 ml of dimethylformamide containing 10 mg EEDQ were added to the biotinylated glucose oxidase. The mixture was left for 1 hour at room temperature and chromatographed on a Sephadex G25 column. The yellow fraction was collected and mixed with 1 mg sheep anti-rabbit IgG antibodies diluted in 0.2 ml 0.15 M NaCl, and then the pH was adjusted to 8.0 with 1 M K_2HPO_4. The mixture was left at 4°C for 24 hours and then dialyzed against PBS. The double labelled antibody preparation was mixed with an equal volume of bidistilled glycerol and kept at -20°C.

BSA antibody conjugate

Sheep anti-human IgE antibodies (2 mg) were mixed with 5 mg BSA in 0.9 ml of 0.1 M phosphate buffer, pH 6.8. 0.1 ml of 1% glutaraldehyde was then added to the mixture. After 3 hours incubation at room temperature, 0.05 ml of 2 M glycine was added. The mixture was then treated as above for enzyme-protein conjugates.

Lectin-antibody conjugates

Lectins were coupled by the one-step glutaraldehyde procedure. Four mg of lectin were mixed with 2 mg of isolated antibodies in a final volume of 0.9 ml of 0.1 M phosphate buffer pH 6.8. Lectin-specific saccharide (0.1 ml) was added to the mixture to obtain a final concentration of 0.1 M. Twenty μl of a 1% aqueous glutaraldehyde solution was added while the mixture was stirred. After 3 hours of incubation at room temperature, 0.05 ml of 2 M glycine solution was added. Two hours later, the mixture was dialyzed against PBS at 4°C and treated as above.

Chimera antibodies

The procedure used to couple sheep anti-human IgE antibodies with monoclonal anti-peroxidase antibodies was as follows. To 0.9 ml of 0.1 M phosphate buffer pH 6.8 containing 0.5 mg of sheep antibodies and 4 mg of monoclonal antibody, 0.1 ml of a 1% aqueous solution of glutaraldehyde was added with gentle stirring. The reaction mixture was allowed to incubate for 2 h at room temperature and then 0.05 ml of 2 M glycine was added. The preparation was further incubated for 2 h and treated as above for the other conjugates.

Enzyme immunoassays

Buffers. The following buffers were used.
PBS : 0.15 M NaCl containing 0.010 M phosphate buffer pH 7.4.
PBS-T : PBS containing 0.1% Tween 20 (washing buffer).
PBS-T-G : PBS-T containing 0.3% gelatin (incubation buffer).

Coating of polystyrene plates. The immobilized antibody or antigen for the various enzyme immunoassays described in this chapter was prepared by passive coating of polystyrene plates. This was done by incubating 0.1 ml of antibodies or antigens diluted in PBS for 2 h at 37°C and then for 18 h at 4°C. After this step, the plates could be kept for several weeks at 4°C. Just before use, coated plates were washed 3 times with PBS-T. Because a prozone phenomenon often occurs, it is important to determine the optimal concentration of the coating antigen or antibody for each new immunological system. This is done by checkerboard titration of antigen or antibody solution using appropriate positive and negative controls. The concentration giving the maximal difference between the positive and the negative control is used to coat the plates.

Assay conditions. All measurements were made in duplicate. Samples and various conjugates were diluted in PBS-T-G, with a final volume of 0.1 ml per well. Between each incubation step the plates were washed three times by emptying and filling with TBS-T from a wash bottle, and placed in an inverted position on clean absorbent paper for 1 minute to remove liquid as completely as possible. Incubation usually lasted 2 hours at room temperature ; in some experiments requiring high sensitivity, the incubation time was increased to 18 hours.

Measurement of β-galactosidase activity.

Galactosidase activity was measured by adding 0.2 ml per well of either chromogenic or fluorogenic substrate. The chromogenic substrate solution comprised : 0.1 M phosphate buffer pH 7.0 containing 2.66 mM ONPG, 0.1 M β-mercaptoethanol, 1 mM magnesium sulphate, 0.2 mM manganese sulphate and 2 mM Mg Titriplex. The fluorogenic substrate solution was prepared by replacing ONPG by MUG. 1 ml of 6.76 mg/ml MUG in dimethylformamide was mixed with 100 ml of the above buffer. The plates were incubated in the dark at 37°C generally for 1-4 h. At an arbitrarily chosen time, the galactosidase was inactivated by adding 0.1 ml 1 M Na_2CO_3. Absorbance of o-nitrophenol was measured by a Titertek multiskan photometer with a 414 nm filter and fluorescence of 4 methyl-umbelliferone was measured by a Titertek Fluoroskan fluorometer (excitation and emission wavelengths were 355 nm and 480 nm respectively).

Measurement of peroxidase activity.

Peroxidase activity associated with the solid phase was measured by adding 0.2 ml substrate solution composed of 0.05 M citrate buffer containing 1 mg o-phenylenediamine/ml and 0.06% hydrogene peroxide. After a 10 minute enzyme reaction, peroxidase was denatured by adding 0.05 ml 2 M H_2SO_4 containing 0.5% Na_2SO_3 and absorbance was read at 492 nm. The tangerine-colored product formed was highly soluble but somewhat light sensitive.

In one experiment, peroxidase activity was determined according to Zaitsu and Ohkura (1980) using 3-(p-hydroxyphenyl) propionic acid, the best fluorogenic substrate available at present to assay this enzyme.

Measurement of glucose oxidase activity.

Glucose oxidase substrate was prepared by adding 250 mg of D-glucose, 2.5 mg of peroxidase (Sigma, type II) and 10 mg o-phenylenediamine in 25 ml 0.01 M phosphate buffer pH 8. The enzyme reaction was started by adding 0.2 ml substrate solution to each well and stopped after 30 minutes incubation at 37°C by adding 0.05 ml 3 N HCl per well. Optical density was read at 492 nm.

Measurement of alkaline phosphatase.

0.2 ml of fluorogenic substrate solution (MUP, 1 mM in 1 M triethanolamine buffer pH 9.8 containing 1 mM $MgCl_2$) was pipetted into each well. The plate was incubated at 37°C for 20-30 minutes and the reaction was stopped by the additon of 0.05 ml of 3 N NaOH. The 4-methylumbelliferone produced was measured on a Fluoroskan fluorometer. The wavelengths used for excitation and emission analysis were 355 nm and 480 nm respectively.

TYPICAL RESULTS

Avidin-Biotin amplified enzyme immunoassay

The determination of mouse IgG was chosen as an example to illustrate the set-up for an Avidin-Biotin amplified enzyme immunoassay.

First, the efficiency of the biotin-labelled anti-mouse IgG antibodies and the enzyme-labelled avidin were tested. In the example given here, the enzyme marker was alkaline phosphatase from calf intestinal mucosa. Biotinylated antibodies and enzyme avidin conjugate may be titrated in several ways ; it is preferable, however, to follow the whole assay procedure, including the incubation of mouse IgG on anti-mouse IgG antibody-coated plates, incubation with biotinylated antibodies, and incubation with enzyme-avidin conjugate.

Indeed, the reactants may be titrated on an IgG-coated plate, but only by carrying out the whole assay procedure can such information as the real background intensity and the value of specific signal/background ratio be obtained.

Figure 1 shows the results of titration tests. For subsequent assays, biotinylated anti-mouse IgG antibodies and alkaline phosphatase-avidin conjugate were diluted 1:1000.

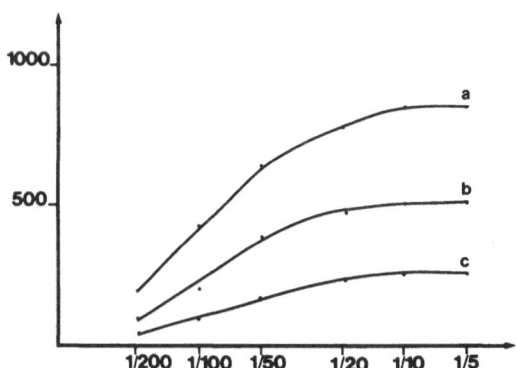

Fig. 1   Titration of biotinylated anti-mouse IgG
          antibodies in microtiter wells coated with
          anti-mouse IgG antibodies, using three
          different concentrations of mouse IgG :
          16 ng/ml (curve a), 8 ng/ml (curve b),
          4 ng/ml (curve c) and using alkaline
          phosphatase-avidin conjugate diluted
          1:1000. Ordinate: Fluorescence intensity
          in arbitrary units ; abscissa: dilution
          of biotin labelled antibodies (x 100).

A comparison of the Avidin-Biotin amplified enzyme immunoassay with a classical enzyme immunoassay is given in Figure 2. Mouse IgG were quantified using a polystyrene plate coated with purified anti-mouse IgG and either biotin-labelled sheep anti-mouse IgG and alkaline phosphatase-labelled avidin or alkaline phosphatase labelled sheep anti-mouse IgG antibodies. Figure 2 shows that amplification was higher with biotin-labelled anti-IgG antibodies than with the enzyme-antibody conjugate.

As part of the study of the molecular amplification for enzyme immunoassay, we tested the possibility of double labelling antibodies using the Avidin-Biotin system. This double labelling was carried out by biotinylating the enzyme, coupling the biotinylated enzyme with an antibody, and adding an enzyme-labelled avidin conjugate to the reaction medium. If antibody and avidin are thus labelled with the same enzyme, the number of enzyme molecules linked to antibody will be increased. The experimental system chosen to test this possibility included : 1) biotinylation of Aspergillus niger glucose oxidase with biotinyl-N-hydroxysuccinimide ester (BNHS), a high BNHS/amino group ratio was used to obtain a substitution greater than 90% ; 2) activation of carboxylic groups present on biotinylated glucose oxidase by using N-ethoxy-carbonyl-2 ethoxy-1,2-dihydro quinoline (EEDQ) ; 3) coupling of sheep anti-rabbit IgG with EEDQ-treated biotinylated glucose oxidase. An enzyme-antibody conjugate used as control was prepared by labelling the same sheep anti-rabbit IgG antibodies with EEDQ treated glucose oxidase.

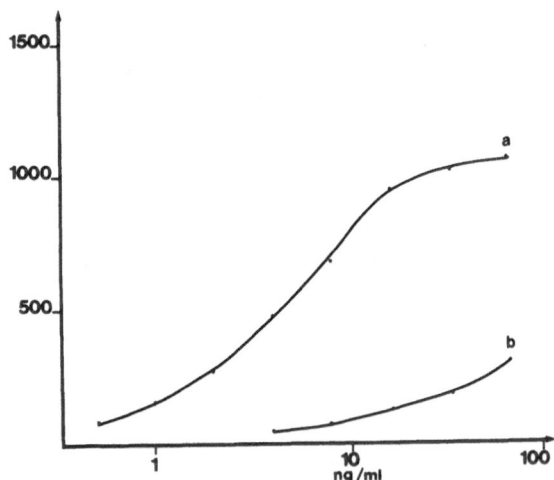

Fig. 2   Quantification of mouse IgG on sheep anti-
         mouse IgG-coated plate. Comparison of
         standard curves obtained using either biotin-
         labelled anti-mouse IgG and phosphatase-
         labelled avidin (a) or phosphatase-labelled
         anti-mouse IgG (b). Ordinate: fluorescence
         intensity in arbitrary units ; abscissa:
         mouse IgG concentration.

The reactivity of both conjugates was tested on a rabbit IgG-coated plate with or without the addition of glucose oxidase-avidin conjugate prior to enzyme determination. Avidin was covalently bound to glucose oxidase using the one-step glutaraldehyde method.

From the results presented in Figure 3, it can be concluded that : 1) glucose oxidase-antibody conjugate prepared using EEDQ as a coupling agent is more efficient when glucose oxidase has been previously biotinylated. The reason may be that after biotinylation, few amino groups remain free and consequently the homopolymerization of glucose oxidase molecules is limited ; 2) the introduction of biotin residues in the glucose oxidase-antibody conjugate allows other glucose oxidase molecules to bind specifically via the avidin-biotin interaction ; 3) double labelled antibody can be used to amplify the specific signal in enzyme immunoassays.

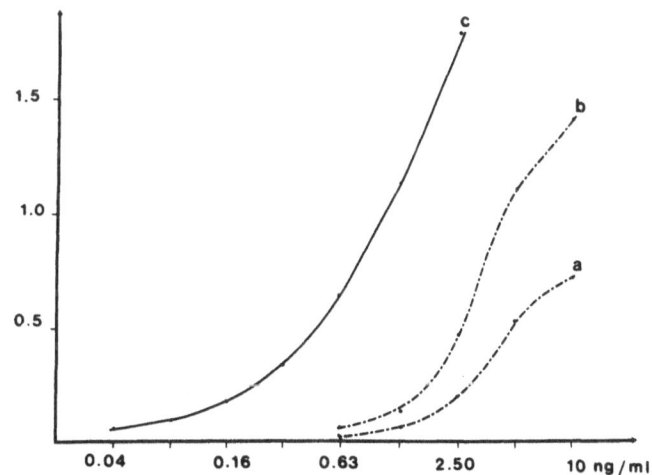

Fig. 3   Amplification by using double labelled antibodies. Sheep anti-rabbit IgG antibodies labelled with biotinylated glucose oxidase were tested in rabbit IgG-coated wells.
Curve a : control with glucose oxidase-labelled antibodies prepared using EEDQ as the coupling agent.
Curve b : testing of antibodies labelled with biotinylated glucose oxidase. Enzyme activity was determined just after conjugate incubation.
Curve c : testing of antibodies labelled with biotinylated glucose oxidase. Enzyme activity was determined after a further incubation with glucose oxidase labelled avidin.

Amplified enzyme immunoassay using albumin-antibody conjugate

We found that whole immune serum can be labelled with an enzyme if a very large excess of enzyme was present. The use of such preparations in routine experiments, although effective, proved very costly and in some cases a high non-specific binding of the conjugate was observed. We therefore examined the possibility of labelling the whole serum by the one-step glutaraldehyde procedure with bovine serum albumin (BSA), because this protein is stable, easy to prepare in pure form and soluble at high concentrations. At the pH used to prepare the conjugate and in contrast to what was often observed with various coupling procedures, no precipitation was observed during the labelling reaction. Moreover, we found that the specific activity of the conjugate was greater with BSA antibody conjugate than with a classical enzyme antibody conjugate. An example is given in Figure 4, where human IgE are quantified in sheep anti-IgE antibody-coated wells, using the same antibodies labelled with either BSA or galactosidase. The BSA used as label is revealed by an anti-BSA antibody labelled with galactosidase. Dose response curves obtained after IgE was quantified with both conjugates under the same experimental conditions show that amplification was greater with the BSA conjugate.

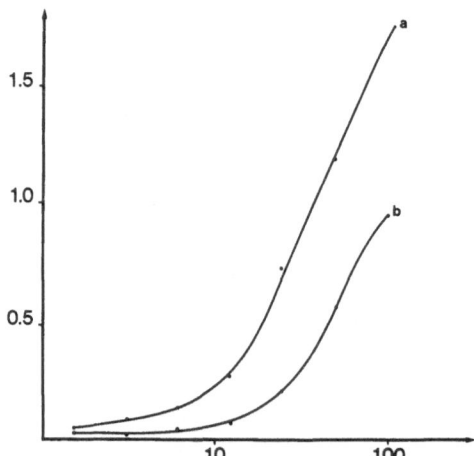

Fig. 4   Human IgE quantification using a plate coated with anti-human IgE antibodies and either BSA-labelled antibodies (a) or β-galactosidase labelled antibodies (b). Ordinate: absorbance at 414 nm. Abscissa: IgE concentration in IU/ml.

Amplified enzyme immunoassay using unmodified enzyme

Two systems in which a native unmodified enzyme served as label in a enzyme immunoassay are described here. The heterogeneous immunoassay, in which an unmodified enzyme is used to determine the quantity of antigen present, is carried out as follows. First, the antigen is allowed to react with its corresponding antibody, immobilized on a solid phase. Then an antibody, directed against the antigen to be quantified, is added. This antibody carries a ligand, which is able to bind the unmodified enzyme marker specifically. The purpose of this procedure is to obtain the fullest enzymatic amplification of the marker. Among the ligands which can be used, we tested lectins and anti-enzyme antibodies.

Lectin-antibody conjugates were prepared using various lectins and glutaraldehyde as the coupling agent. The results obtained by testing lectin-labelled sheep anti-rabbit IgG in rabbit IgG-coated wells with glucose oxidase as marker showed that the various lectins tested, viz Lens culinaris, Canavalia ensiformis, Ricinus communis and Triticum vulgare agglutinins were not equally effective in binding glucose oxidase. A time-course experiment with glucose oxidase at 10 μg/ml demonstrated that the fixation of this enzyme to Canavalia ensiformis agglutinin was completed in 15 minutes (Guesdon and Avrameas, 1980) ; thus, the enzyme incubation step required in this procedure does not considerably increase the whole test time.

The results of labelling the same antibody either with a lectin or an enzyme and comparing the conjugates for their effectiveness in quantifying antibody are shown in Figure 5. The same sheep anti-rabbit IgG antibody, labelled with either Canavalia ensiformis agglutinin (Con A) or horseradish peroxidase (POD), was used to titrate anti-BSA antibodies in rabbit serum. Plates coated with BSA were incubated with increasing dilutions of rabbit immune serum and washed ; 5 μg/ml of the anti-rabbit IgG antibody, labelled with Con A or POD was then added. After further washing the wells that had received enzyme conjugate received enzyme substrate and those which had received Con A conjugate received first POD and then substrate. Whatever the conjugate used, the enzymatic reaction was allowed to proceed for 5 minutes at room temperature. As shown in Figure 5, Con A antibody conjugate was found to amplify the signal to a greater degree than the enzyme-antibody conjugate.

In order to broaden the application of this kind of amplified enzyme immunoassay, we tested the possibility of using chimera antibodies formed by covalently coupling monoclonal or polyclonal anti-enzyme antibodies to monoclonal or polyclonal antigen-specific antibodies (Guesdon et al., 1983a).

Systematic comparison of conventional enzyme antibody conjugates and chimera antibodies showed that the latter always gave higher amplification. Further, the same sensitivity of chimera antibody amplified immunoassay was achieved whether a highly purified enzyme or a crude preparation of enzyme was used (Guesdon et al., 1983a ; Porstmann et al., 1984).

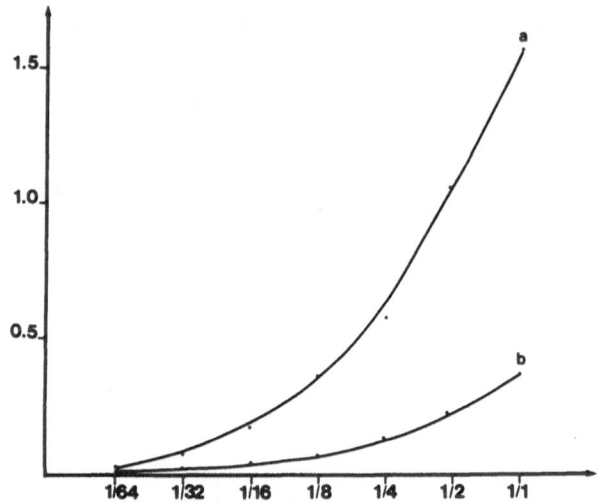

Fig. 5   Titration of anti-BSA antibodies in rabbit immune serum, using plates coated with BSA and sheep anti-rabbit IgG antibodies labelled either with <u>Canavalia ensiformis</u> agglutinin (a) or peroxidase (b). Lectin-antibody conjugate was revealed by using peroxidase (purity number 3, 10 μg/ml). Ordinate: absorbance at 492 nm. Abscissa: rabbit immune serum dilution (x 10^4).

        To illustrate this observation, standard curves obtained by quantifying human IgE are given in Figure 6. In this experiment, a human serum containing a known concentration of IgE was diluted and incubated in sheep anti-IgE antibody-coated wells. After washing, the wells were filled with anti-peroxidase/anti-IgE chimera antibodies, washed and filled again with various preparations of peroxidase. All the preparations tested (purity number from 3.0 to 0.57) were equally effective.

Influence of the substrate on enzyme amplification

        Several enzyme immunoassays have been developed with substrates that generate either light fluorescence or radioactivity which are effective for measuring very small amounts of antigens or antibodies. Fluorogenic substrates appear to be the most attractive because they are safer and the assays are more adaptable to automation. Recent reports indicate that fluorogenic substrates may increase the sensitivity of the enzyme immunoassay. The detection limits of enzyme immunoassays using fluorogenic substrates may even surpass those of the radioimmunoassay (Ishikawa et al., 1983b).

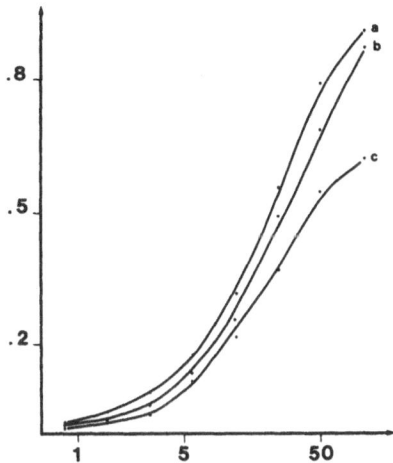

Fig. 6   Quantification of human IgE using chimera antibodies consisting of sheep anti-human IgE antibodies coupled with monoclonal anti-peroxidase antibodies and different preparations of peroxidase, purity number 3 (a), 1.7 (b), 0.57 (c). Ordinate: absorbance at 492 nm. Abscissa: IgE concentration in IU/ml.

Light emission techniques (Puget et al., 1977) and fluorogenic substrates (Zaitsu and Ohkura, 1980) have been used to measure peroxidase activity, but the best signal amplification in the enzyme immunoassay was observed when β-galactosidase and 4-methylumbelliferyl-β-D-galactoside were employed.

However, although the fluorescence amplified enzyme immunoassay appeared more sensitive than the corresponding enzyme immunoassay performed with a chromogenic substrate, the theoretical sensitivity limit expected on the basis of fluorometric measurements of β-galactosidase activity has not yet been achieved. For example, on the one hand, when galactosidase was quantified with a fluorogenic substrate instead of a chromogenic substrate, the sensitivity of the enzyme assay increased 5000-fold, but on the other hand, when human IgE were quantified by a galactosidase immunoassay using an anti-IgE antibody-coated plate, galactosidase labelled anti-IgE antibodies, and either the fluorogenic or the chromogenic substrate, the fluorometric assay was only above 20-fold more sensitive than the colorimetric assay (Figure 7).

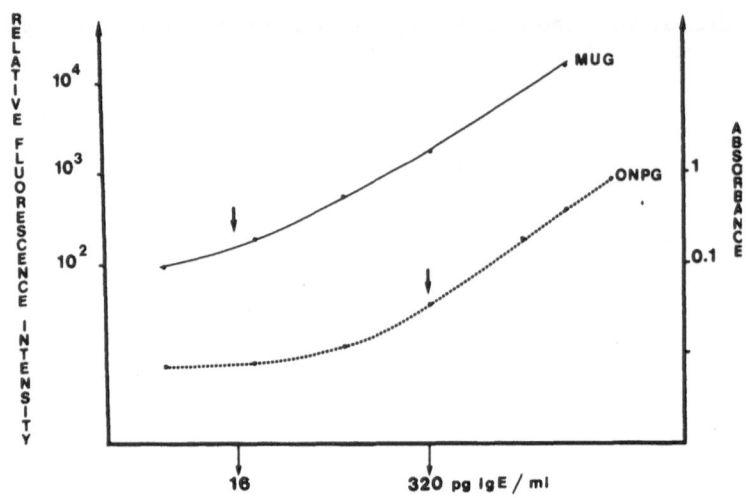

Fig. 7   Human IgE quantification using a plate
coated with anti-human IgE antibodies
and β-galactosidase/anti-IgE antibody
conjugate. Enzyme activity was determined
for 25 h at 37°C using either the fluorogenic substrate, i.e. 4-methylumbelliferyl-
galactoside (MUG) or the chromogenic
substrate, i.e. o-nitrophenyl-galactoside
(ONPG). Arrows indicate the detection limit.

DISCUSSION

Quantitative enzyme immunoassays, originally described
by Avrameas and Guilbert (1971), Engvall and Perlmann (1971)
and Van Weemen and Schuurs (1971), are being used with
increasing frequency in clinical diagnosis and basic
research. The advantages of enzyme immunoassays over the
radio-immunoassays are stability of the reagents, safety and
lower cost. The basic principles of all heterogeneous enzyme
immunoassays are the same : an antibody (or antigen) linked
to an enzyme is allowed to react with the corresponding
immobilized antigen (or antibody), then the enzyme activity
is revealed by specific measurement. The most important
elements in these procedures are the effectiveness of the
antibody (or antigen)-enzyme conjugate employed and the
sensitivity of the enzyme detection method. Generally, the
enzyme-antibody conjugate is prepared by covalently coupling
the enzyme to the antibody (Avrameas et al., 1978). It was
observed, that during the coupling procedure partial
denaturation of the enzyme often occurs. For this reason,
enzyme immunoassays based on a biospecific interaction
between the antibody and the enzyme have been proposed. As
shown here and in previous studies, biospecific interactions,
such as those between avidin-biotin, lectin-saccharide or
antibody-antigen, could lead to a signal amplification not
obtained with chemical binding. This is probably due first to

the conditions used to link the enzyme molecules to the antigen-specific antibody ; indeed in this case, the enzyme marker is used after mild modification (biotinylation) or even without any chemical modification (procedures based on the enzyme/anti-enzyme antibody or enzyme/lectin interactions), and second to the increased number of enzyme molecules linked to each antibody molecule.

The sensitivity of the enzyme immunoassay can be enhanced by using Biotin-Avidin. The biotin-avidin amplification possesses several advantages :

1) Avidin has an extremely high affinity for biotin with an association constant of 10^{15} M^{-1} (Green, 1975).

2) Each avidin molecule can bind four biotin molecules (Green, 1975).

3) Biotin can be covalently bound to antibody or enzymes under relatively mild conditions at a high specific activity without affecting the antigen-binding capacity of the antibody or the activity of the enzyme (Guesdon et al., 1979 ; Kendall et al., 1983).

4) The use of biotinylated antibodies instead of antibody-enzyme conjugates minimizes the problem of steric hindrance.

5) Biotin-avidin interaction is a versatile and specific method when an immunochemical detection method cannot be used (antigen-antibody cross reactions).

6) Many biotinyl derivatives are available to bind biotin to the functional groups of proteins (Bayer and Wilchek, 1980).

In contrast to hybrid antibody molecules with dual specificity, which must be prepared by reduction and subsequent reoxidation of a solution containing a mixture of antigen specific and enzyme-specific antibodies, the chimera antibodies are easily prepared with high yield.

This technique offers several other advantages : in contrast to the classical enzyme immunoassay using enzyme-antibody conjugate, the chimera antibody technique may be performed with crude enzyme preparations, which could be of particular interest when the enzyme used as marker is difficult to obtain in pure form. Moreover the enzyme is used in its native form, i.e. without chemical modification, and thus its turnover rate is unchanged.

A certain background is always seen in the enzyme immunoassay technique. This background, which may be due to nonspecific ionic interactions between the conjugate and the protein-coated plastic, lowers the detection capacity of this kind of assay. In contrast to the techniques using covalently prepared enzyme-antibody conjugate, the assays which employ biospecific interactions to link an enzyme to an antibody make use of homogeneous unmodified proteins. Thus, nonspecific binding, observed during the amplified-immunoassay, may be eliminated by simply changing the pH of the incubation medium so as to neutralize the net electric charge of the protein causing the background. It is not possible to proceed in this way with classical enzyme-antibody conjugates because these preparations are often too heterogeneous. This is a further advantage of using noncovalent binding to link an enzyme to an antibody in comparison with covalent linkage. For example, it was possible to completely eliminate the nonspecific binding of avidin to the solid phase by using solutions of alkaline pH and high ionic strength (Guesdon et al., 1979).

In conclusion, the numerous studies of the enzyme immunoassay demonstrate the potential of this method for ultrasensitive immunoassays. Recent advances in biological technology have enabled substantial improvements to be made, and by intensifying the specific signal, assays even more sensitive than the radioimmunoassay have been designed.

SUMMARY

This chapter deals with amplification systems in enzyme immunoassays. Several methods to enhance the efficiency of the enzyme-antibody linkage and to amplify the specific signal are discussed. These methods involve biospecific interactions such as those between antigen-antibody, avidin-biotin and lectin-saccharide. The performance characteristics and the limitations of the amplified immunoassays are discussed and compared with those for conventional enzyme immunoassays employing covalently prepared enzyme-antibody conjugates. It appears that signal amplification in enzyme immunoassays can be increased by using a coupling procedure which does not affect the enzyme and antibody activities, by increasing the number of enzyme molecules specifically bound to the antibody, and by using a substrate with which a minimal amount of enzyme may be detected.

REFERENCES

Adler-Storthz, K., Dreesman, G.R., Graham, D.Y., and Evans, D.G., 1985, Biotin-Avidin amplified ELISA for quantitation of human IgA, J. Immunoassay, 6:67-77.
Ali, A., and Ali, R., 1983, Enzyme-Linked-Immunosorbent Assay for anti-DNA antibodies using fluorogenic and colorigenic substrates, J. Immunol. Methods, 56:341-346.
Avrameas, S., 1969, Indirect immunoenzyme techniques for the intracellular detection of antigens, Immunochemistry, 6:825-831.
Avrameas, S., and Guilbert, B., 1971a, Dosage enzymo-immunologique de protéines à l'aide d'immunoadsorbants et d'antigènes marqués aux enzymes, C. R. Acad. Sci. (Paris), 273:2705-2707.
Avrameas, S., and Guilbert, B., 1971b, A method for quantitative determination of cellular immunoglobulins by enzyme-labelled antibodies, Europ. J. Immunol., 1:394-396.
Avrameas, S., Ternynck, T., and Guesdon, J.-L., 1978,

Coupling of enzymes to antibodies and antigens, in: "Quantitative Enzyme Immunoassay", E. Engvall and A.J. Pesce, eds, Blackwell Scientific Publications, Oxford.

Avrameas, S., Druet, P., Masseyeff, R., and Feldmann, G., 1983, Immuno-enzymatic Techniques, Elsevier Science Publishers, Amsterdam.

Bacquet, C., and Twumasi, D.Y., 1984, A homogeneous enzyme immunoassay with Avidin-Ligand conjugate as the enzyme-modulator, Anal. Biochem., 136:487-490.

Bayer, E.A., Wilcheck, M., and Skutelsky, E., 1976, Affinity cytochemistry : the localization of lectin and antibody receptors on erythrocytes via the avidin-biotin complex, FEBS Lett., 68:240-244.

Bayer, E.A., and Wilcheck, M., 1980, The use of the Avidin-Biotin Complex as a tool in molecular biology, Meth. Biochem. Anal., 26:1-45.

Berman, J.W., and Bash, R.S., 1980, Amplification of the biotin-avidin immunofluorescence technique, J. Immunol. Methods, 36:335-338.

Butler, J.E., McGivern, P.L., and Swanson, P., 1978, Amplification of the enzyme-linked immunosorbent assay (ELISA) in the detection of class-specific antibodies, J. Immunol. Methods, 20:365-383.

Childs Moriarty, G., and Unabia, G., 1982, Application of the Avidin-Biotin-Peroxidase Complex (ABC) method to the light microscopic localization of pituitary hormones, J. Histochem. Cytochem., 30:713-716.

Costello, S.M., Felix, R.T., and Giese, R.W., 1979, Enhancement of immune cellular agglutination by use of an Avidin-Biotin system, Clin. Chem., 25:1572-1580.

Engvall, E., and Perlmann, P., 1971, Enzyme-linked immunosorbent assay (ELISA). Quantitative assay of immunoglobulin G, Immunochemistry, 8:871-874.

Engvall, E., and Pesce, A.J., 1978, Quantitative Enzyme Immunoassay, Blackwell Scientific Publications, Oxford.

Feldmann, G., Druet, P., Bignon, J., and Avrameas, S., 1976, Immunoenzymatic Techniques, North-Holland Publishing Company, Amsterdam.

Green, N.M., 1975, Avidin, Adv. Protein Chem., 29:85-133.

Guesdon, J.-L., and Avrameas, S., 1976, Polyacrylamide-agarose beads for the preparation of effective immunoadsorbents, J. Immunol. Methods, 11:129-133.

Guesdon, J.-L., Thierry, R., and Avrameas, S., 1978, Magnetic enzyme immunoassay for measuring human IgE, J. Allergy Clin. Immunol., 61:23-27.

Guesdon, J.-L., Ternynck, T., and Avrameas, S., 1979, The use of avidin-biotin interaction in immunoenzymatic techniques, J. Histochem. Cytochem., 27:1131-1139.

Guesdon, J.-L., and Avrameas, S., 1980, Lectin-immunotests : quantitation and titration of antigens and antibodies using lectin-antibody conjugates, J. Immunol. Methods, 39:1-13.

Guesdon, J.-L., and Avrameas, S., 1981, Solid phase Enzyme Immunoassays, in: "Applied Biochemistry and Bioengineering", vol. 3, L.B. Wingard, E. Katchalski-Katzir, L. Goldstein, eds, Academic Press, New York, 207-232.

Guesdon, J.-L., Naquira Velarde, F., and Avrameas, S., 1983a, Solid phase immunoassay using chimera antibodies prepared with monoclonal or polyclonal anti-enzyme and anti-erythrocyte antibodies, Annal. Immunol. (Inst.Pasteur), 134C:265-274.

Guesdon, J.-L., Jouanne, C., and Avrameas, S., 1983b, An amplification system using BSA-antibody conjugate for sensitive enzyme immuno-assay, J. Immunol. Methods, 58:133-142.

Habeeb, A.F.S.A., 1966, Determination of free amino groups in proteins by trinitrobenzenesulfonic acid, Anal. Biochem., 14:328-336.

Harris, C.C., Yolken, R.H., Krokan, H., and Ih Chang Hsu, 1979, Ultrasensitive enzymatic radioimmunoassay : application to detection of cholera toxin and rotavirus, Proc. Natl. Acad. Sci. USA, 76:5336-5339.

Heggeness, M.H., and Ash, J.F., 1977, Use of the avidin-biotin complex for the localization of actin and myosin with fluorescence microscopy, J. Cell Biol., 73:783-788.

Heitzmann, H., and Richards, F.M., 1974, Use of the avidin-biotin complex for specific staining of biological membranes in electron microscopy, Proc. Natl. Acad. Sci. USA, 71:3537-3541.

Hsu, S.M., Raine, L., and Fanger, H., 1981, Use of avidin-biotin-peroxidase complex (ABC) in immunoperidase techniques : a comparison between ABC and unlabeled antibody (PAP) procedures, J. Histochem. Cytochem., 29:577-580.

Ishikawa, E., and Kato, K., 1978, Ultrasensitive Enzyme Immunoassay, Scand. J. Immunol., Vol. 8, Suppl. 7:43-55.

Ishikawa, E., Kawai, T., and Miyai, K., 1981, Enzyme Immunoassay, Igaku-Shoin, Tokyo.

Ishikawa, E., Imagawa, M., Hashida, S., Yoshitake, S., Hamaguchi, Y., and Ueno, T., 1983a, Enzyme-labeling of antibodies and their fragments for enzyme immunoassay and immunohistochemical staining, J. Immunoassay, 4:209-327.

Ishikawa, E., Imagawa, M., and Hashida, S., 1983b, Ultrasensitive enzyme immunoassay using fluorogenic, luminogenic, radioactive and related substrates and factors to limit the sensitivity, in: Immunoenzymatic Techniques, S. Avrameas, P. Druet, R. Masseyeff and G. Feldmann, eds, Elsevier Science Publishers, Amsterdam, 219-232.

Kendall, C., Ionescu-Matin, I., and Dreesman, G.R., 1983, Utilization of the biotin/avidin system to amplify the sensitivity of the enzyme-linked immunosorbent assay (ELISA), J. Immunol. Methods, 56:329-339.

Labrousse, H., Guesdon, J.-L., Ragimbeau, J., and Avrameas, S., 1982, Miniaturization of β-galactosidase immunoassay using chromogenic and fluorogenic substrates, J. Immunol. Methods, 48:133-147.

Leipold, B., and Remy, W., 1984, Use of Avidin-Biotin-Peroxidase Complex for measurement of UV lesions in human DNA by MicroELISA, J. Immunol. Methods, 66:227-234.

Malvano, R., 1980, Immunoenzymatic Assay Techniques, Martinus Nijhoff Publishers, The Hague.

Mason, T.E., Phifer, R.E., Spicer, S.S., Swallow, R.A., and R.B. Dreskin, 1969, An immunoglobulin-enzyme bridge method for localizing tissue antigens, J. Histochem. Cytochem., 17:563-569.

Metzger, W.J., Butler, J.E., Swanson, P., Reinders, E., and Richerson, H. B., 1981, Amplification of the enzyme-linked immunosorbent assay for measuring allergen-specific IgE and IgG antibody, Clin. Allergy, 11:523-531.

Mohr, J., and Franz, H., 1981, Preparation of antibody-lectin conjugate using reversibly inactivated Concanavalin A, Ann. Immunol. (Inst.Pasteur), 132D:89-96.

Neurath, A.R., and Strick, N., 1981, Enzyme-linked fluorescence immunoassays using β-galactosidase and antibodies covalently bound to polystyrene plates. J. Virol. Methods, 3:155-165.

Ogata, K., Arakawa, M., Kasahara, T., Shioiri-Nakano, K., and Hiraoka, K., 1983, Detection of Toxoplasma membrane antigens transferred from SDS-polyacrylamide gel to nitrocellulose with monoclonal antibody and avidin-biotin, peroxidase anti-peroxidase and immuno-peroxidase methods, J. Immunol. Methods, 65:75-82.

Porstman, B., Avrameas, S., Ternynck, T., Portsman, T., Micheel, B., and Guesdon, J.-L., 1984, An antibody chimera technique applied to enzyme immunoassay for human alpha-1-fetoprotein with monoclonal and polyclonal antibodies, J. Immunol. Methods, 66:179-185.

Puget, K., Michelson, A.M., and Avrameas, S., 1977, Light emission techniques for the microestimation of femtogram levels of peroxidase. Application to peroxidase (and other enzymes) coupled antibody-cell antigen interactions. Anal. Biochem., 79:447-456.

Self, C.H., 1985, Enzyme amplification. A general method applied to provide on immunoassisted assay for placental alkaline phosphatase, J. Immunol. Methods, 76:389-393.

Sternberger, L.A., Hardy, Jr., P.H., Cuculis, J.J., and Meyer, H.J., 1970, The unlabelled antibody-enzyme method of immunohistochemistry. Preparation and properties of soluble antigen-antibody complex (horseradish peroxidase-anti-horseradish peroxidase) and its use in identification of spirochetes, J. Histochem. Cytochem., 18:315-333.

Subba-Rao, P.V., McCartney-Francis, N.L., and Metcalfe, D.D., 1983, An Avidin-Biotin MicroELISA for rapid measurement of total and allergen-specific human IgE, J. Immunol. Methods, 57:71-85.

Talwar, G.P., 1983, Non-Isotopic Immunoassays and their Applications, Vikas Publishing House, New Dehli.

Ternynck, T., Gregoire, J., and Avrameas, S., 1983, Enzyme anti-enzyme monoclonal antibody soluble immune complexes (EMAC) : their use in quantitative immunoenzymatic assays, J. Immunol. Methods, 58:109-118.

Van Weemen, B., and Schuurs, A., 1971, Immunoassay using antigen-enzyme conjugates, FEBS Lett., 15:232-236.

Yolken, R.H., Leister, F.J., Whitcomb, L.S., and Santosham, M., 1983, Enzyme immunoassays for the detection of bacterial antigens utilizing biotin-labeled antibody and peroxidase biotin-avidin complex, J. Immunol. Methods, 56:319-327.

Yorde, D.E., Sasse, E.A., Wang, T.Y., Hussa, R.O., and Garancis, J.C., 1976, Competitive enzyme-linked immunoas

say with use of soluble enzyme/antibody immune complexes
for labeling. Measurement of human chloriogonadotropin,
Clin. Chem. (Winston-Salem-N.C.), 22:1372-1377

Zaitsu, K., and Ohkura, Y., 1980, New fluorogenic substrates
for horseradish peroxidase : rapid and sensitive assays
for hydrogen peroxide and the peroxidase, Anal.
Biochem., 109:109-113.

ENZYME IMMUNOASSAY BASED ON PARTITION AFFINITY LIGAND ASSAY (PALA)
SYSTEM

Bo Mattiasson

Department of Biotechnology
Chemical Center, P.O.B. 124, S-221 00 Lund, Sweden

INTRODUCTION

A crucial step in most immunoassays is the separation of free from
bound reactant. The assays based on solid phase technology which were
developed to overcome this problem are characterized by elaborate
washing procedures. In order to avoid such steps and make immunoassays
easier to perform, certain homogeneous assays have been developed
(Ullman and Maggio 1980). The present paper deals with yet another
approach to immunoassays where the binding takes place in free solution
and separation is achieved by an extraction step using an aqueous
two-phase system (Mattiasson 1983). The assay procedure may be designed
in such a way that all steps are carried out in one and the same
vessel.

Aqueous two-phase systems.

When aqueous solutions of two different water-soluble polymers are
mixed with each other, the mixture will often be turbid. When left for
some time, the mixture separates into two phases. This is a well
documented phenomenon that is explained by incompatibility of the
polymers. The phase separation has been used in biochemistry for quite
some time (Albertsson 1971), but has, during recent years, become of
interest also to biotechnologists (Mattiasson 1983). The phase systems
are characterized by the presence of a high water content, each phase
having 85-95 % water. The surface tension between the two phases has
been found to be extremely low as compared to that found in oil-water
systems. Hence, a very gentle mixing results in a fine emulsion, which
in turn forms the basis for the rapid separation that is attainable in
aqueous two-phase systems.

Biochemists have shown that structures like enzymes and other
biological macromolecules maintain their biological properties when
suspended in such phase systems (Mattiasson 1980). Furthermore, in most
cases a certain partition pattern is observed. The partition of a
substance in a two-phase system is described in terms of the partition
constant, K_{part}, which is the ratio of the concentration in the
top phase to that in the bottom phase, i.e.
$K_{part} = C_{top}/C_{bottom}$.

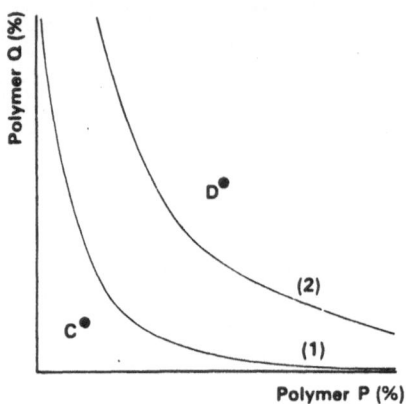

Fig. 1 Mixtures of two water-soluble polymers, P and Q, respectively,
        represented by points above the curves (binodials) such as point
        D, give two liquid phases, whereas mixtures below the binodials,
        such as point C, result in a homogeneous solution. The two
        binodials shown differ in molecular weight in such a way that
        mol.wt(2)<mol.wt(1). (Reproduced with permission).

    Partitioning of molecules also depends on the composition of the
two phase systems. Many polymers have been studied either in
combination with each other, or with a salt for constituting the two
phases. The most well known system is however, the
poly(ethyleneglycol)/dextran system (Albertsson 1971). Systems composed
of two polymers in solution may be represented by a phase diagram
similar to that shown in Fig. 1.

Binding assays in aqueous two-phase systems.

    If aqueous two-phase systems are to be used for carrying out
separations in the binding assay, the molecular entities taking part in
the binding assay must exhibit different partition constants. This
means that when binding has taken place, the formed complex is to
partition in a different way to one of the entities. This change in
partition is the basis for quantifying the assay. In a competitive
assay, where native and labelled antigen are competing for binding to
the antibodies, then the degree of transfer of labelled antigen to the
other phase gives the measure of the assay. The technique of performing
binding assays in aqueous two-phase systems has been called Partition
Affinity Ligand Assay (PALA), and the general principle assay is
illustrated in Fig 2.

Assay of digoxin

    A radioimmunoassay of digoxin in aqueous two-phase system is
described as a typical assay according to the general scheme given
above.
In many phase systems proteins are seen to partition to the bottom
phase. Assays of haptens favouring the top phase are therefore
facilitated. In the case of digoxin assay, for example, the partition
coefficient for digoxin is 4.0, whereas that for the antibody is 0.30
(Mattiasson 1980). Since the reactants are also different in size, a
rather simple assay situation is at hand, in which the complex formed
partitions like a native antibody. By measuring the decrease in label
in the top phase and/or the concomitant increase in the bottom phase at
constant concentrations of antibodies and labelled digoxin, a value for

the concentration of the native digoxin can be obtained from a calibration curve.

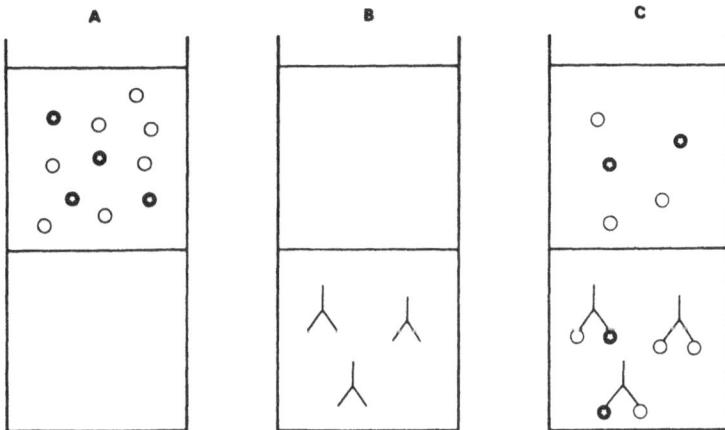

Fig. 2 Schematic presentation of a partition affinity ligand assay
where fixed amounts of labelled antigen and varying amounts of
the native antigen to be quantified are allowed to compete for
the binding positions on a limiting amount of specific
antibodies.
(A) Distribution of labelled (starred circles) and unlabelled
antigen in the partition system.
(B) Distribution of antibodies (inverted Y-shapes) in the
partition system.
(C) Distribution of the complex of antibody-antigen complexes in
the partition system. By measuring standard solutions of native
antigen in this competitive manner, a calibration curve is
obtained. The more native antigen, the less labelled antigen is
bound, and vice versa. Assays of unknown samples are performed
identically and the outcome of the binding assay is then used to
read the concentration of native antigen from the calibration
curve. (Reproduced with permission).

Thus, the experimental procedure involves mixing of labelled antigen
with the sample to be analyzed, addition of antiserum, incubation for
a predetermined period of time, addition of phase system, phase
separation for a certain time period and finally pipetting of aliquots
of either of the phases, or both, and measuring the label in a gamma
counter. A typical calibration curve is shown in Fig. 3. Several
haptens behaving like digoxin have been analyzed, some of which are
listed in Table I.

Digoxin assays on serum samples from a local hospital were
analyzed using the PALA-technique and compared to the results of
conventional solid-phase RIA techniques. The correlation between the
methods was 0.979.

When a fixed amount of native digoxin was added to the sample and
analyzed, recovery values between 95 and 115% were obtained, mostly
within a few percent from unity. The procedure can be performed in just
one test tube except for the counting. In order to investigate whether
it was possible to also eliminate the potential risk of introducing
errors by pipetting of the viscous phases, an analytical system was
investigated where all steps were carried out in a small test tube

Table 1 Partition-affinity ligand assay – immunoassays of different antigens

Type of antigen	Antigen	Separator	Other reactant	Label	Type	Time, min I	S	Range
Hydrophobic hapten	Digoxin	(Ab)	Ab/digoxin*	^{125}I	Comp	5	5	1–8nM
Hydrophobic hapten	T$_3$	(Ab)	Ab/T$_3$*	^{125}I	Comp	30	30	1–6nM
Hydrophobic hapten	T$_4$	(Ab)	Ab/T$_4$*	^{125}I	Comp	120	15	50–200nM
Hydrophobic hapten	Digoxin	Dextran Ab	Digoxin*	^{125}I	Comp	120	39	1–8nM
Hydrophilic hapten	Glucose	PEG–Con A	Glucose*	HRP	Comp	10–30	10	20–1000µM
Protein	β$_2$-Micro-globulin	PEG–Ab	β$_2$m*	^{125}I	Comp	20	15	3–96µg/l
Protein	β$_2$-Micro-globulin	PEG–Staph.	β$_2$m*/Ab	^{125}I	Comp	20	15	3–96µg/l
Cells	Staphylococci	(Staph)	IgG*	^{125}I	Dir	30	120	10^6–10^7
Cells	Staphylococci	PEG(Staph)	IgG*	^{125}I	Comp	30	120	10^5–10^7
Cells	Streptococci	(Ab*)	Ab*	HRP	Dir	30,60	60	2,5 x 10^3–10^5
Cells	Streptococci	PEG–Strep	Ab*	^{125}I	Comp	30,60	60	2,5 x 10^4–10^5

*Labelled reactant. Label: HRP = Horseradish Peroxidase. Type: Comp = competitive assay; Dir = direct binding assay. Time: I = reaction period (at room temperature); S = time allowed for phase separation.

fitting the gammacounter (a LKB Minigamma). Fig. 4 shows an example of changes in sensitivity of the assay by just counting at varying levels of the bottom phase. By introducing a large bottom phase it was possible to take the top phase out of the counting zone, and thus to count only the lower part of the bottom phase. Assays of triiodothyronine (T3) and thyroxine (T4) in serum samples by this approach gave a correlation of 0.97 and 0.93 respectively with the conventional methods (Eriksson, Nilsson and Mattiasson 1983; Mattiasson, Eriksson and Nilsson 1983).

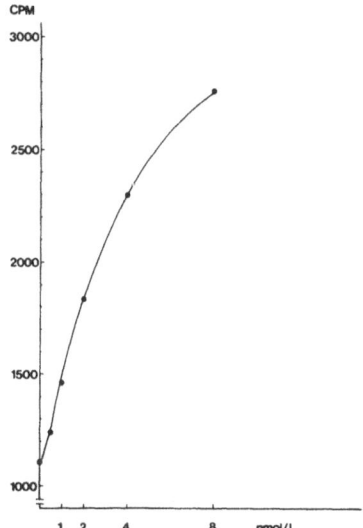

Fig. 3 The cpm in the top phase as a function of digoxin content in the sample. Phase system used: 1ml PEG (30% w/w) + 3ml $MgSO_4$ · $7H_2O$ (30% w/w). Prior to partitioning, the digoxin and the Sephadex-bound antidigoxin antibodies were in contact for 10 min. Reaction conditions are given in the text. (Reproduced with permission).

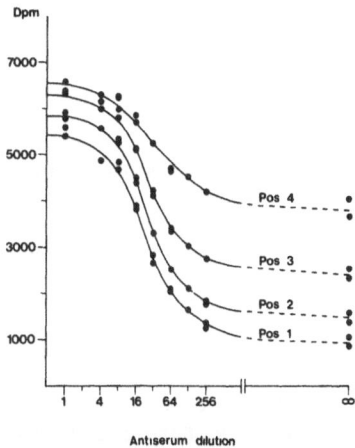

Fig. 4 The influence of the measuring position used in the counter in relation to the radioactivity measured was tested on samples containing 7100 dpm and 4 ml phase system. The position 1,2,3 and 4 refer to the position of the LKB minigamma counter used, where 1 stands for 0 mm up from the bottom; 2 for 10, 3 for 20 and 4 for 30 mm, respectively. (Reproduced with permission).

Improvement of partition by chemical modification

However, sometimes the spontaneous partition pattern of the two reactants was not good enough for setting up a sensitive assay. In such cases one of the reactants had to be modified. Since the haptens are very small and a chemical modification probably would disturb their ability to bind to the antibodies, the only alternative was to modify the antibodies. There have been innumerable reports on chemical modification of proteins, especially immobilization of proteins. In order to improve the tendency of the antibody to stay in the bottom phase, it was made more hydrophilic by coupling it to small particles of carbohydrate origin, either Sephadex or very small agarose beads. Coupling was performed according to the BrCN method (Axén, Porath and Ernback 1967). Empirically, it has been shown that partition constants for proteins in most cases lie in the region of 1-10, or of 1- 0.1 if the bottom phase is preferred.

Partition of particles in aqueous two-phase systems is more easily predicted, since in most cases, all the particles are recovered from either one of the phases, or from one phase and the interface; Hence, coupling the antibody to a particle, an extreme partition behaviour could be predicted. In the case of digoxin assay, however, the use of particle bound antibodies did not show much improvement in the performance of the system.

Binding assays involving Concanavalin A

The lectin-carbohydrate interactions provide a good model for studying binding reactions. Lectins are protein molecules with a specific binding of carbohydrates. They are normally much cheaper than antibodies, and thus constitute a suitable model system for binding studies. One of the best known lectins is Concanavalin A which binds α -D-mannopyranoside and α-D-glucopyranoside residues. It has, further, been shown to bind to several of the most commonly used glycoenzymes, which in this system may be regarded as enzyme labelled carbohydrates. The system we chose to study was the binding between Concanavalin A and horse radish peroxidase (E.C.1.11.1.7). Several pure carbohydrates were used as competing ligands in the binding assay.

When working with lectins and, also, glucopyranoside in an assay system, it is not recommendable to use dextran as the bottom phase component. Instead a salt-PEG phase system was set up. The partition constants for Con A and peroxidase were measured and found to be 0.031 and 0.063, respectively. Obviously, such a reactant pair could not be used in a binding assay before modifying the partitioning of one of them. It should also be borne in mind that in this system there is no significant difference in size between the participating entities. Thus, covalent modification was the only choice. We initially chose to modify Con A to make it more hydrophobic and partition more favourably to the top phase (Mattiasson and Ling 1980). In theory, several groups of substances may be used to achieve such effects (Mattiasson and Ling 1982). It is known from the literature on affinity partitioning that modification of ligands with poly(ethyleneglycol) is useful in changing the partition behaviour (Flanagan and Barondes 1975). In the cases reported, only small molecules had been modified. When working with proteins, several reactive groups are available for modification and this may very well be essential, since there may be a requirement for strong lifting power to transfer a protein across the phase boundary. By using monomethoxypoly(ethyleneglycol) (MPEG) all risks for crosslinking during the modification process were eliminated.

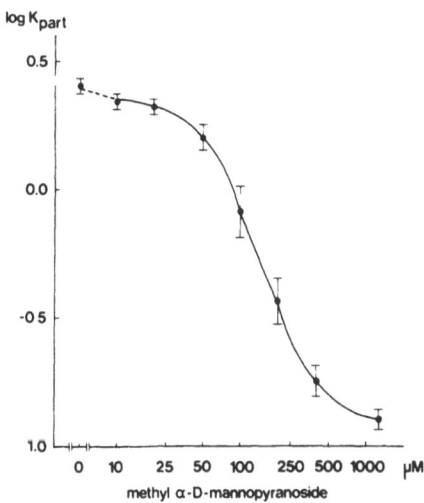

Fig. 5 Calibration curve for methyl-α-D-mannopyranoside obtained. Carbohydrate, enzyme (horseradish peroxidase) and modified
concanavalin A were incubated for 10 min in a total volume of
100 μl. The two phase constituents were added to make a final
volume of 1 ml and after mixing on a Vortex mixer the phases
were allowed to separate for 10 min. The 200 μl was taken from
each phase for determination of enzyme activity. Log K_{part}
of the enzyme activity is plotted versus the logarithm of the
carbohydrate concentration. (Reproduced with permission).

Modification with MPEG (M_r 5000) was seen to change the
partition constant of Concanavalin A from 0.031 to 80, i.e. a change in
partition constant of approx. 2,500 times. Furthermore, when peroxidase
was bound to the modified Con A, it partitioned to the top phase. Thus,
the conditions for setting up a binding assay involving Con A and
peroxidase were met. The chemistry behind PEG-modification of proteins
has attracted quite a lot of interest in recent years and several
methods are now available (Harris 1985).

In a competitive binding assay between a fixed amount of
peroxidase and varying amounts of the ligand, methyl-α-D-mannopyranoside, a calibration curve as depicted in Fig 5 was obtained (Mattiasson
and Ling 1980). The enzymatic activity was measured in both phases and
the partition constant was plotted as a function of ligand
concentration. In order to avoid the pipetting of the viscous aliquots
of the phases, an experiment to measure directly in the phase system
was tried. After binding had taken place a phase system containing
substrate for peroxidase was added so that enzyme activity in the top
phase could be read directly after separation had taken place. Fig. 6
illustrates such an experiment. It was, in this case, important to
consider the partitioning of the substrate as well and hence, the added
amounts had to be corrected so that the final concentration in the top
phase should be optimal.

The Con A -- peroxidase system was also used to quantitate
macromolecular carbohydrates. Furthermore, when operating with small
sugars at a fixed concentration it was possible to get a good relation
between the amount of binding and the binding constant.

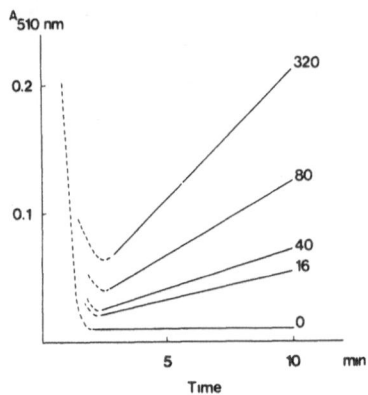

Fig. 6 Time course for a system analogous to the one used in Fig. 5, where all reactants, including phase components and substrates, were mixed. Immediately afterwards, the enzyme activity in the bottom phase was measured. The different concentrations of methyl-α-D-mannopyrannoside are indicated on the graphs (in μM). The initial period of decreasing absorbance is due to phase separation. (Reproduced with permission).

The PALA-approach was tried when setting up an immunoassay for β_2-microglobulin. It is well-known from literature that IgG and β_2-microglobulin have structural similarities (Petersson, Cunningham, Berggård and Edelman 1972). It was thus not unexpected when it turned out to be almost impossible to get the molecules to partition to different phases. When the conditions were changed so as to move one of the entities to the other phase, the other partitioned in a similar way. However, by modification of the antibody preparation it was possible to set up an assay. The operational range was not satisfying since, during the modification step, the antibodies were partially inactivated. A very low degree of modification gave antibodies with satisfying binding properties but insufficient partition behaviour (Ling and Mattiasson 1983).

A strategy employed to solve this problem was to attach the antibody with a "separator", i.e. a molecule or particle with appropriate partitioning behaviour, which, in turn, was modified to partition to the desired phase, Staphyloccus aureus cells and avidin were used as separators. Protein A on the cell surface of S. aureus was used to bind the antibodies through their Fc-fragments. An attractive feature of this system is that each bacterial cell carries approximately 10^6 molecules of protein A, and if a portion of these is destroyed in the modification process, it will not affect the partition behaviour. Moreover, it is quite easy to create systems with desirable partitioning because of the extreme partition behaviour of the bacterial cells.

An alternative to S. aureus as a separator is the avidin/biotin system. In this case, the IgG-preparation was conjugted with the biotin whereas the MPEG-modified avidin served as a separator for lifting biotinylated antibodies to the top phase. In table 2 is seen that the modification of both the bacterial cells and the avidin molecules dramatically changed their partition behaviour. It should be stressed in this context that both the separators could be used as general systems for achieving good separations without interfering much with the binding properties of the antibodies. The cell concept involves just the modification of the bacterial cells and no chemical modifica-

Table 2.

DISTRIBUTION IN TWO-PHASE SYSTEMS FOR THE SEPARATORS AVIDIN AND
Staphyloccoccus aureus.

A phase system consisting of 0.15 g/g PEG-4000 and 0.15 g/g MgSO$_4$ ·
7H$_2$O was used.

	Staphylococcus aureus		Avidin M-PEG-modified
	Native	M-PEG-modified	
Top phase	0 %	80%	90%
Interface	10%	20%	0%
Bottom phase	90%	0%	10%

tion of the antibodies, whereas the avidin-biotin systems requires the
covalent binding of one or a few biotin molecules to each antibody.
Loading of the S. aureus cells with high amounts of IgG molecules was,
however, avoided, which otherwise might have altered the preferential
partitioning to the top phase. The risk for such disturbances is, of
course, much smaller when operating the avidin-biotin system
(Mattiasson, Ling and Ramstorp 1981).

Using the approach of such secondary separators it was possible to
set up a useful analysis. The alternative system involved the
acceptance of a poor difference in partition behaviour and instead use
a mathematical treatment of the data. A computer program was set up to
evaluate the binding reactions taking care of the partition between the
phases, and it turned out that using an iterative process it was quite
easy to obtain good binding data. The method was also used for
evaluating binding constants and stoichiometry in binding reactions
(Ling And Mattiasson 1982).

An advantage of aqueous two-phase systems over many other
separation systems used in biochemical work is that they can be used
for the partitioning of small molecules as well as macromolecules, and
even particles. This property has been exploited for quantification of
bacterial and yeast cells. Direct as well as competitive binding assays
have been set up. In the direct binding assay, a system was selected
where the labelled antibodies were found in a different phase than the
cells. When binding took place the antibodies were transferred to the
phase containing the cells. Alternatively, the competitive assay was
set up when antibodies and cells partitioned to the same phase. By
modifying a preparation of cells with MPEG it was possible to have
modified cells in one phase and the native cells in the other. The
partition of the antibodies in such a system depended of course very
much on the binding of the cells. More the native cells present, lesser
was the amount of antibodies bound to the modified cells (Mattiasson,
Ling and Ramstorp 1981; Ling, Ramstorp and Mattiasson 1982; Mattiasson,
Ling, Nilsson and Dürholt 1982).

In conclusion; Partition Affinity Ligand Assay (PALA) may be used
for quantifying haptens, macromolecules and cells. Table 2 gives some
of the analyses that have been set up with this technique. PALA
involves very few experimental steps, and the whole procedure may be
carried out in only one vessel. These features make PALA a good
candidate when automating immunological binding assays.

Acknowledgements: Part of this work was supported by The National Swedish Board for Technical Development.

References

Albertsson, P.-Å. (1971) Partition of Cell Particles and Macromolecules. 2nd Ed. Almqvist and Wiksell, Stockholm.

Axén, R., Porath, and Ernback, S. (1967) Chemical Coupling of Peptides and Proteins to Polysaccharides by Means of Cyanogen Halides. Nature 214:1302.

Eriksson, H., Nilsson, J. and Mattiasson, B. (1983) Radioimmunoassay of triiodothyronine (T_3) and thyroxine (T_4). As assay with both the bound and free fraction of the hormone present in the counting vessel. Appl. Biochem. Biotechnol. 8,1.

Flanagan, S.D. and Barondes, S.H. (1975) Affinity Partitioning J. Biol. Chem., 250:1484.

Harris, J.M. (1985) Laboratory Synthesis of Polyethylene Glycol Derivatives. J. Macromol. Sci. C25, 325.

Ling, T.G.I. and Mattiasson, B. (1982) Comparison between binding analyses performed by equilibrium dialysis and partitioning in aqueous two-phase systems exemplified by the binding of cibacron blue to serum albumin. J. Chromatography 252, 159.

Ling, T.G.I., Ramstorp, M. and Mattiasson, B. (1982) Immunological quantitation of bacterial cells using a partition affinity ligand assay: A model study on the quantitation of Streptococci B. Anal. Biochem. 122, 26.

Ling, T.G.I. and Mattiasson, B. (1983) A general study of the binding and separation in partition affinity ligand assay. Immunoassay of $\beta 2$-microglobulin. J. Immunol.Methods 59, 327.

Mattiasson, B. (1980) Partition affinity ligand assay (PALA): Radioimmunoassay of digoxin. J. Immunol. Methods 35, 137

Mattiasson, B. and Ling, T.G.I. (1980) Partition affinity ligand assay (PALA). A new approach to binding assays. J. Immunol. Methods 38, 217.

Mattiasson, B., Ling, T.G.I. and Ramstorp, M. (1981) Application of partition affinity ligand assay (PALA) in a quick test for quantitation of Staphylococcus aureus bacterial cells. J. Immunol. Methods 41, 105

Mattiasson, B and Ling, T.G.I. (1982) US. Pat. 4.312.944.

Mattiasson, B, Ling, T.G.I., Nilsson, J. and Dürholt, M. (1982) Lectin – carbohydrate interactions as a tool to quantify microbial cells, in: "Lectins – Biology, Biochemistry, Clinical Biochemistry, Vol II, (ed. T.C. Bøg-Hansen) Walter de Gruyter, Berlin, pp. 573-581.

Mattiasson, B. (1983a) Partition Affinity Ligand Assay (PALA) for quantifying haptens, macromolecules and whole cells. Methods Enzymol. 92, 498.

Mattiasson, B. (1983b) Applications of aqueous two-phase systems in biotechnology. Trends in Biotechnol. 1, 16.

Mattiasson, B., Eriksson, H. and Nilsson, J. (1983) A modified partition affinity ligand assay (PALA) for direct reading of γ -labels without further separation or washing. Clin. Chim. Acta 127, 301.

Petersson, P.A., Cunningham, B.A., Berggård, I., and Edelman, G.M. β_2-Microglobulin - A Free Immunoglobulin Domain (light and heavy chain/amino-acid sequence) (1972). Proc. Natl. Acad. Sci. 69, 1697.

Ullman, E.F. and Maggio, E.T. (1980) Principle of homogeneous enzyme-immunoassay. In "Enzyme Immunoassay" ed. E.T. Maggio, CRC-Press, Boca Raton, Fl, pp 105-134.

IMMUNOENZYMATIC ASSAYS FOR PEPTIDE HORMONES

Y. Amir-Zaltsman, B. Gayer, E.A. Bayer*, M. Wilchek* and F. Kohen

Departments of Hormone Research, and of *Biophysics, The Weizmann Institute of Science, Rehovot, 76100, Israel

INTRODUCTION

"Two-site" sandwich-type assays, exemplified by immunoradiometric assays (IRMA), have been developed originally with a solid-phase capture antibody and a radiolabeled antibody, each of which recognizes a different antigenic determinant of the analyte (Miles and Hayes, 1968). Since this technique is based on an excess reagent system, it offers several potential advantages over conventional competitive radioimmunoassay methods in terms of speed, sensitivity, precision, and range of the standard curve.

However, isotopically labeled antibodies may pose problems associated with radioactive waste disposal and radiolysis of the labeled marker. Recent studies from several laboratories and ours (Amir-Zaltsman et al., 1986) indicate that substitution of ^{125}I by nonisotopic labels such as enzymes (Wada et al., 1982), chemiluminescent (Weeks et al., 1984; Barnard et al., 1984) or fluorescent (Pettersson et al., 1983) markers or sol particles (Leuvering et al., 1983) may be a feasible alternative to radiolabeling of proteins.

In the immunoenzymatic methods that we developed, we investigated two different approaches, direct and indirect. In the direct approach, monoclonal or polyclonal antibodies were covalently linked to the enzyme penicillinase derived from Bacillus cereus using glutaraldehyde as the cross-linking agent. The penicillinase-labeled conjugates convert the substrate penicillin to the corresponding penicillinoic acid, which in turn forms a colorless complex with iodine (present in the reaction mixture in the form of the dark blue starch-iodine complex) (see Fig. 1). The time required for forming a colorless solution from the initial dark blue solution is inversely proportional to the amount of enzyme (conjugate) present in the sample. The end-point in the assay can thus be detected independent of instrumentation.

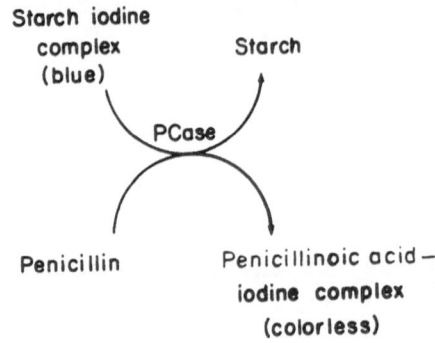

Fig. 1. A schematic presentation of the determination of penicillinase activity.

In the indirect approach, advantage was taken of the four binding sites of the bacterial protein streptavidin for the vitamin biotin which results in amplifying the sensitivity of the assay (Bayer and Wilchek 1980; Wilchek and Bayer, 1984). In this system, a monoclonal or polyclonal antibody preparation, which recognizes a different epitope of the antigen, is biotinylated and serves as the primary probe. After the immunological reaction with immobilized antibody and sample, secondary probes consisting of streptavidin and biotinylated enzyme are added sequentially. The biotinylated enzymes which we have examined include horseradish peroxidase, NAD^+-dependent enzymes (e.g. glucose-6-phosphate dehydrogenase) (Kohen et al., 1984), ATP-dependent enzymes (e.g. myokinase and pyruvate kinase), penicillinase and alkaline phosphatase. Depending on the particular enzyme used, the end point is determined either by spectrometry, luminometry or by measurement of the time required for disappearance of color. We report here the results obtained using the above-described approaches.

MATERIALS AND METHODS

Chemicals

Penicillin G, bovine serum albumin (BSA, fraction V, RIA grade), Tween-20, alkaline phosphatase (from bovine intestine), p-nitrophenyl phosphate, penicillinase (from Bacillus cereus) were obtained from Sigma; 8% glutaraldehyde from Ted Bell Inc. (Tustin, A); L-lysine:L-phenylalanine copolymer (1:1) from Bio-Yeda (Rehovot, Israel), Ficoll-70 from Pharmacia (Uppsala, Sweden), and streptavidin from Cell-Tech (London, U.K.).

Reagents for immunoassay

Etched polystyrene balls (0.5 cm in diameter) were obtained from Northumbria Biologicals (U.K.), immunobead reagent, 5-10 u, from Bio-Rad Laboratories (Richmond, A). Polyclonal and monoclonal antibodies to hCG were generated in the Hormone Research Department. Monoclonal antibodies to hGH (clone #518 and #69) were a gift from Dr. N. Moav (Inter-Yeda, Rehovot).

Reagent solutions

Assay buffer: PBS, pH 7.2 containing 0.1% BSA (unless otherwise noted).

Wash solution: Tween-saline containing, per liter, 9 g of NaCl, and 0.5 ml of Tween-20.

Enzyme substrates: The substrate for penicillinase was freshly prepared, and had the following composition: 3.0 ml of 0.5% gelatin in water; 0.5 ml of 25 mM iodine in 125 mM potassium iodide; 0.1 ml of soluble 1% (w/v) starch solution; and 1 ml of benzyl penicillin (3 mg) in 0.1 M phosphate buffer, pH 7.0.

The substrate for alkaline phosphatase, p-nitrophenyl phosphate (Sigma) in a tablet form of 5 mg, was dissolved in 5 ml of 10% diethanolamine buffer, pH 9.

Procedures

Preparation of immunoadsorbents

Conjugation of antibodies to immunobeads. Purified IgG fractions of specific polyclonal antibodies (1 mg protein/ ml, 2 ml) were dialyzed against 3 mM potassium phosphate, pH 6.3 (coupling buffer) overnight. This solution was added to the immunobead matrix (40 mg in 2 ml of coupling buffer). To this suspension 1-ethyl-3-[3-dimethylamino-propyl]- carbodiimide (15 mg) was added, and the suspension was left at 4°C overnight. The immunobeads were then washed according to the instructions provided by the supplier (Bio-Rad). The antibodies prepared in this way were stored at 4°C in a 5 mM potassium phosphate solution (pH 7.1), containing 1% BSA, 1% sodium azide, at a concentration of 10 mg of immunoadsorbant per ml.

Conjugation of antibodies to polystyrene balls. Polystyrene balls (1000) were activated with 200 ml of an aqueous solution of 0.0025% Phe-Lys for 3 days at room temperature, according to the procedure of Wood and Gadow (1983). The balls were then washed once with 0.15 M NaCl and twice with double distilled water and then immersed in 0.5% solution of glutaraldehyde in 50 mM phosphate buffer, pH 8.0, for 30 min at room temperature. The balls were washed again twice with water and once with 0.15 M NaCl. The antibody solution was then added (2.5 ug protein/ball) in 50 mM phosphate buffer, pH 8:0. The coupling reaction was allowed to proceed for 30 min at room temperature. The resulting Schiff's bases were reduced with 75 mM $NaBH_4$ for 20 min at room temperature. Non specific binding sites on the ball surfaces were saturated with 2.5% BSA in 10 mM Tris-HCl, pH 7.2, for 96 h at 4°C. The antibody-coupled balls were stored in 50 mM Tris HCl, pH 7.2, containing 1% BSA and 0.01% azide.

Preparation of labeled probes

Preparation of enzyme-labeled antibodies. Purified IgG fractions of monoclonal antibodies (1 mg) were dissolved in 0.8 ml PBS, pH 7.4. The enzyme penicillinase (1000 U) was added, and the solution was dialyzed three times against 500 ml PBS, pH 7.4, containing 2 mM $MgCl_2$. After dialysis, the solution was brought to 0.04% with glutaraldehyde and stirred for 2 h at room temperature. The reaction was stopped by adding 1 M glycine (0.1 vol/vol reaction), and then dialyzed against PBS containing 2 mM $MgCl_2$. This reaction mixture was then applied on a Sephacryl S-300 column. The high-molecular-weight conjugates were collected and checked for enzymatic and binding activity. Fractions showing good enzymatic and binding activity were considered to be labeled conjugates and were stored at 4°C in the presence of 0.5% Ficoll and 0.02% azide. A typical chromatogram is shown in Fig. 2.

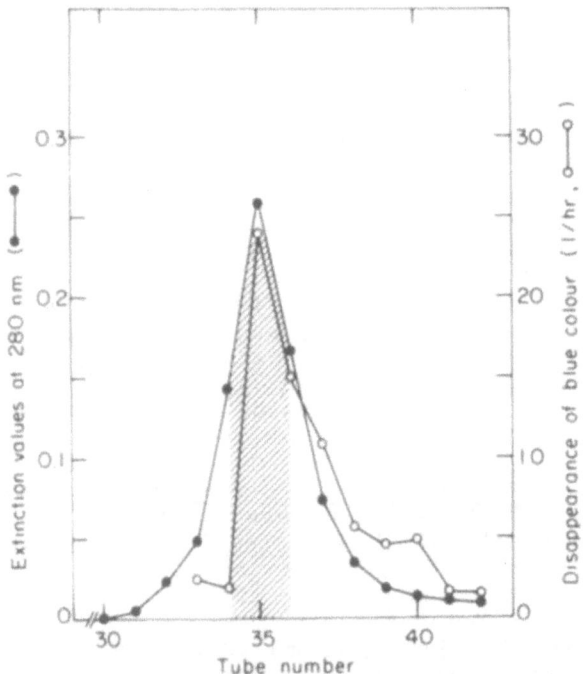

Fig. 2. Chromatogram of the reaction mixture of monoclonal anti-β-hCG-penicillinase conjugate. The shaded area represents the fractions possessing both binding and enzymatic activity.

120

Preparation of biotinylated proteins. Biotin N-hydroxy-succinimide ester (BNHS, 250 ug in 10 ul of dimethylformamide solution) was added to a solution containing purified monoclonal antibody or enzyme (e.g. alkaline phosphatase) at a concentration of 1 mg protein/ml buffer (20 mmol/l phosphate, pH 7.4). The solution was stirred for six hours at room temperature, and then dialyzed exhaustively against PBS. The biotinylated antibody or enzyme (ca. 1 mg/ml) was stored in aliquots at -20°C or 4°C respectively.

Immunoassay methods

Two types of format were used, (a) one independent of instrumentation, and (b) and another based on avidin-biotin technology.

(a)   Immunoenzymatic method independent of instrumentation. hCG was measured in plasma or in urine, by an immunoenzymatic method which utilizes disappearance of color as an end point. The stock solution of polyclonal rabbit anti-hCG, bound to immunobeads, was diluted 50-fold with a suspension of 1 mg unconjugated immunobead matrix per ml of assay buffer, and 0.1 ml of this suspension was added per tube. Standards of hCG (prepared in stripped plasma), and plasma or urine samples were diluted 1:1 with assay buffer and 0.1 ml of this solution was added to each tube. The tubes were incubated for 1 hour at 30°C. One ml of wash solution was then added, and the tubes were centrifuged (10 min, 2000xg, 25°C). The supernatant fluid was decanted and 0.1 ml of

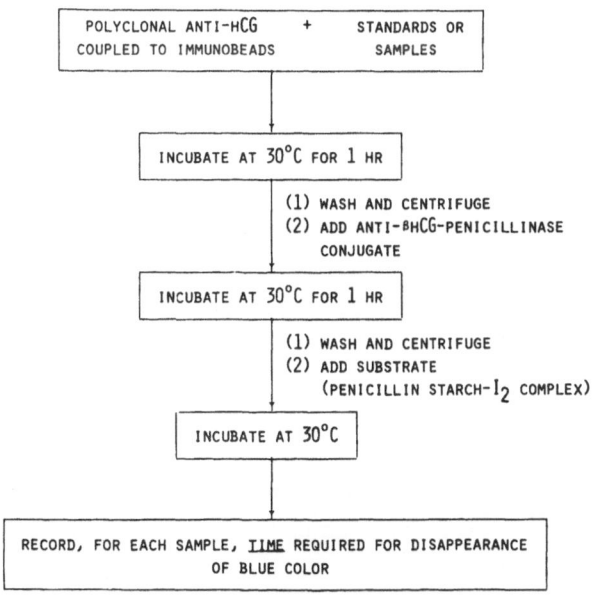

Fig. 3. A procedure for an immunoenzymatic assay for hCG based on determination of penicillinase activity.

anti-β-hCG-penicillinase conjugate (1:50 dilution of 0.074 O.D/ml) was added to the pellet. The tubes were incubated for another hour at 30°C, followed by a washing and centrifugation step.

A freshly prepared penicillinase substrate solution (0.2 ml) was added to each tube. The tubes were incubated at 30°C for the enzymatic hydrolysis. The time required for the enzyme-mediated disappearance of blue color was noted. Fig. 3 shows a flow diagram of the method.

(b)   Immunoenzymatic method based on avidin-biotin interaction. hGH was measured in plasma by an immunoenzymatic method using goat anti-mouse IgG coupled to polystyrene balls as a solid-phase matrix (Amir-Zaltsman et al., 1986). Monoclonal antibodies to hGH [clone #518, 0.1 ml at a dilution of 1:10,000 in PBS pH 7.2 containing 0.25% BSA (Buffer A)], and plasma samples (0.1 ml) or hGH standards (0.1 ml) were added to test tubes containing polystyrene balls coupled to goat anti-mouse IgG. The balls were incubated for two hours at 37°C and washed three times with wash solution. Biotinylated anti-hGH (clone #69, 0.3 ml at a dilution of 1:4000 in Buffer A) was added, and the balls were incubated for 90 min at 37°C. Following aspiration of the reaction mixture and washing, the balls were incubated for 30 min at 25°C with streptavidin (0.5 ug/ball), dissolved in a PBS solution (0.3 ml), containing 10% horse serum and 1% mouse serum (Buffer B). The balls were then washed and incubated (30 min at 25°C), with biotinyl-alkaline phosphatase (0.1 mU/ball) in 0.3 ml of Buffer B. After washing, the balls were incubated with p-nitrophenyl phosphate for 1 h at 37°C. The appearance of the yellow color was measured at 405 nm. A flow diagram of the method is shown in Fig. 4.

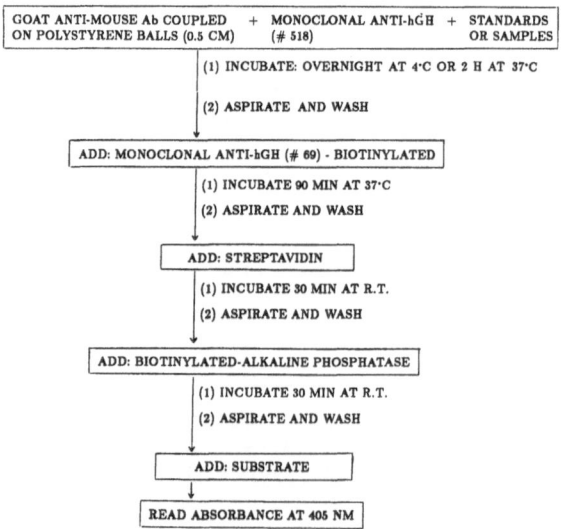

Fig. 4. A procedure for "two-site" immunoenzymatic assay for hGH.

RESULTS

An immunoenzymatic assay for hCG independent of instrumentation

hCG in urine or in plasma was measured using the method (a) described under Immunoassay Methods in the Materials and Methods Section (see Fig. 3). In this technique, the starch-iodine substrate was added to the immobilized fraction containing the penicillinase-labeled monoclonal anti-β-hCG. The time required for decoloration of the starch-iodine substrate, was directly proportional to the amount of hCG present in the sample over the range of 5-200 mIU/ml, and was recorded with the naked eye (Fig. 5).

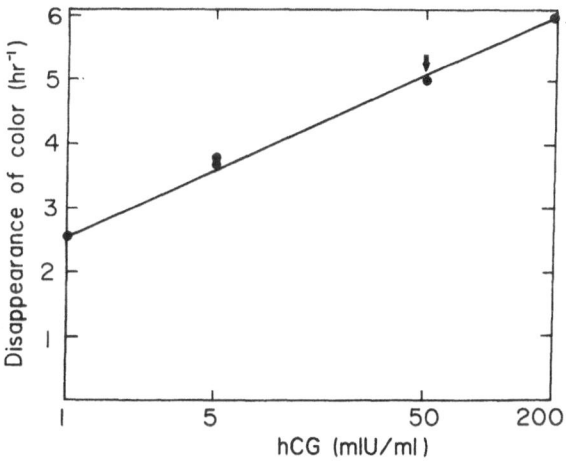

Fig. 5. A dose-response curve for hCG using the procedure described in Fig. 3. The arrow represents the threshold level for pregnancy (50 mIU/ml).

Evaluation of the method

The sensitivity of this technique was found to be 5 mIU hCG/ml (defined as zero dose minus 2 S.D.). This sensitivity was satisfactory for detecting pregnancy in plasma or in urine since the threshold value for pregnancy is 50 mIU hCG/ml.

Measurement of hCG in urine or in plasma samples from pregnant and non-pregnant women was determined by immunoenzymatic assay and by conventional radioimmunoassay. A good correlation was obtained (n=55; r=0.92). Table 1 shows typical results obtained using this method.

Table 1. Plasma hCG determinations using an immunoenzymatic method based on determination of penicillinase activity.

Samples	Disappearance of Color (H^{-1})	Diagnosis
5230	6.6	Pregnant
5597	7.5	"
6003	6.6	"
6858	8.6	"
5735	6.7	"
5303	3.3	Non-pregnant
5274	4	"
5765	3.3	"
5706	4	"
5203	3	"
5255	4	"

The intra-assay (C.V.) precision was 9.2% (dose 50 mIU hCG/ml).

The inter-assay CV (five assays) of the same sample was 12.5%.

Remarks

The use of penicillinase as a label in the immunoenzymatic assay offers several advantages. Penicillinase is a relatively inexpensive and stable enzyme. This enzyme exhibits a high turnover rate (2000 units/mg protein), and it possesses several lysine residues that can be utilized for conjugation to proteins. The resulting conjugates are stable and retain a substantial portion of the enzymatic and antigen-binding activities compared to those of the original reagents. In addition, the substrates do not have potential mutagenic or carcinogenic activity. The end-point determination (see Fig. 3) can be recorded with the naked eye, avoiding the use of colorimetric instrumentation. Thus, this assay can be particularly useful in small clinical laboratories and in developing countries. Although several papers have been published on the use of penicillinase in the detection of microbial antigens (Yolken and Wee, 1984) and in competitive type immunoassays (Joshi et al., 1983), to our knowledge this is the first report utilizing an inexpensive instrument - a watch - for end point determination.

An immunoenzymatic assay for hGH based on avidin-biotin interaction.

Using the procedure described in Fig. 4, an immunoenzymatic assay for hGH was developed. A dose-response curve was obtained by plotting the absorbance readings at 405 nm against log-mass of added hGH over the range 0.3-100 ng hGH/ml (see Fig. 6). The amount (ng) of hGH present in each assay tube was read directly from this curve.

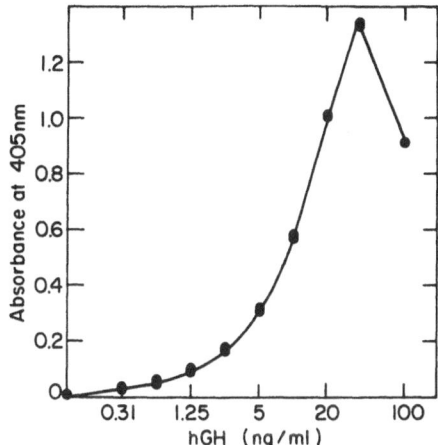

Fig. 6. A dose-response curve for hGH as measured by the two-site immunoenzymatic method.

Evaluation of the method

The specificity of the enzyme immunoassay for hGH, as performed in Fig. 3, was tested by examining alkaline phosphatase activity bound in the presence of various other human pituitary hormones. The observed color-production was not significantly different from that of the background in the presence of 500 ng of hPRL, 500 mIU of hLH and 500 mIU of hFSH.

The sensitivity of the assay was comparable to that obtained by the radioimmunoassay procedure (see Fig. 2). The least amount of hGH that could be distinguished from zero ($p < 0.08$) was 0.3 ng hGH/ml.

The precision of the assay was determined by performing (a) replicate samples of the same plasma pool simultaneously in the same assay; and (b) by evaluating replicate samples of this pool on five different occasions. The results for intra-assay precision were: mean 3.8 ± 0.247 ng hGH/ml and CV 6.5%. The inter-assay CV for this pool was 7.4%.

Comparison with radioimmunoassay. The dynamics of hGH concentrations in basal plasma levels or under stimulation tests in normal patients and in patients with acromegaly and dwarfism were determined by radioimmunoassay and two-site enzyme immunoassay. The regression and correlation coefficients were y(EIA) = 1.149 (RIA) + 0.02; r = 0.974 (n = 60). (Fig. 7).

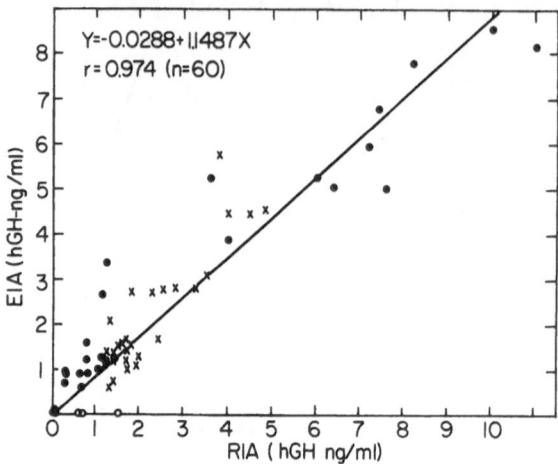

Fig. 7. Correlation between RIA and EIA for the measurement
of hGH.

Remarks

The work described here indicates that a "two-site" immunoenzymatic assay for hGH can be used for the determination
of the dynamics of hGH concentrations in basal and in stimulated conditions (Amir-Zaltsman et al., 1986). The method
is simple, and does not require centrifugation. In addition, some of the reagents are universally applicable. For
instance, the second antibody coupled to polystyrene balls
can be utilized in a number of "sandwich-type" assays. The
use of avidin-biotin interaction as probes in immunoassay
systems has several advantages (Wilchek and Bayer, 1984):
(i) biotinylation of antibodies or enzymes can be achieved
under mild conditions and the immunological or enzymatic activity of the conjugates are only nominally affected; (ii)
biotinylated enzymes can be used as universal reagents;
(iii) in many cases, mediation via the avidin-biotin interaction enables an amplified signal and increased sensitivity; and (iv) streptavidin and biotin, as well as biotinylated enzymes, are available commercially.

In this procedure we describe the stepwise addition of
the various reagents (see Fig. 4). Premade complexes, which
comprise the appropriate probes (i.e. streptavidin:biotinyl
antibody and/or biotinyl enzyme premixed in appropriate ratios) can also be used in order to minimize the number of
steps necessary for the assay.

ACKNOWLEDGEMENTS

We are grateful to Mrs. M. Kopelowitz for excellent secretarial assistance.

REFERENCES

Amir-Zaltsman, Y., Gayer, B., Bayer, E.A., Wilchek, M., Za-
dik, Z., Kostyo, J., and Kohen, F., 1986, A two-site
immunoenzymatic assay for hGH based on the avidin-biotin interaction. J. Clin. Endocrin. Metab., submitted.

Barnard, G.J., Kim, J.B., Borckelbank, J.L., Collins, W.P.,
Gaier, B. and Kohen, F., 1984, Measurement of choriogonadotropin by chemiluminescence immunoassay and immuno-
chemiluminometric assay: 1. Use of isoluminol derivatives. Clin. Chem., 30/4:538.

Bayer, E.A., and Wilchek, M., 1980, The use of the avidin-
biotin complex as a tool in molecular biology. Methods
Biochem. Anal., 26:1.

Joshi, U.M., Shah, H.P., and Sankolli, G.M., 1983, Penicillinase as a marker in enzyme-linked immunoadsorbent assays for steroid hormones. J. Steroid Biochem.,
19:419.

Kohen, F., Bayer, E.A., Wilchek, M., Barnard, G., Kim, J.B.,
Collins, W.P., Beheshti, I., Richardson, A., and McCapra, F., 1984, Development of luminescence-based immunoassays for haptens and for peptide hormones. in:
"Analytical Applications of Bioluminescence and Chemiluminescence", L. Kricka, and T.P. Whitehead, eds.,
Academic Press, Inc., pp. 149-158.

Leuvering, J.H.W., Goverde, B.C., Thal, P.J.H.N., and Scu-
urs, A.H.W.M., 1983, A homogeneous sol particle immunoassay for human chorionic gonadotrophin using monoclonal antibodies. J. Immunol. Meth., 60:9.

Miles, L.E.M., and Hales, C.N., 1968, An immunoradiometric
assay of insulin. In: Protein and Polypeptide Hormones, Part I (M. Margoulies, ed.), Excerpta Medica,
Amsterdam, pp. 61-70.

Petterson, K., Siitari, H., Hemmila, I., Soini, E., Lovgren,
T., Hanninen, V., Tanner, P., and Stenman, U.-H., 1983,
Time-resolved fluoroimmunoassay of human choriogonado-
tropin. Clin. Chem., 29:60.

Wada, H.G., Danisch, R.J., Baxter, S.R., Federici, M.M.,
Fraser, R.C., Brownmiller, L.J., and Lankford, J.C.,
1982, Enzyme immunoassay ofthe glycoprotein tropic hormones - CGH, lutropin, thyrotropin - with solid-phase
monoclonal antibody for the α-subunit and enzyme
coupled monoclonal antibody specific for the β-subunit.
Clin. Chem., 28:1862.

Weeks, I., Sturgess, M., Siddle, K., Jones, M.K., and Woodhead, J.S., 1984, A high sensitivity immunochemilumino-
metric assay for human thyrotrophin. Clin. Endocrinol., 20:489.

Wilchek, M., and Bayer, E.A., 1984, The avidin-biotin complex in immunology. Immunology Today, 5:39.

Wood, W.G., and Gadow, A., 1983, Immobilization of antibodies and antigens on macro solid phases. A comparison
between adsorptive and covalent binding. A critical
study of macro solid phases for use in immunoassay systems, Part 1. J. Clin. Chem. Clin. Biochem., 21:789.

Yolken, R.H. and Wee, S-B., 1984, Enzyme immunoassays in
which biotinillated beta-lactamase is used for the detection of microbial antigens. J. Clin. Microbiol.,
19:356.

APPLICATION OF ENZYME IMMUNOASSAY IN VETERINARY MEDICINE: SERODIAGNOSIS OF BOVINE BRUCELLOSIS

P.F. Wright and K.H. Nielsen

Agriculture Canada, Animal Diseases Research Institute
NEPEAN, P.O. Box 11300, Station H, Nepean, Ontario
Canada K2H 8P9

INTRODUCTION

Enzyme immunoassay (EIA) techniques have been applied in two principle areas of veterinary medicine; 1) antigen detection for direct demonstration of disease causing agents and 2) antibody detection for indirect or presumptive diagnosis of infection. EIA techniques have been chosen over other primary binding immunoassays (radio- and fluoro-) to replace many of the 'conventional' serological assays. EIA techniques offer many advantages; 1) the inherent dangers, waste disposal problems and limited shelf life associated with radioisotopes are not encountered; 2) whereas most fluoroimmunoassay techniques still require subjective or operator judgement, enzymic reactions (chromogenic) may be objectively quantitated with 96 well plate photometers; 3) although 96 well plate fluorimeters are now available, in general photometers are less expensive and far more flexible with respect to the range of chromogens which can be quantitated; 4) the 96 well plate format of EIA techniques has been the major focus for development of computer controlled, automated in-strumentation and as a result of these technological advances, accuracy, precision and quality control has been greatly enhanced; 5) electronic data acquisition and processing of EIA results is now well established and supported by commercial or custom microcomputer software, is an important aspect of overall quality assurance in the diagnostic labor-atory; 6) EIA techniques in general are faster and less expensive to perform and are technically less demanding than conventional assays such as serum neutralization and complement fixation; 7) EIA techniques, as primary binding assays, are not dependent on antibody specific functions which occur secondarily after binding to antigen (e.g. complement fixation, virus neutralization, bacterial agglutination, precipitation, etc.); 8) the immunological specificity and sensitivity of EIA techniques is adjustable through selection of appropriate antigens and detection reagents; 9) the use of monoclonal detection reagents in EIA techniques has greatly enhanced assay specificity and standardization and 10) the same EIA protocol and test sample may be used for simultaneous detection of antibody to a number of different diagnostic antigens from one or more infectious agents. For the reasons stated above and constant pressures to tailor diagnostic specificity· and sensitivity to meet eradication, control and surveillance requirements for infectious diseases, EIA tech-niques have found wide application in veterinary medicine as applied to both food producing and companion animals. However, it is not our intention to review all of the applications of EIA techniques in veterinary medicine in this chapter, as an extensive bibliography has been recently published (Charan and Gautam, 1984).

Perhaps the most widely used application of EIA techniques is the detection of antibody to infectious agents. The indirect EIA, based on antigen immobilization on a solid matrix, the subsequent binding of test sample antibody and finally detection by the addition of an anti-species globulin-enzyme conjugate, has been applied to a considerable number of different antigens. The competition type of assay and the class capture principle have been used to some extent in other areas of medicine but only few applications have been described in veterinary diagnostics. As well, various EIA techniques have been applied to the detection of antigens and infectious agents. An overview of both antibody and antigen detection techniques may be found in Maggio (1980). For reasons of popularity and because our laboratory has had considerable experience in the application of the indirect EIA, this assay will be described in detail. For the purposes of specific discussions, we have chosen the EIA for detection of bovine antibody to Brucella abortus as an example of a test which has been developed and is currently being implemented for diagnostic use on a relatively large scale in Canada.

In this chapter, the performance of the indirect EIA is compared to the currently used conventional assays with respect to the reactivity of the 4 major bovine antibody isotypes. The serodiagnostic use of these assays is explained with reference to screening and confirmatory test application. An indirect EIA protocol developed for use in centralized diagnostic laboratories is presented with special emphasis given to assay standardization and quality control. As diagnostic performance will ultimately determine the usefulness of any serological test, a strong emphasis has been placed on diagnostic evaluation.

SEROLOGICAL DIAGNOSIS OF BOVINE BRUCELLOSIS

Brucella abortus infection is enzootic in many countries. It is a disease of considerable economic importance in cattle and has important public health implications. Canada and some other countries have eliminated brucellosis from their cattle populations. However, continuous surveillance is still required to ensure that disease is not reintroduced. Serological detection is just as essential in maintaining Brucellosis-free status as it is in eradication and control programs in countries where B. abortus still occurs.

An approach taken by many countries to protect cattle from B. abortus infection has been to use live or killed vaccines. The most widely used vaccine is B. abortus strain 19 (attenuated live bacteria), usually given to calves or to adult cattle in reduced dosages in order to minimize serum antibody that would subsequently interfere with the serological diagnosis of field infection. The serological response to the vaccine strain has not been distinguishable from that produced in response to field infection by conventional serological tests except that the response to vaccine in the great majority of cases diminishes with time. In contrast, the antibody response to field infection remains at a high level and involves all of the four major antibody isotypes.

Several serological procedures have been and are being used extensively for antibody detection. All of the so-called 'conventional techniques' rely on the bovine antibody being capable of performing a secondary function such as agglutination or fixation of complement.

TABLE 1

REACTIVITY OF THE FOUR MAJOR BOVINE IMMUNOGLOBULIN ISOTYPES
IN SEROLOGICAL TESTS (Nielsen et al., 1984)

Assay*	IgM	IgG_1	IgG_2	IgA
STAT	20**	–	125	650
BPAT	–***	550	8500	–
Card	–	600	7500	–
Riv	–	1550	2750	–
CFT	6500****	290	–	–
EIA 1	155	190	220	700
EIA 2	78	112	126	92

*   STAT   – standard tube agglutination test (neutral pH)
     BPAT   – buffered plate antigen test (acid pH)
     Card   – Card test (buffered antigen)
     Riv   – rivanol agglutination test
     CFT   – complement fixation test
     EIA   – enzyme immunoassay – 1. using a commercial, polyclonal
               anti-bovine IgG (H+L chain) detecting agent and 2.
               using a monoclonal anti-bovine L-chain reagent
               (unpublished data).

**   based on the minimum weight of affinity purified antibody (ng)
     required to agglutinate 50% of a bacterial cell suspension.
***   negative using 20,000 ng/test.
****   based on the minimum weight of affinity purified antibody (ng)
     required to fix 50% of $3CH_{50}$ units of guinea pig complement.

Thus as can be seen in Table 1, the standard tube agglutination tests (STAT) detects IgM, IgG_2 and IgA antibody but not IgG_1 while the buffered plate agglutination test (BPAT), Card and rivanol (Riv) tests only detect IgG_1 and IgG_2. The complement fixation test (CFT) can only detect IgG_1 antibody with any sensitivity. It is therefore clear that results observed for one serum where a given antibody isotype may predominate may differ from a second serum containing a different predominant antibody isotype. For example, if large amounts of IgG_2 antibody are produced, the STAT will be positive and a reaction may occur in the Riv test, while the Card, BPAT and CFT may be negative.

Table I compares the immunological sensitivities of serological tests for bovine brucellosis including the EIA, using a commercially available polyclonal, anti-bovine IgG (H+L chain)-enzyme conjugate (EIA 1) or a monoclonal anti-bovine L chain-enzyme conjugate (EIA 2) to detect antibody of the four major isotypes. It is apparent that EIA is an excellent assay for estimating antibody of all the isotypes and its sensitivity with two exceptions is higher than that of the other tests. In addition, EIA allows manipulation via the anti-species reagent to exclude one or more isotypes should it be necessary for reasons of specificity.

Although all of the serological tests just described have been effectively employed in the diagnosis of bovine brucellosis, the choice of testing strategies will be greatly influenced by laboratory resources, prevalence of disease and action(s) (e.g. slaughter, quarantine, depopulation and/or vaccination) to be initiated as a result of detection of serological reactors. Depending on the intent of the program (e.g. control, eradication or surveillance), the requirements for diagnostic sensitivity and specificity will vary. These parameters will be discussed in detail with respect to diagnostic evaluation. In general, tests of high diagnostic sensitivity are required when disease prevalence is high and control and eradication programs are initiated. As prevalence declines there is a simultaneous need for improved diagnostic specificity. Increased diagnostic specificity is most often accomplished through use of a battery of tests. When a program requires that extremely large numbers of sera must be 'screened' either on site (e.g. farm, market or abattoir) or in a laboratory, a simple, rapid, robust and economical screening test of high diagnostic sensitivity and moderate specificity is required. In many countries, card or plate-based agglutination tests are employed as screening tests for bovine brucellosis. Although the diagnostic specificity of a given screening test is quite variable amongst different countries and cattle populations, a false positive rate of less than 5% is usually encountered. In areas of low disease prevalence, reactor sera are then subjected to assays of higher diagnostic specificity for confirmation of reactor status. In general, confirmatory tests are performed in the laboratory and are more labour intensive, more expensive and have a much slower 'turn-around time'. The STAT and the CFT are currently the most widely used confirmatory tests. Reactor status may be based on either series or parallel interpretation of confirmatory test results and is dependant on serum dilution and strength of reaction, age and vaccination status and often herd history. Import/export certification often requires that one or both confirmatory tests be performed. In the case of the STAT, a bank of international reference serum has been established and results are expressed in international units of agglutinating antibody activity per milliliter (IU/ml).

EIA PROTOCOL

The EIA for bovine brucellosis has been used in many research settings with milk or serum (or for antigen detection) in various formats, using several antigen preparations and many anti-species reagents of different specificities. Although a large number of reports have been published, they are too numerous to discuss or reference within the confines of this chapter; however, a bibliography may be found in Nielsen and Wright (1984). Presently no concerted attempts have been made to internationally standardize the EIA and as a result its acceptance in the diagnostic laboratory has not been established.

The essential features of a protocol developed in our laboratory and to be implemented in 8 Agriculture Canada laboratories are described below. A more detailed description of the protocol and equipment used may be found in Nielsen and Wright (1984).

1. 96 well polystyrene plates are rinsed in distilled water immediately prior to use. Excess water is removed by sharply striking the plate, upside down, onto absorbent, lint-free material.

2. The plate wells are coated with 200 µl of antigen solution per well. The antigen solution contains 1.0 µg/ml of smooth lipopolysaccharide (S-LPS) prepared from B. abortus strain 413 by the method of Baker and Wilson (1965) in 0.06 M carbonate buffer pH 9.6. The plates are sealed with an adhesive plastic sheet and incubated at room temperature for approximately 18 hours.

3. Serum samples to be tested are diluted 1:100 in 0.01 M phosphate, pH 7.2, containing 0.15 M NaCl and 0.05% Tween 20 (PBS/T).

4. Immediately prior to use the antigen coated plates are washed four times with PBS/T and excess wash buffer is removed as in Step 1. The samples and controls are dispensed in 200 µl volumes in a quadrant

pattern (Figure 1). The plates are resealed and incubated for 3 hours at room temperature.

5. Unbound serum proteins are removed by washing as above and 200 µl of appropriately diluted detecting reagent is added to each well. The detection reagent used initially was a commercially prepared horse-radish peroxidase (HRPO)-conjugated rabbit anti-bovine IgG (H+L). An HRPO-conjugated mouse monoclonal anti-bovine L chain (prepared from ascites fluid and conjugated by the periodate method of Nakane and Kawai (1974) using ascorbic acid instead of sodium borohydride) is likely to replace the polyclonal conjugate for reasons discussed below. The plates are resealed, incubated at room temperature for 1 hour followed by wash-ing in PBS/T.

6. Substrate (4 mM hydrogen peroxide) and chromogen (1 mM ABTS (2,2'-azino-bis(3-ethylbenzthiazoline sulfonic acid)) in solution in 0.05 M citrate, pH 5.0 is dispensed in 200 µl amounts into each well. The plate is maintained at room temperature. Timing of the reaction begins immediately and the plate is placed on a plate shaker to assure uniform enzyme-substrate contact and formation of a uniform meniscus.

7. At exactly 4 min of development, the plate is assessed for development in a photometric 96-well plate scanner pre-blanked on 200 µl of undeveloped substrate/chromogen solution in a separate plate. The average optical density (OD) of a 'target' reference serum placed in 4 wells (Figure 1) is used to extrapolate the time required for the target serum to achieve an OD of 1.0 using the equation (from Wright et al., 1985a)

$$\text{Final development time} = \frac{\text{OD at 4 min} + 0.1394}{4.5939}$$

Given an average development of about 10 min, OD units derived in this manner will not require between plate or day-to-day variability corrections.

8. The results are interpreted for diagnostic relevance using an O.D. threshold as established in the following discussions on diagnostic evaluation.

STANDARDIZATION AND QUALITY CONTROL

To minimize error in diagnostic interpretation of the EIA, a number of steps can be taken to facilitate standardization. Plasticware (96 well plates) vary from manufacturer to manufacturer, from batch to batch and with the manufacturer's post-molding treatment. Generally speaking, more antigen added up to a limit increases sensitivity. Beyond this limit, loss occurs (Cantarero, Butler and Osborne, 1980). Thus to avoid assay discrepancies due to antigen adsorption/desorption, it might be useful to test each batch of 96-well plates for their capacity to immobi-lize antigen. This is fairly simply accomplished with a radiolabelled antigen preparation added at various levels (Nielsen and Wright, 1984). A second variable in antigen binding to plastic is the antigen itself. With crude LPS extracts, standardization is difficult due to multiple components; however, with purified LPS preparations, molecular size and chemical nature can be ascertained. The O-chain of S-LPS has been chemically defined (Caroff et al., 1984a,b) and its monoclonal antibody (Bundle et al., 1984) makes standardized antigen titration and quantita-tion more feasible. A third variable is the coating buffer, 0.06 M car-bonate, pH 9.6, of which the pH is important; however, minor variations

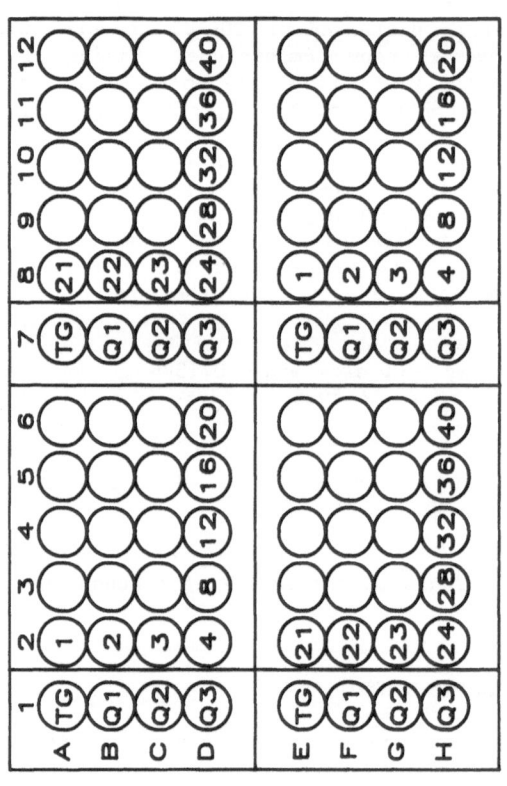

Figure 1: Sample placement configuration used in the indirect EIA; TG, target serum; Q1, positive quality control serum; Q2, negative quality control serum and Q3, buffer control (no serum).

in pH (+ 0.1 pH units) are acceptable. For maximum binding of antigen, the carbonate buffer is prepared freshly each day for plate preparation. A final variable in antigen binding is the incubation period of the antigen in the plate. With protein antigens maximum binding is reached in about 6 hours (Cantarero, Butler and Osborne, 1980). While with S-LPS, 90% of maximum binding is achieved in 2 hours at 37°C, generally, 18 hours at room temperature (overnight) is convenient in preparing plates for the next day. It is necessary with each batch of antigen to establish the exact quantity required per test by titration with sera demonstrating various levels of previously characterized antibody activity.

Test sample dilution was established (at 1:100) by using two-fold serum dilutions from 1:25 to 1:51200 of sera with a range of antibody activities and some were negative. The 1:100 dilution was found to consistently give low non-specific reactions and acceptable sensitivity in detecting antibody in low titered sera. Using the above guidelines, no blocking agent is required, however, the Tween 20 component of the wash buffer is essential to prevent non-specific interactions. Maximum primary immune complex formation is achieved in 3 hours. Antibody dilutions are in PBS/T to prevent non-specific binding.

The polyclonal rabbit anti-bovine IgG H+L chain HRPO conjugate previously described was used initially as a detection agent. However, due to limited availability and potential batch variation, a mouse monoclonal anti-bovine L chain HRPO conjugate is currently being introduced. The use of a monoclonal antibody assures a continuous supply of a reagent which is identical from batch to batch not only in its ability to bind to its antigen but also with respect to conjugation uniformity. The polyclonal conjugate was titrated directly using bovine gammaglobulin (predominantly IgG) coated directly on the polystyrene plate in amounts of 500 ng/ml followed by dilutions of the conjugate. A working dilution of conjugate resulting in an OD of 1.0 after about 10 minutes of development was chosen. A second method which perhaps more accurately reflects the detection of antibody in the form of antigen-antibody complexes was used to titrate the monoclonal conjugate. In this method S-LPS coated plates were reacted with an excess of affinity purified antibody to B. abortus of each isotype. An OD of 1.0 after approximately 10 minutes of development was observed with approximately half as much antiglobulin-conjugate as the former method. The working concentration of the monoclonal anti-L chain was determined to be 15 ng per well. The choice of titration method will ultimately depend on the availability of purified gammaglobulins and/or antibody. Both methods are useful and diagnostic OD thresholds, as will be discussed, would likely differ slightly depending on the method chosen.

Optimization and standardization of the EIA requires other essential considerations as well. All buffers should be fresh and at room temperature for use. The pH should be within 0.1 units of the optimal pH described above. The ambient temperature should be monitored as it has an effect particularly on the enzyme reaction. Incubation times should be established and used. Washing procedures are important to minimize non-specific interactions. Washed plates should not be allowed to air dry as this increases non-specific binding at the next step.

Automation is one of the key features of the 96 well plate-based, indirect EIA. Microprocessor-controlled instrumentation is now available for virtually every operational aspect of the assay. This instrumentation has become an integral part of overall quality assurance. These instruments are not without inherent limitations and therefore should be chosen carefully. In addition, they must be properly maintained to ensure uniform function and minimize 'down time'.

Plate washers are available which will independantly wash all 96 wells simultaneously. Pump-driven washers can deliver with uniform efficiency, specific volumes of wash buffer across all wells. Because of the high salt content in many of these buffers, it is essential that the washer head be thoroughly rinsed with distilled water after use to prevent salt 'build up' which may block delivery and aspiration tubes. Despite preventative measures, it will periodically be necessary to dismantle and clean the washer head to remove foreign residues. If the washer is not to be used for prolonged periods of time, the entire system should be rinsed and all internal compartments subsequently dried or alternatively a non-corrosive, anti-microbial agent should be added to the rinse. Prior to use (i.e. daily), washer efficiency should be visually checked to ensure uniform delivery and aspiration.

Several types of single and multiple channel dispensing systems are available. We currently use a system which employs an eight piston cartridge for reagent delivery. As the cartridges are removable, a separate cartridge can be reserved exclusively for use with one reagent (e.g. antigen, conjugate or substrate). Periodically, the efficiency of delivery is checked by dispensing 200 µl of 0.5 M $NiSO_4$ across all 96 wells and scanning in a plate photometer at a wavelength of 414 nm.

Liquid handling technology has advanced to the point where test sample dilutions can be made directly into the microtiter plate wells. In our case, a 2 µl serum sample must be aspirated and diluted 100 fold in the well. Most diluters employ a double syringe system, one sample and one reagent syringe. If samples of less than 5 µl are required for dilution, the choice of sample syringe becomes critical. The volume of sample to be aspirated should be no less than 5% of the total syringe volume. We have chosen a 10 µl syringe for our application. The sample probe is also a critical aspect when handling small volumes. The tip of the probe, whether it be metal or plastic, should be of sufficiently fine bore as to minimize capillary action.

The choice of plate photometer will depend greatly on a number of factors; assay protocol (e.g. multiple vs single point reading), number of samples, interfacing options, elecronic data processing requirements, etc. Generally all plate photometers are designed around a fiber optic system in which the appropriate absorbance wavelength is selected by means of interference filters. As these filters can deteriorate over time, their efficiency must be checked periodically according to the manufacturer's instructions. The light source can also be a source of variability and should be replaced should reading become eratic. Photometers vary with respect to the number of wells which are read at any one time (e.g. single, eight or twelve wells) and consequently vary with respect to blanking operation. As a general rule, the number of blanking wells should correspond to the number and order of optic channels (e.g. 8 channels/8 blanking wells). In order to preserve wells, we have chosen a photometer which can be blanked on one plate while test sample reactions are read on a separate plate. Only one set of blanks is then required for a day's run of six or more plates and the blanking plate may be reused after washing with distilled water. Most EIA procedures require the presence of detergent in reagent buffers. This induces the formation of a meniscus in the plate wells. The lens effect of meniscus formation causes light to scatter as it passes vertically through the substrate solution to the photoreceptor and it also decreases the pathlength of the sample (Wright et al., 1985b). It is important that each optic channel be blanked on wells exhibiting the same degree of meniscus formation as the sample wells. If not, significant between channel variability may result. To ensure uniform meniscus formation in all wells of both the sample and blanking plate, the plates should be shaken immediately prior to reading. Most photometers now offer a single or dual wavelength reading mode. The dual wavelength mode is intended to remove the effects of optical aberrations not attributable to specific absorbance. In general, we have found that this feature is not an essential aspect of reading given the quality and optical uniformity of most plates and reagents.

Given that an EIA protocol has been highly standardized with respect to buffers, reagents, physicochemical parameters and instrumentation, there is still a certain degree of inherent variability which must be monitored. Acceptance criteria must be established for both within and between assay variation. The indirect EIA is semi-quantitative at a single dilution but replicate samples must be tested in order to establish acceptance criteria. Although confidence increases proportionately with the number of replicates tested, so does the cost per sample. We have chosen for economic reasons to test all samples in duplicate. The duplicate samples are tested within one plate as to avoid the complication of having one or the other plate rejected for other reasons as discussed below. The spatial placement of the duplicates within one plate should also be considered. There are a number of factors which can affect intra-plate variability (Wright et al., 1985b). The pattern as originally described by Stemshorn et al., 1983 is best suited for duplicate or quadruplicate sample placement. In brief, the plate is subdivided horizontally and vertically into four equal quadrants as illustrated in Figure 1. Diagonally opposed quadrants each receive one of two replicates. The mean OD and percent co-efficient of variation of each duplicate is determined and only those duplicates which fall within predetermined limits of variability are accepted. Acceptable limits will tend to narrow as OD values increase; therefore, one must establish a sliding scale by repeated testing of a cross section of samples. As variability is normally distributed around the mean OD, acceptance limits can be set within plate at two standard deviations. The most critical area of acceptability will involve OD values at or near the established diagnostic threshold. We have determined for our assay using the polyclonal conjugate, given a critical diagnostic threshold range of 0.340 ± 0.040 OD units, that duplicate assay variability must fall below 15% to be acceptable.

Although duplicate sample variability may fall within acceptable limits, this does not necessarily indicate that the assay itself is in control. The four left hand wells of each quadrant are reserved for a 'target' reference serum and three quality control reagents. We have developed a timing protocol to monitor chromogen conversion in our assay (Wright et al., 1985a). Inter-plate variability is an inherent problem of the indirect EIA. In this protocol a high titered reference serum is used to monitor the enzyme kinetics. A reference serum and dilution is chosen such that an average OD of 1.0 can be expected at an optimal development time (e.g. 10 minutes for our assay). An OD of 1.0 then becomes the 'target' and the development time required to achieve this value is calculated from a hyperbolic relationship as a function of the OD at four minutes of development. Acceptance limits are established for between plate variability. For our assay, we calculate the mean quadruplicate target OD and standard deviation on a continuous basis for the last 50 plates. The mean target value of each new plate must fall within two standard deviations of the grand mean. This represents an acceptable between plate variability of less than 15%. The target time is also monitored; however, acceptability is not defined using a mean and standard deviation. Development times are considered to be acceptable only if the target reference and internal quality control reagents fall within acceptable limits. An upper and lower range of times may then be established; however, this range is quite variable and will likely change from one operator to another within a given laboratory. We have observed that approximately 95% of final development times fall within a range of 5 to 15 minutes for acceptable assays given a mean development time of 10.2 minutes for 84 test plates (Wright et al., 1985a). The distribution tends to be biased towards shorter development times (approximately 60% of values fall below the mean).

The four quality control wells per quadrant also include a low titer positive control serum pool, a negative control serum pool and a buffer control (no serum). The positive control pool should be chosen such that reactivity at a standard test dilution (1/100) is slightly higher than the upper limit of the diagnostic threshold OD range. The negative control (1/100) should represent a pool of healthy cattle from herds of minimal infectious disease status. As for the target serum, the mean OD of the positive and negative controls is calculated on a continuous basis for the last 50 plates. The mean quadruplicate value of positive and negative serum controls for each new plate must fall within two and three standard deviations of their grand means, respectively. This represents an acceptable between plate variability of less than 25% for the positive control and less than 75% for the negative control. Although on a percent basis, the variability of the negative control may appear to be extreme, in terms of OD units it is much less than that of the positive control. The OD values of the buffer control are less normally distributed about the mean. Because the lower limit of acceptance is delineated by zero absorbance, the distribution tends to be skewed towards the high end. The upper limit of acceptance should therefore be defined as the 95th percentile of the OD distribution.

Within plate variability is also monitored but is not necessarily required for determination of assay acceptability. For both the target and quality controls, the within plate variability should be much lower (in the order of 25 to 50%) than the between plate equivalent. Only if a plate is rejected on the basis of between plate variability should the within plate variability be checked to determine whether or not the rejection was due to an aberrant value in the quadruplicate OD's. It is likely that this type of rejection will be based on one of either the target or quality controls. We have observed a positive correlation between mean target and positive control (R = 0.73 for 89 plates) and independently, between mean negative and buffer control values (R = 0.88) in any given plate. This would suggest that the majority of plates will be rejected on the basis of at least two control points being unacceptable as defined by the above correlations. In such cases there would be no need to examine the quadruplicate values for outliers.

DIAGNOSTIC EVALUATION

The indirect EIA, as a single dilution assay, is quantitative, in terms of antibody activity, over a relatively limited range. In our system, end point titers as great as 1:800 or 1:1600 can be predicted from single dilution OD values. Beyond this, the antigen/antibody system approaches saturation and the OD value is no longer a quantitative indication of antibody activity (e.g. OD values greater than approximately 1.2). However, single dilution OD values are valid criteria for determining seropositive and seronegative status. In addition, they are readily applicable to seroepidemiological studies.

Proper evaluation of diagnostic performance is a critical aspect of assay development. The two principle parameters which must be defined are diagnostic specificity and diagnostic sensitivity. Specificity is an estimate of the probability that an animal will be correctly classified as seronegative in the absence of infection with Brucella. Sensitivity is an estimate of the probability that an animal will be correctly classified as seropositive given that it has been infected. It is very important to define, at the outset, criteria by which cattle are chosen to establish these estimates. The samples collected should preferably represent a cross-section of the target population with respect to breed, age and sex. To evaluate diagnostic specificity, serum samples were selected from Ontario dairy herds as they were submitted for annual certification of brucellosis-free status (Dohoo et al., in press). Only those herds, declared negative on the basis of a combination of conventional serological tests in the current and previous two annual submissions, were sampled. By strict definition, samples used to estimate diagnostic specificity should be chosen by means not involving serological reactor status, conventional or otherwise. However, this is often next to impossible when dealing with infectious diseases for which serologically-based eradication, control or surveillance programs have been in existence for a number of years.

The selection of sera for estimation of specificity may be complicated by vaccination practices. Presently in Canada, less than 2% of calves are vaccinated annually with a standard dose of B. abortus strain 19 for purposes of exportation. Vaccinal antibody activity in the sera of both calfhood and adult vaccinates generally declines to very low levels over a period of approximately one year. By definition, these cattle are not considered to be infected (note: infection and bacterial shedding may occur in a very small percentage of vaccinated animals); however, they have immunologically responded to the vaccine strain. Therefore, vaccinated cattle from herds considered to be brucellosis-free (and may include herds not necessarily certified by serological tests) should be considered separately from non-vaccinated cattle and estimates of diagnostic specificity should be established for both groups if necessary.

To evaluate diagnostic sensitivity, sera must be obtained from cattle of known positive, Brucella culture status. From actively infected cattle, isolations are often made from tissues and fluids following slaughter or abortion or from milk and these cattle are often strong serological reactors. However, these cattle are not representative of the entire spectrum of infection. In many cases, bacteria may only be isolated after slaughter and exhaustive culture attempts. In an eradication program, an entire herd may be slaughtered on the basis of serological detection and positive culture from one or a few serological reactors within the herd. In such an event, bacteriological culture should be attempted on all cattle and sera should then be matched with positive culture results. Only these sera may be used for estimation of diagnostic sensitivity. It is important to note that negative culture status does not preclude infection and sera from these cattle within infected herds should not be used for estimation of diagnostic specificity.

Once the selection criteria have been defined, it is equally important to define appropriate sample sizes for each estimate. The sample size for estimating a binomial proportion will be determined by three parameters; the expected sensitivity or specificity (i.e. as a proportion), the allowable error in the estimation of the expected sensitivity or specificity and the confidence that the error will not exceed allowable limits. For 95% confidence, the sample size (n) may be determined from the following equation:

$$n = 4 \ pq/e^2$$

where p = expected proportion of false positives (1- specificity) or false negatives (1- sensitivity)
q = expected specificity or sensitivity
e = allowable error of the estimate

Because the diagnostic threshold of the EIA need not be established prior to evaluation, the expected sensitivity and specificity should reflect the diagnostic requirements as related to a particular prevalence of disease and eradication, control or surveillance programs. Common sense should prevail in determining expected sensitivity and specificity and allowable error. For our study, given that brucellosis has been eradicated from cattle herds in Canada, major emphasis was placed on diagnostic specificity. Based on the performance of the conventional tests, a diagnostic specificty of at least 0.995 would have to be demonstrated by the EIA if it is to be effectively used in our surveillance program. In order to accurately estimate specificity, an allowable error of 0.005 was considered to be tolerable. Given these criteria, it was determined from the preceding formula that at a 95% confidence level, approximately 800 sera would be required. A decrease in specificity (0.960) was expected in the testing of vaccinated cattle from brucellosis-free herds and the allowable error margin was increased to 0.025 because of the unpredictable nature of the immune response to vaccination and its subsequent decline. It was determined, as above, that approximately 250 sera would be required. A diagnostic sensitivity in the range of 0.960 was expected and a relatively large allowable error margin (0.030) was considered to be acceptable because of the heterogeneous nature of the immune response of cattle recently infected under natural conditions. Approximately 170 sera would be required to meet these criteria. Note that the 95% confidence intervals (for binomial distributions) must be redefined once estimates of sensitivity and specificity have been calculated from actual test results.

After collection and testing of the appropriate number of samples for each group, the mean OD values for each sample should be sorted in ascending order. Results of such an evaluation, using the rabbit polyclonal anti-bovine IgG (H+L) are presented in Figure 2. In order to evaluate the diagnostic performance of the EIA, a base OD threshold was defined as that OD value resulting in a minimum acceptable specificity estimate of 0.995. This value corresponds to the median OD value (0.260) of the 100th percentile of the non-vaccinated, brucellosis-free group. Four thresholds (in arbitrary increments of 0.04 OD units) are compared in Table II. As can be seen from this Table, there is an inverse relationship between diagnostic specificity and sensitivity.

The choice of threshold will not only depend on the estimates of specificity and sensitivity but the predictive value of the assay as well. The predictive value of a negative test is an estimate of the probability that a cow is brucellosis-free given a seronegative test and a particular prevalence of disease. The predictive value of a positive test is an estimate of the probability that a cow is infected given a seropositive test and a disease prevalence. As can be seen in Table II, the predictive value of a positive test is greatly influenced by both diagnostic specificity and prevalence of disease. Therefore, the choice of diagnostic threshold will depend very much on the economic impact of action(s) to be initiated as a result of a positive test. From the data presented in Table II, a diagnostic OD threshold in the range of 0.340

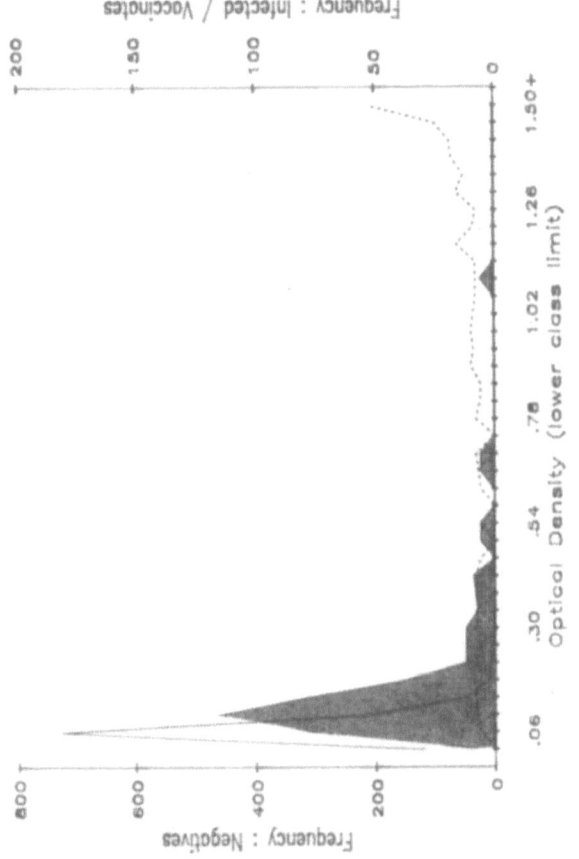

Figure 2: Frequency distribution of OD values used to estimate diagnostic specificity and sensitivity; solid line, non-vaccinated, negative cattle sera; shaded area, vaccinated cattle sera and dotted line, culture positive, infected cattle sera.

TABLE II

DIAGNOSTIC PERFORMANCE OF THE INDIRECT ENZYME IMMUNOASSAY INTERPRETED AT VARIOUS OPTICAL DENSITY THRESHOLDS

Positive[a] Threshold	Specificity[b] NV	Specificity[b] V	Sensitivity[c]	Disease[d] Prevalence	Predictive Value[e] +'ve NV	Predictive Value[e] +'ve V	Predictive Value[e] −ve Either
≥0.260	0.9947	0.9144	0.9479	.0001	0.02	0.00	1.00
				.001	0.16	0.01	1.00
				.01	0.66	0.10	1.00
				.1	0.95	0.54	0.99
≥0.300	0.9982	0.9358	0.9427	.0001	0.09	0.00	1.00
				.001	0.49	0.01	1.00
				.01	0.91	0.12	1.00
				.1	0.99	0.60	0.99
≥0.340	0.9991	0.9572	0.9271	.0001	0.09	0.00	1.00
				.001	0.51	0.02	1.00
				.01	0.91	0.16	1.00
				.1	0.99	0.67	0.99
≥0.380	1.000	0.9633	0.9115	.0001	1.00	0.00	1.00
				.001	1.00	0.03	1.00
				.01	1.00	0.24	1.00
				.1	1.00	0.77	0.99

[a] mean $OD_{414\ nm}$ using a rabbit anti-bovine IgG (H+L) conjugate
[b] specificity based on 1128 individual non-vaccinated (NV) cattle sera and 327 individual vaccinated (V) cattle sera
[c] sensitivity based on 192 individual sera from infected cattle
[d] theoretic disease prevalence in either a non-vaccinated or vaccinated population
[e] predictive value of positive (+ve) or negative (−ve) tests from Galen and Gambino (1975)

to 0.380 would provide an appropriate specificity for use of the EIA in a surveillance program where the prevalence of brucellosis is extremely low or absent in a non-vaccinated cattle population. The upper limit of this threshold range would minimize the number of field investigations attributable to false positive test results. However, when the same threshold range is applied to a vaccinated cattle population, it is clear that the lack of diagnostic specificity will present major problems especially when disease prevalence is very low. From the frequency distribution of OD values for vaccinated cattle (Fig. 2) and from the skewness of the distribution, it is apparent that the specificity amongst vaccinates must be improved by methods other than increasing the OD threshold. The diagnostic specificity could be improved if the results of the EIA were to be interpreted in series with the results of one or the other of the conventional tests (i.e. the STAT or CFT). Alternatively, the specificity of either the antigen or the conjugate or both could be improved. Initial comparisons of the rabbit polyclonal anti-bovine IgG (H+L) and mouse monoclonal anti-bovine light chain conjugates indicate that the latter conjugate has a much higher diagnostic specificity in the testing of a limited number of field sera from B. abortus strain 19 calfhood vaccinated cattle. Both conjugates, however, demonstrated comparable specificities in the testing of non-vaccinated cattle and comparable sensitivities (data unpublished).

Estimates of diagnostic specificity and sensitivity do not directly relate to the performance of the EIA as a confirmatory test. As a confirmatory test, the EIA would only be applied to serum samples which are screening test positive. Of the 192 sera from infected cattle, 171 (sensitivity = 0.8906) were BPAT screening test positive and all 171 were EIA positive at an OD threshold of 0.340. Therefore, at this threshold, the diagnostic sensitivity of the EIA relative to the BPAT is 1.000. Of 108 BPAT positive sera from brucellosis-free herds, serologically certified or otherwise, only 8 were EIA positive. Given an estimate of specificity, for the BPAT, of 0.9887 amongst non-vaccinated and 0.9210 amongst vaccinated cattle, the predictive value of a positive test at a low prevalence of disease (0.0001) would be less than 0.01. Therefore, if we assume that the 108 BPAT positives are actually false positives, then the specificity of the EIA as a confirmatory test can be estimated. Although the number of sera in this sample is low from a statistical standpoint and includes 23 sera from vaccinated cattle, it illustrates that the specificity of the EIA as a confirmatory test (0.9259) is likely to be lower than the estimate amongst either non-vaccinated or vaccinated cattle which have not previously been screened.

It is interesting to note that of the 8 cattle positive on both the BPAT screening test and the EIA as a confirmatory test, 4 were vaccinated cattle and were decidedly seropositive in these and other conventional tests as well. The other 4 cattle had not been vaccinated and were also strongly positive. This illustrates yet another complicating factor in that exposure of cattle to some other bacteria, unrelated to Brucella, may induce antibody which strongly cross reacts with the S-LPS used in the EIA and other antigens of the whole cell bacteria used in the conventional tests (Corbel, 1985). These cattle are difficult to differentiate from Brucella-infected cattle. The relative importance of this type of reactor becomes more acute as the prevalence of disease becomes extremely low.

Once the EIA is implemented and has been established in a diagnostic routine, it would be necessary to create a data-base from which the diagnostic performance may be monitored. If the assay is to be run in more than one laboratory as in our situation, it may be necessary to establish optimal thresholds for each laboratory. This may reflect differences in cattle populations (e.g. within or between dairy or beef populations) or perhaps laboratory differences.

DISCUSSION

In the preceding review, consideration was given to some of the factors that influence the performance and application of an EIA applied to veterinary medicine. Thus the example of an indirect EIA for detection of antibody to B. abortus may be used as a functional model for further test development as most of the criteria for proper conduct of testing and its standardization should be directly applicable, at least to the diagnosis of other cattle diseases. While at present this test offers distinct advantages over conventional procedure, application of further research could undoubtedly improve both performance and standardization.

Aspects of quality control have been discussed as this is an essential feature of any assay which is to be put into routine diagnostic use in one or more laboratories. Quality control is 'all encompassing' and must include analytical and operational aspects as well as instrumentation maintenance. This is ultimately reflected in assay performance and acceptance criteria must be defined to ensure that the assay is in control.

Proper diagnostic evaluation is a critical step in determining the ultimate usefulness of a test. Diagnostic performance should be assessed with careful consideration being given to the prevalence of disease and to the type of program in which the test is to be used. This will impact on the minimum acceptable performance criteria required of an assay with respect to its use either alone or in combination with other assays. The EIA as presented in this review may be used as either a screening test or a confirmatory test. This choice will depend on the numbers of sera requiring testing.

Presently a semi-automated S-LPS antigen is used; however, other antigens should be considered. Of particular interest in this regard is the chemically defined O-chain which may be synthesized and therefore eliminates variability encountered with extraction of antigens from bacteria. However, as presently understood, the immune response to B. abortus strain 19 and to pathogenic field strains cannot be distinguished by serological means. Thus an alternative choice of antigen may be the polysaccharide 'poly-B' which in gel diffusion has been shown to react with sera from field infected cattle but not with sera from vaccinated animals (Diaz et al., 1979 and Jones et al., 1980). Its application as an antigen to EIA awaits further study and its chemical characterization is incomplete.

With the advent of highly refined antigens, the EIA procedure itself may be revised to a much more manageable 'homogeneous assay' in which an epitope of the antigen is labelled with enzyme, the function of which may be disrupted by binding of antibody to the epitope. This one-step assay can be performed quickly and appears to be highly amenable to 'barn-side' use.

Application of monoclonal anti-species antibody-enzyme conjugates to EIA procedures are essential for standardization on a national or world wide basis. In addition, with the advent of across species fusion procedures, bovine x mouse hybridomas which secrete bovine anti-Brucella antibody may be produced. Such hybridomas, with desired specificities will provide the final link in the total standardization of the brucellosis EIA.

Utilization of microprocessors and robotics will presently allow semi- or complete automation of EIA procedures including sample identification and print-out of the final report and with the appropriate software, the only human requirement is in the final data evaluation. Thus EIA of any type, whether for antigen or antibody detection, would achieve prominence in mass testing procedures frequently required in a veterinary diagnostic environment.

ACKNOWLEDGEMENTS

The authors wish to thank Dr. Bob Duncan and Dr. Rich Jacobson for their encouragement and constructive criticism. The technical support of David Gall, Sofija Balsevicius and Diane Henning is also appreciated. The word processing skills and patience of Joan Graham is also greatly appreciated.

REFERENCES

1. P. J. Baker and J. B. Wilson, Hypoferremia in mice and its application to the bioassay of endotoxin, J. Bacteriol. 90:903-910 (1965).
2. D. R. Bundle, M. A. J. Gidney, M. B. Perry, J. R. Duncan and J. W. Cherwonogrodzky, Serological confirmation of Brucella abortus and Yersinia enterocolitica 0:9 O-antigens by monoclonal antibodies, Infect. Immunity 46:389-393 (1984).
3. L. A. Canterro, J. E. Butler and J. W. Osborne, The adsorptive characteristics of proteins for polystyrene and their significance in solid-phase immunoassays, Analyt. Biochem. 105:375-382 (1980).
4. M. Caroff, D. R. Bundle and M. B. Perry, Structure of the O-chain of the phenol-phase soluble cellular lipopolysaccharide of Yersinia enterocolitica serotype 0:9, Eur. J. Biochem. 139:195-200 (1984).
5. M. Caroff, D. R. Bundle, M. B. Perry, J. W. Cherwonogrodzky and J. R. Duncan, Antigenic S-type lipopolysaccharide of Brucella abortus 1119-3, Infect. Immun. 46:384-388 (1984).
6. S. Charan and O. P. Gautam, Applications of enzyme-linked immunosorbent assay in veterinary medicine: A bibliography, Vet. Res. Com. 8:255-267 (1984).
7. M. J. Corbel, Recent advances in the study of Brucella antigens and their serological cross-reactions, Vet. Bull. 55:927-941 (1985).
8. A. Diaz, P. Garatea, L. M. Jones and I. Moriyon, Radial immunodiffusion test with a Brucella polysaccharide antigen for differentiating infected from vaccinated cattle, J. Clin. Microbiol. 10:37-41 (1979).
9. I. R. Dohoo, P. F. Wright, G. M. Ruckerbauer, B. S. Samagh, F. J. Robertson and L. B. Forbes, A comparison of five serological tests for bovine brucellosis. Can. J. Comp. Med. (in press).
10. R. S. Galen and S. R. Gambino, 1975, Beyond normality: the predictive value and efficiency of medical diagnosis, John Wiley and Sons, New York, N.Y., pp. 167-221.
11. L. M. Jones, D. T. Berman, E. Moreno, B. L. Devoe, M. J. Gilsdorf, M. J. Huber and P. Nicoletti, Evaluation of a radial immunodiffusion test with polysaccharide B antigen for diagnosis of bovine brucellosis, J. Clin. Microbiol. 12:753-760 (1980).
12. E. Maggio (ed.), Enzyme immunoassay: principles and practices, CRC Press, Boca Raton, Florida, pp. 1-295.
13. P. K. Nakane and A. Kawaoi, Peroxidase-labelled antibody: a new method of conjugation, J. Histochem. Cytochem. 22:1084-1091 (1974).
14. K. Nielsen, F. C. Heck, G. G. Wagner, J. Stiller, B. Rosenbaum, R. Pugh and E. Flores, Comparative assessment of antibody isotypes to Brucella abortus by primary and secondary binding assays, Prev. Vet. Med. 2:197-204 (1984).
15. K. Nielsen and P. F. Wright, 1984, Enzyme immunoassay and its application to the detection of bovine antibody to Bovine abortus, Agriculture Canada/Animal Diseases Research Institute (publisher), Nepean, Ontario, pp. 1-121.
16. B. W. Stemshorn, D. J. Buckley, G. St. Amour, C. S. Lin and J. R. Duncan, A computer-interfaced photometer and systematic spacing of duplicates to control within-plate enzyme-immnoassay variation, J. Immunol. Meth. 61:367-375 (1983).
17. P. F. Wright, W. A. Kelly and D. E. J. Gall, Application of a timing protocol to the reduction of inter-plate variability in the indirect enzyme immunoassay for detection of anti-Brucella antibody, J. Immunoassay 6:189-205 (1985).

18. P. F. Wright, D. E. J. Gall and W. A. Kelly, Effect of meniscus formation and duplicate sample placement configuration on the variability of measurement by three microtiter plate photometers, J. Immunol. Methods 81:83-93 (1985).

ENZYME IMMUNOASSAY OF PLANT CONSTITUENTS

Richard L. Mansell and Cecilia A. McIntosh

Biology Department
University of South Florida
Tampa, FL 33620

The use of immunoassay to quantitate plant derived compounds was first developed in clinical laboratories in the early 1970's. These first assays were directed toward small organic molecules, e.g., colchicine (Boudene et al., 1975), nicotine (Langone et al., 1973), and morphine (Spector, 1971), which have medicinal or pharmaceutical value. These assays were based on the radioimmunoassay (RIA) and were developed for compounds whose small molecular size renders them non-immunogenic. Assays for low molecular weight compounds of animal origin had been developed earlier (see Erlanger, 1980, and references therein) using synthesized protein-hapten conjugates for immunization. The immunoassays mentioned above demonstrated that the hapten-protein conjugate principle (Landsteiner, 1945) also could be used to induce animals to produce antibodies against compounds of plant origin.

The pioneering work of Weiler and Zenk (1976) demonstrated that highly specific and sensitive radioimmunoassays could be developed and used in plant science and it was suggested that this technique could be utilized in a wide variety of investigative systems. In addition to the specificity and sensitivity of these assays, it was also discovered that this method could be utilized in unpurified extracts and several hundred samples could be processed by a single person per day. The impact of this first report has resulted in a great expansion of the immunoassay technique in plant science and at present it is clear that the only limit to the utilization of such an assay is one's own creativity and imagination.

Most of the early immunoassays were based on the use of radioisotopes since these are easily detected, are relatively stable and are inexpensive. However, over the past five years or so it has become increasingly obvious that for the use of immunoassay in research to reach its full potential, other detection molecules or systems must be utilized. Non-isotopic assays are proving to be less expensive, faster and less hazardous than isotopic assays. In addition, the sensitivities and specificities of these non-isotopic assays equals or exceeds those of radioimmunoassays. As early as 1971, it had been demonstrated (Engvall and Perlmann; van Weemen and Schuurs) that immunoassays utilizing an enzyme rather than an isotope would be of practical use in laboratories not equipped to handle radiation. These enzyme based assays are most often referred to as enzyme immunoassay (EIA) or enzyme-linked immunosorbent assay (ELISA) and like RIAs, their utilization is rapidly gaining wide-spread acceptance. The surge in the use of these enzyme-based assays has been founded on a number of technological advances in spectrophotometric instrumentation, enzyme isolation and detection, and the result has been a rapid advancement in employment of these systems (see Engvall and Pesce, 1978; Maggio, 1980 for discussions of approaches and applications of enzyme immunoassays in general). It will be the purpose of this chapter to explore some of the recent uses which EIA and ELISA have found in plant science.

In our search of the literature since 1976, we have found that there are a number of widely different uses for which the principles of the EIA have been employed. There are numerous manuscripts on the subject of plant diseases, viral detection and other applications to plant pathology and agriculture, all of which are covered elsewhere in this volume. This chapter will concentrate mainly on plant hormones, proteins, enzymes, secondary metabolites, and polysaccharides. To this end, we encourage the reader to obtain copies of the original articles so that the unique uses of each of these assay systems can be fully appreciated.

In addition to the use of enzymes as tracers in these assays, it must also be pointed out that the immunoassay systems being utilized in plant science are making wide use of both mono- and polyclonal antibodies. Although it is not our goal to discuss the merits and idiosyncratic characteristics of these different types of antibody preparations, it is important to emphasize that each of these is unique and must be carefully analyzed prior to use in any assay system. The researcher will be justly served by these different antibody preparations as each of them can provide different, yet important, pieces of information (see Vora, 1985 for a discussion of the use of monoclonal antibodies in enzyme research). The strategy which the researcher wishes to employ will dictate which antibody system will function most effectively.

Theoretically, the most straightforward plant immunoassay system is one which uses a protein as the antigen, however, the literature contains many more reports describing assays which have been developed against low molecular weight compounds; less than 1000 daltons (see Weiler, 1983 for a comprehensive review). Since it is recognized that these small molecules must be conjugated to protein before they will elicit an immunogenic response, a useful and functional set of assays for a specific molecule or related compounds can be developed through the careful choice of the hapten conjugate as well as the tracer molecule (Jourdan et al., 1985; Weiler, 1983; Weiler et al., 1986).

Table 1 presents a list of representative studies which are currently involving the use of EIA. As can be seen, there is great diversity in the type of compounds which are under investigation and some of these assays have been adapted to commercial use. It is also important to note that many of these assays have several applications within the same plant system, and for some specialized uses, e.g. immunohistology, the enzyme is routinely substituted with a fluorescent molecule (Saunders et al., 1983).

HORMONES

One of the major areas which is currently receiving a great deal of attention is that of plant hormones. Until the development of immunoassays for these compounds, it was generally agreed that the knowledge of the occurrence and levels of hormones in plants was surprisingly lacking. The data available was mainly restricted to average levels which were determined in whole shoots, seeds or even whole plants. In addition, since it was necessary to extract and process large amounts of tissue in order to obtain measurable amounts of compounds, it was virtually impossible to do studies

Table 1. Enzyme immunoassays currently being utilized to measure and analyze plant constituents.

Antigen	Application of Assay	Anti-sera(1)	Assay Type(2)	Detection Limit	References
Quassin (terpene)	Detection, isolation, culture screening, distribution, quant.	pc	ic	5 ng, 5 pg	Robins et al., (1984a,b)
Phytochrome	Association with membranes	pc	ic	--	Jordan et al., (1984)
	Quant., R/FR form specificity	pc/mc	ic	0.2ng, --	Thomas et al., (1984a,b)
	Quant. in crude extracts	pc/mc	nc	100 pg	Shimazaki et al. (1983)
Lectins	Quant., distrib. developmental changes, plant part analysis	pc	nc	0.3 ng	Borrebaeck et al., (1983), Borrebaeck, (1984)
	Structural studies	mc	nc	--	Borrebaeck & Etzler, (1981)
Glyco-alkaloids	Quant., screening, qual. control	pc	ic	2 pg	Morgan et al., (1983)
Cell wall carbohydrates	Localization, pattern spec., changes during growth and development	mc	nc	--	Vreeland et al., (1982, 1984)
Cell surface molecules	Identification of mating type antigens	mc	nc	--	Kosfiszer et al., (1982)
Limonin (triterpene) U.S. pat.no. 4,305,923	Quant. in juice, plant extracts, localization, industrial quality control biosynthesis	pc	c	0.15 pmol	Weiler et al., (1984), Jourdan et al., (1984), Mansell et al., unpub. res.
Paeoniflorin	Screen biological fluids	pc	c	1 ng	Kanoaka et al., (1984)
Glycyrrhizin	Study metab. by humans	pc	c	0.2 ng	Kanoaka et al., (1983)

(continued)

Table 1. (continued)

Antigen	Application of Assay	Anti-sera(1)	Assay Type(2)	Detection Limit	References
Glycyrretic acid	Screen biological fluids	pc	c	2.5 ng	Kanoaka et al., (1981)
Naringin (flavonoid)	Quant. in juice, plant extracts, localization	pc	c	--	Mansell et al., unpub. res.
Soya protein	Quant. in food, quality control	pc	nc	0.1 ug	Hitchcock et al., (1981) Griffiths & Hitchcock, (1984)
NADP-Malate Dehydrogenase	Localization, culture screening	pc	nc	--	Perrot-Rechenmann et al., (1983)
Sorbital-6-P-Dehydrogenase	Physiological, field studies.	pc	nc	0.5 ng	Hirai (1983)
Nitrate Reductase	Quant., determination of rates of biosyn. and turnover; char. of mutants	pc	nc	1 ng	Campbell & Ripp, (1984), Narayan et al., (1983)
Zein	Quant., deter. structural relatedness	pc	nc	1 ng	Conroy & Esen (1984)
L-Phenylalanine ammonia lyase	Culture screening	pc	nc	--	Bolwell et al., (1985)
Phosphoglucose Isomerase	Deter. struct. relatedness, species specificity	pc	c	--	Higgens & Gottlieb, (1984)

Antigen	Application of Assay	Anti-sera(1)	Assay Type(2)	Detection Limit	References
Lipoxygenase	Test activ. and specificity of IgG fractions, local. in tissues of germ. seeds	pc	nc	--	Vernooy-Gerritsen et al., (1983)
HORMONES					
Abscisic acid	Quant., culture screening	pc	c	50 fmol	Weiler, (1982a); Daie & Wyse (1982)
Brassinolide	Quant., distrib. in plants	mc	nc	--	Horgen et al., (1984)
Cytokinins	Quant., develop. changes distrib.	pc mc pc	c c c	50 pg 0.1 pmol 0.3 pmol	Barthe & Stewart, (1985) Eberle et al, (1986) Hansen et al., (1983)
Gibberellins	Detection, quant.	pc	c	0.5 fmol	Atzorn & Weiler, (1983)
Indole-3-acetic acid	Detection, quant., distrib. physiol. changes	pc mc	c c	3 pg 0.5 pmol	Weiler et al., (1981) Weigel et al., (1984) Mertens et al., (1985)

(1) pc = polyclonal, mc = monoclonal
(2) c = competitive, nc = noncompetitive, ic = indirect competitive

151

on short distance transport, daily or hourly concentration fluctuations and physiological changes which triggered or accompanied the subsequent morphological responses.

Since the late 1970ᶠs there has been rapid development of immunoassays directed toward plant growth hormones and most of this work has been accomplished in relatively few laboratories (see Weiler, 1984 and Weiler et al., 1986 for comprehensive reviews). The early assays were radiolabel-based but over the past several years a superior EIA has also been developed for each of these compounds. The sensitivity of the EIAs for these compounds (currently in the fmol range) enables them to be used to study the distribution in plants, changes in levels through development, physiological changes, and levels and changes in plant tissue cultures. Importantly, it must also be realized that the development of these non-isotopic assays has made this type of technology available to nearly every research laboratory which, for one reason or another, was unable to utilize the radiation based assays.

In several of the immunoassays developed for plant hormones, some degree of sample purification is necessary to remove cross-reacting or interfering compounds which are extracted along with the hormone in question. However, the sensitivity of the EIA coupled with its processing capacity, speed, and cost-effectiveness far outweighs the effort expended in sample preparation. In some of these assays, prepurification of samples by HPLC or TLC enables more than one substance to be quantitated by the same antibody preparation (Barthe and Stewart, 1985; Hansen et al., 1984). In these situations, it is quite likely that the separation step is both easier and more practical than trying to develop separate assays specific for each compound (Weiler, 1982b; Jourdan et al. 1985). In addition, as more and more monoclonal antibody based assays are developed, the problems associated with these interfering substances should diminish.

On the other hand, Arens and Zenk (1980) succeeded in the production of antisera either highly selective for lysergic acid or reactive with a whole range of lysergic acid derivatives and clavines. This was done by immunizing against lysergic acid-protein conjugates differing in the sites chosen for attachment of the antigen. The latter of these has proven to be useful in selection of cell lines which synthesize at least some derivative of this compound. Thus, by appropriate antibody choice, the researcher can choose whether to examine an entire molecule, part of a molecule or a group of related substances.

The region of the antigen used for conjugation to the carrier protein (normally BSA or HSA) can have a profound effect on the specificity of the antisera produced and therefore on the assay itself. For example, an RIA for abscisic acid developed by Weiler (1980) utilized a conjugate made through the carbonyl group of ABA and the resulting assay recognized only the free form of abscisic acid. Subsequent assays developed by Weiler (1982a) and Daie and Wyse (1982) were based on an immunogen conjugate in which ABA was linked through its carboxyl group. The resulting serum yielded an assay which could be used to measure both free cis-(+)-abscisic acid and its naturally occurring C-1 conjugates (esters). As a consequence, this assay permits the measurement of both free ABA and the glucosyl ester without the need for hydrolysis of the latter. If each compound is to be measured separately, then a chromatographic resolution of the two is needed.

PHYTOCHROME

The utilization of EIA for the study of phytochrome is another rapidly expanding research area. Studies on phytochrome have been greatly hampered by the limitations of spectrophotometric assays, eg. chlorophyll interference (Pratt, 1978). Radioimmunoassays developed for phytochrome partially alleviated these problems, however, isotope based assays were not totally ideal for many types of studies since there was often interference (quenching and/or chemi-luminescence) from other compounds found in crude extracts (Hunt and Pratt, 1979; Thomas et al., 1984a).

Both monoclonal and polyclonal antibody-based EIA systems have been developed to measure phytochrome and each has proved to be useful for specific strategies. In 1983, Shimazaki et al. reported on the development of an ELISA for the quantification of phytochrome in crude extracts of Avena. This assay utilized polyclonal antibodies produced in rabbits. The serum was first purified by affinity chromatography on a column to which phytochrome had been bound. The antibodies were eluted and used to coat plastic wells. The phytochrome samples were then immunoabsorbed and after washing, four monoclonal antibodies (produced in mice and specific for some part of the phytochrome molecule) were applied. Phytochrome was then quantitated by the addition of rabbit anti-mouse IgG-alkaline phosphatase conjugate and a correction factor for interference from other compounds in the crude extracts was determined.

Thomas et al. (1984a) developed an ELISA for Avena phytochrome which utilized polyclonal antibodies and the antibody population revealed that there is an immunological difference between phytochromes from dark-grown and light-grown plants. This property allowed the authors to "identify antigenic heterogeneity in phytochrome from different sources". This same assay was also utilized to study phytochrome-membrane associations (Jordan et al., 1984). It was found that membrane-associated and soluble phytochrome react differently with the assay (give different immunological responses) and that dissociation of phytochrome from the membranes results in a response identical to that of soluble phytochrome. This ELISA, based on a polyclonal antibody preparation, was not able to differentiate between the red (Pr) and far-red (Pfr) absorbing forms of phytochrome even though an elaborate sample purification protocol was used for this work.

An ELISA using monoclonal antibodies to phytochrome was subsequently developed which does discriminate between Pr and Pfr (Thomas et al., 1984b). Eight monoclonal antibody lines (produced in rats) were established of which one was specific for Pfr, two were specific for Pr, and the remaining five did not discriminate between the two forms. This type of analytical work has great potential and it is likely that these types of assays will ultimately lead to the resolution of the area(s) of the phytochrome molecule affected by its photoconversion. In a related study, the relatedness of phytochrome in eight different plant species is being studied using the cross-reactivity values of monoclonal antibodies raised against either pea or rye phytochrome (Saji et al., 1984). Importantly, these monoclonal based systems distinguish between different surface structures of phytochrome and thus can be successfully employed as molecular probes (Nagatani et al., 1984). Studies such as these should provide important phylogenetic information about this physiologically important molecule.

COMMERCIAL ADAPTATION

Phytohormones

EIA kits for indole-3-acetic acid, abscisic acid, trans-zeatin riboside, and dihydrozeatin riboside are presently available commercially (IDETEK, Inc., San Bruno). These assays utilize monoclonal antibody systems developed by Prof. E.W. Weiler (W.Germany) and can be used for the quantification and localization of these plant growth regulating compounds. Bulk antibody is also available and can be utilized to isolate (via immunoaffinity chromatography) these hormones from low-level sources (eg. culture media, diffusates, etc.) prior to quantification by EIA. Other monoclonal antibody EIA kits for the quantification of gibberellins are being developed and hopefully these will be available in the not-too-distant future.

Since so much confusion and controversy exists in the plant hormone field, it must be considered a major achievement to have assay systems which are based on the same principle, e.g. immunoassay, and moreover it is critical that the antibodies used in these tests be derived from the same source. This will permit the direct comparison of data between laboratories and it will provide an assay system to which future studies can be standardized worldwide. Since the cost of these immunoreagents is quite low, more laboratories should be able to utilize these systems and thus results and new discoveries should come more rapidly.

Limonin

Limonin is a tetracyclic triterpenoid dilactone which occurs in Citrus sp. and is of importance because of its extremely bitter taste. The chemistry of this molecule makes it impossible to quantitate by standard laboratory techniques, e.g. spectrophotometry. In such cultivars as Navel and Shamouti orange, the presence of this compound causes many economic and organoleptic problems and greatly affects the taste quality of processed fruit (Maier et al., 1980). Limonin is also prevalent in grapefruit but the intrinsic quality of this fruit is further complicated by the presence of naringin, a bitter flavanone neohesperidoside (Rouseff, 1980).

The problems associated with the lack of adequate measuring methods were partially resolved with the development of an RIA for this compound (Mansell and Weiler, 1980a; Weiler and Mansell, 1980), however, since this assay was based on the use of isotopes, it proved to be inappropriate for many of its intended uses in juice production and quality control. Food processing plants are not able to employ radiation techniques as standard protocols and thus we developed two different EIA systems, one of which can be used in a quality control laboratory (Jourdan et al., 1984) and the other which is more appropriate for a research and development facility (Weiler et al., 1984). For a comprehensive discussion on immunological tests for the evaluation of citrus quality see Mansell and Weiler (1980b).

Since the organoleptic response to limonin levels in juice is characterized by a very narrow concentration range [e.g., non-bitter = 1 ppm; bitter = 5 ppm; very bitter = 10 ppm (Maier et al., 1977)], the assay developed for quality control is characterized by having a measuring range of between two and 15 parts per million. This test is being produced by IDETEK, Inc. and is presently being utilized for juice blending, quality control and in the determination of product contamination. The assay is constructed in such a way that all dilutions are done in a single step. The use of a forced air incubator and vertical light path microstrip reader allows the assay to be semi-automated. Forty-plus samples can be analyzed per plate and the entire assay can be done in about 90 minutes. Results can be calculated directly from the linear standard curve and software exists for automated analyses.

The second EIA for limonin utilizes the longer measuring range for which immunoassays are noted and this assay has also been characterized for both short and long time periods. Both assays have nearly identical characteristics and since they are developed from the same antibody preparation, the results are comparable.

154

DISCUSSION AND CONCLUSION

In addition to the EIA for limonin we have also developed several radioimmunoassays for naringin (Jordan et al.,1982; Jordan et al., 1983; Jordan et al., 1985) the other bitter substance which is abundant in grapefruit. An EIA for this compound has also been developed (see Table I), but as yet, has not been fully characterized and optimized for commercial use. The first enzyme-linked assay was done with a second antibody precipitation but during our early solid-phase experimentation, we encountered technical problems based on the interaction of the antigen with the plastic surface. Studies to circumvent this problem are in progress but this illustrates again that immunoassays must be carefully designed and manipulated in order to maximize their use. Care must be taken to assure that all reagents of the assay will function appropriately and extraordinary effort must be given to avoid cross-reactivity and false positives. Problems with binding to solid surfaces and stability of both the antibody and antigen must be anticipated, however, each of these can be overcome by the use of creative assay design.

One additional feature of the use of EIA for the study of plant constituents is the fact that many of these assays can be conducted in the presence of relatively harsh solvents. Robins et al. (1984a,b) noted that quassin, a triterpene, could be assayed in the presence of 20 per cent methanol without affecting antibody behavior. Conroy and Esen (1984) found that the maize prolamin, zein remained an active antigen after solubilization in 6-8 M urea and they suggested that many proteins or molecules which are insoluble or rendered insoluble in aqueous buffers might still react in an antigen-antibody complex.

The use of EIA in plant research is a rapidly growing field which has yet to achieve its potential. Numerous areas await future utilization of antibodies and the application of this technology is restricted only by the imagination of those involved. Early indications point to the use of EIA based systems as an aid to taxonomy of plant families, genera and even species (Aldwell et al., 1985; Schneider and Leidgens, 1981) and eventually it should be possible to develop an assortment of antibodies which will permit the immuno-typing of taxa. Additional uses include the localization of genes on chromosomes (Rennard et al., 1981) and an analysis of membranes and organelles. These latter studies should evolve to the point where they can be used to study the synthesis and assembly of proteins and other molecules into membranes (Gold et al., 1985). Such information would greatly clarify the nature of membrane architecture and the dynamics of turnover.

Another current research area is the characterization of receptors for plant hormones. These studies should yield important information which will clarify our understanding of the regulation and mode of action of hormones which takes place during plant development. While some receptor work has been done on plant material, this research has not progressed as far as that of animal studies. Consequently, this area of plant science lacks the level of understanding which animal hormone research has produced. Studies currently being pursued (Weiler, 1986) include an analysis of the specific location of receptors, analysis of the nature of the hormone binding site and comparative analyses of different membrane systems. By combining the EIA with immunofluorescence, a combination of parameters can be monitored during the binding and release of physiologically active molecules and membranes.

Regardless of the utilization of any of these assay systems, it must be fully realized that plant chemistry is exceedingly diverse, therefore, each of the assays developed must be fully characterized relative to the plant being studied. The presence of interfering or cross-reacting materials must be thoroughly investigated and this requirement becomes more critical if the assay developed for one plant species is to be utilized with unrelated taxa (McLaughlin and Barnett, 1979; Sandberg et al., 1985).

REFERENCES

Aldwell, F.E.B., Hall, I.R. and J.M.B. Smith. 1985. Enzyme-linked immunosorbent assay as an aid to taxonomy of the Endogonaceae. Trans. Br. Mycol. Soc. 84: 399-402.

Arens, H. and M.H. Zenk. 1980. Radioimmunoassays for the determination of lysergic acid and simple lysergic acid derivatives. Planta Medica 39: 336-347.

Atzorn, K. and E.W. Weiler. 1983. The immunoassay of gibberellins. II. Quantitation of GA3, GA4, and GA7 by ultra-sensitive enzyme immunoassays. Planta 159: 7-11.

Barthe, G.A. and I. Stewart. 1985. Enzyme immunoassay (EIA) of endogenous cytokinin in Citrus. J. Agric. Food Chem. 33: 293-297.

Bolwell, G.P., Bell, J.N., Cramer, C.L., Schuch, W., Lamb, C.J. and R.A. Dixon. 1985. L-phenylalanine ammonia lyase from Phaseolus vulgaris - characterization and differential induction of multiple forms from elicitor-treated cell suspensions. Eur. J. Biochem. 149: 411-419.

Borrebeack, C.A.K. 1984. Detection and characterization of a lectin from non-seed tissues of Phaseolus vulgaris. Planta 161: 223-228.

Borrebaeck, C.A.K. and B. Mattiasson. 1983. Distribution of a lectin in tissues of Phaseolus vulgaris. Physiol. Plant. 58: 29-32.

Borrebeack, C.A.K. and M.E. Etzler. 1981. Production and characterization of a monoclonal antibody against the seed lectin of the Dolichos biflorus plant. J. Biol. Chem. 256: 4723-4725.

Boudene, C., Duprey, F.and C. Bohuon. 1975. Radioimmunoassay of colchicine. Biochem. J. 151: 413-415.

Campbell, W.H. and K.G. Ripp. 1984. An elisa for higher plant nitrate reductase. Annals N.Y. Acad. Sci. 435: 123-125.

Conroy, J.M. and A. Esen. 1984. An enzyme-linked immunosorbent assay for zein and other proteins using unconventional solvents for antigen adsorption. Anal. Biochem. 137: 182-187.

Daie, J. and R. Wyse. 1982. Adaptation of the enzyme-linked immunosorbent assay (ELISA) to the quantitative analysis of abscisic acid. Anal. Biochem. 119: 365-371.

Eberle, J., Arnscheidt, A., Klix, D. and E.W. Weiler. 1986. Monoclonal antibodies to plant growth regulators III. zeatinriboside and dihydrozeatinriboside. Plant Physiol. In Press.

Engvall, E. and P. Perlmann. 1971. Enzyme-linked immunosorbent assay (ELISA) quantitative assay of immunoglobulin G. Immunochem. 8: 871-874.

Engvall, E. and A.J. Pesce. 1978. Quantitative enzyme immunoassay. Suppl. #7, Scan. J. Immunol. Blackwell Scientific Publications, (Oxford). 129 pp.

Erlanger, B.F. 1980. The preparation of antigenic hapten-carrier conjugates: a survey. in Methods in Enzymology, ed. by H. van Vunakis and J.J. Langone. Academic Press, N.Y. vol 70(a): 85-104.

Gold, P., Lewis, M., Mazzarella, R. and M. Green. 1985. An enzyme-linked immunoassay for the detection of antibodies to endoplasmic reticulum. Anal. Biochem. 146: 82-89.

Griffiths, N.M., Billington, M.J., Crimes, A.A. and C.H.S. Hitchcock. 1984. An assessment of commercially available reagents for an enzyme-linked immunosorbent assay of soya protein in meat products. J. Sci. Food Agric. 35: 1255-1260.

Hansen, C.E., Wenzler, H. and F. Meins, Jr. 1984. Concentration gradients of trans-zeatin riboside and trans-zeatin in the maize stem. Plant Physiol. 75: 959-963.

Higgins, R.C. and L.D. Gottlieb. 1984. Subunit hybridization and immunological studies of duplicated phosphoglucose isomerase isozymes. Biochem. Genet. 22: 957-979.

Hirai, M. 1983. Seasonal changes in sorbitol-6-phosphate dehydrogenase in loquat leaf. Plant & Cell Physiol. 24: 925-931.

Hitchcock, C.H.S., Bailey, F.J., Crimes, A.A., Dean, D.A.G. and P.J. Davis. 1981. Determination of soya proteins in food using an enzyme-linked immunoassay procedure. J. Sci. Food Agric. 32: 157-165.

Horgen, P.A., Nakagawa, C.H. and R.T. Irvin. 1984. Production of monoclonal antibodies to a steroid plant growth regulator. Can. J. Biochem. Cell Biol. 62: 715-721.

Hunt, R.E. and L.H. Pratt. 1979. Phytochrome radioimmunoassay. Plant Physiol. 64: 327-331.

Jordan, B.R., Partis, M.D. and B. Thomas. 1984. A study of phytochrome-membrane association using an enzyme-linked immunosorbent assay and western blotting. Physiol. Plant. 60: 416-421.

Jourdan, P.S., E.W. Weiler and R.L. Mansell. 1985. Naringin levels in citrus tissues. I. Comparison of different antibodies and tracers for the radioimmunoassay of naringin. Plant. Physiol. 77: 896-902.

Jourdan, P.S., Mansell, R.L., Oliver, D.G. and E.W. Weiler. 1984. Competitive solid phase enzyme-linked immunoassay for the quantification of limonin in Citrus. Anal. Biochem. 138: 19-24.

Jourdan, P.S., Weiler, E.W. and R.L. Mansell. 1983. Radioimmunoassay for naringin and related flavanone 7-neohesperidosides using a tritiated tracer. J. Agric. Food Chem. 31: 1249-1255.

Jourdan, P.S., Mansell, R.L. and E.W. Weiler. 1982. Radioimmunoassay for the citrus bitter principle, naringin, and related flavonoid-7-O-neo-hesperidosides. Planta Med. 44: 82-86

Kanoaka, M., Yano, S., Kato, H., Nakanishi, K. and M. Yoshizaki. 1984. Studies on the enzyme immunoassay of bio-active constituents contained in Oriental medicinal drugs. III. Enzyme immunoassay of paeoniflorin, a constituent of Chinese paeony root. Chem. Pharm. Bull. 32: 1461-1466.

Kanoaka, M., Yano, S., Kato, H., Nakao, N. and E. Kinoshita. 1983. Studies on the enzyme immunoassay of bio-active constituents contained in Oriental medicinal drugs. II. Enzyme immunoassay of glycyrrhizin. Chem. Pharm. Bull. 31: 1866-1873.

Kanoaka, M., Yano, S., Kato, H. and N. Nakano. 1981. Glycyrrhetylamino acids: synthesis and application to enzyme immunoassay for glycyrrhetic acid. Chem. Pharm. Bull. 29: 1533-1538.

Kosfizer, M., Clausell. A., Imam, S., Lavery Jr., B., Border, B., Witte, P. and W.J. Snell. 1982. A monoclonal antibody that blocks adhesion of mt+ gametes of Chlamydomonas. J. Cell Biol. 95: 71a.

Landsteiner, K. 1945. The specificity of serological reactions. Harvard University Press,(Cambridge, MA).

Langone, J.J., Gjika, H.B. and H. Van Vunakis. 1973. Nicotine and its metabolites. Radioimmunoassays for nicotine and cotinine. Biochem. 12: 5025-5030.

Maggio;, E.T. 1980. Enzyme-immunoassay. CRC Press, Inc.(Boca Raton, FL). 295 pp.

Maier, V.P., Hasegawa, S., Bennett, R.D., and L.C. Echols. 1980. Limonin and limonoids: chemistry, biochemistry, and juice bitterness. in Citrus nutrition and quality, ed. by S. Nagy and J.A. Attaway. ACS Symposium Series 143: 63-82.

Maier, V.P., Bennett, R.D. and S. Hasegawa. 1977. Limonin and other limonoids. in Citrus science and technology, ed. by S. Nagy, P.E. Shaw and M.K. Veldhuis, vol. 2: 482. Avi Publishing, Westport.

Mansell, R.L. and E.W. Weiler. 1980a. Radioimmunoassay for the
    determination of limonin in Citrus. Phytochem. 19: 1403-1407.
Mansell, R.L. and E.W. Weiler. 1980b. Immunological tests for the
    evaluation of citrus quality, in Citrus nutrition and quality, ed.
    by S. Nagy and J.A. Attaway. ACS Symposium Series 143: 341-359.
McLaughlin, M.R. and O.W. Barnett. 1979. The influence of plant sap and
    antigen buffer additives in the enzyme-immunoassay of two plant
    viruses. Phytopath. 69: 1038.
Mertens, R., Eberle, J., Arnscheidt, A., Ledebur, A. and E.W. Weiler. 1985.
    Monoclonal antibodies to plant growth regulators. II. Indole-3-acetic
    acid. Planta 166: 389-393.
Morgan, M.R.A., McNerney, R., Matthew, J.A., Coxon, D.T. and H.W.S. Chan.
    1983. An enzyme-linked immunosorbent assay for total glycoalkaloids
    in potato tubers. J. Sci. Food Agric. 34: 593-598.
Nagatani, A., Yamamoto, K.T., Furuya, M., Fukumoto, T. and A. Yamashita.
    1984. Production and characterization of monoclonal antibodies which
    distinguish different surface strutures of pea (Pisum sativum cv.
    Alaska) phytochrome. Plant & Cell Physiol. 25:1059-1068.
Narayanan, K.R., Somers, D.A., Kleinhofs, A. and R.L. Warner. 1983. Nature
    of cytochrome c reductase in nitrate reductase-deficient mutants in
    barley. Mol. Gen. Genet. 190: 222-226.
Perrot-Rechenmann, C., Jacquot, J.P., Gadal, P., Weeder, N.F., Cseke, C. and
    B.B. Buchanan. 1983. Localization of NADP-malate dehydrogenase of
    corn leaves by immunological methods. Plant Sci. Lett. 30: 219-226.
Pratt, L.H. 1978. Molecular properties of phytochrome. Photochem.
    Photobiol. 27: 81-105.
Rennard, S.I., Church, R.L., Rohrbach, D.H., Shupp, D.E., Abe, S., Hewitt,
    A.T., Murray, J.C. and G.R. Martin. 1981. Localization of the human
    fibronectin (FN) gene on chromosome 8 by a specific enzyme
    immunoassay. Biochem. Genet. 19: 551-566.
Robins, R.J., Morgan, M.R.A., Rhodes, M.J.C. and J.M. Furze. 1984a. An
    enzyme-linked immunosorbent assay for quassin and closely related
    metabolites. Anal. Biochem. 136: 145-156.
Robins, R.J., Morgan, M.R.A., Rhodes, M.J.C. and J.M. Firze. 1984b.
    Determination of quassin in picogram quantities by an enzyme-linked
    immunosorbent assay. Phytochem. 23: 1119-1123.
Rouseff, R.L. 1980. Flavonoids and citrus quality. in Citrus nutrition and
    quality, ed. by S. Nagy and J.A. Attaway. ACS Symposium Series 143:
    83-108.
Saji, H., A. Nagatani, K.T. Yamamoto, M. Furuya, T. Fukumoto and A.
    Yamashita. 1984. Cross-reactivity of monoclonal antibodies against
    rye and pea phytochrome with phytochromes extracted from eight
    different plant species. Plant Sci. Lett. 37: 57-61.
Sandberg, G., Ljung, K. and P. Alm. 1985. Precision and accuracy of
    radioimmunoassay in the analysis of endogenous 3-indoleacetic acid
    from needles of Scots pine. Phytochem. 24:1439-1442.
Saunders, M.J., Cordonnier, M.M., Palevitz, B.A. and L.H. Pratt. 1983.
    Immunofluorescence visualization of phytochrome in Pisum sativum L.
    epicotyls using monoclonal antibodies. Planta 159: 545-553.
Schneider, H.A.W. and W. Leidgens. 1981. An evolutionary tree based on
    monoclonal antibody-recognized surface features of a plastid enzyme
    (5-aminolevulinate dehydratase). Z. Naturforsch. 36: 44-50.
Shimazaki, Y., M.-M. Cordonnier and L.H. Pratt. 1983. Phytochrome
    quantitation in crude extracts of Avena by enzyme-linked
    immunosorbent assay with monoclonal antibodies. Planta 159: 534-544.
Spector, S. 1971. Quantitative determination of morphine in serum by
    radioimmunoassay. J. Pharmacol. Exp. Ther. 178: 253-258.
Thomas, B., Crook, N.E. and S.E. Penn. 1984a. An enzyme-linked
    immunosorbent assay for phytochrome. Physiol. Plant. 60: 409-415.

158

Thomas, B., Penn, S.E., Butcher, G.W. and G. Galfre. 1984b. Discrimination between the red- and far-red-absorbing forms of phytochrome from Avena sativa L. by monoclonal antibodies. Planta 160: 382-384.

van Weemen, B.K. and A.H.W.M. Schuurs. 1971. Immunoassay using antigen-enzyme conjugates. FEBS Lett. 15: 232-236.

Vernooy-Gerritsen, M., Bos, A.L.M., Veldink, G.A. and J.F.G. Vliegenthart. 1983. Affinity chromatography of antibodies directed against soybean lipoxygenase-1 and -2 and an enzyme-linked immunosorbent assay (ELISA) for antibodies and lipoxygenases. Biochim. Biophys. Acta 748: 148-152.

Vora, S. 1985. Monoclonal antibodies in enzyme research: present and potential applications. Anal. Biochem. 144: 307-318.

Vreeland, V., Slomich, M. and W.M. Laetsch. 1984. Monoclonal antibodies as molecular probes for cell wall antigens of the brown alga, Fucus. Planta 162: 506-517.

Vreeland, V., Larsen, B. and W.M. Laetsch. 1982. Monoclonal antibodies to Fucus cell wall antigens: localization and specificity patterns. J. Cell Biol. 95: 127a.

Weigel, U., Horn, W. and B. Hock. 1984. Endogenous auxin levels in terminal stem cuttings of Chrysanthemum morifolium during adventitious rooting. Physiol. Plant. 61: 422-428.

Weiler, E.W. 1986. Personal communication.

Weiler, E.W. 1984. Immunoassay of plant growth regulators. in Annual Review of Plant Physiology ed. by W.R. Briggs, R.L. Jones and V. Walbot. Annual Reviews (Palo Alto) pp. 85-95.

Weiler, E.W. 1983. Immunoassay of plant consituents. Biochem. Soc. Trans. 11: 485-495.

Weiler, E.W. 1982a. An enzyme immunoassay for cis-(+)-abscisic acid. Physiol. Plant. 54: 510-514.

Weiler, E.W. 1982b. Plant hormone immunoassay. Physiol. Plant. 54: 230-234.

Weiler, E.W. 1980. Radioimmunoassay for the differential and direct analysis of free and conjugated abscisic acid in plant abstracts. Planta 148: 262-272.

Weiler, E.W., Eberle, J., Mertens, R., Atzorn, R., Feyerabend, M., Jourdan, P.S., Arnscheidt, A. and U. Wieczorek. 1986. Antisera- and monoclonal antibody-based immunoassays of plant hormones. in Immunology in Plant Science ed. by T.L. Wang. Cambridge University Press, pp 27-58.

Weiler, E.W., Jourdan P.S. and R.L. Mansell. 1984. Peroxidase-linked, solid-phase enzyme immunoassay for the determination of picomole levels of limonin. Plant Sci. Lett. 35: 159-167.

Weiler, E.W., Jourdan, P.S. and W. Conrad. 1981. Levels of indole-3-acetic acid in intact and decapitated coleoptiles as determined by a specific and highly sensitive solid-phase enzyme immunoassay. Planta 153: 561-571.

Weiler, E.W. and R.L. Mansell. 1980. Radioimmunoassay of limonin using a tritiated tracer. J. Agric. Food Chem. 28: 543-545.

Weiler, E.W. and M.H. Zenk. 1976. Radioimmunoassay for the determination of digoxin and related compounds in Digitalis lanata. Phytochem. 15: 1537-1545.

SECTION II

LUMINESCENCE IMMUNOASSAY

PHYCOBILIPROTEINS AS LABELS IN IMMUNOASSAY

Mel N. Kronick

Applied Biosystems

850 Lincoln Centre Drive, Foster City, CA   94404

INTRODUCTION

Fluorescence has for many years been viewed as a potential means of
providing high sensitivity in a non-isotopic test format (Soini and
Hemmila, 1979; Hemmila, 1985). As Jolley et al. (1984) have clearly
pointed out, the number of photons emitted from a sample of
fluorescently tagged molecules can exceed the number of gamma ray
photons emitted from the same number of gamma-ray-emitting tagged
molecules. The reason for this is two fold: each radioactive molecule
can emit only once and it emits its photon only when it naturally and
spontaneously decays. Nevertheless, high sensitivity assays using
fluorescence have not replaced radioactive assays to any great extent.
Non-isotopic versions of most high sensitivity assays are run today
using enzymes as labels. Fluorescence immunoassay has penetrated the
marketplace but only in certain applications. One technique,
fluorescence polarization (Dandliker and Saussure, 1970; Jolley et al.,
1981), has proven very useful, but only for small molecules such as
drugs which are present at concentrations from approximately 10^{-6} to
10^{-10} M. The dynamic range inherent in fluorescence polarization assays
in a competition assay configuration result in this limitation and the
physics of fluorescence polarization prevents its being adapted easily
to immunometric assays or even to competition assays for larger (≥ 1000
molecular weight) molecules. Assays based upon fluorescence quenching
or fluorescence energy transfer (Ullman et al., 1976; Ullman and Khanna,
1981) similarly have proven most useful to date only in competition
assays for analytes of concentration greater than about 10^{-10} M.

The primary reason for the limited application for fluorescence
immunoassay has been outlined in the often cited review article by Soini
and Hemmila (1979): Conventional fluorescent probes emit photons in the
same spectral region where most endogenous materials fluoresce. Thus,
even though the fluorescently tagged molecules of interest may emit a
very large and measureable number of photons, the background does also,
and discrimination of signal from background becomes difficult or
impossible. Efforts to address this shortcoming have taken on a variety
of forms. The use of time-resolved fluorescence attempts to minimize
background by using fluorescent probes with long excited state lifetimes
so that the time at which the signal appears can be used to discriminate
signal from background (Soini and Hemmila, 1979; Hemmila, 1985). Much

work has gone into the chemical synthesis of dyes which emit in the red end of the spectrum (Hemmila, 1985) although few developments in this field have made significant inroads into solving the background problem. Hirschfeld and others have attempted to use synthetic polymers of dyes to increase the effective extinction coefficient but in doing so have had to contend with significant non-specific binding and quenching problems (Hirschfeld, 1976; Hassan et al., 1979). A discussion of the drawbacks and limitations of current techniques for marking cell surface molecules with red-emitting dyes led to a collaboration of workers in several different fields to develop a new set of tags based upon certain molecules, the phycobiliproteins, that have existed in nature for millions or possibly billions of years (Oi et al., 1982; Stryer et al., 1985). The realization and subsequent demonstration that phycobiliproteins could be used as fluorescent tags has opened up new opportunities for fluorescence techniques in immunoassay. As discussed in detail below, the extremely high extinction coefficients and quantum yields of phycobiliproteins, along with their other chemical and physical properties, have made possible several applications that just were not possible using conventional fluorescent tags.

Properties of Phycobiliproteins

Phycobiliproteins are a class of proteins containing bilin prosthetic groups which act as efficient light harvesting antennae in a wide variety of red algae, blue-green algae, and cryptomonads (Glazer, 1981). In their native environments, the phycobiliproteins are grouped together in particles called phycobilisomes which are physically located near chlorophyll reaction centers (Gantt, 1980; Glazer, 1984; Glazer, 1985). In the phycobilisomes, the light absorbed is efficiently transferred, through a series of tightly coupled intermolecular energy transfers, to the chlorophyll and hence there is no significant fluorescence. When released from the phycobilisomes and purified, however, phycobiliproteins are highly fluorescent and can exhibit quantum yields in excess of 0.8 (Oi et al., 1982).

There are three main classes of phycobiliproteins: phycoerythrin, phycocyanin, and allophycocyanin. Variant forms are denoted with prefix letters that have some historic roots, (e.g., R-phycoerythrin for the variant from red algae) but are now somewhat arbitrary. Examples of the spectral and physical properties of a representative sample of these proteins are shown in Figure 1 and Table 1. Comparison of these properties with the data on conventional dyes (Hemmila, 1985) demonstrate that the phycobiliproteins can exhibit an extinction coefficient as much as thirty times that of a good dye like fluorescein with no sacrifice in quantum yield. Many of the phycobiliproteins have been studied extensively. The complete amino acid sequences of allophycocyanin (Sidler et al., 1981), C-phycocyanin (Frank et al., 1978), phycoerythrocyanin (Fuglistaller et al., 1983), and C-phycoerythrin (Sidler et al., 1985) have been determined and a 3 Å resolution X-ray crystallographic structure for C-phycoerythrin has been reported (Schirmer et al., 1985). For R-phycoerythrin from Gastroclonium coulteri and B-phycoerythrin from Porphyridium cruentum, the amino acid sequences about all of the chromophore attachement sites have been determined and the exact structures of the peptide-linked bilins established (Klotz and Glazer, 1985; Lundell et al., 1984; Nagy et al., 1985; Schoenleber et al., 1984a; Schoenleber et al., 1984b). In all phycobiliproteins examined to date, the bilins are linked to the polypeptide through thioether bonds to cysteinyl residues. In general, phycobiliproteins are acidic and carry a negative charge at neutral pH. This negative charge and their generally hydrophilic character minimize non-specific binding that is often associated with the use of fluorescein as a label. The proteins typically consist of several

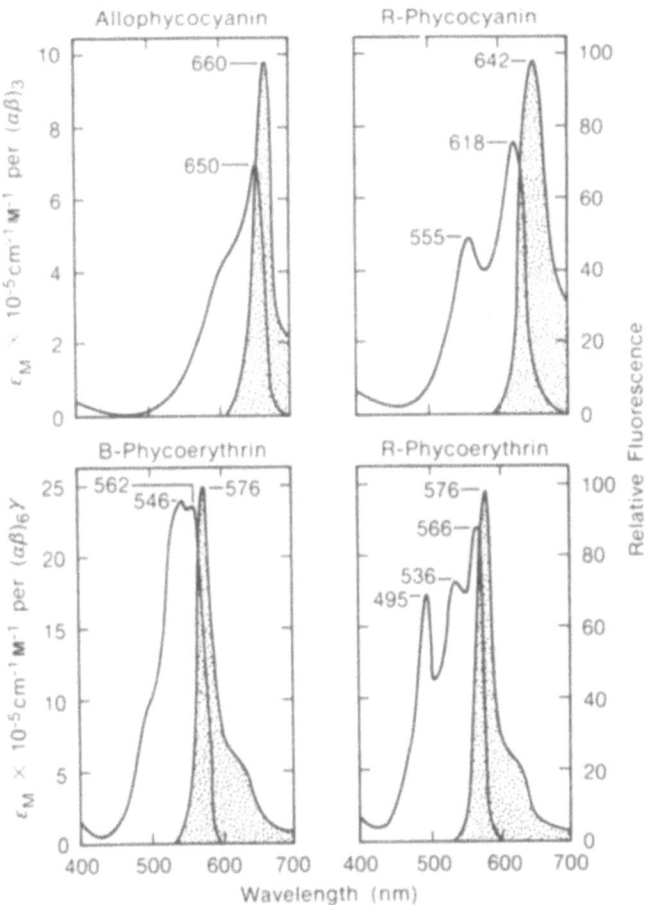

Fig. 1   Absorption and fluorescence emmission spectra of four re-
          presentative phycobiliproteins: allophycocyanin from the
          filamentous cyanobacterium, <u>Anabaena variabilis</u>, R-phycocyanin
          and B-phycoerythrin from the unicellular red alga, <u>Porphyridium</u>
          <u>cruentum</u>, and R-phycoerythrin from the higher red alga,
          <u>Gastroclonium coulteri</u>. Structural properties of these proteins
          are listed in Table 1. From Glazer and Stryer, (1984), by
          permission.

polypeptide chains with an exact structure which is very dependent on the particular protein, algal source, and the specific purification protocol used (Glazer, 1981). As an example, R-phycoerythrin from Gastroclonium coulteri consists of 13 separate amino acid chains of three distinct types as noted in Table 1.

Table 1. Properties of some major phycobiliproteins. From Glazer and Stryer (1984), by permission.

Protein	Subunit composition	Approx. mol. wt.	Bilins per subunit	Total bilins
Allophycocyanin	$(\alpha\beta)_3$	100,000	α 1 PCB; β 1 PCB	6
R-Phycocyanin	$(\alpha\beta)_3$	110,000	α 1 PCB; β 1 PCB; 1 PEB	9
B-Phycoerythrin	$(\alpha\beta)_6\gamma$	240,000	α 2 PEB; β 3 PEB; γ 2 PEB, 2 PUB	34
R-Phycoerythrin	$(\alpha\beta)_6\gamma$	240,000	α 2 PEB; β 2 PEB, 1 PUB; γ 1 PEB, 3 PUB	34

PCB = phycocyanobilin, PEB = phycoerythrobilin, PUB = phycourobilin.

The structures of the three major bilin pigments are shown in Figure 2 which also illustrates the most common mode of linkage to the polypeptide chains. The bilins are brought together in an extremely small volume but nature has controlled the distance, angle, and physical environment of each bilin such that the quantum yield is not lost to excitation traps. As noted by Glazer and Stryer (1984), when multiple conventional fluorescent molecules are brought together synthetically into clusters, they exhibit a large decrease in quantum yield. The bilins in phycobiliproteins are isolated from their external environment by virtue of the surrounding polypeptide chains. This isolation results in phycobiliprotein fluorescence being independent of pH over a broad pH range (approximately pH 5 to pH 9) and being immune to collisional quenching from most molecules or ions found in biochemical environments.

In addition to enormous molar extinction coefficients and high quantum yields, several qualitative features of the spectra of phyco-biliproteins have proven extremely useful in the utilization of phycobiliproteins as labels. Phycoerythrin, for example, has a very wide absorption band. This allows a significant Stokes shift, the separation of excitation and emission wavelengths, which can be very useful in discriminating phycoerythrin fluorescence from that of other dyes (fluorescein, for example) or from endogeneous background fluorescers. The ability to excite at the same wavelength two dyes which emit at two very different wavelengths allows the measurement of two independent molecular species on cell surfaces (Oi et al., 1982) or dissolved in solution (Houghton, 1985).

As has been pointed out by Sepaniak (1985) and others, laser excitation provides an exquisite means for excitation of a small number of molecules present in a small volume. Here again, the phycobili-proteins possess some advantageous features. Mathies and Stryer (1986) have, in fact, reported observation of just three molecules of phycoery-thrin. Phycoerythrin absorbs very efficiently at the output wavelengths of the argon ion laser (488 and 514 nm), the laser used in most flow cytometry systems. The new, green-emitting, helium-neon lasers at 543 nm

166

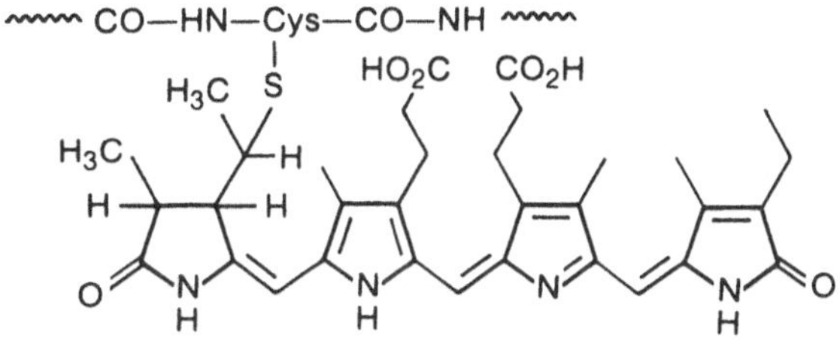

Peptide-linked PHYCOCYANOBILIN (PCB)

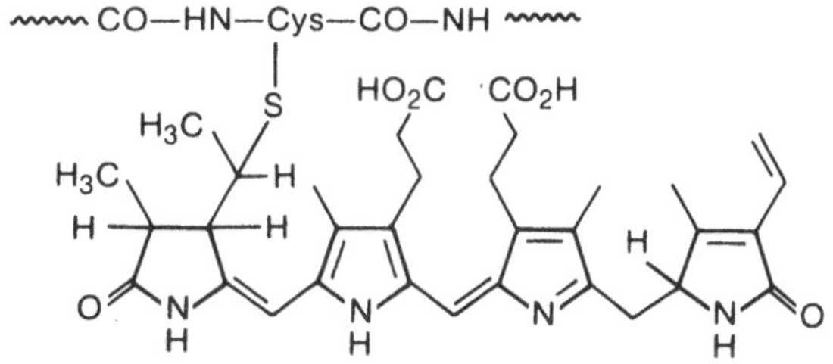

Peptide-linked PHYCOERYTHROBILIN (PEB)

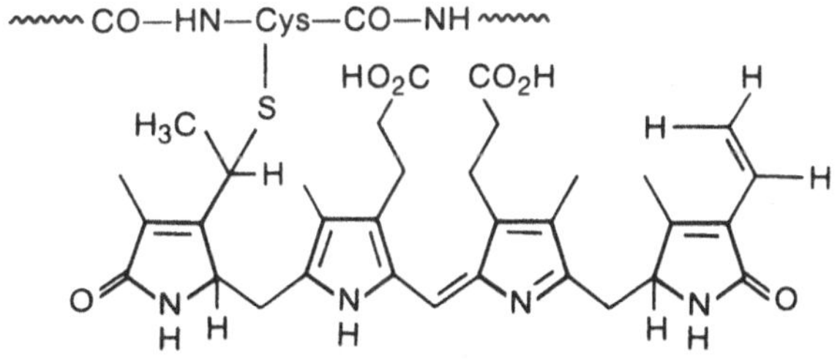

Peptide-linked PHYCOUROBILIN (PUB)

Fig. 2   Structures of the three most common phycobiliprotein prosthetic groups, phycocyanobilin, phycoerythrobilin, and phycourobilin, as linked to their respective peptide chains. Thioether linkages to ring A are the only ones shown. Bilins are also linked by thioether linkages to ring D, and by two such linkages to rings A and D in the case of certain phycoerythrobilin and phycourobilin groups (Glazer, 1985).

as well as the mercury arc emission line at 546 nm represent additional excitation sources for phycoerythrin. Allophycocyanin, similarly, shows great promise as a label for laser excitation because it will absorb efficiently at the 633 nm emission line of the inexpensive and reliable red-emitting helium-neon laser. Shapiro et al. (1983) have already demonstrated the use of allophycocyanin for flow cytometry using 633 nm helium-neon excitation. The use of allophycocyanin should also result in much less background in many applications because, as Soini and others have observed, very few endogenous molecules are excited or emit in the red end of the spectrum (Soini and Hemmila, 1979). In addition, the red absorbance and emission of allophycocyanin have been used for multi-color flow cytometry to measure as many as four independent cell parameters (Hardy et al., 1984). In this particular example, fluorescein and phycoerythin are excited by the 488 nm line of an argon ion laser and Texas Red and allophycocyanin are excited by the use of a rhodamine dye laser at 605 nm.

Finally, as demonstrated by Glazer and Stryer, the complementary absorption and emission characteristics, which are a consequence of the natural function of phycobiliproteins, allow the formation of synthetic species with an extremely long Stokes shift (Glazer and Stryer, 1983; Stryer and Glazer, 1985b). Glazer and Stryer used a covalent complex of phycoerythrin and allophycocyanin to produce a label that could be excited with an argon ion laser at 488 nm and emit at 660 nm.

Phycobiliproteins as Labels

Oi et al. (1982) performed a series of experiments which proved that phycobiliproteins could be used very successfully as covalently-coupled labels. These researchers demonstrated that the crosslinking chemistry did not destroy the spectral properties of the phycobili-proteins and that the labeled molecule, e.g., an antibody, avidin, a lectin or a drug, did not lose its particular chemical binding specificity when it was linked to a phycobiliprotein. Their work was accomplished with conventional heterobifunctional chemical crosslinking reagents (see below) that were already in common use for protein chem-istry. The phycobiliproteins were typically coupled through the ϵ-amino group of their many lysine side chains: B-phycoerythrin has an esti-mated 85 lysines per molecule and allophycocyanin has approximately 36 (Glazer and Stryer, 1984). The concept of phycobiliproteins as valuable fluorescent labels was thus established.

The widest and most immediate acceptance of phycobiliproteins as labels has been in cell labeling for fluorescent microscopy and flow cytometry. Flow cytometry, a technique complementary to microscopy, rapidly characterizes cells in suspension by flowing a stream of cells, substantially in a single file manner, past one or more interrogating light sources (Parks and Herzenberg, 1984; Parks et al., 1986). The use of phycobiliproteins in soluble antigen or hapten assays has also occurred. Most of these applications, however, will require a much longer development time to format the phycobiliprotein label in a particular instrument/reagent system and then receive the appropriate regulatory approval. Nevertheless, clinical assays utilizing phycobiliproteins are already commercially available (Winfrey and Wagman, 1984, Khanna, 1985). Following a discussion of the materials and methods for the formation and purification of phycobiliprotein conjugates below, each of these application areas will be reviewed in detail.

168

MATERIALS AND METHODS

Sources of Purified Phycobiliproteins

Many species of red and blue-green algae contain significant amounts of the various phycobiliproteins. Many different purification protocols have been described in the literature. For unicellular sources the purification begins by rupture of the cell envelope. This is usually accomplished by use of a French press, a homogenizer, or, for simple blue-green algae (cyanobacteria), by using enzymes such as lysozyme that dissolve the cell wall. For red macroalgae, homogenization in a high speed blender is adequate. Following centrifugation to remove cell debris, the phycobiliproteins are precipitated with ammonium sulfate. Subsequent purification steps involving ion exchange (on resins or hydroxylapatite) and usually gel filtration or crystallization will bring the phycobiliprotein of interest to near homogeneity. The unicellular red alga, Porphyridium cruentum, has proven a favorite source because of the high content of B-phycoerythrin present. Glazer and Hixson (1977) have described one technique for this algae source and Gantt and Lipschultz (1974) have described another. Purification of R-phycoerythrin from red macroalgae is discussed in many papers including those of Oi et al. (1982), Klotz and Glazer (1984) and Hardy (1986). Purification of C-phycocyanin and allophycocyanin is discussed by many authors including Bryant and Glazer (1976) and Hardy (1986). Earlier workers used purification procedures that were designed to produce extremely pure material since the original papers were attempting to describe properties of pure phycobiliproteins. The somewhat simpler techniques such as Hardy used are probably sufficient for most routine labeling applications. High quality purified materials are now available from many commercial sources and thus purification should no longer be required by most users.

Crosslinking Procedures

Most of the initial experiments performed by Oi, Stryer and Glazer utilized the reagent SPDP (N-succinimidyl 3-(2-pyridyldithio) propionate), a heterobifunctional reagent first described by Carlsson et al. (1978). SPDP, as well as other heterobifunctional reagents such as SMCC (succinimidyl 4-(N-maleimidomethyl) cyclohexane-1-carboxylate) or SMPB (succinimidyl-4-(p-maleimidophenyl) butyrate), are commercially available from several sources and allow the separate activation of the two molecules of interest, the phycobiliprotein and the specific-binding molecule, using reagents under mild aqueous conditions at near neutral pH. SPDP does result in a disulfide linkage between the coupled molecules. Thus, some workers who were concerned about the possible reduction of the disulfide bonds have preferred to use maleimide-containing heterobifunctional reagents such as SMCC or SMPB to produce thioether linkages (Hardy, 1986). The SH group that is used to react with the maleimide is produced by a reaction with iminothiolane (Oi et al., 1982) or S-acetyl succinic anhydride (Houghton, 1985), reduction of SPDP (Oi, et al., 1982), or even by mild reduction of endogenous disulfides (Parks et al., 1984). Homobifunctional reagents such as glutaraldehyde could in principle be used, as is often done with enzyme labeling, but the loss of control resulting in the formation of homopolymers has discouraged most investigators from taking this approach.

Small molecules such as biotin have been attached to phycobiliproteins by reaction with an active ester or sulfonyl halide derivative (Oi et al., 1982). In principle, any chemistry used to conjugate a hapten to an enzyme or a carrier protein could be used so long as the conjugation conditions do not denature the phycobiliproteins.

Detailed protocols for conjugation may vary depending upon the particular molecule to be coupled to the phycobiliprotein of interest. Different proteins and even different monoclonal antibodies can give very different results with the same protocol. General references on protein modification chemistry should be consulted for special cases (Means and Feeney, 1971; Glazer, 1976). The example which follows is for the coupling of rabbit IgG to B-phycoerythrin using SPDP (Kronick and Grossman, 1983). Initial concentrations of 2.0 mg/ml for the B-phycoerythrin and 1.0 mg/ml for the IgG are assumed and should result in an approximate 50% yield of useful conjugate. Changing molar ratios and concentrations may increase total yield but may also affect parameters like non-specific binding. Some optimization is almost always necessary for any specific conjugate pair. In general, it is recommended to start with conditions that err on the side of inefficient coupling but do result primarily in well characterized conjugates of one molecule of phycobiliprotein to one specific-binding molecule. Higher efficiency procedures will necessarily involve mixtures of conjugates containing three or more molecules because of the statistical nature of the coupling process.

The SPDP disulfide linking procedure described by Kronick and Grossman is begun by dialyzing separately the two proteins to be linked, B-phycoerythrin and rabbit IgG, into a coupling buffer of 0.1 M sodium phosphate, 0.1 M sodium chloride, at pH 7.4. A fresh solution of 1.3 mg/ml of SPDP in anhydrous methanol or ethanol is prepared just prior to the conjugation procedure. To 1.0 ml of the 2.0 mg/ml solution of the B-phycoerythrin solution, 20 ul of the SPDP solution is added while gently vortexing to give an approximate 10:1 molar excess of SPDP in the mixture. The reaction is allowed to proceed at room temperature in the dark for 30 minutes and is then terminated by the addition of 30 ul of a 0.5 M aqueous solution of dithiothreitol to create free sulfhydryl groups on the B-phycoerythrin. After 20 minutes more at room temperature, the solution is applied to a 1.2 cm by 6 cm column of Sepadex G-25 gel filtration media that has been pre-equilibrated with a low phosphate buffer, 0.001 M sodium phosphate, 0.1 M sodium chloride, pH 7.0. The colored fraction is collected and then applied to a 1 ml bed volume column of hydroxylapatite that has been previously equilibrated with same low phosphate buffer. After rinsing the column with a few column volumes of the low phosphate buffer, the red material is eluted in a minimal volume using the coupling buffer. This procedure concentrates the derivatized B-phycoerythrin and removes any residual dithiothreitol. While the above reactions are occurring, the rabbit IgG is similarly derivatized: 10 ul of the fresh SPDP solution added to 1.0 ml of a 1.0 mg/ml solution of the IgG in coupling buffer (to give a 6:1 molar excess). The mixture is then allowed to react for forty minutes and then is transferred to a low phosphate buffer and concentrated on hydroxylapatite exactly as above. Finally the two, separately derivatized protein solutions are mixed together in appropriate volumes to give a molar ratio of 1.2 B-phycoerythrin molecules to each 1.0 molecules of IgG. After an 18 hour incubation in the dark at room temperature, the reaction is quenched with a stoichiometric amount of N-ethyl maleimide and the coupling is complete. Purification of the conjugated material is then usually performed as described below.

Other groups have used very similar conjugation procedures. Hardy (1986) describes the use of SMPB to form thioether links. Houghton (1985) has described the use of SMCC as a crosslinker. The general principles are all the same. The details will differ depending upon the specific molecules involved.

Purification of Conjugates

In most applications it is desirable to purify the conjugates away from any unconjugated label or specific-binding molecule that may remain. This purification serves to remove label that conceivably could cause background or non-specific binding and it eliminates any unlabeled specific-binding molecules which could compete for target molecules and thus reduce the amount of signal one hopes to measure. Gel filtration chromatography has proven to be the most common approach to conjugate purification and is especially applicable when large molecules such as antibodies are being coupled to phycobiliproteins and the resulting conjugates are thus much larger in size than either of the starting molecules. In the particular conjugation example discussed above, a 1.5 by 70 cm column of Bio-Gel A-1.5m from Bio-Rad, Richmond, CA, works extremely well. The conjugation reaction mixture is applied to the gel filtration column and after a few hours two well-separated red bands will appear. The faster band is comprised of the B-phycoerythrin-rabbit IgG conjugates which have molecular weights of at least 400,000 and can easily be resolved from the unconjugated B-phycoerythrin at 240,000 molecular weight and the 160,000 molecular weight IgG. The fractions of purified conjugates can, if desired, be grouped into pools that consist of 1:1 conjugates, 1:2 and 2:1 conjugates, etc. on the basis of the retention time and corresponding molecular weight of the particular fractions. Figure 3 shows a typical chromatogram found when gel filtration is used to purify conjugates utilizing the conditions described above (Robison and Kronick, 1985). Absorbance at both 546 nm and 280 nm is plotted to show content of both label (at 546 nm) and protein (at 280 nm) in each fraction. The characteristic two peaks of conjugated and unconjugated material are seen. It can be seen that unconjugated IgG, which runs just behind the unconjugated phycoerythrin, is clearly separated from the conjugates.

Hydroxylapatite chromatography has been used to purify conjugates based upon the fact that the conjugates will have different retention properties from the unconjugated starting materials. Resin-based ion exchange chromatography is also possible. Results using hydroxylapatite chromatography for the separation of conjugates were described by Glazer and Stryer, (1983) for "tandem" conjugates and Hardy (1986) for antibody conjugates. In both cases, a low phosphate buffer rinse at approximately 0.02 M sodium phosphate, 0.1 M sodium chloride, removes unconjugated phycoerythrin and then a step change to higher phosphate removes the conjugates which have an intermediate affinity for the hydroxylapatite between that of phycoerythrin and the other protein used (allophycocyanin for the "tandem" and antibody for the antibody conjugates). These methods are obviously highly tailored to the particular substances being used because retention properties are so sensitive to chemical composition.

Gel filtration can be seen to be more general than ion exchange but may take longer and will result in considerable dilution of the conjugates. Affinity chromatography in conjunction with gel filtration may be useful especially when dealing with purification of conjugates of low molecular weight molecules like drugs or hormones with phyco-biliproteins.

When the final purified conjugate has been isolated, it is desirable to store the material in the dark at 2 to 8 degrees Celsius. 0.02-0.1% sodium azide is desirable as an anti-microbial agent and a stabilizing agent like gelatin or bovine serum albumin is useful, especially if the conjugate is fairly dilute. Conjugates purified as

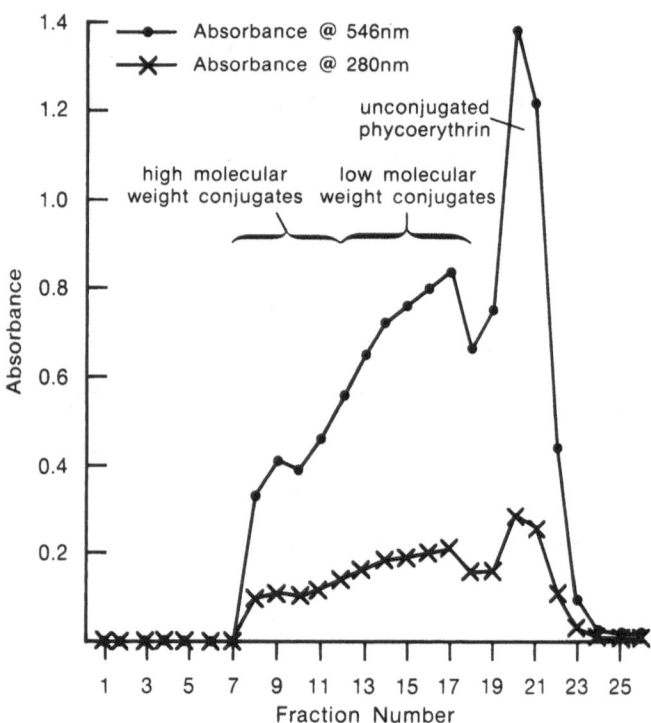

Fig. 3   Chromatogram of the separation of conjugates of phycoerythrin
          with rabbit IgG from unconjugated starting materials. Un-
          conjugated IgG elutes just after the unconjugated phycoerythrin.
          Separation is done by gel filtration using a 1.5 by 70 cm column
          of Bio-Gel A-1.5m (Bio-Rad Laboratories, Richmond, CA).

above and stored as recommended are certified by some commercial manufacturers to be stable for at least three months and stability for over a year has been observed (Kronick, 1985).

RESULTS

Cell Staining and Analysis

Cell staining with phycoerythrin conjugates for flow cytometry has been described by many researchers including Oi et al. (1982), Hardy (1986), and Parks and Herzenberg (1984). Shapiro et al. (1983) have described similar procedures for allophycocyanin conjugates. Particular detailed methodologies for use with fluorescent microscopy may be found in product inserts from various manufacturers of cell labeling reagents (Becton Dickinson, 1985; Ortho Diagnostic Systems, 1985). The typical procedure involves incubating cells on ice with a dilution of the conjugate followed by extensive rinsing. Sometimes fixation with formaldehyde is used before the final rinse.

The original paper of Oi et al. (1982) clearly demonstrated the advantages of phycobiliprotein conjugates in flow cytometry. The simultaneous measurement of two independent cell parameters allowed the definition of cell subsets that absolutely could not be established before without the use of two lasers. The two lasers were necessary so that two different dyes with ·different fluorescent emission peaks could be simultaneously excited. The use of phycoerythrin makes the use of one laser sufficient. Data is most often displayed in the form of a two-parameter contour plot such as shown in Figure 4 for human peripheral blood lymphocytes. Peak height on the contour plot is a measure of the number of cells with a particular red and green fluorescence. The fluorescein-labeled Leu-4 antibody stains all human peripheral T-cells. The phycoerythrin labeled Leu-3a stains only helper T-cells. The contour plot thus shows three populations: The cells in the lower left hand corner have very little red fluorescence (from the Leu-3a) or green (from the Leu-4) and are thus not T-cells. The population in the bottom right are T-cells that are not helper cells since they bear a green stain but no red stain. The population in the upper right are thus helper T-cells since they bear both a red and green stain.

Two color flow cytometry using a combination of fluorescein- and phycoerythrin-labeled antibodies has already become a standard technique in flow cytometry. The technique has widespread use in the identification and tracking of lymphocyte subsets in both research and clinical applications. Phillips and Lanier, (1985) for example, used the technique to establish a model for the differentiation of human natural killer cells. Lewis and coworkers (1985) applied it to the study of T-cell subsets in patients with lymphadenopathy syndrome and AIDS. Ryan (1985) has used the technique to monitor the remission of pediatric patients with certain leukemias. Ledbetter et al. (1985) have used two-color flow cytometry to correlate lymphocyte activation and differentiation with the presence of certain cell surface markers and IL2 receptors. There are many other examples.

The ability to monitor two independent cell parameters with one laser was quickly perceived as a means to allow the measurement of three or four independent cell parameters with two lasers. Lanier and Loken (1984) used an argon ion laser at 488 nm to excite antibodies labeled with fluorescein and phycoerythrin in combination with a dye laser at 600 nm to excite biotinylated antibodies indirectly labeled with Texas Red-avidin in order to determine the co-expression of cell surface

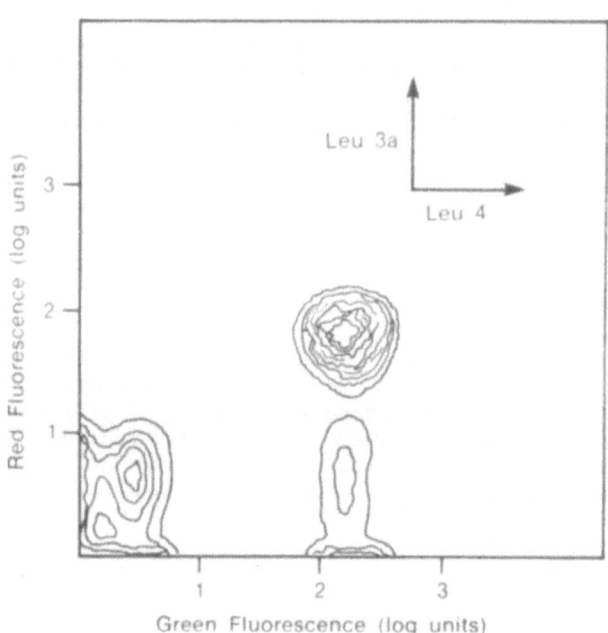

Fig. 4   Typical two-dimensional contour plot obtained from two-color fluorescence-activated cell sorting anaylsis of human peripheral blood lymphocytes. The green channel detects the fluorescence from the fluorescein-labeled Leu-4 antibody specific for all T-cells. The red channel detects the fluorescence from the phycoerythrin-labeled Leu-3a antibody specific for T-helper-inducer cells. From Oi et al. (1982), by permission.

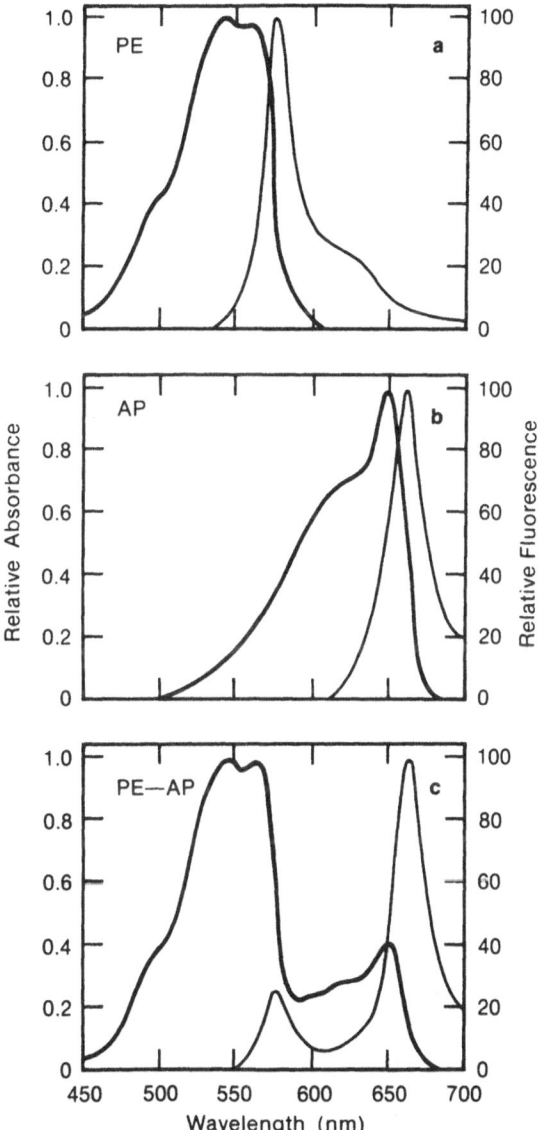

Fig. 5   Absorption (▬) and fluorescence (—) spectra of (a) phycoery-
         thrin (PE), (b) allophycocyanin (AP), and (c) the phycoerythrin-
         allophycocyanin conjugate (PE-AP). The fluorescence spectra of
         PE and PE-AP are excited at 500 nm and the spectra of AP is
         excited at 600 nm. Reproduced from Glazer and Stryer, the
         Biophysical Journal, 1983, 43, 383-386 by copyright permission
         of the Biophysical Society.

antigens on human peripheral lymphocytes. Hardy, Parks and coworkers similarly used an argon ion laser at 488 nm to excite fluorescein and phycoerythrin labeled reagents and a dye laser operating at 615 nm to excite an allophycocyanin labeled reagent to provide information on relationships of B-cell populations in normal and immunodeficient mice (Hardy et al., 1983; Parks et al., 1984). Finally, the two techniques were combined by Hardy and coworkers to do four-color flow cytometry to measure four independent parameters in a study of murine B-cell differentiation linkages (Hardy et al., 1984). In this study, the dye laser was operated at 605 nm and used to excite simultaneously Texas Red and allophycocyanin labeled reagents while the argon ion laser was used as before. Still others have used a similar three-color trick to correlate two cell surface markers with the amount of DNA in a cell as a function of cell cycle (Rabinovitch, 1985).

As discussed above, Glazer and Stryer created "tandem" conjugates of phycoerythrin and allophycocyanin to produce labels with extremely long Stokes shifts. Figure 5 shows the absorption and emission spectra of such conjugates. Chen and coworkers (1985) have recently utilized such labels to do three-color flow cytometry on human lymphocyte populations. The results are similar to those obtained by Loken and Lanier but the use of the "tandem" conjugates allowed all three independent labels to be excited by the same argon ion laser being operated at 488 nm.

When a sophisticated flow cytometer is not available, the fluorescence microscope can be a very powerful tool for examining cell populations when used with phycobiliprotein-labeled reagents. The distinct color of the phycoerythrin reagents in combination with the large extinction coefficient relative to fluorescein offers the potential for increased sensitivity for fluorescence microscope assays. Also, since fluorescein and phycoerythrin can simultaneously be excited with the same wavelength and yet emit very different color fluorescence, two distinct cell populations in a single sample can clearly be seen in the microscope. With appropriate filters, the population stained with the fluorescein reagent appears green and the population stained with the phycoerythrin reagent appears yellow-orange. This technique has very successfully been applied to microscopic determination of T-cell subset ratios which are useful clinical parameters in immune system disorders such as AIDS (Becton Dickinson, 1985). Pizzolo and Chilosi (1984) applied this two-color microstaining method to lymph node sections to investigate cellular heterogeneity in normal and patho-logical conditions. Oliver and coworkers, in a different application, took advantage of the size of phycoerythrin, in addition to its spectroscopic properties, to cause and subsequently follow degranulation in basophil cells (Deanin et al., 1984; Oliver, 1985).

Fluorescence Quenching Assays

The principles of fluorescence quenching immunoassays were first described by Ullman et al. (1976). The binding of quencher-labeled antibody to fluorescer-labeled hapten or antigen is monitored by the extent of fluorescence detected and can be used to monitor concentra-tions of hapten or antigens if the appropriate conditions for a competition assay are set up. Kronick and Grossman (1983) first described the uses of phycobiliproteins in such fluorescence quenching assays and Winfrey and Wagman (1984) and Khanna (1985) have reported use of phycoerythrin conjugates in a clinically useful assay for digoxin. The use of phycoerythrin in such assays parallels its use in nature as an energy donor in an intermolecular energy transfer of high efficiency. Figure 6 shows the results of the fluorescence quenching assay for human

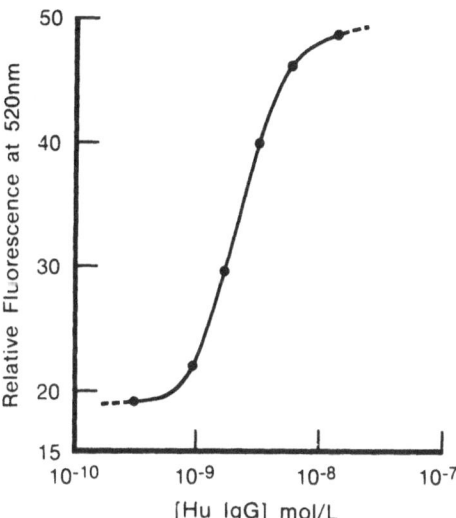

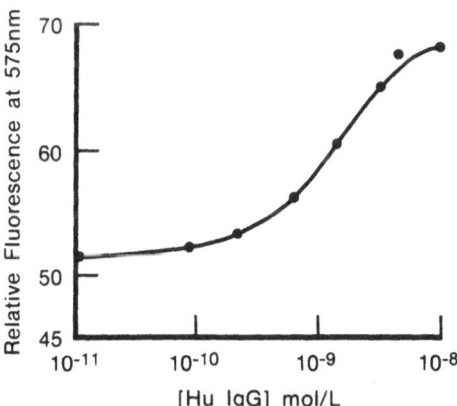

Fig. 6   Fluorescence intensity versus antigen (human IgG) concentration
in a fluorescence excitation transfer immunoassay. The upper
plot shows results from competition between unlabeled IgG in the
sample and fluorescein-labeled IgG for binding sites on
phycoerythrin-labeled antibody. Phycoerythrin quenches the
fluorescein fluorescence. The lower plot shows results from the
competition between unlabeled IgG in the sample and
phycoerythrin-labled IgG for binding sites on antibody labeled
with Texas Red. Texas Red quenches the fluorescence of
phycoerythrin. From Kronick and Grossman (1983), by permission.

IgG performed by Kronick and Grossman (1983). The assay was run using a competition for Texas Red labeled antibodies by phycoerythrin-labeled and unlabeled human IgG or analogously, competition for phycoerythrin-labeled antibodies by fluorescein-labeled IgG and unlabeled IgG. Sensitivity levels of $< 10^{-9}$ M were obtained. A commercially available assay for the cardiac drug digoxin with sensitivity of close to 10^{-10} M serum digoxin concentration has subsequently been described (Winfrey and Wagman, 1984; Khanna, 1985). In the digoxin assay, digoxin-labeled phycoerythrin competes with digoxin in the serum sample for the binding sites of an antibody labeled with a dye that quenches the fluorescence of phycoerythrin very effectively. In this particular homogeneous assay format in which no separation steps are performed, phycoerythrin's high extinction coefficient in the 540 nm region of the spectrum helps to discriminate signal fluorescence from that of endogenous materials in the clinical sample being analyzed. Digoxin is present in clinical samples at very low levels and the spectral properties of conventional fluorescent dyes are not adequate for the performance of these quenching assays at such high sensitivity. The use of phycoerythrin resulted in an order of magnitude increase in sensitivity of the assay.

Sandwich Assays

Kronick and Grossman (1983) first reported the use of phycobili-proteins in sandwich or immunometric assays. The techniques utilized were identical to those employed in conventional latex bead assays using fluorescein-labeled antibodies as reported by Curry et al. (1979). Kronick and Grossman substituted a phycoerythrin-labeled rabbit anti-human IgG antibody for a fluorescein-labeled one in a sandwich assay on latex beads and found an increase in sensitivity. They were able to obtain sensitivity of about 10^{-11} M in the sample with a commercial fluorimeter. This sensitivity was somewhat less than was expected based on spectroscopic differences alone. The discrepancy was probably due to scattering from the beads which was picked up as fluorescence due to inadequate filtering and possibly due to some increase in non-specific binding. In addition, there was some evidence that the large phycoery-thrin tag caused some steric interference with the antibody binding sites and thus lowered the effective concentration of labeled antibody.

This sandwich methodology, as well as competition assays on latex beads, can easily be applied to the techniques of particle concen-tration fluorescence immunoassay (PCFIA) developed by Jolley and coworkers (1984). PCFIA extends the sensitivity and ease of use of latex bead assays by concentrating the specifically-labeled latex beads on filters in the bottom of small wells, rinsing the beads that have been so concentrated, and then reading the fluorescence of the concentrated beads in situ. Francis et al. (1985) has recently described the use of phycobiliproteins as labels for antibodies in PCFIA assays. Francis and coworkers developed an immunoassay for human IgG run as a sandwich assay using particle concentration fluorescence immunoassay on the Pandex Screen Machine^{TM}. An enhancement in sensitivity of four times was observed with the phycoerythrin-labeled reagent compared to a similar fluorescein-labeled reagent. As with the results of Kronick and Grossman, this was a significant increase although somewhat less than expected. The reasons for the shortfall are probably one or more of the same factors: light scattering picked up as fluorescence, non-specific binding, and steric interference. It should be noted that the measurements with the phycoerythrin-labeled antibodies were also done at significantly lower antibody concentration than with the fluorescein-labeled antibodies. In principle, multi-color assays analogous to multi-color cell sorting could be run in the PCFIA format with the appropriate optical filters. Bethell and Buck (1986) and

Francis et al. (1985) have also discussed measurements of cell staining using PCFIA and note significant increases in sensitivity with the use of phycoerythrin as a label.

A novel patented separation scheme (Nowinski and Hoffman, 1985) based on polymer chemistry has been utilized in two-color fluorescence assays of soluble antigens by Houghton and coworkers (1985). The phycoerythrin label permitted simultaneous analysis of two independent parameters exactly as in flow cytometry. Independent sandwiches were formed in solution and then separated by means of initiation of polymerization of the monomer-labeled antibody used in each sandwich. Resulting polymer aggregates were measured in a flow cytometer for two colors of fluorescence, resulting in the simultaneous measurement of both human IgG and IgM in a single sample. In monochromatic assays, phycoerythrin labels again gave a significant (about a factor of five) increase in sensitivity over that obtained with fluorescein labels but the increase again was slightly less than expected, probably for the same reasons as discussed above.

DISCUSSION

Sensitivity

As discussed in the results section above, the use of phycoerythrin labels results in significant increases in sensitivity of anywhere from two to ten times over that obtained with the use of conventional labels like fluorescein. The factors discussed above, scattering, non-specific binding, and steric interference, may be responsible for the fact that the theoretically possible improvements have not been fully realized. These factors, however, are related to the specific instrumentation and methodology employed in the particular examples described and further improvements are very likely as various groups become more familiar with the idiosyncracies of working with phycoerythrin and the other phyco-biliproteins. Scattering can be reduced by changes in the optical design and filters. Non-specific binding may be mitigated with changes in materials, surfactants, bulking agents, conjugation techniques, and coating procedures. Steric interferences could be addressed by changes in cross-linking procedures and adjustment of concentrations could compensate for inactive species. In certain applications, allophyco-cyanin, in spite of the fact that its spectral properties are not as impressive as those of phycoerythrin, may result in some of the most sensitive measurements because it is used in a spectral region where there is so little endogenous background interference. In fact, Loken et al. (1985) have reported on flow cytometric measurements that demonstrate the potential for very sensitive measurements using allophycocyanin. In their experiments, the signal to background ratio for alveolar macrophage cells stained indirectly with allophycocyanin-streptavidin conjugates and excited with a helium-neon laser at 633 nm gave results at least as good as those obtained using phycoerythrin conjugates and argon ion laser excitation at 514 nm. Autofluorescence was almost non-existent with 633 nm excitation of the alveolar macrophage cells which normally give high background with conventional 488 or 514 nm excitation. The increased reliability and decreased cost of helium-neon excitation make this result extremely significant.

Multicolor Measurements

The ability to make multicolor measurements that has been made possible by phycobiliproteins has already transformed the field of flow cytometry and cell labeling. The results of Houghton indicate similar improvements are possible in the assay of substances in solution.

Further improvements in the use of "tandem" conjugates should result in even simpler configurations for three parameter experiments.

Areas for Potential Improvement

It was noted above that the exact structure of the individual phycobiliproteins is a function of the purification procedure utilized. All the major phycobiliproteins consist of assemblies of non-covalently linked polypeptide chains and are thus liable to dissociate into their constituent subunits. Phycoerythrin, with its $(\alpha\beta)_6\gamma$ structure, is actually quite stable because of the presence of the γ subunit which appears to act to stabilize the molecule in its aggregated form (Glazer, 1984; Glazer, 1985). Dilutions of phycoerythrin appear linear to less than 10^{-13} M, the limit of sensitivity of most conventional fluorometric measurements (Oi et al., 1982). The fluorescence excitation and emission spectra of phycocyanin and allophycocyanin can be seen to change at concentrations as high as 10^{-8} M corresponding to the dissociation of these proteins into their constituent subunits (Robison and Kronick, 1985). One observes a decrease in quantum efficiency and a blue shift in the fluorescence. The formation of a conjugate of phycocyanin or allophycocyanin with an antibody has been observed by several workers to increase the stability of the native phycocyanin or allophycocyanin molecule but it does not necessarily guarantee stability (Kronick, 1985). Ong and Glazer (1985), in an attempt to understand interactions among the subunits of allophycocyanin, have developed a technique for stabilization of allophycocyanin that results in linear fluorescence for dilutions to the limits of sensitivity of about 10^{-13} M (Robison and Kronick, 1985). This should be very helpful in the use of allophycocyanin for high sensitivity measurements and similar techniques should be possible for other phycobiliproteins.

Photobleaching of fluorescence is always of concern, especially when using a fluorescence microscope in which a significant light flux is concentrated in a very small area. Mathies and Stryer (1986) have measured the photodestruction rate for phycoerythrin and found it to be about two times slower than fluorescein per photon of light absorbed per molecule. Further means of stabilization of all the phycobiliproteins against photodestruction would obviously be desirable. A fruitful area of investigation would be the relationship between photodestruction and subunit stability discussed above.

Finally, in spite of the relatively straightforward methods employed for the formation of conjugates, it would certainly be desirable to develop even simpler conjugation techniques with higher degrees of control and efficiency. Because phycobiliproteins are proteins, the use of heterobifunctional reagents seems to be the only logical method for forming conjugates. Some truly creative insights here would appear necessary to simplify things so that conjugation would be easy as coupling with fluorescein isothiocyanate.

SUMMARY

In less than four years since the first description in the literature of their use as labels, phycobiliproteins have become an important tool in the use of fluorescence in immunoassay. More than half a dozen companies are already marketing fluorescent phycobili- protein conjugates including some that are already being used for measurements of clinical significance. The field of flow cytometry has already been transformed by the two-, three-, and even four-color cell sorting that has been made possible by the use of phycobiliproteins.

The use of phycobiliproteins in flow cytometry has become so routine, in fact, that often the original work of Oi et al. (1982) is no longer even referenced. In the future, it can be expected that phycobiliproteins will make even further inroads to clinical and research immunoassay as their advantages, limitations, and idiosyncracies become more understood and appreciated.

ACKNOWLEDGMENTS

Applied Biosystems has provided a nurturing environment in which to develop an appreciation for the physical and practical beauty of the phycobiliproteins. Paul Grossman and Debra Robison have put in many long hours conducting the many experiments that have made the purification and utilization of phycobiliproteins almost commonplace at Applied Biosystems. David Parks, Randy Hardy, and Lubert Stryer at Stanford University have been most helpful in many discussions. Finally, Alex Glazer of the University of California at Berkeley has been a continuing source of inspiration and enthusiasm and has continued to amaze me with his profound knowledge, intuition, and love of the phycobiliproteins.

REFERENCES

Becton Dickinson, 1985, Product Insert for Kit 95-1005, T Helper/ Supressor Ratio Test, Becton Dickinson Immunocytometry Systems, Mountain View, California

Bethell, D.R., and Buck, D.W., 1986, The use of PCFIA for detecting lymphocyte cell surface antigens using Leu monoclonal antibodies, Research Report No. 13 (January 1986), Pandex Laboratories, Mundelein, Illinois

Bryant, D.A., Glazer, A.N., and Eiserling, F.A., 1976, Characterization and structural properties of the major biliproteins of Anabaena sp. Archives of Microbiology, 110, 61-75.

Carlsson, J., Drevin, H., and Axen, R., 1978, Protein thiolation and reversible protein-protein conjugation. N-succinimidyl 3-(2-pyridyldithio) propionate, a new heterobifunctional reagent, Biochemical Journal, 173, 723-737.

Chen, C.H., 1985, Application of phycobiliprotein tandem conjugates for cell surface immunofluorescence, presentation at Conference on Phycobiliproteins in Biology and Medicine, Seattle, September 9-10, 1985.

Curry, R.E., Heitzman, H., Reige, D.H., Sweet, R.V., and Simonsen, M.E., 1979, A system approach to fluorescent immunoassay, general principles and representative applications, Clinical Chemistry, 25, 1591-1595.

Dankliker, W.B., and Saussure, V.A., 1970, Fluorescence polarization in immunochemistry, Immunochemistry, 7, 799-828.

Deanin, G.G., Davis, B.H., Haugland, R.P., Seagrave, J.C., Stewart, C.C., Steinkamp, J., and Oliver, J.M., 1984, The DNP-phycobiliproteins, a new family of fluorescent ligands to study IgE-mediated degranulation in rat basophilic leukemic cells, Journal of Cell Biology, 99, 331a.

Francis, B., Buck, D.W., Lanier, L.L., LaFoe, M.F., Bethell, D.R., and Jolley, M.E., 1985, Enhanced sensitivity using B-phycoerythrin conjugates to antibodies in particle concentration fluorescence immunoassay, poster at Conference of Phycobiliproteins in Biology and Medicine, Seattle, September 9-10, 1985.

Frank, G., Sidler, W., Widmer, H., and Zuber, H., 1978, The complete
amino acid sequence of both subunits of C-phycocyanin from the
cyanobacterium Mastigocladus laminosus, Hoppe-Seyler's Zeitschrift
fur Physiologische Chemie 359, 1491-1507.

Fuglistaller, P., Suter, F., and Zuber, H., 1983, The complete amino
acid sequence of both subunits of phycoerythrocyanin from the
thermophilic cyanobacterium Mastigocladus laminosus.
Hoppe-Seyler's Zeitschrift fur Physiologische Chemie 364, 691-712.

Gantt, E., and Lipschultz, C.A., 1974, Phycobilisomes of Porphyridium
cruentum: Pigment analysis, Biochemistry, 13, 2960-2966.

Gantt, E., 1980, Structure and function of phycobilisomes; Light
harvesting pigment complexes in red and blue-green algae.
International Review of Cytology 66, 45-80.

Glazer, A.N., 1976, Chemical modification of proteins by group-specific
and site-specific reagents, The Proteins, volume IIA, 3rd edition,
ed. by Neurath, H. and Hill, R.L., Academic Press, New York, 1-103.

Glazer, A.N., and Hixson, C.S., 1977, Subunit structure and chromophore
compostion of rhodophytan phycoerythrins, Journal of Biological
Chemistry, 252, 32-42.

Glazer, A.N., 1981, Photosynthetic accessory proteins with bilin
prosthetic groups, The Biochemisty of Plants, Academic Press, New
York, 8, 51-96.

Glazer, A.N., and Stryer, L., 1983, Fluorescent tandem phycobiliprotein
conjugates, Biophysical Journal, 43, 383-386.

Glazer, A.N., 1984, Phycobilisome: A macromolecular complex optimized
for light energy transfer, Biochimica et Biophysica Acta, 768,
29-51.

Glazer, A.N., and Stryer, L., 1984, Phycofluor probes, Trends in
Biochemical Sciences, 9, 423-427.

Glazer, A.N., 1985, Light harvesting by phycobilisomes, Annual Reviews
of Biophysics and Biophysical Chemistry, 14, 47-77.

Hardy, R.R., Hayakawa, K., Parks, D.R., and Herzenberg, L.A., 1983,
Demonstration of B-cell maturation in X-linked immunodeficient mice
by simultaneous three-colour immunofluorescence, Nature, 306,
270-272.

Hardy, R.R., Hayakawa, K., Parks, D.R., and Herzenberg, L.A., 1984,
Murine B cell differentiation linkages, Journal of Experimental
Medicine, 159, 1169-1188.

Hardy, R.R., in press 1986, Purification and coupling of fluorescent
proteins for use in flow cytometry, Handbook of Experimental
Immunology, 4th edition, ed. by Weir, D.M., Herzenberg, L.A.,
Blackwell, C.C., and Herzenberg, L.A., Blackwell Scientific
Publications, Edinburgh.

Hassan, M., Landon, J., and Smith D.S., 1979, Multiple fluorescein-
substitued polymers as labels in fluoroimmunoassay, FEBS Letters,
103, 339-341.

Hemmila, I., 1985, Fluoroimmunoassays and immunofluorometric assays,
Clinical Chemistry, 31, 359-370.

Hirschfeld, T., 1976, Optical microscopic observation of single small
molecules, Applied Optics, 15, 2965-2966.

Houghton, R., 1985, A dichromatic polymerization-induced separation
immunoassay for the simultaneous measurement of human serum IgG and
IgM, presentation at Conference on Phycobiliproteins in Biology and
Medicine, Seattle, September 9-10, 1985.

Jolley, M.E., Stroupe, S.J., Schwenzer, K.S., Wang, C.J., Lu-Steffer,
M., Hill, H.D., Popelka, S.R., Holen, J.T., and Kelso, D.M., 1981,
Fluorescence polarization immunoassay III. An automated system for
therapeutic drug determination, Clinical Chemistry, 27, 1575-1579.

182

Jolley, M.E., Wang, C.J., Ekenberg, S.J., Zuelke, M.S., and Kelso, D.M., 1984, Particle concentration fluorescence immunoassay (PCFIA): a new, rapid immunoassay technique with high sensitivity, Journal of Immunological Methods, 67, 21-35.

Khanna, P., 1985, Energy transfer immunoassays using phycobiliproteins, presentation at Conference of Phycobiliproteins in Biology and Medicine, Seattle, September 9-10, 1985.

Klotz, A.V., and Glazer, A.N., 1985, Characterization of the bilin attachment sites in R-phycoerythrin, Journal of Biological Chemistry, 260, 2856-2863.

Kronick, M.N., and Grossman, P.D., 1983, Immunoassay techniques with fluorescent phycobiliprotein conjugates, Clinical Chemistry, 29, 1582-1586.

Kronick, M.N., 1985, unpublished observations.

Lanier, L.L., and Loken, M.R., 1984, Human lymphocyte subpopulations identified by using three-color immunofluorescence and flow cytometry analysis: correlation of Leu-2, Leu-3, Leu-8, and Leu-11, cell surface antigen expression, Journal of Immunology, 132, 151-156.

Ledbetter, J., Rose, L.M., Spooner, C.E., Beatty, P.G., Martin, P.J., and Clark, E., 1985, Antibodies to common leukocyte antigen p220 influence human T cell proliferation by modifying receptor expression, Journal of Immunology, 135, 1819-1825.

Lewis, D.E., Puck, J.M., Babcock, G.F., and Rich, R.R., 1985, Disproportionate expansion of a minor T cell subset in patients with lympadenopathy syndrome and acquired immunodeficiency syndrome, Journal of Infectious Diseases, 151, 555-559.

Loken, M.R., Keij, J., and Kelly, K., 1985, Comparison and helium-neon and dye laser excitation of allophycocyanin, poster at Analytical Cytology XI, Hilton Head Island, November 17-22, 1985.

Lundell, D.J., Glazer, A.N., DeLange, R.J., and Brown, D.M., 1984, Bilin attachment sites in the α and β subunits of B-phycoerythrin. Amino acid sequence studies, Journal of Biological Chemistry, 259, 5472-5480.

Mathies, R.A., and Stryer, L., in press 1986, Single molecule fluorescence detection: a feasibility study using phycoerythrin, in Fluorescence in the Biological Sciences, edited by Taylor, D.L., Waggoner, A.S., Lanni, F., Murphy, R.F., and Birge, R., Alan R. Liss, Inc., New York.

Means, G.E., and Feeney, R.E., 1971, Chemical Modification of Proteins, Holden Day, San Franicsco, 254 pages.

Nagy, J.O., Bishop, J.E., Klotz, A.V., Glazer, A.N., and Rapoport, H., 1985, Bilin attachment sites in the α, β, and γ subunits of R-phycoerythrin. Structural studies on singly and doubly linked phycourobilins, Journal of Biological Chemistry, 260, 4864-4868.

Nowinski, R.C., and Hoffman, A.S., April 16, 1985, U.S. Patent # 4,511,478, Polymerizable compounds and methods for preparing synthetic polymers that integrally contain polypeptides.

Oi, V.T., Glazer, A.N., and Stryer, L., 1982, Fluorescent phycobiliprotein conjugates for analyses of cell and molecules, Journal of Cell Biology, 93, 981-986.

Oliver, J.M., 1985, DNP-phycobiliproteins as novel antigens for IgE-mediated basophil reactions, presentation at Conference on Phycobiliproteins in Biology and Medicine, Seattle, September 9-10, 1985.

Ong, L.J., and Glazer, A.N., 1985, Crosslinking of allophycocyanin, Physiologie Vegetale, 23, 777-787.

Ortho Diagnostic Systems, 1985, Product insert for Orthomune products OKT*3, OKT4, OKT8, OKB*7, OKDR phycobiliprotein conjugates, Ortho Diagnostic Systems, Raritan, New Jersey.

Parks, D.R., Hardy, R.R., and Herzenberg, L.A., 1984a, Three-color immunofluorescence analysis of mouse B-lymphocyte subpopulations, Cytometry, 5, 159-168.

Parks, D.R., and Herzenberg, L.A., 1984b, Fluorescence activiated cell sorting; theory, experimental optimization, and application in lymphoid cell biology, Methods in Enzymology, 108, 197-241.

Parks, D.R., Lanier, L.L., and Herzenberg, L.A., in press 1986, Flow cytometry and fluorescence activated cell sorting (FACS), Handbook of Experimental Immunology, 4th edition, ed. by Weir, D.M., Herzenberg, L.A., Blackwell, C.C., and Herzenberg, L.A., Blackwell Scientific Publications, Edinburgh.

Phillips, J.H., and Lanier, L.L., 1985, A model for the differentiation of human natural killer cells, Journal of Experimental Medicine, 161, 1464-1482.

Pizzolo, G., and Chilosi, M., 1984, Double immunostaining of lymph node sections by monoclonal antibodies using phycoerythrin labeling and haptenated reagents, American Journal of Clinical Pathology, 82, 44-47.

Rabinovitch, P., 1985, Multicolor cell cycle and immunofluorescence analysis, presentation at Conference of Phycobiliproteins in Biology and Medicine, Seattle, September 9-10, 1985.

Robison, D., and Kronick, M., 1985, unpublished results.

Ryan, D., 1985, Detection of rare leukemic cells in acute lymphoblastic leukemia using two color immunofluorescence, presentation at Conference on Phycobiliproteins in Biology and Medicine, Seattle, September 9-10, 1985.

Schirmer, T., Bode, W., Huber, R., Sidler, W., and Zuber, H., 1985, X-Ray crystallographic structure of the light-harvesting biliprotein C-phycocyanin from the thermophilic cyanobacterium Mastigocladus laminosus and its resemblence to globin structures, Journal of Molecular Biology 184, 257-277.

Schoenleber, R.W., Lundell, D.J., Glazer, A.N., and Rapoport, H., 1984, Bilin attachment sites in the α and β subunits of B-phycoerythrin. Structural studies on a doubly peptide-linked phycoerythrobilin, Journal of Biological Chemistry, 259, 5481-5484.

Schoenleber, R.W, Lundell, D.J., Glazer, A.N., and Rapoport, H., 1984, Bilin attachment sites in the α and β subunits of B-phycoerythrin. Structural studies on a singly linked phycoerythrobilins, Journal of Biological Chemistry, 259, 5485-5489.

Sepaniak, M.J., 1985, The clinical use of laser-excited fluorimetry, Clinical Chemistry, 31, 671-678.

Shapiro, H.M., Glazer, A.N., Christenson, L., Williams, J.M., and Strom, T.B., 1983, Immunofluorescence measurement in a flow cytometer using low power helium-neon laser excitation, Cytometry, 4, 276-279.

Sidler, W., Gysi, J., Isker, E., and Zuber, H., 1981, The complete amino acid sequence of allophycocyanin, a light-harvesting protein-pigment complex from the cyanobacterium Mastigocladus laminosus, Hoppe-Seyler's Zeitschrift fur Physiologische Chemie 362, 611-628.

Sidler, W., Kumpf, B., Zuber, H., and Rudiger, W., 1985, The complete amino acid sequence of C-phycoerythrin from Fremyella diplosiphon, V International Symposium of Photosynthetic Prokaryotes, Grindelwald, Switzerland, October 1985, Abstracts, p. 303.

Soini, E. and Hemmila, I., 1979, Fluoroimmunoassay: Present Status and key problems, Clinical Chemistry, 25, 353-361.

Stryer, L., Glazer, A.N., and Oi, V.T., May 28, 1985, U.S. Patent #4,520,110 Fluorescent immunoassay employing a phycobiliprotein labeled ligand or receptor.

Stryer, L., and Glazer, A.N., September 17, 1985, U.S. Patent #4,542, 104, Phycobiliprotein fluorescent conjugates.

Ullman, E.F., Schwartzberg, M., and Rubenstein, K.E., 1976, Fluorescent excitation transfer immunoassay, Journal of Biological Chemistry, 251, 4172-4178.

Ullman, E. and Khanna, P., 1981, Fluorescent excitation transfer immunoassay, Methods in Enzymology, 74, 28-60.

Winfrey, L., and Wagman, B.S., 1984, Sensitive homogeneous fluorescence immunoassay for serum digoxin determinations, American Clinical Products Review, 3, 31-35.

CHELATED TERBIUM AS A LABEL IN FLUORESCENCE IMMUNOASSAY

M.P. Bailey, B.F. Rocks and C. Riley

Biochemistry Department
Royal Sussex County Hospital
Eastern Road
Brighton BN2 5BE
Sussex UK

The use of fluorescent substances to covalently label biological molecules was described as early as 1941 by Coons and co-workers (Coons et al., 1941), who used anthracene isocyanate to label bacterial proteins. Later Coons and Kaplan (1950) introduced fluorescein isothiocyanate (FITC) as a more effective label. FITC has since become widely used as a fluorescent label in immunoassay. Its disadvantages are well recognised, however, and stem principally from the overlap of its emission spectrum with the endogenous fluorescence of plasma, and from its very small Stokes shift (only about 25 nm), which can lead to severe interference from scattered light.

Fluorescence assay is potentially capable of spectacular sensitivity (Wieder, 1978), but such sensitivity can only be realised if the background luminescence is extremely low. Blood plasma, irradiated at around 350 nm, has a broad fluorescence spectrum with a maximum at around 400 nm, and a tail extending to very long wavelengths. Many other materials, notably the plastics often used for solid-phase supports or for disposable cuvettes, also have appreciable fluorescence. This background fluorescence is not particularly intense, but nevertheless leads to a very poor signal-to-noise ratio if very high sensitivity is sought. This is particularly a problem in homogeneous fluorescence immunoassays, where fluorescence measurement is required on a solution which may contain a high proportion of plasma.

These difficulties have ensured that, until recently, fluorescence immunoassay has not achieved the sensitivity of which it is theoretically capable.

TIME RESOLUTION.

Fluorescence involves a sequence of processes with widely differing timescales. An incident photon will be scattered or absorbed by a molecule in about 1 femtosecond - the period of oscillation of the light. If absorption takes place, an electron in the molecule is raised to one of the vibrational energy levels of the first singlet excited state, from where it rapidly relaxes to the lowest vibrational energy level of that state. The excited state persists for about 5 to 50 nanoseconds, and may release its energy by transition to the ground state with emission of a fluorescent photon.

187

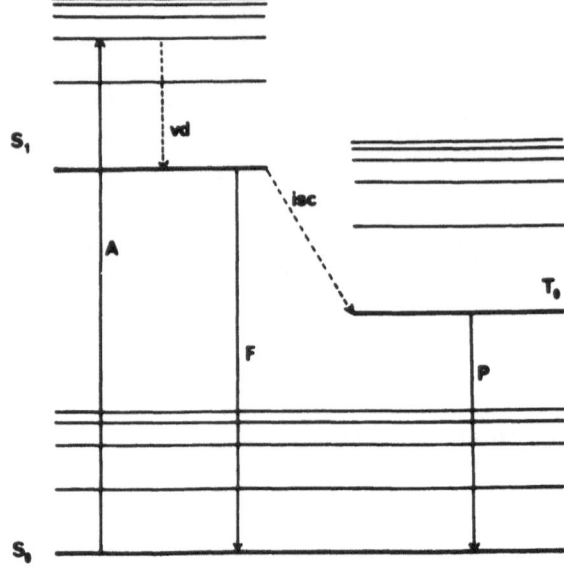

Figure 1. Simplified energy-state diagram
   illustrating the processes occurring in
   fluorescence and phosphorescence. S_0 is the
   singlet electronic ground state, S_1 is the
   first singlet excited state, and T_0 is the
   triplet state. The processes illustrated are:
   A - absorption; vd - vibrational deactivation;
   F - fluorescence; isc - intersystem crossing;
   P - phosphorescence.

Alternatively, intersystem crossing to the triplet state may occur. The probability of singlet - triplet transitions is low, and consequently the triplet state is formed only in low yield; however, the reverse transition also has low probability, and the triplet state therefore persists for much longer (milliseconds to seconds). Phosphorescence arises when the triplet - ground state transition occurs with the emission of a photon. Fluorescence and phosphorescence, when excited by a light pulse, exhibit exponential decay with the characteristic lifetimes of the excited singlet and triplet states respectively. Fluorescence (and phosphorescence) lifetime is normally expressed as the time taken for the emission to fall to $1/e$ of its initial intensity after excitation by a pulse of light ($e=2.71828....$).

   Wieder (1978) noted that most organic fluorophores have fluorescence lifetimes of less than 50 ns, and suggested that gated detection would allow the unwanted emission from plasma constituents and cuvettes to be eliminated. Long-lived fluorescence could then be measured against a very low background, and the resultant improvement in signal-to-noise ratio should allow increased sensitivity.

LONG-LIFETIME LABELS.

   Very few organic fluorophores have fluorescence lifetimes longer than about 50 ns. Pyrene (figure 2a) is one such compound, and was originally suggested for use as a label in time-resolved fluorescence immunoassay. Its derivatives have fluorescence lifetimes in the region of 100 ns, and

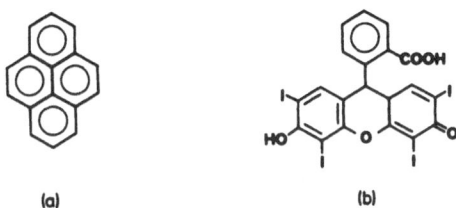

Figure 2. Structural formulae of (a) pyrene; (b) erythrosine.

it can readily be coupled to protein: however, effective discrimination against even short-lived background luminescence requires a label with a considerably longer fluorescence lifetime, and pyrene has not so far proved useful.

Sidki and Landon (1985) removed oxygen from the assay solution and were able to use erythrosine (figure 2b) as a phosphorescent label. Phosphorescence lifetimes are typically much longer than fluorescence lifetimes (10 ms to 100s), but phosphorescence has also typically much smaller quantum yield than fluorescence.

RARE-EARTH CHELATES.

The use of europium tris-diketonate complexes as labels was suggested by Wieder (1978), and has subsequently been developed by others (Hemmilä et al, 1984). The possibility of using terbium chelates was also considered, but europium chelates, having higher intensities of fluorescence and lower limits of detection, were seen as more promising.

The chelates of europium and terbium (and also of samarium and dysprosium), have certain properties which make them extremely attractive as potential labels for fluorescence immunoassay. As well as unusually long fluorescence lifetimes (1 μs to over 1 ms), they also have very large Stokes shifts (over 200 nm), and exhibit emission spectra consisting of sharp lines.

The lanthanides, as aqua ions, are feebly fluorescent. The ions have very low absorptivity, and the excited states are deactivated by interaction with solvent water molecules, causing a reduction in both the intensity and the lifetime of the fluorescence. Formation of chelate complexes with certain organic ligands leads to enhancement of the lanthanide ion fluorescence by many orders of magnitude: the presence of the chelating ligands shields the metal ion from interaction with the solvent, and energy from light absorbed by the ligand is efficiently transferred to the metal ion, effectively increasing its absorptivity.

A major difficulty with all the lanthanide chelates has been that those chelates which are most highly fluorescent are rather loosely bound, and dissociate readily on dilution, whilst those which are very stable (notably the chelates with polyaminopolycarboxylate ligands) are feebly fluorescent. The strongly fluorescent europium tris-diketonate chelates suffer the additional disadvantage of extremely low water solubility. Terbium forms strongly fluorescent chelates with a wider variety of organic ligands, and many of these complexes are freely soluble in water. Terbium fluorescence is also less sensitive to quenching by coordinated water molecules, although

Table 1. Analytical reagents for terbium.

Reagent	Sensitivity	Reference
Bis[1-(2-pyridyl)-3-methyl-5-pyrazolonyl]-4,4'-methane	30 nmol/L	Butter et al., 1968
Dipicolinic acid	10 nmol/L	Barela & Sherry, 1976
Dipicolinic acid	600 pmol/L	Miller & Senkfor, 1982
Hydroxyphenyliminodiacetic acid	60 pmol/L	Lyle & Za'tar, 1983
Ethylenediamine-bis-hydroxyphenyliminodiacetic acid	6 nmol/L	Lyle & Za'tar, 1983
EDTA + sulphosalicylic acid	4 nmol/L	Lyle & Za'tar, 1983
EDTA + Tiron	120 nmol/L	Lyle & Za'tar, 1983
EDTA + 2,3-dihydroxynaphthalene	6 nmol/L	Haddad, 1977
EDTA + salicylate	4 nmol/L	Haddad, 1977
Antipyrine + tetraphenylborate	300 nmol/L	Haddad, 1977
Iminodiacetic acid + Tiron	25 pmol/L	Haddad, 1977
EDTA + acetylacetone	12 nmol/L	Haddad, 1977
Trifluoromethyldimethylpentanedione	20 pmol/L	Hemmila, 1985

some quenching does occur. Terbium chelates typically show maximum
fluorescence at strongly alkaline pH.

ANALYTICAL REAGENTS FOR TERBIUM.

Terbium forms fluorescent chelates with a large number of ligands, and
many of these have been used for the determination of terbium in solution.
Table 1 shows some of the reagents described in recent literature. For such
analytical applications, a moderate to large excess of reagent is used.
Consequently, a high binding affinity has not been a particular consideration
when evaluating these reagents. Indeed, the binding of the ligands in, for
example, the terbium-EDTA-salicylic acid ternary complex is sufficiently weak
that a solution containing equimolar amounts of the three components shows
only feeble fluorescence if it is diluted below about 10 micromolar. In
practice this low binding affinity also limits the sensitivity of the method,
since the large excess of salicylate required to ensure complete formation of
the complex at very low terbium concentrations also causes a marked
inner-filter effect.

We have investigated a series of reagents in which an aromatic group is
covalently linked to a diethylenetriamine pentaacetic acid structure (Bailey
et al., 1985); one of these compounds has proved useful as a label for
fluorescence immunoassay (Bailey et al., 1984).

PREPARATION OF LIGANDS

Diethylenetriamine pentaacetic acid anhydride (DTPA anhydride; DTPAA)
was obtained commercially (Sigma, Fancy Road, Poole, Dorset, UK.). DTPAA was
dissolved in dimethyl sulphoxide (DMSO), previously dried over molecular
sieve 4A, at a concentration of 7.2 mg/ml (200 mmol/L). This solution
develops a brown colour on storage, and was prepared fresh for each use.
Solutions of a number of aminoaromatic compounds were prepared in DMSO at 200
mmol/L, and mixed in 1:1 mole ratio with the DTPAA solution. The reaction is
rapid, and we used a reaction time of, typically, ten minutes. For
investigation of the fluorescence properties of the terbium chelates of the

reaction products, one equivalent of terbium chloride solution (50 mmol/L in 10 mmol/L hydrochloric acid) was added to the DMSO solution, and the mixture was diluted in 0.1 mol/L triethanolamine buffer, pH 7. Fluorescence intensities were measured using a Perkin-Elmer MPF3-L spectrofluorimeter, and fluorescence lifetimes on a Perkin-Elmer LS2 filter fluorimeter.

LABELLING OF HUMAN SERUM ALBUMIN

Human serum albumin was labelled with the reaction product formed from DTPAA and 4-aminosalicylic acid (DTPA-pAS). A solution of DTPA-pAS in DMSO was prepared as described above, and 0.2 mL of this solution, containing approximately 0.4 mmol of DTPA-pAS, was added dropwise with continuous vortex mixing to a solution of 50 mg of human serum albumin in 5 mL of 0.1 mol/L phosphate buffer, pH 7. The excess DTPA-pAS was removed by dialysis against three 1L volumes of saline (9g/L). The dialysate was stored at -20°C in 0.4 mL aliquots.

FLUOROIMMUNOASSAY OF ALBUMIN

To an aliquot of DTPA-pAS albumin was added 0.1 mL of 50 mmol/L terbium chloride solution. This mixture was diluted 1 in 10 with 0.1 mol/L phosphate buffer, pH 7. The excess terbium is precipitated as phosphate under these circumstances and can be removed by centrifugation. The supernatant liquid was diluted a further 1 in 10 in phosphate buffer for use.

The assay was performed in 10x50 mm polystyrene tubes. Plasma samples were prediluted 1 in 200 for assay. Standard or diluted sample (0.1 mL) was mixed with diluted label (0.1 mL), and 0.1 ml of a dilution of antiserum to human albumin was added to each tube. The tubes were incubated at room temperature for 15 to 30 minutes and the globulin fraction precipitated by the addition of 1 mL of a solution of polyethylene glycol (200g/L) in phosphate buffer. The tubes were centrifuged for 20 minutes at 1500 g, and the supernatant liquid aspirated and discarded. The precipitates were resuspended in 1.4 mL of phosphate buffer and the fluorescence intensity measured In the absence of a time-resolving instrument, background rejection was achieved by scanning the 545 nm emission peak.

RESULTS

Most of the aromatic compounds we tested gave some enhancement of terbium fluorescence when treated as described. A few, notably the halogenated compounds gave fluorescence which changed rapidly with time in the very small flowthrough cuvette of the LS2. It was impossible to measure the fluorescence lifetime for these compounds. Table 2 shows the relative intensities of terbium fluoresence for the compounds investigated, and the fluorescence lifetimes where these could be measured. The greatest enhancement was given by the product of reaction between DTPAA and 4-aminosalicylic acid: terbium fluorescence in the presence of this reagent was approximately 100,000 fold greater than that of the aqua ion. Most of the compounds tested did, however, give terbium complexes with long fluorescence lifetimes compared with, for example, those of the terbium-diketone complexes (Hemmilä, 1985).

The fluorescence emission spectrum of DTPA-pAS-Tb is shown in figure 3. The major peaks are at 488 and 545 nm, and of these the 545 nm peak was better resolved from the background at low concentrations. The excitation spectrum, which is closely similar to the absorption spectrum of the uncomplexed ligand, has a relatively broad maximum at about 312 nm. The

Table 2. Fluorescence characteristics of terbium chelates
of ligands derived from DTPA and aminoaromatic
compounds.

DTPAA reacted with:	Excitation maximum (nm)	Fluorescence lifetime (ms)	Relative fluorescence intensity
4-aminosalicylate	312	1.58	100
2-aminobenzoic acid	280	not measured	23
5-aminosalicylic acid	307	not measured	18
5-aminoisophthalic acid	295	1.63	10
5-aminomethylsalicylic acid	309	1.64	9
4-amino-2-chlorobenzoic acid	280	not measured	5.5
3-aminobenzoic acid	285	1.25	5
4-aminobenzoic acid	284	not measured	4
2-aminobenzophenone-2'-carboxylic acid	300	1.42	2.5
3-aminophthalic acid	300	1.56	2
4-aminophthalic acid	290	1.48	2
4-aminophenazone	300	1.35	1
3-amino-2,5-dichlorobenzoic acid	290	not measured	1
3,4-dihydroxybenzylamine	285	1.49	1
5-aminophenanthroline	294	0.59	1
aniline	285	1.1	0.5

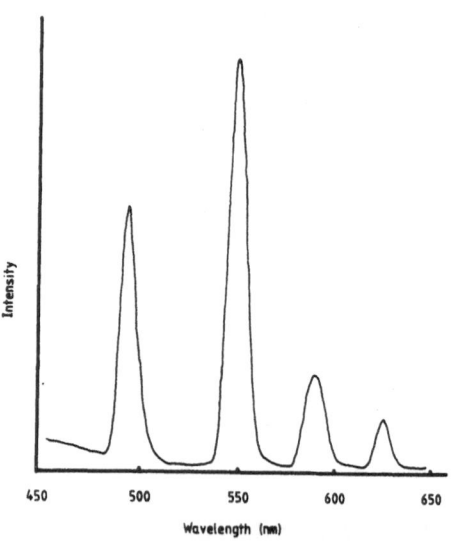

Figure 3. Fluorescence emission spectrum of the terbium
complex of DTPA-pAS.

complex has an extinction coefficient of about 7500 in a 1 cm cuvette. Attempts to isolate the product have been unsuccessful; evaporation of the solvent under vacuum yielded a brown gum which could not be induced to crystallise, and which probably contains much residual DMSO.

The fluorescence intensity of the chelate was relatively insensitive to variations in pH in the range 4 to 11, and showed a broad maximum at pH 8 to 9. Dilution of a mixture containing equimolar proportions of DTPA-pAS and terbium showed a linear relationship between terbium concentration and fluorescence intensity from 0.1 μmol/L to 5 nmol/L (figure 4). Below this concentration it was impossible to accurately measure the height of the 545 nm emission peak against the background without the use of time-resolved detection.

Labelled albumin showed similar fluorescence characteristics, and could be diluted without loss of specific fluorescence to the limit of detection. As little as 250 ng (5 pmol) of labelled albumin could be detected in a 2 mL sample volume. Terbium chloride added to unlabelled human albumin showed negligible fluorescence.

The fluorescence immunoassay for human albumin gave a usable standard curve between 19 and 1250 mg/L; correlation was good with an automated dye-binding method using bromocresol purple (r=0.96, n=10), and the within-batch coefficient of variation for 20 replicate samples with a concentration of 30 g/L (150 mg/L after dilution) was 4.9%.

NATURE OF THE PRODUCT

We were not able to confirm the identity of the terbium-binding ligand by independent means. The structure shown in figure 5 is consistent with the

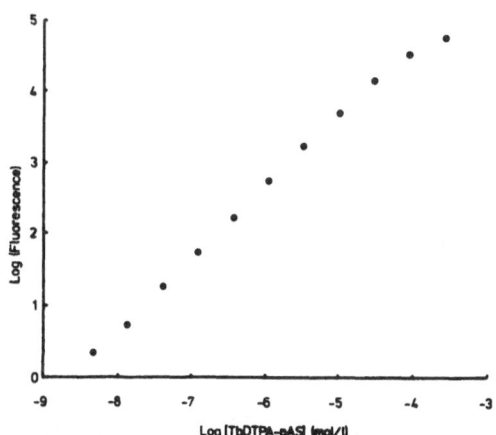

Figure 4. Dependence of fluorescence intensity on concentration for the terbium complex of DTPA-pAS.

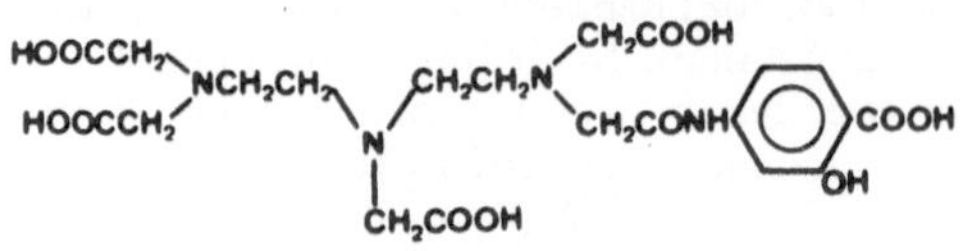

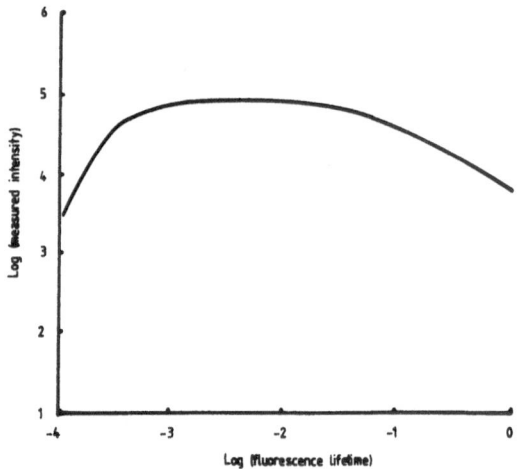

Figure 5. Presumed structure of DTPA-pAS.

nature of the reactants, and offers a plausible explanation for the
enhancement of terbium fluorescence. Molecular models of this structure
readily adopt conformations in which a cavity in the centre of the molecule,
of appropriate size to accommodate a terbium atom, is bounded by the
carboxylate groups of the acetic acid side chains and by the hydroxyl or
carboxylate groups of the aromatic moiety. The molecular model of phenyl-EDTA
has a similar cavity, but bounded only by aliphatic carboxyl groups. The
aromatic ring in this molecule is constrained to lie behind the "spine" of
the EDTA structure, and away from the metal atom.

Figure 6. Calculated dependence of measured fluorescence
intensity on fluorescence lifetime for a
hypothetical fluorophore, using xenon flash
illumination at 1000 pulses per second.

THEORETICAL CONSIDERATIONS

Our choice of terbium over europium as a possible label for fluorescence
immunoassay was determined largely by practical considerations: terbium forms
fluorescent chelates with a variety of reagents, and many of these chelates
are soluble in water, and fluorescent in aqueous solution. However, the
theoretical constraints on the lifetime of a label for time-resolved
applications have also interested us. It is readily apparent that a very
short-lived fluorescence will be difficult to resolve from the background
luminescence; and that a very long-lived emission will require long measuring

194

times. We performed computer simulations using a simplified mathematical model for the excitation/emission process, to determine the optimum value for the fluorescence lifetime (Bailey et al., 1986).

Using a xenon flash source, our calculations indicate that, for otherwise identical fluorophores, the measured intensity of fluorescence emission will rise with increasing fluorescence lifetime and reach a plateau, before decreasing with further increase in lifetime (figure 6). The plateau is quite broad, with an optimum lifetime of the order of 1 ms. Labels with lifetimes of less than about 300 μs show considerable reduction in measured intensity, and at still shorter lifetimes the measured intensity falls very rapidly. This loss occurs because most of the label emission occurs in the interval before measurement begins.

With increasing label lifetime, the output falls less precipitately; an increased excited state population tends to counteract the slower rate of excited state decay. However, the time required for the excited state population, and therefore the fluorescence output, to reach a steady value, becomes appreciable (several seconds for a fluorophore with a fluorescence lifetime of a few hundred milliseconds). Eventually, saturation of the excited state population will occur, and the fluorescence emission will fall with increasing lifetime: under these conditions, non-radiative processes will become increasingly significant, and it is likely that our simple model overestimates the output from very long-lived fluorophores.

Several other groups of workers have made some use of terbium chelates. Miller and Thirkettle (1975) found that the terbium complex of transferrin was highly fluorescent, and were able to develop an assay for transferrin in plasma based on the enhancement of terbium fluorescence. The terbium-transferrin complex was subsequently used to label gentamicin (Wilmott et al., 1984) and the conjugate was shown to react with anti-gentamicin antibodies. Unfortunately the macromolecular label appears to have caused problems with the development of a fluorescence immunoassay, and the authors suggest that labels based on iminodiacetic acid derivatives will eventually prove more useful.

A patent granted to Kodak (Hinshaw et al., 1982) describes one group of iminodiacetic acid derivatives of phenols, aromatic ketones, and coumarins. The use of some of these substances in solid-phase immunoassay systems is described.

Derivatives of phenyl-EDTA, carrying a diazo- or isothiocyanato-substituent, have been used for the attachment of both europium and terbium (inter alia) to macromolecules (Sundberg et al., 1974). The use of the terbium complex of phenyl-EDTA as a label in time-resolved fluorescence immunoassay was described by Kuo et al (1985). These authors used laser excitation and photon-counting detection to achieve ng/mL sensitivity in the assay of immunoglobulin G.

Recently Hemmilä (1985) has described the use of fluorinated β-diketone chelates of terbium in fluorescence immunoassay, using an approach analogous to that previously described for the corresponding europium chelates (Lövgren et al., 1984). One notable characteristic of these terbium chelates is their very short fluorescence lifetime (ca 100 μs) compared with that of many other terbium chelates (typically around 1 ms).

HOMOGENEOUS ASSAY

One of the attractions of conventional fluorescence assay has been the ability to develop homogeneous assays (Ullman et al., 1976; Nargessi et al.,

1978). The separation step in radioimmunoassay is a major source of imprecision, and has also been a serious obstacle to the development of completely automated systems. So far, no homogeneous time-resolved fluorescence immunoassays have been described. We have attempted without success to develop an energy-transfer immunoassay (Ullman et al., 1976) using a terbium-DTPA-pAS label: generally, no significant non-trivial quenching occurs. It seems likely that the resistance to quenching of this chelate is a consequence of the extent to which the emitting centre is protected from its environment.

An alternative approach would be to place the terbium atom on, say, the antigen, and a suitable enhancing ligand on the antibody. The close approach of the two species on antibody binding would then result in enhancement of the terbium fluorescence. We have been able to observe enhancement of the fluorescence of aqueous terbium-EDTA in the presence of albumin labelled with 4-diazosalicylic acid; and enhancement by salicylate ion of the fluorescence of terbium bound to albumin with 2-carboxyethyl-EDTA. Attempts to extend this approach, using albumin labelled with the terbium chelate of 2-carboxyethyl-EDTA and anti-albumin labelled with 4-diazosalicylic acid, have so far given results which, whilst encouraging, have not permitted the development of an assay. It seems probable, however, that a homogeneous assay will eventually be possible based on this approach. Such an assay should be easily automated using a flow system: ultimately we envisage an automated time-resolved fluorescence immunoassay system based on flow injection analysis (Riley et al., 1984), using pulsed laser excitation to achieve very high sensitivity.

SUMMARY

Terbium forms strongly fluorescent chelates with a variety of ligands containing aromatic groups. These chelates absorb at wavelengths characteristic of the aromatic group, and have emission spectra characteristic of terbium, consisting of a few narrow lines. The fluorescence also has an unusually long lifetime (typically around 1 ms). Reagents which incorporate aromatic groups combined with polyaminopolycarboxylate structures can form terbium complexes which are water soluble and highly fluorescent in aqueous solution, and do not dissociate at high dilution. This is in contrast to the fluorescent chelates of europium. These terbium chelates are suitable for the direct labelling of immunoassay reagents. Homogeneous assays have not yet been described, but it is likely that the development of such assays will be possible using terbium labels. The combination of a homogeneous assay using a terbium chelate label with time-resolved detection and flow injection technology offers the prospect of a serious alternative to radioimmunoassay.

ACKNOWLEDGEMENTS

We are grateful to the South-East Thames Regional Health Authority for financial support: to Perkin-Elmer Ltd. for the loan of an LS2 filter fluorimeter: and to Professor J.N. Miller of the Chemistry Department, University of Loughborough, for fluorescence lifetime measurements.

REFERENCES

Bailey, M.P., Rocks, B.F. and Riley, C., 1986, Theoretical Considerations on the Lifetimes of Labels for Time-Resolved Fluoroimmunoassay, J. Pharm. Biomed. Anal., in press.
Bailey, M.P., Rocks, B.F. and Riley, C., 1984, Terbium Chelate for Use as a Label in Fluoroimmunoassay, Analyst, 109: 1449-1450.
Bailey, M.P., Rocks, B.F. and Riley, C., 1985, Terbium Chelates for Fluorescence Immunoassay, Analyst, 110: 603-604.

Barela, T.D. and Sherry, D.A., 1976, A Simple, One-Step Fluorometric Method for Determination of Nanomolar Concentrations of Terbium, Anal. Biochem., 71: 351-357.

Butter, E., Kolowos, I. and Holzapfel, H., 1968, Ein Neues Reagenz zur Fluorometrischen Bestimmung von Terbium(III) und Dysprosium(III) in Wässriger Lösung, Talanta, 15: 901-911.

Coons, A.H., Creech, H.J. and Jones, R.N., 1941, Immunological Properties of an Antibody Containing a Fluorescent Group, Proc. Roy. Soc. Exp. Biol. (N.Y.), 47: 200-202.

Coons, A.H. and Kaplan, M.H., 1950, Localization of Antigen in Tissue Cells.-II. Improvements in a Method for the Detection of Antigen by Means of Fluorescent Antibody, J. Exp. Med., 91:1-13.

Haddad, P.R., 1977, The Application of Ternary Complexes to Spectrofluorometric Analysis, Talanta, 24: 1-13.

Hemmilä, I., Dakubu, S., Mukkala, V-M., Siitari, H. and Lövgren, T., 1984, Europium as a Label in Time-Resolved Immunofluorometric Assays, Anal. Biochem., 137: 335-343.

Hinshaw, J.C., Toner, J.L. and Reynolds, G.A., 1982, Eur. pat.appl. 82303380.8, Fluorescent Chelates and Labelled Specific Binding Reagents prepared therefrom.

Lövgren, T., Hemmilä, I., Pettersson, K., Eskola, J.U. and Bertoft, E., 1984, Determination of Hormones by Time-Resolved Fluoroimmunoassay, Talanta, 31: 909-916.

Lyle, S.J. and Za'tar, N., 1983, A Comparative Study of Some Methods for the Spectrophotometric Determination of Terbium in Aqueous Solutions Containing Other Lanthanides and Yttrium, Anal. Chim. Acta, 153: 229-236.

Miller, J.N. and Thirkettle, C., 1975, The Fluorimetric Determination of Transferrin in Blood Serum, Biochem. Med., 13: 98-100.

Nargessi, R.D., Landon, J. and Smith, D.S., 1978, Non-separation Fluoroimmunoassay of Human Albumin in Biological Fluids, Clin. Chim. Acta., 89: 461-467.

Sidki, A.M. and Landon, J., 1985, Fluoroimmunoassay and Phosphoroimmunoassay, in "Alternative Immunoassays", W.P. Collins, ed., J. Wiley, Chichester, 185-201.

Sundberg, M.W., Meares, C.F., Goodwin, D.A. and Diamanti, C.I., 1974, Selective Binding of Metal Ions to Macromolecules Using Bifunctional Analogs of EDTA, J. Med. Chem., 17: 1304-1307.

Ullman, E.F., Schwarzberg, M. and Rubenstein, K.E., 1976, Fluorescent Excitation Transfer Immunoassay. A General Method for Determination of Antigens, J. Biol. Chem., 251: 4172-4178.

Wieder, I., 1978, Background Rejection in Fluorescence Immunoassay, in: "Immunofluorescence and Related Staining Techniques. Proceedings of the VIth International Conference on Immunofluorescence and Related Staining Techniques.", W. Knapp, K. Holubar and G. Wick, eds., Elsevier/North-Holland Biomedical Press, Amsterdam and New York, 67-80, and references therein.

Wilmott, N.J., Miller, J.N. and Tyson, J.F., 1984, Potential Use of a Terbium-Transferrin Complex as a Label in an Immunoassay for Gentamicin, Analyst, 109: 343-345.

FLUORESCENCE POLARIZATION IMMUNOASSAY

THEORY AND APPLICATION

Fukuko Watanabe

Clinical Chemistry Laboratory
Kobe Women's College of
Pharmacy
Kobe, Japan

Kiyoshi Miyai

Department of Laboratory
Medicine
Osaka University Medical
School
Osaka, Japan

INTRODUCTION

The principles of fluorescence polarization were first developed by Perrin (1926). About 30 years later the technique was applied to biological system by Weber (1953), and its application to the antigen-antibody reaction was first described by Dandliker and Feigen (1961). Since then, the principles and practice of fluorescence polarization and fluorescence polarization immunoassay have been described in a number of review articles (Dandliker et al., 1964; Parker, 1973; Dandliker and de Saussure, 1970).

Fluorescence polarization immunoassay makes use of competitive-binding assay, measuring tracer binding directly, without the need for a separation procedure (homogeneous method). The specificity of immunoassay is thereby combined with the speed and convenience of a homogeneous method, providing a precise and reliable procedure for determining the concentration of biologically interesting substances in serum and plasma. Fluorescence polarization immunoassay has been successfully applied to the measurement of such compounds as insulin, cortisol, thyroxine, human chorionic gonadotropin, and various drugs present in serum.

Several experimental studies on fluorescence polarization immunoassay have been reported during the past 10 years, but with little clinical application because of some major drawbacks existing, that is, the lack of simple and high performance instrumentation, non-specific binding of tracer to serum proteins, and the intrinsic fluorescence of serum.

Recently, much interests have been focused on monitoring the serum concentrations of drugs such as antibiotics and anticonvulsants, since monitoring these drugs has improved their therapeutic usefulness be allowing adjustment of their dosages. Several different instrument systems have been developed, some of which are now available commercially, to overcome many of the problems encountered in this field.

Principles of fluorescence polarization immunoassay

The principles of fluorescence polarization is as follows (Fig. 1). Light is resolved with a fixed polarizing lens or prism into rays with their electrical vectors in a single plane (e.g., plane Z-O-Y). This vertical polarized light is vibrating in the plane along the OZ axis and is propagated in direction OY axis. When fluorescence substances in solution are excited with the polarized light, they emit partially polarized fluorescence as viewed at right angles (OX axis) to the incident beam. The emission is determined by a rotating polarizer. On the contrary, in a totally automated system for measuring polarization, the plane of polarization of the exciting light is rotated from vertical to horizontal, and the emission optics contain a fixed vertical polarizer. The polarization (P) is calculated by the following formula:

$$P = (F_{/\!/} - F_{\perp})/(F_{/\!/} + F_{\perp})$$

where $F_{/\!/}$ is the fluorescence intensity with the analyzer prism positioned vertically (Parallel to OZ) and $F_{\perp}$ is the fluorescence intensity in the prism positioned horizontally (parallel to OY).

Polarization of fluorescence is due to the relationship between molecular orientation and the absorption and emission of fluorescence. In a solution of randomly oriented fluorescent molecules, polarized light will preferentially excite molecules positioned so that their absorption oscillators are parallel to the plane of polarization. Provided the position of the moelcule at the time of excitation is partially retained at emission and fluorescence will be partially polarized. The extent to which fluorescent molecules polarize fluorescence is influenced by a number of factors. The fluorescence polarization is expressed as a function of the fluorescence molecule, the lifetime of its excited state and solvent temperature and viscosity as follows (Perrin, 1926).

$$\left(\frac{1}{P} - \frac{1}{3}\right) = \left(\frac{1}{P_0} - \frac{1}{3}\right) \cdot \left(1 + \frac{3\tau}{Q}\right), \qquad Q = \frac{3\eta V}{RT}$$

P = observed polarization of fluorescence
P_0= the limiting polarization of fluorescence
τ = the fluorescence lifetime of molecule
Q = the rotational relaxation time of molecule
η = viscosity
V = the molar volume of the fluorescent molecule
R = the gas constant
T = absolute temperature

The rotational relaxation time of molecule (Q) is defined as the time required for a molecule to rotate through an angle of approximately 68.5°. The rotational relaxation time which depend on the volume of the fluorescence molecule (V), is small (less than 1nsec) for small molecules (e.g. fluorescein) and large (1 – 100nsec) for large molecules(e.g. immunoglobulin). If the fluorescence lifetime (τ), the temperature (T) and viscosity (η) of the solution are constant, the observed polarization of fluorescence (P) gives to direct indication of the size of molecule (V).

The principle of a fluorescence polarization immunoassay is as follows. If a small fluorescent hapten or antigen (mol wt 1,000 – 10,000) with rapid movement become bound to antibody (mol wt 160,000), rotational relaxation time will be increased as a result of formation of the antigen-antibody complex with slow movement. Thus the polarization is increased (Fig. 1A). When unlabeled antigen is added to this system, it competes with fluorescent antigen for binding to the antibody and the amount of free fluorescent antigen with rapid movement is increased (Fig. 1B). Thus the rotational relaxation time is decreased and the polarization is decreased. If the amount of antibody is constant, the degree of polarization will be inversely correlated with the amount of added unlabeled antigen. In other words, the added unlabeled antigen can be determined from the degree of decreased polarization.

Based on the above principle, fluorescence polarization immunoassay is not stuitable for measurement of large molecular antigens, since Brownian motion of these molecules are originally inactive and is not largely affected by the binding of the antibody, producing no marked change in polarization. It is more useful for measurement of small molecule antigen.

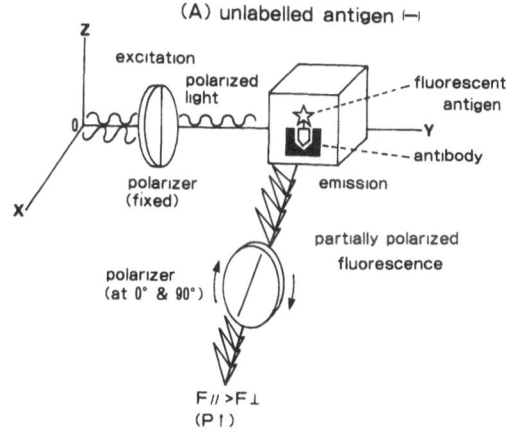

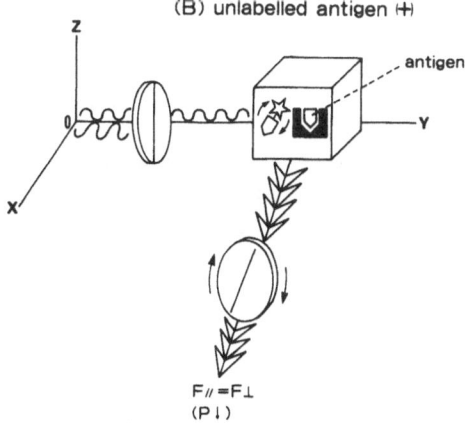

Fig. 1 Schematic representation of principle of
fluorescence polarization immunoassay

MATERIALS AND METHODS

Fluorescence labels

Fluorescence labels used in fluoroimmunoassay should fulfill the following conditions.
1. The molecular absorbance should be as great as possible.
2. The quantum yield in the buffered aqueous solution in which immuno-reaction occurs should be as great as possible.
3. Since serum or antiserum shows the maximum fluorescence near the excitation wave length at 340 nm and the emission wave length of 470 nm, it produces background fluorescence. To avoid this background, therefore, substances with an emission wave length of 500-700 nm and a Stokes' shift of 50 nm or greater are desirable as fluorescence labels.
4. The substances must by highly soluble in water, and solubility should not decrease when the fluorescence label-antigen ratio is increased.
5. They must be stable, and their conjugates with the antigens must also tolerate long-term storage.
6. They should not interfere with the antigen-antibody reaction.
7. For fluorescence polarization immunoassay, fluorescence labels with a relatively long lifetime (20 ns) are required.

Table 1 List of fluorescence polarization immunoassay

Analyte	Parent compound	Fluorescent probe	Reference
Gentamicin	Gentamicin	5-[(4,6-Dichlorotriazin -2-yl)-amino] fluorescein	a, b c, d
Phenobarbital	5-(3'-Aminophenyl)-5 -ethyl-2,4,6 (1H,3H, 5H)pyrimidinetione	"	b, d
Theophyllin	8-Aminoethyl 1-3,7 -dihydro-1,3-dimethyl -1H-purine-2,6-dione	"	b, d
Tobramicin	Tobramicin	"	c
Amicacin	Amicacin	"	c
Streptomycin	Streptomycin	"	e
Phenytoin	2-(2'-Aminoethyl) -5,5-diphenyl-2,4 -imidazolidinedione	5-Carboxyfluorescein -N-hydroxy-succinimide ester	d, f g
Theophyllin	8-(3-Carboxypropyl) -theophyllin	Cumarin-3-carboxylic acid	h
Cortisol	21-Amino-21 -deoxycortisol (Cortisol 21-amine)	Fluorescein isothiocyanate (FITC)	i, j
Thyroxine	Thyroxine	"	k, l
Insulin	Insulin	"	m
Human chorionic gonadotropin (HCG)	HCG	"	n

a:Watson et al., 1976, b:Jolley, 1981, c:Jolley et al., 1981a, d:Jolly et al., 1981b, e:Schwenzer et al., 1983, f:Lu-Steffes, 1982, g:McGregor et al., 1978, h:Li et al., 1981, i:Kobayashi et al., 1979, j:Kobayashi et al., 1979a,k:Nakajima et al., 1979, l:Muratsugu et al., 1982, m: Yamaguchi et al., 1982, n:Urios et al., 1978

The most commonly used fluorescent labels in fluorescence polarization immunoassay is fluorescein isothiocyanate (FITC). Fluorescent labeled analytes are prepared by conjugation the analyte, or a derivative thereof, with a fluorescent molecule such as fluorescein (Table 1).

Antibody

It is desirable to use purified globulin fractions of antibody rather than whole serum. Fluorescent molecules bind non specifically to serum albumins, requiring the use of correction factors in calculating binding data if serum is used. Satisfactory globulin fractions have been obtained by precipitation with ammonium sulfate or by DEAE-cellulose fractionations.

Apparatus

Several fluorimeters have been used for fluorescence polarization immunoassay. In our previous experiments (Kobayashi et al., 1979; Kobayashi et al., 1979a), we used a model IBF-129 polarization fluorimeter (Kowa Kizai Ltd., Japan). The excitation and emission wave lengths were 490 nm and 520 nm, respectively. During measurement of polarization, the temperature changes of cell chamber was controlled to within $\pm 0.5^{o}$C. In the reports by Watson (1976) and McGregor (1978), fluorescence polarization measurements were made with a model RS-3 polarization fluorimeter (SLM Instruments, 2403 Hathaway Drive, Champaigen, Ill. 61820) which was connected to a pen recorder.

The instrument, developed by Abbott Laboratories (North Chicago, Ill. 60064) for quantitating fluorescence polarization immunoassays, can measure automatically polarization components and compute a polarization value after correcting for background and optical bias. Furthermore, a fully automated system for performing fluorescence polarization immunoassay has been developed by other laboratories.

TYPICAL RESULTS

(A) CORTISOL

Fluorescence polarization immunoassay of cortisol was developed in our laboratory (Kobayashi et al., 1979; Kobayashi et al., 1979a)

Materials.

Conjugation of fluorescein isothiocyanate (FITC) and 21-amino-21-deoxy-cortisol (cortisol 21-amine): A method for the preparation of cortisol 21-amine was described elsewhere (Kobayashi et al., 1978). All the procedures for the conjugate preparation were carried out away from daylight. A solution of 4 mg of cortisol 21-amine·HCl in 0.5 ml of methanol was adjusted to pH 9.0 with 0.1 mol/l NaOH and mixed with 20 mg of FITC and then 0.4 ml of dioxane. The mixture was stirred at room temperature for 4 h and applied to thin layer chromatography for purification. The developing solvent systems consist of ethyl acetate/chloroform/methanol/acetic acid (5:3:1:1, v/v) and benzene/ethyl acetate/ethanol (8:2:1, v/v). The conjugate thus prepared is stable for at least one year when stored at 4°C.

Preparation of antiserum to cortisol: Cortisol-21-hemisuccinate was conjugated with bovine serum albumin (BSA) by a mixed anhydride reaction, and the antibody to the conjugate was raised in rabbits. The globulin fraction of antiserum was obtained by precipitation with 50% saturation of ammonium sulfate, and the IgG fraction (anti-cortisol IgG fraction) was prepared by DEAE-Sephadex column chromatography.

Procedure

Samples of 10 μl of standard serum or test serum were mixed with 50 μl of 1% sodium dodecyl sulfate (SDS) solution and the mixtures were left at room temperature for 30 min. Then, 1 ml of FITC-cortisol solution in phosphate buffered saline (0.01 mol/l, pH 7.4) was added and the polarization (A) before the antigen antibody reaction was measured with a model IBF-129 fluorescence polarization fluorimeter using quartz cells with excitation and emission wave lengths at 490 nm and 520 nm, respectively. Then, 10 μl of anti-cortisol IgG fraction was added and the mixture was incubated at constant temperature of 20-25°C for 60 min, and the polarization (B) was measured. A correction was made for minimizing interference due to serum by subtracting the polarization value (A) from the value (B).

Results and discussion

A typical standard curve is shown in Fig.2. Serum concentration of 1 to 100 μg/100 ml (0.027 to 2.7 μmol/l) of cortisol could be measured. Therefore, the minimal amount of cortisol detectable was estimated as 0.1 ng (0.27 pmol)/assay tube.

When serum samples were diluted serially with cortisol free serum and their cortisol levels were determined by the present method, a linear plot passing through the zero point was obtained. The precision was estimated as the coefficient of variation (CV) from 6 determinations. The intra-assay CV was 7.2 to 11.6% (mean concentration of cortisol 34.5 to 4.3 μg/100 ml) and the inter-assay CV was 4.6 to 10.6% (32.1 to 17.2 μg/100 ml), respectively.

Cortisol levels in 20 serum samples were determined by radioimmunoassay (x) and the present method (y). There was a good correlation between the values obtained by the two methods ($y = 1.08x - 0.13$, $r = 0.95$).

In our previously experiments (Kobayashi et al., 1979), purification procedures were required to eliminate the non-specific binding of fluorescein-labeled hapten to serum protein. But in the present method (Kobayashi et al., 1979a), the non-specific binding was successfully eliminated by the addition of SDS to serum samples.

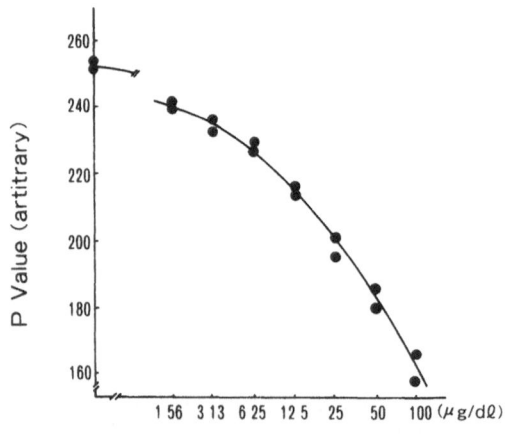

Fig. 2   Standard curve for fluorescence polarization
immunoassay of cortisol

(B) THYROXINE

Fluorescence polarization immunoassay was developed by Nakajima et al. (1979).

Materials
FITC-thyroxine: FITC-labeled thyroxine was prepared according to the method of Smith (1977).

Antiserum to thyroxine: Thyroxine was conjugated with BSA by carbo-diimide reaction, and the antibody to thyroxine-BSA was produced in rabbits. The globulin fraction of antiserum was obtained by the same as the preparation of cortisol IgG fraction.

Procedure
Serum sample was shaken with two volumes of methanol, and the mixture was centrifuged for 10 min at 1000 g. Fifty µl of the super-natant and 50 µl of anti-thyroxine IgG fraction were added to 2 ml of 6 mmol/l phosphate buffered saline (pH 7.2). The mixture was allowed to stand at 25°C for 30 min. After incubation, 50 µl of FITC-thyroxine was added, and the fluorescence polarization was immediately measured with a model IBF-129 fluorescence polarization fluorimeter using quarz cells.

Results and discussion
A typical standard curve and the specificity of the anti-thyroxine IgG fraction to thyroxine are presented in Fig. 3. Monoiodotyrosine (MIT), diiodotyrosine (DIT), and 3,5,3'-triiodothyronine (T$_3$) tested showed almost no detectable cross-reaction.

Thyroxine levels in 63 samples were determined by radioimmunoassay (x) and the present method (y). There was a good correlation between the values obtained by the two methods (y = 1.03x + 0.18, r = 0.96).

For clinical purpose, the present method requires only a small amount of serum (less than 100 µl is available).

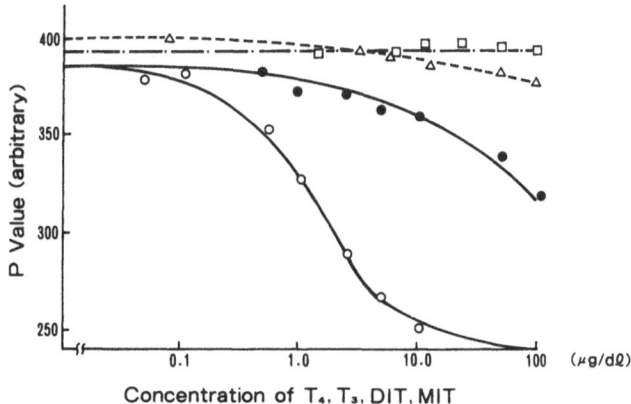

Fig. 3 Standard curve for fluorescence polarization immunoassay of thyroxine and cross reactivity test with other thyroid compounds

thyroxine (T$_4$) (o), 3,5,3'-triiodothyronine (T$_3$) (●)
monoiodotyrosine (MIT) (Δ), diiodotyrosine (DIT) (□)

(C) THERAPEUTIC DRUGS

A fully automated system for performing fluorescence polarization immunoassay for therapeutic drugs has been developed by Abbott Laboratories (Jolley et al., 1981a). All reagents for each assay are contained in coded reagent packs.

Materials

Buffer solution: 0.1 mol/l sodium phosphate solution, pH 7.5 containing 0.1 g of bovine gamma-globulin and 0.1 g of sodium azide per liter.

Pretreatment reagent (for releasing bound drug from serum protein): 0.1 mol/l Tris buffer solution, containing 0.1 g of sodium azide per liter.

Antiserum: antiserum diluted with pH 7.5 phosphate buffer solution containing 0.1 g of bovine gamma-globulin and 0.1 g of sodium azide, and 50 mg of benzalkonium chloride per liter.

Tracer reagent: Tracers diluted with the Tris buffer solution (0.1 mol/l, pH 7.5) containing 1.2 to 12 g of SDS, 0.1 g of bovine gamma-globulin and 0.1 g of sodium azide per liter. The concentration of tracer was approximately 100 nmol/l for each.

Pure drugs were obtained from the U.S. Pharmacopoeial Convention, Inc., Rockville, MD 20852 and were used according to the supplier's instructions. Pooled drug-free normal human serum was used throughout. The standards and controls were prepared by dilution from a concentrated stock solution.

Apparatus

The analyzer was developed by Jolley et al.(1981b) and the following explanation is cited from their manuscript.

A diagram of major components of the apparatus is shown in Fig. 4. The apparatus consists of the reagent-reagent sampling component, carousel component, and optic module component. The determination is performed with a microprocessor-based electronic system that has the following features: sample dilution, supplying of each reagent, rotation of the carousel, polarization fluorimetry, and temperature control.

The carousel can be mounted with 20 sample cups and cuvettes. The tracer, antibody, and pretreatment reagent are kept in a three-chamber pack, identified by a bar-code.

The buffer solution is added during the assay with a pump. The sample (50 µl) is placed on the sample compartment of the dual-chamber sample cup. The reagent pack for assay is stored at 4-8°C and is directly mounted on the instrument. The carousel and the reagent pack are placed at assigned positions, the door is closed, and the run button is pressed.

The bar-code on the reagent pack, which indicates the item to be assayed is read automatically, and the determination is started. Each reagent is dispenses and diluted according to the protocol for the substance to be measured, and the volumes of the reagent and the sample are checked by the liquid-level sensor at the tip of the pipettor. The temperature of the cuvettes is controlled by an infrared sensor in the optics module and an air heater at 35 ± 0.5°C.

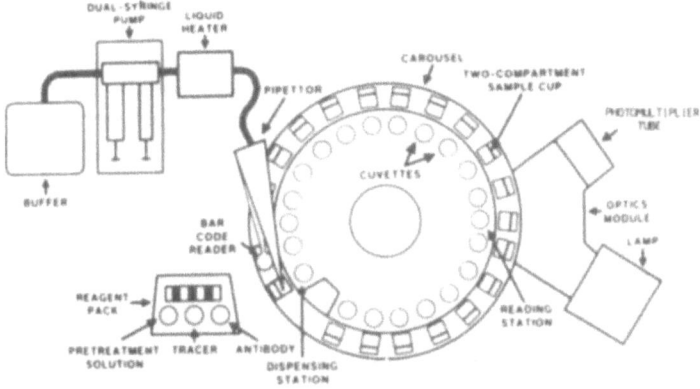

Fig. 4   The major components of the totally automated, bench-top, fluorescence polarization analyzer. (cited from Jolley et al, 1981b)

Procedure

A typical protocol is as follows.

1. During the first revolution of the carousel, the sample (20 μl) and the buffer solution (700 μl) are mixed and diluted in the dilution well of the sample cup.
2. During the second revolution, the pretreatment solution (25 μl) is added to the diluted sample (10-25 μl), and the total volume is adjusted by simultaneously adding the buffer solution to 975 μl. When the cuvette arrives at the fluorimeter, the blank reading is obtained.
3. During the third revolution, the tracer (25 μl), the antibody (25 μl), and the same volume of the diluted sample as was added during the second revolution are added, and the total volume is adjusted to 1.95 ml with the buffer solution.
4. During the fourth revolution after a 3-minute incubation, fluorescence polarization is determined. The blank-corrected polarization value is calculated, and the concentration of the drug is determined from the standard curve.

The entire process from the pretreatment of samples to the end of fluorimetry is completely automated. The time required from mounting the samples to printing out the results is less than 10 min for 20 samples and less than 5 min for a single determination.

DISCUSSION AND SUMMARY

The limited acceptance of fluorescence polarization immunoassay to date may be ascribed to the delay in the development of the polarization fluorimeter and the difficulty of its access for general use. The development of reagents in kit forms and completely automated assay apparatus as described here is considered to be of great important to the future spread of fluorescence polarization immunoassay.

Furthermore, laser excited fluorescence polarization immunoassay, in which laser instead of a tungstic lamp is used, is under consideration at our laboratory in an attempt to achieve higher sensitivity.

Fluorescence polarization immunoassay is useful when examination of a large number of samples is required as in a) drug monitoring in which serial samples must be assayed rapidly and b) neonatal mass-screening for such disorders as congenital adrenal hyperplasia and hypothyroidism. It is expected to be used more widely in preventive medicine in the future.

REFERENCES

Dandliker W.B. and Feigen G.A., 1961, Quantification of the antigen-antibody reaction by the polarization of fluorescence, Biochem. Biophys. Res. Commun. 5: 299-304.

Dandliker W.B., Schapiro H.C., Meduski J.W., Alonso R., Feigen G.A. and Hamrick J.R., 1964, Application of fluorescence polarization to the antigen-antibody reaction, Theory and experimental method, Immunochemistry 1: 165-191.

Dandliker W.B. and de Saussure V.A., 1970, Fluorescence polarization in immunochemistry, Immunochemistry 7: 799-828.

Jolley M.E., 1981, Fluorescence polarization immunoassay for the determination of therapeutic drug levels in human plasma, J. Anal. Toxicol. 5: 236-240.

Jolley M.E., Stroupe, S.D., Wang J., Panas H.N., Keegan C.L., Schmidt R.L. and Schwenzer K.S., 1981a, Fluorescence polarization immunoassay I. Monitoring aminoglycoside antibiotics in serum and plasma, Clin. Chem. 27: 1190-1197.

Jolley M.E., Stroupe S.D., Schwenzer K.S., Wang C-H. J., Lu-Steffes M., Hill H.D., Popelka S.R., Holen J.T. and Kelso D.M., 1981b, Fluorescence polarization immunoassay III. An automated system for therapeutic drug determination, Clin. Chem. 27: 1575-1579.

Kobayashi Y., Ogihara T., Amitani K., Watanabe F., Kiguchi T., Ninomiya I. and Kumahara Y., 1978, Enzyme inmunoassay for cortisol in serum using cortisol 21-amine, Steroids 32: 137-144.

Kobayashi Y., Amitani K., Watanabe F. and Miyai K., 1979, Fluorescence polarization immunoassay for cortisol. Clin. Chim. Acta 92: 241-247.

Kobayashi Y., Miyai K., Tsubota N. and Watanabe F., 1979a, Direct fluorescence polarization immunoassay of serum cortisol, Steroids 34: 829-834.

Li T.M., Benovic J.L. and Burd J.F. 1981, Serum theophyllin determination by fluorescence polarization immunoassay utilizing an umbelliferone derivative, Anal. Biochem. 118: 102-107.

Lu-Steffes M., 1982, Fluorescence polarization immunoassay IV. Determination of phenytoin and phenobarbital in human serum and plasma, Clin. Chem. 28: 2278-2282.

McGregor A.R., Prookall-Greening J.O., Landon J. and Smith D.S., 1978, Polarization fluoroimmunoassay of phenytoin, Clin. Chim. Acta 83: 161-166.

Muratsugu M. and Makino M., 1982, Evaluation of fluorescence polarization immunoassay for serum thyroxine determination, J. Clin. Chem. Clin. Biochem. 20: 567-570.

Nakajima K., Mochida H., Ueno T., Suzuki M., Kobayashi I., Matsuda I. and Miyai K., 1979, A simple, rapid method of blood thyroxine assay using a fluorescence polarization technique, Clin. Chem. Symposium (in Japanese) 18: 30-35.

Parker C.W., 1973, Spectrofluometric methods, in "Hand Book of Experimental Immunology", Weir D.M., ed., Blackwell Scientific Pub., Oxford.

Perrin F., 1926, Polarization de la lumiere de fluorescence. Vie moyenne de molecules dans l'etat excite, J. Phys. Radium. 7: 390-401.

Schwenzer K.S. and Ahalt J.P., 1983, Automated fluorescence polarization
        immunoassay for monitoring streptomycin, Antimicrobial Agents &
        Chemotherapy, 23: 683-687.
Smith D.S., 1977, Enhancement fluoroimmunoassay of thyroxine, FEBS Lett.
        77: 25-27.
Urios P., Cittanova N. and Jayle M-F., 1978, Immunoassay of the human
        chorionic gonodotropin using fluorescence polarization, FEBS
        Lett. 94: 54-58.
Watson R.A.A., Landon J., Shaw E.J. and Smith D.S., 1976, Polarization
        fluoroimmunoassay of gentamicin, Clin. Chim. Acta 73: 51-55.
Weber G., 1953, Rotational Brownian motion and polarization of the fluo-
        rescence of solutions, Adv. Protein Chem. 8: 415-459
Yamaguchi Y., Hayashi C. and Miyai K., 1982, Fluorescence polarization
        immunoassay for insulin preparation, Anal. Lett. 15: 731-737.

FLUORESCENCE ENERGY TRANSFER IMMUNOASSAYS

Pyare L. Khanna

Syva

900 Arastradero, Palo Alto, California 94304*

INTRODUCTION

Homogeneous immunoassays have gained broad acceptance during the past
decade and are being widely utilized for both quantitative and qualitative
analysis of a wide variety of analytes in the clinical laboratory. Concur-
rent to the development of enzyme immunoassays, homogeneous fluorescence
immunoassays have played an equally important role in the development of a
wide variety of immunochemical methods. One of the earliest procedures of
histochemical staining using fluorescent labels (Coones, Creech and Jones,
1941) is still a method of choice for immunochemical localization of tissue
antigens. After incubation of tissue antigens with fluorescer-labeled
primary or secondary antibody, the excess of the labeled reagent is washed
away and the resulting stained immune complexes can be visualized under a
dark field microscope. This method, although cumbersome as an immunoassay
technique, does provide useful qualitative data. Such use of receptor
groups in competitive immunoassays was not further exploited until 1959
when the technique of radioimmunoassays was introduced (Berson and Yallow,
1959). The basic technique involves competition of the analyte with a
radio-labeled antigen for a limited number of binding sites. After separation of free and bound antigen and measurement of the radioactivity of

* Present address: Microgenics, 2380A Bisso Lane, Concord, CA 94520

these fractions, the quantitation can be achieved by comparison with measurements obtained from a set of known standards. This method provided the ability to determine antigens in solutions quantitatively. Although various modifications have been developed during the past years, the basic technique suffers from the short shelf life of the radioisotopes and the complex labor intensive protocols involving separation of the bound and free fractions prior to measurement. More recently, there has been considerable interest in the use of nonisotopic labels for immunoassays. These have included enzymes (Scharpe, Cooneman and Bloome, 1976; Wisdom, 1976; Van Weeman and Schuurs, 1971), metals (Cais, Dari, Eder, et al, 1977), chemiluminescent groups (Simpson, Campbell and Ryall, 1979), and fluorescent groups (Soini and Hemmila, 1979).

Although these nonradioactive labels offer the basic advantages of safety and required sensitivity, the concept of separation of free and bound fractions (heterogeneous immunoassays) still makes the overall assay protocol less attractive. In the last 20 years, new immunochemical procedures have been introduced in which separation of free from bound label was eliminated. These assays have been defined as homogeneous immunoassays. They are competition procedures that characteristically avoid the separation step and hence are fast, more reliable and convenient to be applied on common instruments.

The first homogenous method (Dandliker and Geiger, 1961) was based on the change of fluorescence polarization of fluorescein-labeled penicillin upon binding by antipenicillin antibodies. This change in tumbling rate associated with antibody binding, leading to increased polarized fluorescence, has been used for assay of a variety of haptens. This basic concept, although not exploited commercially for 20 years due to lack of availability of appropriate instrumentation, has recently been fully developed and is widely used in clinical laboratories for quantitative determination of antigens. The method, however, suffers from the limitation of

212

being applicable to only small molecules and is inherently less sensitive, since only a small percentage of the total fluorescence intensity can be modulated. In addition, weak nonspecific interactions of fluorescer-labeled antigens with serum proteins can change its tumbling rate and produce anomalous results.

A more general homogenous method, Fluorescence Energy Transfer Immunoassay (FETI) (Ullman, Schwarzberg and Rubenstein, 1976) is described in greater detail in this chapter. This method offers greater simplicity and generality for both small molecules and proteins.

ASSAY PRINCIPLE

The FETI method is based on the principle of dipole-dipole coupled energy transfer from an electronically excited fluorescent dye to an acceptor dye when they are in close proximity to each other. Fluorescence decay of an electronically excited molecule results in a change in its electric dipole moment. In the proximity of a second dye molecule, a nonradiative energy transfer between the molecules can occur through dipole-dipole coupling. An antigen labeled with an energy donor (fluorescer), when brought in close proximity to an antibody labeled with an energy acceptor (quencher), can result in energy transfer through space by inductive resonance over substantial distances. The efficiency of energy transfer E is inversely proportional to the sixth power of the distance r between the donor and acceptor molecules (Förster, 1948). The term E is defined as:

$$E = \frac{\emptyset_o - \emptyset_q}{\emptyset_o} = \frac{r^{-6}}{r^{-6} + R_o^{-6}}$$

Where $\emptyset_o$ and $\emptyset_q$ are the donor fluorescence quantum yields in the absence and presence of the acceptor, respectively, and r is the distance in angstroms between the centers of the donor and acceptor chromophores, the term R_o is distance in angstroms at which 50% transfer efficiency occurs and is represented by:

$$R_o = \frac{9.7 \times 10^3}{\sqrt[6]{J\, k^2\, \emptyset_o\, n^{-4}}} \; A^o$$

Where k^2 is the orientation factor for dipole-dipole transfer (which for randomly oriented dyes is normally $\sqrt{2/3}$), J is the spectral overlap integral of the donor emission and acceptor absorption, and n is the index of refraction of the medium. The spectral overlap integral J is dependent on the molecular extinction coefficient E of the acceptor at each emission wavelength. For efficient energy transfer, the relative orientations of both donor and acceptor oscillations must permit strong interactions.

The value of R_o depends primarily on the quantum yield of the donor and the extinction coefficient of the acceptor, and will vary with donor-acceptor pairs. For commonly used dye pairs like fluorescein-rhodamine, R_o has been calculated to be 58 A^o. The value of R_o could approach around 84 A^o for a hypothetical perfect-dye pair with a molar extinction coefficient of 3×10^5 throughout the emission wavelength of the perfect donor $\emptyset_o = 1$ at 500 nm in water (n = 1.33). Since distances within immune complexes can be of the order of 100 A^o, efficient energy transfer could, in principle, take place with R_o values mentioned above. This basic phenomenon of energy transfer (Gantt, 1981) occurs very efficiently in nature in a number of photosynthetic light harvesting systems.

A most desirable, practical FETI configuration employs a relatively pure antigen labeled covalently with an energy-donor (fluorescer) and specific antibodies labeled with an energy-acceptor (quencher). Mixing of the labeled reagents and an unknown sample results in a competition of labeled antigen and sample antigen for a limited number of antibody binding sites. In the presence of sample antigen, some of the quencher-labeled

214

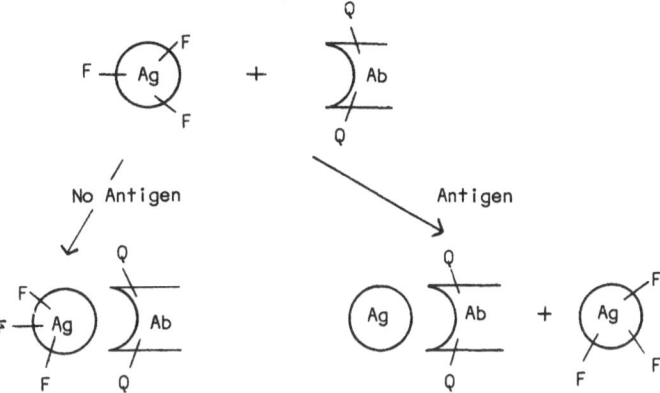

Fig. 1. Principle of FETI

antibody will be used up and be unavailable for binding to the fluorescer-labeled molecules, thereby resulting in a proportional increase in total fluorescence intensity. This is schematically represented in Fig. 1.

The FETI technique has found general applicability both for small molecules (haptens) and macromolecules (proteins). An alternate assay configuration for multivalent antigens employs the use of a mixture of reagents comprising an antibody labeled with a fluorescer and an antibody labeled with a quencher. The resulting immune complex (Fig. 2. Sandwich Assay) renders the fluorescer and quencher in close proximity for significant energy transfer.

The method can have more practical utility if the fluorescer-labeled antibody is either highly purified or monoclonal, in order to avoid large, immodulable fluorescence background. Assays constructed in this way offer exceptionally simple protocols. This method is particularly desirable since the two labeled antibodies can be combined into a single reagent. Fluorescence measurement, after mixing of this single reagent with patient sample, results in the measurement of antigen in the sample. However, in the presence of excess antigen, biphasic response in the standard curve is observed.

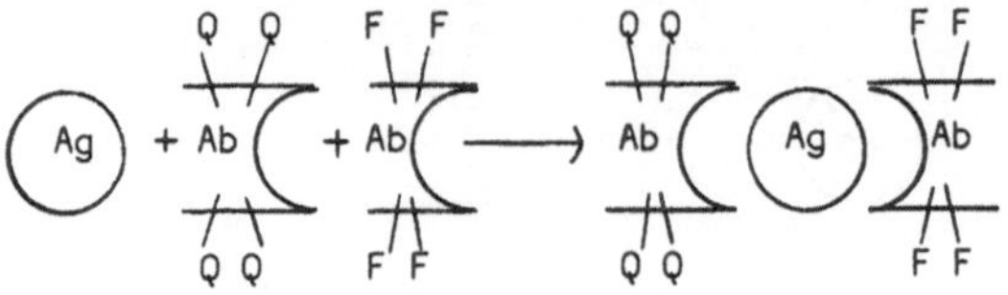

Fig. 2.   Sandwich FETI assay

Other variants of FETI methodology have also been explored where a
fluorescer-labeled antibody and a quencher-labeled antigen are employed.
This assay configuration can offer higher sensitivity due to multi-
attachment of fluorescers to a single antibody molecule. However, sensitivity is limited in a practical sense from two major factors: (a) fluorescence background by direct excitation of the fluorescer-labeled nonspecific
IgG, thereby increasing the intensity of unmodulated fluorescence signal,
and (b) lack of a suitable quencher which can efficiently quench the
fluorescence from an immune complex where an antibody is randomly labeled
with fluorescent molecules. The random labeling would place some of the
fluorescent tags on the F_c fragment, which is usually too far away from the
binding sites to permit efficient quenching.

MATERIALS AND METHODS

Some of the common fluorescer- and quencher-activated derivatives,
like fluorescein isothiocyanate and tetramethylrhodamine isothiocyanate,
are available commercially (Molecular Probes Catalog, Oregon, CA). Preparation of a few new fluorescein derivatives, which could serve as fluorescers or quenchers, depending on the nature of substitution, is described
elsewhere (Ullman and Khanna, 1981). During the last five years, a new
class of naturally-occurring macromolecular fluorescers, phycobiliproteins,
have been described (Oi, Glazer and Stryer, 1982), and are also available
commercially.

Preparation of fluorescer-labeled haptens is normally carried out by
reaction of a suitable hapten derivative with an activated fluorescer

216

molecule. An example is the preparation of a morphine-fluorescein conjugate from 6-carboxyfluorescein, and O^3-aminoethyl morphine (Ullman, Schwarzberg and Rubenstein, 1976). General procedures for the preparation of a hapten-macromolecular fluorescer conjugate, and quencher-labeled anti-hapten antibodies conjugate, are described below.

Preparation of Digoxin-B-Phycoerythrin Conjugate   A solution of N-hydroxy succinimide ester of 3-O-carboxymethyl digoxin (2 mg in 0.25 ml DMF) is added drop-wise to a purified B-phycoerythrin solution (5 mg/5 ml in 0.05 M PO_4^{3-} buffer pH 8.0) over a 30-minute period at 4°C. After an additional two-hour stirring, the resulting conjugate is purified over a sephadex G-50 column using 0.05 M tricine buffer, pH 7.5. The resulting conjugate is used directly in the assay with appropriate loading for adequate assay performance in the desired sensitivity range.

Preparation of Quencher Dye Labeled Antidigoxin Antibody Conjugate
The following is a typical procedure employed for the preparation of a protein-dye conjugate, as exemplified by the preparation of anti-digoxin antibodies labeled with a quencher dye. The antidigoxin antibodies are dialyzed against 0.05M phosphate buffer, pH 8, before conjugation. A solution of the activated dye (N-hydroxysuccinimide ester derivative of [VIII], 6 mg in 0.25 ml DMF prepared using equimolar amounts of N-hydroxysuccinimide and dicyclohexyl carbodiimide in dry DMF) was added over a period of twenty minutes to 7.5 mg of sheep antidigoxin γ-globulin in 0.5 mL of 0.05M phosphate buffer, pH 8.0, at 0-4°C. After stirring overnight, the solution was centrifuged for two minutes and the deep purple solution filtered on a Sephadex G-50 column in the same buffer. A single high molecular weight, nonfluorescent fraction was obtained having λ_{max}^{abs} 558 nm. Typical efficiency of labelling is around 80-90%.

Dye-to-protein ratios can be calculated spectroscopically. The protein content of the conjugate was determined from absorption at 280 nm after subtracting the contribution due to free dye at this wavelength. The dye concentration can be calculated by using the extinction coefficient of the dye at 558 nm to be 89,000 M^{-1} cm^{-1}.

Fluorescence Quenching Experiments   Fluorescence quenching experiments are carried out after incubating quencher-labeled antidigoxin antibodies and digoxin-phycoerythrin conjugate in 0.01M phosphate, pH 8.0, 0.15 M NaCl, 2% PEG for 10-30 minutes at room temperature. The fluorescence background of the buffer or unknown sample can be eliminated by subtraction (in end-point quenching). Alternatively, the rate of change in fluorescence can also be measured over a specific period of time and this change in rate of fluorescence quenching can be used to calculate concentration of the analyte. Since the background signal can be kept relatively static, fluorescence rate measurements have been successfully used to reduce serum-to-serum variations in a number of commercial assays, thereby eliminating the need to make a separate blank measurement.

Selection of Probes   The basic requirements for the FETI assay are the selection of appropriate fluorescer and quencher molecules and their mode of attachment to antigens and antibodies. A somewhat detailed description of the selection of fluorescer-quencher pair and the factors governing their effect on the efficiency of FETI are described below.

Fluorescent Label (Energy Donor) Selection   For an efficient energy transfer, the fluorescent molecule should have the following characteristics: (a) high extinction coefficient, (b) high quantum yield in aqueous medium, (c) water solubility, (d) suitable linking group to proteins and haptens, (e) absorbance and emission at longer wavelengths, and (f) minimal effects by environments. Further, the binding of probe to an antibody or antigen must not have any adverse effects on their properties. Since no separation step is involved in FETI assays, an additional consideration that influences the selection of a suitable fluorophore is the endogenous fluorescence background arising from serum components. In addition to serum's intrinsic fluorescence, its background is affected by strong scattering, which is further increased by proteins and other macromolecular compounds (e.g. fat content). A close look at the excitation and emission spectra of serum samples indicates that interference from serum occurs over a broad range of wavelengths, and the intensity of background signal decreases at excitation wavelengths longer than 500 nm, the major component contributing to fluorescence background arising from albumin-bound bilirubin.

Among the most commonly employed fluorescent labels in aqueous medium are derivatives of dansyl, umbelliferone, fluorescein and rhodamine. Although dansyl [I] derivatives have high quantum yield when bound to proteins, this susceptibility to environmental effect makes them less attractive for use as a label in FETI assays. Use of umbeliferone ([II], λ_{max}^{abs} 360 nm and λ_{max}^{em} 447 nm) or fluorescein ([III], λ_{max}^{abs} 490 nm and λ_{max}^{em} 520 nm) as a fluorescent label for FETI has limited application because of serum fluorescence interference described above. Tetramethylrhodamine [VI] and its other derivatives, although absorbing and emitting at longer wavelengths, do have a potential problem with the spectrum change, due to environmental effects. During the past few years, a number of fluorescent derivatives, having absorption and emission wavelengths greater than 500 nm, have been synthesized. Two such fluorescein derivatives, [IV] & [V], have been successfully used for FETI assays for haptens like thyroxine (Ullman and Khanna, 1981) and proteins like IgG (Rodgers, Schwarzberg, Khanna and Chang, 1978).

Fluorescent metal chelates have also been used (Weider, 1978) as fluorescent labels. The chelates of rare earth metals have unique emission characteristics in that upon excitation of aromatic portions of the ligand of the lanthanide complex, the energy of excitation is efficiently transferred to the metal ion. These probes possess high quantum yields in combination with very large Stokes shift and hence offer higher sensitivity. With the proper combination of a rare earth chelate bound antigen and quencher-labeled antibodies, sensitive immunoassays can be developed. An additional advantage with these probes is in the use of time-delayed

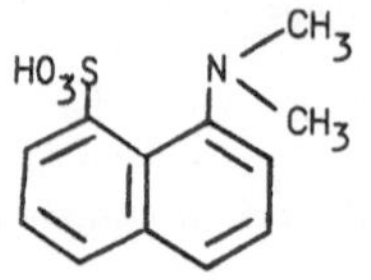

HO$_3$S

N—CH$_3$
CH$_3$

[I] Dansyl

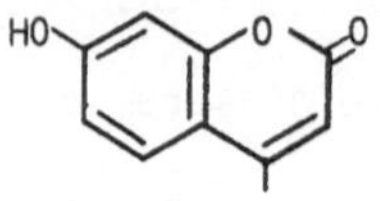

HO

O   O

[II] Umbelliferrone

R$_2$ R$_2$

HO   O

R$_1$ R$_1$

R$_3$ R$_7$

R$_4$ R$_6$

R$_5$

N   O   N$^+$

CO$_2$H

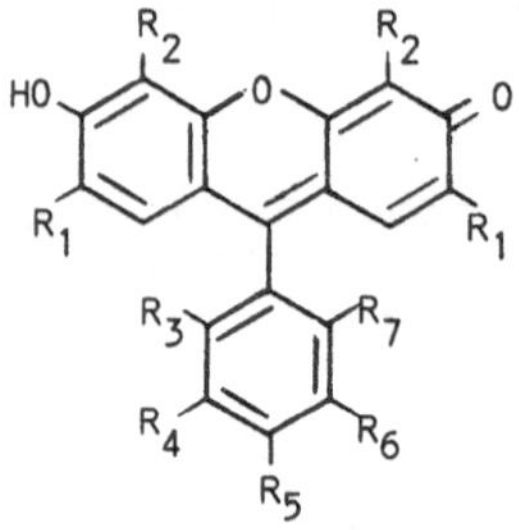

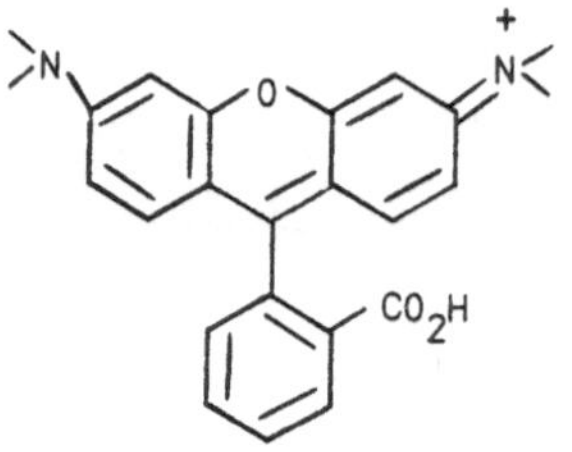

[VI]

[III]   R$_1$=R$_2$=R$_3$=R$_4$=R$_5$=R$_6$=H, R$_7$=CO$_2$H

[IV]   R$_1$=CH$_3$, R$_2$=H, R$_3$=R$_4$=R$_7$=Cl, R$_5$=R$_6$=CO$_2$H

[V]   R$_1$=OCH$_3$, R$_2$=R$_3$=R$_6$=Cl, R$_4$=H, R$_5$=R$_7$=CO$_2$H

fluorescence measurement, which further helps minimize the problems associated with fluorescence background.

In addition to the synthetic fluorophores described above, some of the
naturally occurring fluorescent macromolecules called phycobiliproteins
have also been used (Oi, Glazer and Stryer, 1982) as labels for FETI
assays. These dyes are isolated from red algae, and exhibit excitation and
emission at the red end of the spectrum. Since the phycobiliproteins are
stable and hydrophilic, they can easily be attached to proteins and haptens. Among the most widely used phycobiliproteins, B-phycoerythrin has
λ_{max}^{abs} at 545 nm and λ_{max}^{emis} 565 nm with a high extinction coefficient of
2.4 x 10^6 and $\emptyset_o$ = 0.8 . Among the other phycobiliproteins, allophycocyanin,
although not exploited commercially, has an absorption maximum at 633 nm
and emission around 645 nm, and offers a great future potential, as the
excitation maximum wavelength matches nicely with He-Ne laser wavelength.
Some of the advantages of phycobiliproteins include high extinction coefficient over a wide spectral range, and high quantum yields not affected by
most biomolecules over a broad pH range. Use of these dyes as fluorescent

220

labels for FETI assays will be described later in the chapter. More recently, the possibility of use of porphyrin and chlorophylls as fluorescent
probes has also been mentioned (Hendrix, 1983).

   Quencher Label (Energy Acceptor) Selection   As described earlier,
the efficiency of the FETI technique depends not only on the energy donor,
but also on the efficiency with which the quencher can accept energy. In
general, an ideal quencher among other factors should have the following
characteristics: (a) high extinction coefficient at all emission wavelengths of the donor, (b) very poor quantum yield approaching zero when
excited at donor absorption maximum, (c) high solubility in water so that a
few quencher molecules could be attached to a single antibody molecule, and
(d) an absorption spectrum independent of environmental effects.

   Commonly employed quenchers include tetramethylrhodamine derivatives,
which show good overlap with emission spectra of fluorescein isothiocyanate
conjugates. However, rhodamine conjugates have some undesirable characteristics, like (a) change in absorption spectra due to environmental
effects and degree of conjugation, (b) significant fluorescence emission
from direct excitation of rhodamine, which causes huge background problems,
and (c) poor solubility of rhodamine protein conjugates.

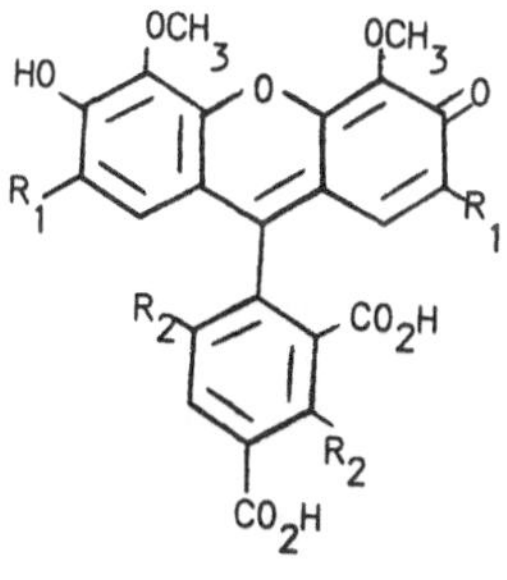

[VII] $R_1 = R_2 = H$

[VIII] $R_1 = R_2 = Cl$

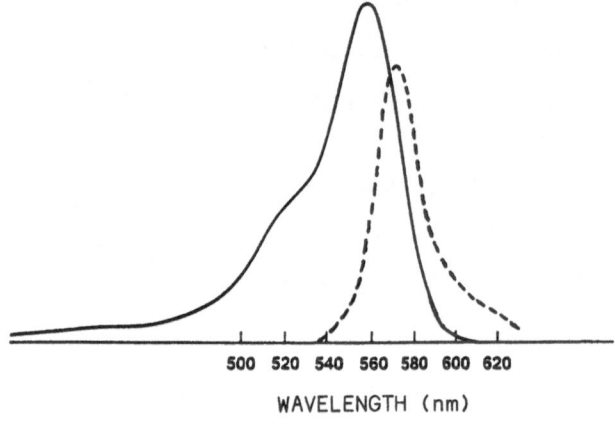

WAVELENGTH (nm)

Fig. 3.   Absorption spectrum of antidigoxin-VIII conjugate
             (————)

             Emission spectrum of digoxin-B-phycoerythrin
             conjugate (— — —)

To overcome some of these problems, a few nonfluorescein fluorescein
derivatives have been synthesized. Compounds of particular interest include [VII] and [VIII], which form very effective pairs with fluorescein
derivatives [III] and [V], respectively. Substitution of methoxy groups at
positions 4 & 5 of fluorescein derivatives results in almost complete
elimination of fluorescence emission.

The absorption spectrum of [VIII] also matches fairly well (Fig. 3.)
with the emission spectrum of B-phycoerythrin, and hence forms a very
effective energy transfer pair. This has been used for development of an
immunoassay for digoxin, described later.

Compound [VII] has λ_{max}^{abs} 510 nm and very poor emission with $\emptyset \leq 10^{-4}\%$.
In addition, [VII] is quite hydrophilic and mimics fluorescein very well in
molecular shape, except for being totally nonfluorescent. Compound [VIII]
has λ_{max}^{abs} 550 nm with practically no fluorescence. By varying the nature
of substituents on the fluorescein nucleus, a variety of quencher molecules
have been synthesized having λ_{max}^{abs} in the 500 to 600 nm region to match
with emission from appropriate fluorescers. R_o values for some selected

fluorescer- quencher pairs are tabulated below:

Donor-Acceptor Pairs	R_o
[I] [III]	38 A^o
[III] [VII]	61 A^o
[V] [VIII]	62 A^o

A macroscopic chromophoric particle, like finely divided charcoal, has
also effectively been used (Ullman and Khanna, 1981) as a kind of universal
quencher.

Preparation of Fluorescer Labeled Antigen Reagents   Preparation of
fluorescer-labeled haptens is normally carried out by reacting a suitable
hapten derivative with an activated fluorescer labeling reagent. Normally,
the point of attachment is to the same site as the protein carrier in the
immunogen that is employed for antibody preparation. The size and specificity constraints usually require that only one fluorescent label be bound.
The nature of the hapten and mode of attachment can sometimes affect the
overall quantum yield of the resulting dye-hapten conjugate. Two examples
include fluorescein derivatives of gentamicin and thyroxine. In both
cases, fluorescence is nonspecifically enhanced by antibody binding, presumably due to relief of an internal quenching effect, caused by either
electron transfer or heavy atom induced quenching, respectively. In the
case of thyroxine, this nonspecific antibody binding induced quenching has
been controlled by synthesis of compound [IX] where the thyroxine part and
fluorescein moeity have been physically kept apart by a rigid piperazine
linking group. The compound [IX] not only has a rigid chain, but the two
iodines that can approach the dye most closely have been replaced by
bromine atoms, which is a less effective quencher, but still permits
binding by anti-T_4.

In the case of macromolecular fluorescers, like B-phycoerythrin,
suitable activated hapten derivatives, can be reacted with the surface
amino groups of the fluorescent dye to yield highly fluorescent conjugates.

[IX]

Similar methodology can be employed for the preparation of protein-dye conjugates. In this case, many labels can be found without directly affecting the efficiency of quenching. However, the net fluorescence signal from a fluorescer-labeled protein will depend on the nature and purity of the protein, selection of the fluorophore, and the degree of labeling. In general, the purity of multivalent antigen is an important requirement for an efficient FETI. This helps in avoiding a large immodulatable fluorescence background, which results in loss of assay sensitivity. The problem of pure antigen requirement puts a limitation for FETI for those macromolecular analytes which have not been isolated or chemically characterized. A variety of methods have been developed to solve this problem. These include: (a) employing a mixture of fluorescent-labeled monoclonal antibodies and quencher-labeled antibodies, and performing a sandwich kind of assay, (b) use of purified immune complexes derived from a mixture of fluorescent-labeled Fab and impure multivalent antigen. The resulting material still containing some free antigenic sites can be directly used in FETI assays using quencher-labeled antibody, and (c) use of BioRad polyacrylamide immunobeads prepared by preincubation with impure fluorescent labeled antigen, followed by extensive washing to remove nonspecific absorption. Such bead suspension (carrying pure labeled antigen) on incubation with quencher-labeled antibody and sample could lead to appreciable quenching so that a standard curve could be modulated by free sample.

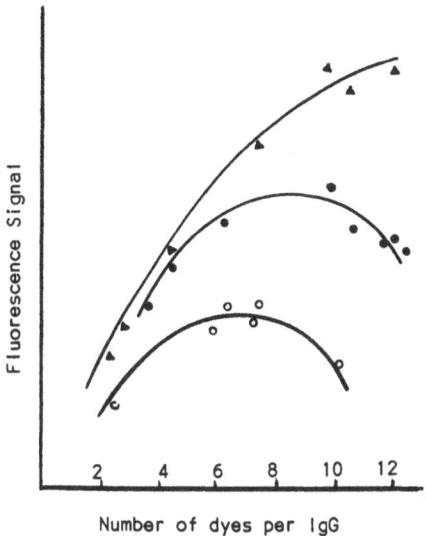

Fig. 4.   Effect of the dye-to-protein ratio on the fluorescence intensity of human IgG with [III] (●),
[IV] (▲) and [V] (o).

The nature of fluorophore and the degree of labeling also affect the
total fluorescence intensity of the labeled protein. In general, the
quantum efficiency of fluorescence decreases with increasing dye to protein
ratios due to self-quenching of dyes. Fig. 4. indicates that, using human
IgG as an example, there is little increase in fluorescence intensity upon
labeling more than eight fluoresceins per IgG molecule. On the other hand,
less self-quenching is observed with dye IV-protein conjugates. By contrast, fluorescent dye [V] is very susceptible to self-quenching.

Quencher-Labeled Antibody Reagent Preparation   The efficiency of
quenching obtained using quencher-labeled antibodies depends not only on
the structure of the quencher label, but also on the degree of labeling and
mode of conjugation to antibodies. In general, the quenching efficiency is
increased by increasing the number of quencher molecules bound to the
antibody. However, the solubility of quencher labeled protein will limit
the number of dye molecules that can be attached to a protein. In addi-

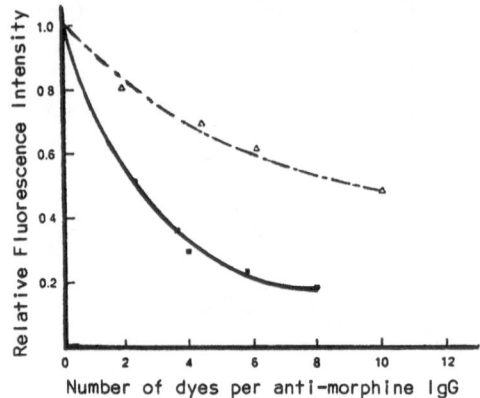

Fig. 5.   Fluorescence efficiency of 10 nM morphine-fluorescein in the presence of excess anti-morphine rhoda-
mine-6-carboxylic acid (————) and tetramethylrhodamine isothiocyanate (— · — · — · —).

tion, higher quencher labeling could lead to: (a) interference observed due
to trivial quenching (energy transfer observed as a result of absorption of
emitted light by an internal filter effect), (b) partial antibody inactivation, and (c) an introduction of excessive fluorescence background caused
by weak quencher emission. The method of linking quencher molecules to
proteins also influences the quenching efficiency. Fig. 5. indicates the
difference in quenching efficiency observed when labeled with rhodamine
isothiocyanate and NHS ester of rhodamine-6-carboxylic acid.

It appears from Fig. 5 that the more random labeling achieved by
labeling an IgG molecule with an active ester derivative of the quencher
molecule results in more efficient fluorescence quenching, as compared to
the use of a less reactive quencher derivative which can result in selective substitution.

RESULTS AND DISCUSSIONS

The principle of FETI has been successfully employed for quantitative
determination of a variety of haptens, as exemplified by the measurement of
serum digoxin concentrations. The FETI assay, based on a competitive

226

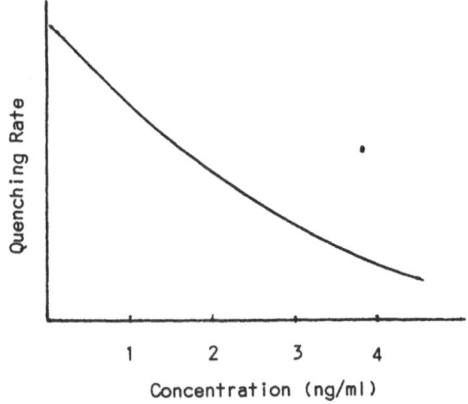

Fig. 6.   Relationship of digoxin concentration
to quenching rate in arbitrary units.

binding immunoassay principle, employs B-phycoerythrin labeled digoxin and
acceptor dye [VIII] labeled antidigoxin antibodies. Since the emission
wavelength of excited fluorescent dye overlaps the absorption wavelength of
the acceptor dye, an energy transfer can occur when the fluorescer and
acceptor are in close proximity (Fig. 3.). During the assay reaction, free
analyte in the sample competes with fluorescent dye-labeled analyte for
binding sites on an antibody. In the resulting immune complex, the energy
transfer from fluorescer to acceptor quenches the fluorescence signal. The
rate of quenching relates inversely to the concentration of analyte in the
sample. Unbound phycoerythrin conjugate in the reaction mixture is excited
by 540 nm. The change in fluorescence over a defined period is measured
during the analysis. The endogenous variable fluorescence among samples is
reduced to a minimum by a homogeneous sample oxidation. A standard curve
illustrating this relationship is shown in Fig. 6.

The assay performance using FETI technology shows an excellent correlation to several other methods in terms of assay precision and assay
accuracy. Rapid assay results obtained using this technology allow the
most timely dosage adjustments during initiation of therapy and in periods

when a patient's health status is changing. Similar assay procedures have been employed for the quantitative determination of proteins, such as IgG, IgM, etc., using this technology.

INSTRUMENTATION

FETI assays can be performed on a standard fluorescence spectrophotometer in either end-point or kinetic mode. By optimizing the instrument for the fluorescent probe to be used, it is possible to use the most suitable lamp as the source of excitation radiation and to remove scattering with accurate filters. This helps in increasing measurement sensitivity considerably. Use of suitable dye lasers can considerably increase the excitation intensity. Sensitivity can also be increased, or background decreased, by using modified optical arrangements. Further improvement in assay sensitivity can be achieved by using the most sensitive photomultiplier tube and single photon counting. For end-point measurements, a quartz cell could be employed and the intensity of emitted light usually measured in a direction perpendicular to the incident light. For rate measurements, the instrument is equipped with a provision to measure change in fluorescence signal with time in a temperature controlled flow cell.

CONCLUSION

The application of FETI technology for the detection and quantitation of a biologically active compound provides a major alternative to isotopic immunoassays and chromatographic methods of analysis. FETI assays offer good sensitivity without the necessity of a separation step. The methodology is convenient to perform, reliable and offers rapid results. Possible interference arising from endogenous serum components and foreign substances can be reduced considerably by the selection of suitable fluorophores and quenchers.

ACKNOWLEDGMENT

The work described in this chapter is a collection of research efforts of a large number of scientists at Syva under the direction of Dr. Edwin Ullman. Special thanks are due to Mosche Schwarzberg, Warren Colvin, Roberta Ernst, Neal Bellet, Laura Winfrey and Bob Zuk for their dedicated efforts to finish various parts of the work described in this chapter.

REFERENCES

Berson, S. A. and Yalow, R. S., 1959, Quantitative aspects of the reaction between insulin and insulin binding antibodies, J. Clin. Invest. 38, 1196-2016.

Cais, N., Dari, S. and Eder, Y., et al, 1977, Metalloimmunoassay, Nature 270, 543-535.

Coons, A. H., Creech, H. J. and Jones, R. N., 1941, Immunological properties of an antibody containing a fluorescent group, Proc. Soc. Exp. Biol. Med. 47, 200-202.

Dandliker, W. B. and Geiger, G., 1961, Quantification of the antigen-antibody reaction to the polarization of fluorescence, Biochem. Biophys. Res. Commun. 5, 299-304

Forster, T., 1948, Zwischenmolekulare energiewanderung und fluoreszenz, Ann. Phys. (Leipzig), 2, 55-75.

Gantt, E., 1981, Phycobilisomes, Ann. Rev. Plant. Physiol. 32, 327-347.

Hendrix, J. L., 1983, Porphyrins and chlorophylls as probes for fluoroimmunoassays, Clin. Chem. 29, 1003.

Molecular Probes Catalogue, Junction City, Oregon 97448.

Oi, V. T., Glazer, A. N. and Stryer, L., 1982, Fluorescent phycobiliprotein conjugates for analytes of cells and molecules, J. Cell Biology 93, 981-986.

Rodgers, R., Schwarzberg, M., Khanna, P. L., Chang, C. H. and Ullman, E., 1978, A fluorescence quenching immunoassay for human IgA, Clin. Chem. 24, 1033.

Scharpe, S. L., Cooneman, W. M. and Bloome, W. J., 1976, Quantitative enzyme immunoassays: current status, Clin. Chem. 22, 733-738.

Simpson, J. S. A., Campbell, A. K. and Ryall, M. E. T., 1979, A stable chemiluminiscent labelled antibody for immunological assays, Nature 279, 646-647.

Soini, E. and Hemmila, I., 1979, Fluoroimmunoassay: present status and key problems, Clin. Chem. 25, 353-361.

Ullman, E. F. and Khanna, P. L., 1981, Fluorescence excitation transfer immunoassay, Methods in Enzymology 74, 28-60.

Ullman, E. F., Schwarzberg, M. and Rubenstein, K. E., 1976, Fluorescence excitation transfer immunoassay, J. Biol. Chem. 251, 4172-4178.

Van Weeman, B. K., and Schuurs, A. H., 1971, Immunoassay using antigen-enzyme conjugates, FEBS. Lett. 15, 232-236.

Weider, I., 1978, Immunofluorescence and related staining techniques, Elsevier/North Holland Biomedical Press, 67.

Wisdom, G. B., 1976, Enzyme immunoassay, Clin. Chem. 22, 1243-1255.

TIME-RESOLVED FLUOROIMMUNOASSAY

Erkki Soini and Timo Lövgren

Wallac Oy
P.O. Box 10
SF-20101 Turku, Finland

INTRODUCTION

The number of published fluoroimmunoassay (FIA) methods and their
applications is increasing continuously and FIA has established its
position as a reliable analytical method and a nonisotopic alternative to
RIA (Hemmilä, 1985). It has been anticipated that RIA methods will be
gradually replaced by nonisotopic methods, first by enzymic methods and
then by FIA. Indeed, EIA methods are widely applied, especially in the
areas of serology, microbiology, and virology. Until now, FIA methods have
had only limited applications - for example in drug monitoring and sero-
logic assays. To be able to replace the traditional RIA methods and the
already well-established EIA methods, higher sensitivity, better precision,
and more widely applicable performance characteristics are required. The
assay principle and the instrument should be suitable for a very wide range
of analytes (from small haptens to large microbes), which may be present in
either high or low concentrations (from millimolar to picomolar).

The progress in the development of FIA methods has been made primarily
in two fields: rapid and simple homogeneous assays, and solid-phase-based
immunofluorometric assays. A large variety of homogeneous assay principles
has been published, but only a few of them have found wider application.
Homogeneous assays are simple to perform and rapid (minimum assay time in
some assays is less than 1 min), but are mostly limited to haptenic molecules present in the sample in relatively high concentrations (drugs,
certain hormones, and some proteins). The sample used (in most cases serum)
limits the sensitivity by its high background fluorescence.

Solid-phase-based immunofluorometric assays (IFMA) have been developed
for proteins and other high molecular weight analytes (peptide hormones,
serum proteins, antibodies, viruses) as the assay principle requires at
least two either similar or different epitopes on the antigen. Various
separating systems and low-background solid-phase matrixes reduce the endogenous background fluorescence, but the assay sensitivity is still often
inferior to that of the corresponding immunoradiometric assays (IRMA). The
benefits of immunofluorometric assays are thus restricted to properties as
long-term stability of reagents, no radioactivity and rapid signal
counting.

Most FIA methods are based on the use of well-known organic fluorescent probes - in most cases fluorescein - and only a few new probes have emerged. New interesting alternatives comprise of fluorescent microparticles, natural and synthetic porphyrins, fluorescent proteins, and especially the lanthanide chelates. In the search for new probes interest has been directed towards fluorescence properties such as long emission wavelength, large Stokes shift, and long fluorescence decay time.

Instrument development has mostly been concerned with adaptation and automation of special techniques, such as measurement of polarization, use of solid surfaces, continuous-flow systems, and time-resolved fluorescence. Unlike higher precision and assay standardisation for certain specified analytes, higher instrument sensitivity has not always been the main objective. Instrument sensitivities have also been increased - for example by single-photon counting and laser excitation. The latter, however, is not very useful for routine instrumentation because of its high purchase and maintenance cost.

Time-resolved fluoroimmunoassay (TR-FIA) is one of the new technologies with the potential to replace isotopic methods. It has high sensitivity, if needed, and is applicable in principle to all known immunoassays in a wide range of concentrations. The combination of the sandwich-type "excess- reagent assay", good-quality monoclonal antibodies, Eu-chelate label, and a time-resolving fluorometer has already resulted in several immunoflurometric assays with higher sensitivities and wider dynamic ranges than those of any other known assays (IRMA or ELISA).

Time-resolved fluoroimmunoassays (TR-FIA)

Time-resolved fluoroimmunoassays (TR-FIA) can be seen as a separate group of fluoroimmunoassays, because the principle of time-resolved measurement and especially the use of lanthanide ions - not necessarily in fluorescent forms - as labels does not fit within any category of conventional FIAs. In TR-FIA the difference between the fluorescence lifetimes of the specific signal and the nonspecific background has been used to increase the signal-to-noise ratio and thus the sensitivity. When the fluorescence decay time of the label is sufficiently longer than the average background decay, the specific signal of the label can be measured after the background signal has decayed. This principle has been used for rejecting background in immunofluorescence (Mueller and Hirschfeld, 1977) and for studying the antigen-antibody reaction with pyrene derivatives as haptens (Lovejoy et al., 1977).

The development of new long-lifetime labels, lanthanide chelates for which the fluorescence decay time ranges from 10 to 1000 µs (more than four decades longer than the average background duration), presents interesting possibilities for increasing the sensitivity of FIA by reducing the background signal to practically zero (Soini and Hemmilä, 1979; Wieder, 1978). This long lifetime also allows the development of simple, inexpensive instrumentation, in which the N_2-laser or simply a pulsed xenon discharge lamp is used for excitation. Among the several published methods for using fluorescent lanthanide chelates as labels in TR-FIA, practically all involve Eu(III)-labelled antibodies in solid-phase-based heterogeneous assays, where the bound Eu has been quantified after its dissociation into solution. TR-FIA with dissociation-based enhancement of Eu fluorescence has been found reliable and practical for a number of clinical routine immunochemistry analytes and has the name DELFIA ("Dissociation Enhanced Lanthanide Fluoroimmunoassay") (Trade mark of LKB-Wallac). Detailed descriptions of various methodological aspects of DELFIA can be found in recent publications (Soini and Kojola, 1983; Hemmilä and Dakubu, 1986; Hemmilä et al., 1984 a); Lövgren et al., 1984).

Background interference is the most important problem of any fluorometric determination in material of biological origin. This is due to the
fact that the analyte concentrations to be measured are often very low and
the sample contains other fluorescent components besides the label itself,
thus background interference often overlaps the analyte-specific signal.
The situation is illustrated in Fig. 1 which shows that the fluorescence
emission of blood serum covers the whole visible wavelength range of the
electromagnetic spectrum including excitation and emission peaks of fluorescein isothiocyanate (FITC) - the most common and traditional fluorescent
label.

Europium and terbium form highly fluorescent chelates with many organic ligands and are thus a potential alternative to radioisotopic and
established non-isotopic labels in immunoassays and other specific ligand
assays. The main advantages of using lanthanide chelates as fluorescent
probes in biological material are their high quantum yield, exceptionally
large Stokes shift, narrow emission peaks, and optimal emission and excitation wavelengths. The emission peak of europium is above the end point of
serum interference. Because of the extraordinarily large Stokes shift the
sharp emission peak of europium (614 nm) can easily be separated from scattering caused by excitation (340 nm) and from the interfering fluorescence
of serum (400-600 nm). The excitation energy is absorbed by the ligand and
then transferred to the chelated metal ion which emits the energy at its
characteristic wavelength. A good fluorescent lanthanide chelate is free
from fluorescence emission of the ligand itself, and the energy is transferred to the lanthanide ion without any radiative transition.

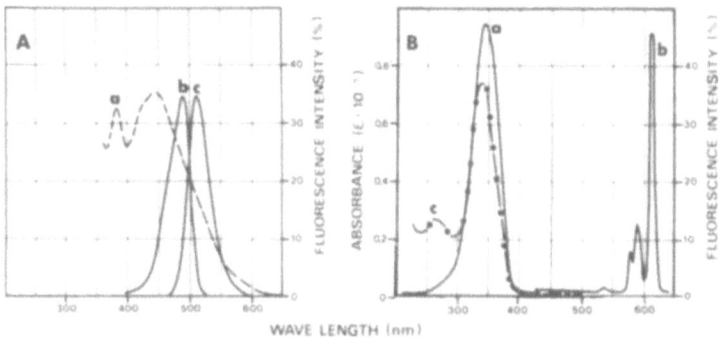

Fig.1. Fluorescence emission spectra of serum (Aa), excitation (Ab) and emission (Ac) spectra of FITC, exci-
(Ba), emission (Bb) and absorption (Bc) spectra of
europium.

The difference between europium and terbium is mainly based on their
different excitation and emission wavelengths and their fluorescent properties in solution. The strongest fluorescent emission peaks of terbium
chelates are at 490 and 545 nm. Terbium as well as europium form stable
chelates in water, but unlike europium, terbium requires no protection
against quenching by water. The use of terbium as a label presents problems
because the excitation maximum of terbium chelates is below 300 nm and
plastics and optical glass cut down the excitation below 320 nm. Europium

233

chelates are particularly suitable as probes as far as their fluorescence spectrum is concerned. Their absorbance at 340 nm is on the same level as that of the best organic fluorochromes although their quantum yield is often lower. Their excitation band is relatively broad and dependent solely on the ligand. Therefore it is possible to use a broad excitation bandwidth and consequently a lower excitation light source intensity. As already mentioned, the advantage lies in the very large Stokes shift (270 nm). In addition, europium chelates have a very narrow light emission peak and that is why background in the region of the fluorescence emission is very low and good separation of the excitation and emission peaks can be achieved with interference filters.

TIME-RESOLVED FLUOROMETRY OF LANTHANIDE CHELATES

Owing to natural fluorescence from various compounds in biological samples such as serum proteins, conventional fluorescent probes suffer from serious limitations of sensitivity. With lanthanide chelates, however, it is possible to considerably reduce the background level by selective detection of long-decay fluorescence. The fluorescence decay time of lanthanide chelates is often in the order of 10-1000 µs, whereas the decay time of natural fluorescence in a typical biological sample is in the order of 1-20 ns. For this reason time-resolved fluorometers used with lanthanide chelates are potentially several orders of magnitude more sensitive than conventional fluorometers (Soini and Hemmilä, 1979). A pulsed light is used to produce an excitation pulse of short duration as compared to the decay time of the lanthanide chelate.

Certain europium chelates have a decay time of 500 µs and the interfering rapid decay of natural fluorescence can be discriminated against by activating the detector after a delay of e.g. 400 µs has elapsed after excitation of the lanthanide chelate (Fig. 2). As a result, the interfering background fluorescence is reduced by 2-3 orders of magnitude (Soini and Kojola, 1983) and a time-resolved fluorometer covers a concentration range of lanthanide chelates which is not measurable using a conventional filter fluorometer or spectrofluorometer.

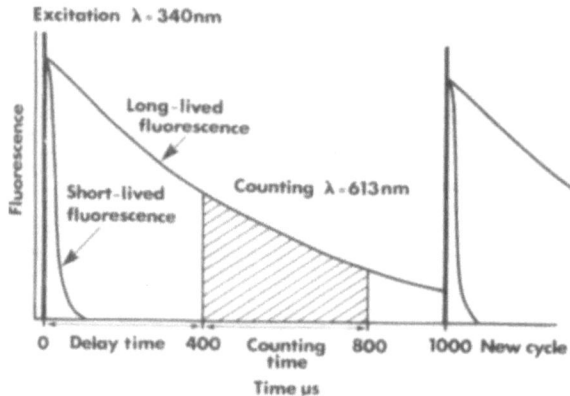

Fig.2. Selected counting time (400-800 µs) of Eu chelate fluorescence emission (1) following an excitation light pulse. Background fluorescence ceases after a few nanoseconds. The excitation pulse is repeated 1000 times during 1 second.

TIME-RESOLVED FLUOROMETER

An automatic time-resolved fluorometer has been developed in our laboratory. This instrument, named the 1230 Arcus Fluorometer, is equipped with a sample changer for sample tube racks and a built-in microcomputer. Two different types of sample holders can be used, either cuvette racks for 10 sample tubes (od. 12 mm, length 80 mm) or racks which can hold one 12-well microtitration strip. There is space for 30 racks on the conveyor table and thus the maximum sample capacity is 300 cuvettes or 360 microtitration wells. The two rack types can be used mixed together.

A xenon flash-lamp is used as the pulsed excitation light source of the Arcus Fluorometer. The flash duration of the lamp is of the order of a few microseconds and a repetition rate of about 1000 flashes per second is achieved (see Fig. 2). The optimal excitation wavelength for europium chelates is 340 nm but in the fluorometer other values are possible with different interference filters. The built-in manually operated filter changer holds two standard 1-inch filter discs. As the excitation of the sample takes place through the side of the tube, the generated fluorescence light is measured through the tube bottom. The right angle between the excitation source and the detector improves the distinction between the source and fluorescent light. A side window-type photomultiplier tube is operated with a red sensitive multialkali cathode in the single-photon mode. The output pulses of the photomultiplier tube are amplified by a fast preamplifier and then fed into the 50 MHz prescaler which divides them by 100 to lower the pulse frequency for the main scaler. After the counting period is completed the microprocessor reads the contents of the scalers and stores the accumulated counts in the memory.

TIME-RESOLVED FLUOROIMMUNOASSAY

TR-FIA is a novel highly sensitive immunoassay technique based on the long fluorescence decay time of lanthanide chelates.

If a europium ion is used as label in immunoassays, it has to be strongly bound to an immunoreactive component. Strong europium binding can be achieved with different polycarboxylic acids such as EDTA, EGTA, HEDTA and DTPA. An additional requirement for labelling with europium is the introduction of a functional group to molecules such as EDTA which allows covalent coupling to the immunoreactive component. Two useful chelate derivatives, diazophenyl-Eu-chelate and isothiocyanatophenyl-Eu-chelate, have been synthesized. The former is coupled mainly to tyrosine residues of the protein (antibody), and the latter to free amino groups of the protein (antibody) (Hemmilä et al., 1984 a). These labelled antibodies are, however, essentially non-fluorescent and must virtually be considered as labelled by a metal ion. In the non-fluorescent form the metal ion is strongly bound to the antibody and it thereby provides the good stability needed during the immunoreaction.

It is evidently difficult to combine strong chelating capacity with good absorption and energy transfer properties. Thus the europium labelling and measurement procedures are separated; an antibody labelled with a non-fluorescent europium chelate is used in the immunoreaction and later the europium is released and bound to another chelating agent which results in intense fluorescence. In the assay the measurement of the amount of label-led antigen-antibody conjugate is carried out after the separation step by adding another chelating agent - called the Enhancement Solution - which dissociates the europium ion from the antibody chelate complex on the solid phase and forms a β-diketone chelate of high fluorescence. As europium quantitation in the immunoassay occurs in aqueous solution, it was necessary to design an enhancement solution in which the europium is measured in order to dissolve the β-diketone required for energy absorption and transfer, as well as to simultaneously protect the ion from quenching by water molecules.

The enhancement solution contains a non-ionic detergent, Triton X-100, which dissolves the sparingly soluble organic component in the micellar phase and excludes the quenching water from the chelated europium ion. Insulation from the aqueous solvent is further improved by adding a synergistic agent, trioctylphosphine oxide, to the europium chelate solution. Fig. 3 presents a schematic outline of the hypothetical micellar structure into which the fluorescent europium chelate, consisting of the europium ion, β-diketone and the trioctylphosphine oxide molecules, is solubilized. The high sensitivity and the long lifetime of europium fluorescence indicate that the quenching effect of water has been almost completely eliminated. When the optimized solution is used for the enhancement of europium fluorescence and a time-resolved fluorometer is used for measurement, europium can be quantitated over a concentration range from 5×10^{-14} to 10^{-7} moles per litre. Thus the immunoassays based on europium chelates as labels provide at least the same order of sensitivity as commonly obtained with [125]I. This unique assay procedure is called DELFIA[TM]* (dissociation-enhanced lanthanide fluoroimmunoassay).

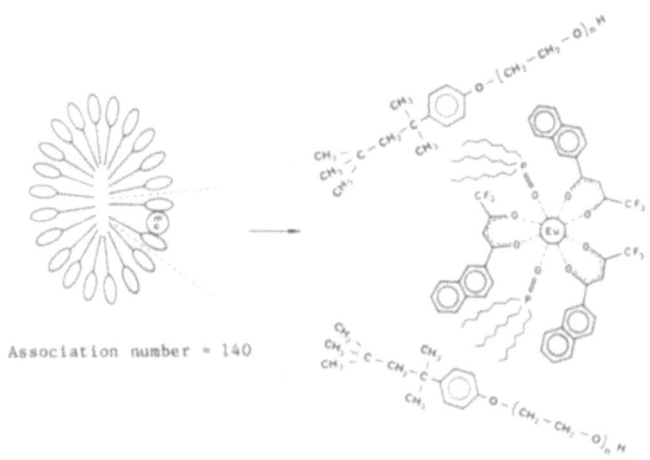

Association number = 140

Fig.3. A hypothetical structure of a fluorescent europium chelate. Europium is chelated with β-diketone and trioctylphosphine oxide within a micellar structure of Triton X-100.

---

*DELFIA is a trade mark of LKB-Wallac

The following label related reasons contribute to the high detection sensitivity obtained in immunoassays:

- The specific activity of the europium chelate is high and it can be increased by using stronger excitation.
- Background fluorescence is avoided by using time-resolved fluorescence detection.
- Lanthanides are not normally present in biological samples.
- Lanthanide labelling can be carried out with minimal interference on immunoreactivity.

APPLICATIONS

A large number of applications have been developed using the DELFIA concept and many of them are now in clinical routine use and available as commercial kits. The analyte range covers the following clinical fields:

- Gynaecology: hCG (Pettersson et al., 1983; Stenman et al., 1983), LH (Stenman et al., 1985), testosterone (Bertoft et al., 1984 a); Bertoft et al., 1984 b)), FSH, prolactin, estradiol, progesterone

- Thyroid disease: TSH (Kaihola et al., 1985; Böttger et al., 1984 a); Han et al., 1985; Böttger et. al., 1984 b); Näntö et al., 1985 a)), T_4 (Näntö et al., 1985 b), neonatal TSH

- Cancer: ferritin, AFP (Suonpää et al., 1985); pancreatic phospholipase A_2 (Eskola et al., 1983; Thuren et al., 1985)

- Viral infections: rubella antibody (Meurman et al., 1982;), rotavirus and adenovirus (Halonen et al., 1984 a); Halonen et al., 1984 b); Halonen et al. 1985, Halonen et al., 1983; Matikainen et al., 1983), tetanus (Hemmilä et al., 1984 b)), HBsAg (Siitari et al., 1983)

- Others: cortisol (Eskola et al., 1985 b)), myelin basic protein (Viljanen et al., 1984), digoxin (Helsingius et al., 1985;), IgG (Hemmilä et al., 1982), C-reactive protein (Hemmilä et al., 1985)

In addition applications have been worked out in cytotoxicity (Blomberg et al., 1986), enzyme activity measurement (Karp et al., 1983), and phospholipid binding (Saris, 1983).

The DELFIA procedure for peptide hormones and proteins, as well as for viral antigens and antibodies is a solid phase, two-site fluoroimmunoassay based on the sandwich technique. Two different antibodies, the first attached by adsorption to the wells of microtitration strips and the second labelled with europium chelate, are directed against two separate antigenic determinants of the analyte molecule. The DELFIA hapten assays are based on a competitive solid-phase procedure employing a labelled antibody.

In general 25-50 µl of undiluted serum is sampled to microtitration strips and the immunoreactions are carried out in one or two steps with 20 min - 4 h incubation periods. In the final step, europium is released from the chelate by the Enhancement Solution and a new europium chelate is formed. The fluorescent signal is measured in the LKB-Wallac 1230 ARCUS time- resolved fluorometer using a 1 second measuring time. In the non-competitive immunometric assays the fluorescence signal is directly proportional to the concentration of the analyte. As an example the standard curve of the HBsAg assay is shown in Fig. 4.

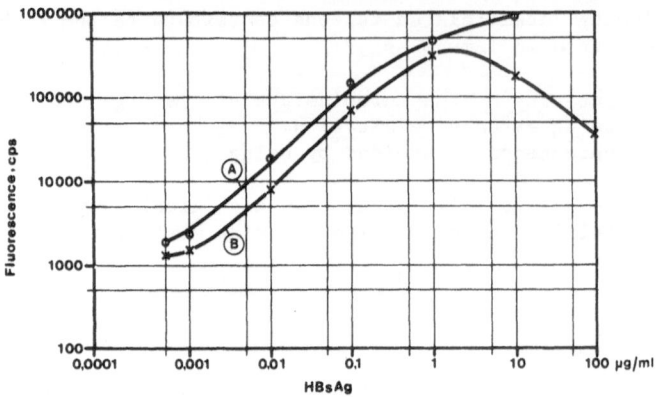

Fig.4. Standard curve of DELFIA assay for HBsAg. Curve A is
the normal two-step assay with 2 h + 2 h incubation
periods and curve B is a 50 min single step assay.

DISCUSSION

Time-resolved fluorometry of europium chelates offers an extraordinarily wide <u>dynamic range</u> and linearity which cover the whole concentration
range of clinical importance without extra dilution of the sample. However,
the large linearity provided by the label can be exploited only in the
immunometric assays because the competitive binding assay principle works
only within a very limited dynamic range.

A unique advantage of time-resolved fluorometry is the extremely high
<u>specific activity</u> of the label in terms of emitted photons per second. This
is due to the fact that the excited states of the fluorescent labels are
created in the measuring position of the sample. The photon emission rate
is a function of excitation intensity and decay time, and excitation itself
can be increased by increasing the intensity of the exciting radiation. The
optimal decay time of fluorescent labels for time-resolved fluorometry is
10-1000 μs. A typical specific activity of fluorescent labels reaches up to
10^7 photons/s for one picomole. The specific activity of radioisotopic
labels is normally in the order of one disintegration/s for one picomole.
Consequently, the specific activity of fluorescent labels is more than a
million times higher than that of radioisotopes and the lowest level of
concentration can be measured in 1 second with satisfactory statistical
precision.

Time-resolved fluoroimmunoassay has thus an extremely high potential
<u>sensitivity</u> and provides a very high count rate with low instrument back-
ground; concentrations down to 10^{-17} Moles per cuvette can be detected.
However, it is important to realize that the limiting factor for the lowest
level of detection in a practical time-resolved fluoroimmunoassay is not
the instrument background or the specific activity of the label, or in
other words the signal to noise ratio but the signal from the blank sample
which is determined in practice only by the non-specific binding effects.
An important prerequisite for the low blank signal is also the purity of
the enhancement solution.

In immunometric assays the first antibody is immobilized and used up to 1 μg per tube in order to achieve the highest speed, linearity and dynamic range. Factors that limit the amount of antibody are the non-specific binding effect and naturally the cost. Both monoclonal and poly-clonal antibodies have been used. When polyclonal antibodies are applied the titer should be as high as possible, if a wide dynamic range has to be achieved. Often affinity purified antibody preparations have been used, but since monoclonal antibodies became available they have gradually replaced the polyclonal antisera. In assay optimization employing a pair of mono-clonal antibodies it is usually desirable to immobilize the one with higher affinity. Of course analyte or antibody specific properties must be con-sidered in each individual case. The final sensitivity of the assay is influenced by the quality of the antibody coating on the solid phase as both the non-specific binding and the precision are affected.

In the reagent excess procedure the second labelled antibody is also critical in respect to the titer of the specific fraction. Polyclonal anti-bodies with low titer involve a problem of using very high total activity of label for achieving high performance. The large non-reactive labelled part of the antibody mixture is then to be washed out. Incomplete washing leads to limited linearity and dynamic range as well as high background. Therefore the use of labelled monoclonal antibodies or affinity purified polyclonal antibodies in high excess up to 100 ng per tube or more is the only practical possibility in immunometric DELFIA.

The sensitivity of an immunometric assay - an idealized sandwich type "excess reagent" assay - is in the first instance dependent on the
- detection limit of the label
- precision of the separation procedure

The detection limit of the radioactive label can be improved by increasing the specific activity and a similar improvement as in case of TR-FIA can be obtained except for the problems related to the radiochemical stability of the tracer and for radiation safety. The specific activity of the DELFIA tracer can be regarded practically "infinite" when the maximum sensitivity of DELFIA is being theoretically considered.

The maximum theoretical sensitivity obtainable in an immunometric non-competitive assay is defined by Jackson et al. (1983):

$$\frac{k \cdot \sigma}{K} = 0.5 \cdot 10^{-15} \text{ moles/L}$$

where

k = fractional non-specific binding of the antibody (10^{-4})
K = affinity constant ($2 \cdot 10^{10}$ L/Mole)
σ = the relative error of the response in the absence of analyte = error of the non-specific bound antibody (10 %)

In practice, sensitivities from 0.01 μIU/mL to 0.001 μIU/mL (zero standard plus one standard deviation) have been reported in TSH-DELFIA, which corresponds to $0.5 \cdot 10^{-13} - 0.5 \cdot 10^{-14}$ moles/L (molecular weight of TSH ~ 30 000). The difference between the theoretical and verified sensi-tivity is still 10-100 fold assuming that the counting error and instrument error = 0 and that an ideal washing procedure is used. Consequently, the outcome from this simplified example is that the most important sensitivity limiting factor in the present DELFIA procedure is the efficiency and precision of separation (including the performance of the solid phase).

Washing off the unreacted fraction in the microtitration wells is a step which requires particular attention in the DELFIA method. Extremely high specific activity and a large excess of the reactant which often contains more than 10^7 counts/sec in one test tube, is used to achieve high sensitivity, large dynamic range and rapid immunoreaction. As a consequence, the washing must be very efficient in order to exploit the potential advantage of a high specific activity label and monoclonal antibodies to reach high sensitivity which requires a stable and low blank level. Normally washing six times is recommended using an appropriate washing/aspirating device.

The extra step in the DELFIA, the addition of the enhancement reagent, is a procedural disadvantage. The enhancement reaction is, however, very rapid and troublefree and undoubtedly presents a simple step as compared to the initiation of the enzyme reaction in photometric methods or the chemiluminescence reaction in luminometric assays.

Coated tube technology is well established in practical immunoassays and in DELFIA the europium-labelled antibody reacts with immobilized antibody-antigen complex on the solid phase. The microtitration strips have proved to be very useful and reliable, and they make sample processing simple and fast. Commonly available microtitration plate compatible sample preparation accessories, dispensors and wash/aspirate devices can be used. The linear standard curve and the stable label provide simple standardization of the assay and in this respect time-resolved fluorometry is superior as compared to any photometric, luminometric or radioisotopic procedure.

Good intra-assay precision can be achieved with many different immunoassays However, in photometric and especially nephelometric assays the quality of cuvettes should be very carefully controlled whereas quality errors which increase the scattered light emission are not important in time-resolved fluorometry. The fluorescent chelate is formed in solution after dissociation of the europium ion. This involves a significant advantage from the fluorometric point of view. The detection of the fluorescence signal in solution is much more stable and reproducible than that on the solid phase coated plastics. Consequently the signal is neither dependent on the homogeneity of the coating nor on the position of the tube in the fluorometer detector. The photon collection geometry is also better and the tube gives a minor contribution to the background.

The temperature coefficient of the fluorescence signal when the enhancement solution is used, is −1 %/ºC. For this reason it is important that the samples are not exposed by a strong temperature change when loaded into the fluorometer. The inside temperature of the Arcus fluorometer is max. 2ºC above the room temperature.

The present DELFIA methodology is simple and straightforward for immunoassays which require high performance. It is highly insensitive to interferences from the sample because the methodology involves three points which as such may be sufficient to eliminate the interference problems:

1. Washing of the solid phase which cleans the cuvette from interfering constituents in the sample.
2. Dissociation of the biochemically inert europium ion from the solid phase bound labelled antibody and formation of a new chelate in solution provides a stable and reproducible fluorescence.
3. The time-resolved fluorescence detection eliminates autofluorescence of the materials necessary in the system as well as possible fluorescence from interfering residues in the sample.

ACKNOWLEDGEMENTS

This paper has reviewed and respectively referred to the results by a
number of our colleagues at the LKB-Wallac Research Laboratories as well as
in our collaborative laboratories. In the field of microbiological applications, the Department of Virology, University of Turku, Finland and the
Finnish Red Cross Blood Transfusion Centre, Helsinki, Finland, have made a
considerable contribution. We gratefully acknowledge the contributions of
the Middlesex Hospital Medical School, Dept. of Molecular Endocrinology,
London, UK; University of Turku, Dept. of Clinical Chemistry and Dept. of
Pathology, Turku, Finland; University of Helsinki, Dept. of Obstetrics and
Gynaecology, Helsinki, Finland; Medix Laboratories Ltd., Kauniainen,
Finland; and Farmos Group, Diagnostics Factory, Oulunsalo, Finland
(formerly Nordiclab).

REFERENCES

Bertoft, E., Eskola, J., Näntö, V., and Lövgren. T., 1984 a), Competitive
    solid-phase immunoassay of testosterone using time-resolved fluores-
    cence, Febs. Lett. 173: 213-216.
Bertoft, E., Mäentausta, O., Lövgren, T., 1984 b), Time-resolved fluoro-
    immunoassay of testosterone, Proceedings: Symposium on the Analysis
    in Steroids, Szeged, Hungary.
Blomberg, K., Granberg, C., Hemmilä, I. , and Lövgren, T., 1986, Europium
    labelled target cells in assay of natural killer cell activity I. A
    novel nonradioactive method based on time-resolved fluorescence, J.
    Immunol. Methods, in press.
Böttger, I., Pabst, H., Serekowitsch, R., and Kriegel, H., 1984 a),
    Performance of two new solid phase ligand assays for TSH resing
    monoclonal antibodies, Brit. J. Nucl. Med., 6: 195-207.
Böttger, I., Sommerfeld, V., Pabst, H., 1985 b), Clinical value of two
    ligand assays for TSH, Proceedings: European Nuclear Medicine
    Congress, Helsinki, Finland, Aug. 14-17, 1985.
Eskola, J.U., Nevalainen, T.J., and Lövgren, T.N-E., 1983, Time-resolved
    fluoroimmunoassay of human pancreatic phospholipase A$_2$, Clin.Chem.,
    29: 1777.
Eskola, J., Näntö, V., Meurling, L., and Lövgren, T., 1985 b), Direct
    solid- phase time-resolved fluoroimmunoassay of cortisol in serum,
    Clin. Chem. 31: 1731-1734.
Halonen, P., Meurman, O., Lövgren, T., Hemmilä, I., and Soini, E., 1983,
    Detection of viral antigens by time-resolved fluoroimmunoassay, In:
    "New Developments in Diagnostic Virology", P.A. Bachman, ed.,
    Springer-Verlag, Berlin, pp. 133-46.
Halonen, P., Meurman, O., Petterson, U., Ranki, M., and Lövgren, T., 1984
    a), New developments in diagnosis of virus infections, in: "Control
    of Virus Diseases", Kurstak, Tijssen, and Kurstak, eds., Marcel
    Dekker, Inc., New York , p. 501.
Halonen, P., Bonfanti, C., Waris, M., Lövgren., T., and Hemmilä, I., 1984
    b), New developments in diagnostic virology, in: "New Horizons in
    Microbiology", Elsevier North Holland Biomedical Press.
Halonen, P., Bonfanti, C., Lövgren, T., Hemmilä, I., and Soini, E., 1985,
    Detection of viral antigens by time-resolved fluoroimmunoassay, in:
    "Rapid Methods and Automation in Microbiology and Immunology",
    K-O.Habermehl, ed., Springer-Verlag, Berlin, p.429.
Helsingius, P., Hemmilä, I., Lövgren, T., 1985, Time-resolved fluoroimmuno-
    assay for digoxin, Abstract, Israel. J. Clin. Biochem. Lab. Sci. 4:
    52.

Hemmilä, I., Dakubu, S., Mukkala, V.-M., Siitari, H., and Lövgren, T., 1984 a), Europium as a label in time-resolved immunofluorometric assays, Anal. Biochem., 137: 335-343.

Hemmilä, I., Viljanen, M., and Lövgren, T., 1984 b), Time-resolved fluoro-immunoassay (TR-FIA) of tetanus antibodies, Anal.Chem. 317: 737-38.

Hemmilä, I., 1985, Fluoroimmunoassays and immunofluorometric assays, Clin. Chem., 31: 359-370.

Hemmilä, I., Pulkki, K., Irjala, K., 1985, Time-resolved immunofluorometric assay for C-reactive protein, Abstract, Israel. J. Clin. Biochem. Lab. Sci., 4: 52.

Hemmilä, I., Dakubu, S., Method for fluorescence spectroscopic determination of a biologically active substance, U.S. patent No. 4,565,790 (issued on January 21, 1986)

Jackson, T., Marshall, N., and Ekins, R., 1983, Optimisation of immuno-radiometric (labelled antibody) assays, in: "Immunoassays for Clinical Chemistry", W. M. Hunter and J.E.T. Corrie, eds., Churchill Livingstone, Edinburgh, p. 557.

Kaihola, H.-L., Irjala, K., Viikari, J. and Näntö, V., 1985, Determination of thyrotropin in serum by time-resolved fluoroimmunoassay evaluated, Clin. Chem. 31: 1706-1709.

Karp, M., Suominen, I., Hemmilä, I. and Mäntsälä, P., 1983, Time-resolved europium fluorescence in enzyme activity measurements: A sensitive protease assay, J. Appl. Biochem. 5: 399-403.

Lovejoy, L., Holowka, D., and Cathou R., 1977, Nanosecond fluorescence spectroscopy of pyrene-butyrate-antipyrene antibody complexes. Biochemistry 16: 3668-3672.

Lövgren, T., Hemmilä, I., Pettersson, K., Eskola, J., and Bertoft, E., 1984, Determination of hormones by time-resolved fluoroimmunoassay, Talanta, 31: 909-916.

Matikainen, M.-T., Halonen, P., Hemmilä, I., and Lövgren, T., 1983, Time-resolved fluoroimmunoassay in the detection of adenovirus hexon antigen, Abstract, Scand. J. Immunol. 18: 77.

Meurman, O., Hemmilä, I., Lövgren, T., and Halonen, P., 1982, Time-resolved fluoroimmunoasay: a new test for rubella antibodies, J.Clin. Microbiol., 16: 920-25.

Mueller, W., and Hirschfeld, T., 1977, Background reduction in fluorochrome staining, Histochem. J. 9: 121-123.

Näntö, V., Kaihola, H.-L., Irjala, K., and Viikari, J., 1985 a), Evaluation of a sensitive assay for thyrotropin in serum, Abstract, Israel. J. Clin. Biochem. Lab. Sci., 4: 36.

Näntö, V., Suonpää, K., Eskola, J., Lövgren, T., 1985 b), Direct solid phase time-resolved fluoroimmunoassay of total thyroxine in serum, Isr. J. Clin. Biochem. Lab. Sci., 4: 52.

Pettersson, K., Siitari, H., Hemmilä, I., Soini, E., Lövgren, T., Hänninen, V., Tanner, P., and Stenman, U.-H., 1983, Time- resolved fluoro-immunoassay of human choriogonadotropin, Clin. Chem. 29: 60-64.

Saris, N.-E., 1983, Europium phosphorescence as a probe of binding to phospholipids, Chemistry and Physics of Lipids, 34: 1-5.

Siitari, H., Hemmilä, I., Soini, E., Lövgren, T., and Koistinen, V., 1983, Detection of hepatitis B surface antigen using time-resolved fluoro-immunoassay, Nature, 301: 258-60.

Soini, E., and Hemmilä, I., 1979, Fluoroimmunoassay: Present status and key problems. Clin. Chem., 25: 353-61.

Soini, E., and Kojola, H., 1983, Time-resolved fluorometer for lanthanide chelates - a new generation of nonisotopic immunoassays. Clin.Chem., 29: 65-68.

Stenman, U.-H., Myllynen, L., Alftan, H., and Seppälä, M., 1983, Ultrarapid and highly sensitive time-resolved fluoroimmunometric assay for chorionic gonadotropin, Lancet, 17: 647-49.

Stenman, U.-H., Alfthan, H., Koskimies, A., Seppälä, M., Pettersson, K., and Lövgren, T., 1985, Monitoring the LH surge by ultrarapid and highly sensitive immunofluorometric assay, In: "In vitro Fertilization and Embryo Transfer", M. Seppälä, R.G. Edwards, eds., The New York Academy of Sciences, New York, p. 544.

Suonpää, M., Lavi, J., Hemmilä, I., and Lövgren, T., 1985, A new sensitive assay of AFP using time-resolved fluorescence and monoclonal antibodies, Clin. Chem. Acta, 145: 341-348

Thuren, T., Virtanen, J., Lalla, M., and Kinnunen, P., 1985, Fluorometric assay for phospholipase A_2 in serum, Clin. Chem., 31: 714-717.

Wieder, I., 1978, Background rejection on fluorescence immunoassay, In: "Immunofluorescence and Related Staining Techniques", W. Knapp, H. Holubar, G. Wick, eds., Elsevier-North Holland Biomedical Press, New York, NY, p. 67.

Viljanen, M., Backman, C., Veromaa, T., Frey, H., Reunanen, M., 1984, Time-resolved fluoroimmunoassay (TR-FIA) of myelin basic protein (MBP), Acta Neurol. Scand., 69: 361-362.

PHASE-RESOLVED FLUOROIMMUNOASSAY

Linda B. McGown

Department of Chemistry
Oklahoma State University
Stillwater, Oklahoma

INTRODUCTION

Immunoassays employ competitive binding equilibria between labelled and non-labelled antigen (or hapten), Ag* and Ag, respectively, for antibodies (Ab) developed to the antigen for the determination of the antigen:

$$Ag^* + Ag + Ab \rightleftharpoons Ag^*\text{-}Ab + Ag\text{-}Ab. \qquad (1)$$

Immunoassay techniques can be described as either heterogeneous or homogeneous. Heterogeneous techniques require the separation of the Ag-Ab and Ag*-Ab from the Ag* and Ag, whereas the homogeneous techniques avoid the separation step. The first immunoassay techniques employed antigens or haptens labelled with radioactive tracers. These techniques are by necessity heterogeneous because there is no way to distinguish between the radioactivity signal due to the Ag*-Ab from that of the Ag*. The use of alternative labels for immunoassays has enabled the development of homogeneous techniques in addition to the traditional heterogeneous techniques. Homogeneous fluoroimmunoassay (FIA) techniques in which fluorescent labels are used can employ differences in fluorescence intensity, spectral characterisitics, polarization, and/or lifetime between the Ag* and the Ag*-Ab.

Prior to the development of phase-resolved fluoroimmunoassays (PRFIA), the use of fluorescence lifetime selectivity had been primarily restricted to time-resolved heterogeneous FIA techniques in which time resolution is used to discriminate between long-lived fluorescence emitted from an appropriate fluorescent label and the generally shorter-lived fluorescence background from the sample. The Ag* and Ag*-Ab are still separated prior to the fluorescence measurements. The PRFIA approach, on the other hand, is homogeneous and exploits the difference between the fluorescence lifetimes of the Ag* and the Ag*-Ab to resolve their individual emission contributions so that the separation step is no longer necessary. The ability to resolve the fluorescence intensity contributions due to a single molecular species in two different microenvironments (free and protein-bound) was demonstrated prior to the development of the PRFIA technique (McGown, 1984).

Phase-Resolved Fluorescence Spectroscopy

Phase-resolved fluorescence spectroscopy (PRFS) (Veselova et al., 1970; Lakowicz and Cherek, 1981; Mattheis et al., 1983) is based on the phase-modulation technique for the determination of fluorescence lifetimes (Gaviola, 1927; Birks and Dyson, 1961). The sample is excited with sinusoidally modulated light which has an angular modulation frequency $\omega=2\pi f$, where f is the linear modulation frequency. The time-dependent intensity of the excitation beam E(t) is given by:

$$E(t) = E_o(1 + m_{ex}\sin\omega t) \qquad (2)$$

where E_o is the wavelength-dependent d.c. intensity contribution and m_{ex} is the modulation depth of the exciting beam (i.e., the ratio of the a.c. amplitude to the d.c. signal). The resulting fluorescence emission function, F(t), for a system exhibiting homogeneous exponential decay will be phase-shifted by an angle ϕ and appear as:

$$F(t) = F_o(1 + m_{ex}m\sin(\omega t-\phi)) \qquad (3)$$

where F_o represents the wavelength-dependent d.c. component of the fluorescence emission function, $\underline{m}$ is the demodulation factor (i.e., the ratio of the modulation depth of the sample relative to the modulation depth of a scattering solution which has a fluorescence lifetime of zero), and ϕ is the phase-delay of the detected emission. The excitation and the emission response functions for emitters with fluorescence lifetimes of 10 and 100 ns are shown in Figure 1 for an excitation modulation frequency of 10 MHz (Demas, 1983). For a mixture of non-interacting fluorescence emitters, the total F(t) function is the sum of the individual contributions from each exponentially decaying emitter.

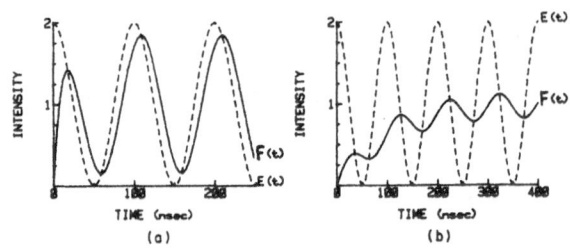

Fig. 1. Transient response F(t) of samples with fluorescence lifetimes of (a) 10 ns and (b) 100 ns to sinusoidal excitation E(t) with a modulation frequency of 10 MHz (Demas, 1983, with permission).

Phase-resolved fluorescence spectroscopy utilizes phase-variable rectifying detection, combining the integration of the a.c. component of F(t) and multiplication by a square wave function which has the value of one for one half-cycle (or π-interval), followed by a half-cycle during which the value is zero (Figure 2). The phase of this "on-off" integration cycle, hereafter refered to as the detector phase angle, ϕ_D, is variable, and can be set to any angle between 0 and 360°. The resulting function, $F(\phi_D)$, is a time-independent phase-resolved fluorescence intensity (PRFI) which is proportional to the cosine of the difference between the relative phases of the detector and the sample emission:

$$F(\phi_D) = F_o(1 + m_{ex}m\cos(\phi_D-\phi))$$ (4)

Therefore, the PRFI is a function of both the fluorescence lifetime and the wavelength-dependent spectral characteristics of the emitter. The PRFI of a solution containing non-interacting emitters is the sum of their individual PRFI contributions.

Selective "nulling" of the PRFI contribution due to a homogeneous (single fluorescence lifetime) component (Figure 2a) is achieved by setting ϕ_D exactly out of phase with the component:

$$\phi_D = \phi_{component} \pm 90^o$$ (5)

thereby eliminating its contribution to the total fluorescence emission of a solution ($F(\phi_D) = 0$ for the component, from Equation 4). The remaining PRFI contributions due to other emitters in the solution can then be measured without interference from the nulled component.

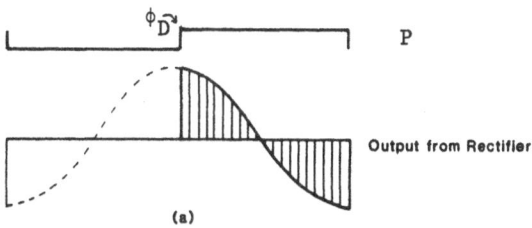

(a)

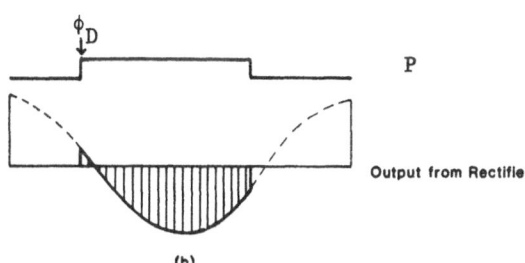

(b)

Fig. 2. Pictorial representation of the output from a phase-variable rectifying detector with ϕ_D (a) 90° and (b) 185° out of phase with the fluorescence signal. $\underline{P}$ represents the integration function. Mattheis et al., 1983, with permission.

Phase-Resolved Fluoroimmunoassay

Phase-resolved fluoroimmunoassay (PRFIA) involves the use of measurements of PRFI at a series of $\underline{n}$ detector phase angle settings (Bright and McGown, 1985), thereby generating a set of $\underline{n}$ independent proportional equations of the form:

$$F(\phi_D)_i = I_{f,i}C_{Ag*} + I_{b,i}C_{Ag*-Ab}$$ (6)

where $F(\phi_D)_i$ is the PRFI of the immunoassay standard or sample solution at the ith detector phase angle, $I_{f,i}$ and $I_{b,i}$ are the molar fluorescence intensities of the free Ag* and the Ag*-Ab, respectively, at the ith detector phase angle, and C_{Ag*} and C_{Ag*-Ab} are the analytical molar concentrations of the two species. An additional equation is added to the series:

$$C_{Ag*} + C_{Ag*-Ab} = C \qquad (7)$$

where the constant C is the known value of total Ag* in the solution. The nx2 matrix formed from Equations 6 and 7 are solved for C_{Ag*} and C_{Ag*-Ab}. If n=1, Equations 6 and 7 form a square 2x2 matrix for which an exact solution is found. If n>1, the matrix is overdetermined and can be solved using a least squares iterative procedure (Hartley, 1961). More detailed descriptions of the data analysis procedures that have been used for PRFS analysis of two-component systems can be found in the literature (McGown, 1984; McGown and Bright, 1985).

The percent of free labelled antigen (%Ag*) is calculated as

$$\%Ag* = C_{Ag*}/C . \qquad (8)$$

The %Ag*-Ab can be similarly calculated if required. The PRFIA calibration curve is generated by plotting %Ag* as a function of concentration of Ag in standard solutions, as in other immunoassay techniques. The analyte Ag concentration of an unknown sample solution is found from the curve using the %Ag* found for the solution.

Due to the relationship expressed in Equation 7, only one additional equation of the type shown in Equation 6 is needed to generate the two equations required to determine the two concentrations. If a sufficient difference exists between the steady-state (d.c.) molar intensities of Ag* and Ag*-Ab, PRFS is not necessary and the immunoassay can be accomplished using conventional steady-state fluorescence measurements. However, the systems studied to date have shown only slight differences or no difference at all in the two molar intensities, and no shifts have been observed between the fluorescence spectra of Ag* and Ag*-Ab. The added dimension of fluorescence lifetime is therefore useful in resolving the intensity contributions of the Ag* and Ag*-Ab.

MATERIALS AND METHODS

THE PRFIA systems described to date have been developed using commercially available reagents. Antibodies have been polyclonal preparations in either dialyzed serum or IgG fractions. Therefore, specificities will be ultimately limited by the specificity of the polyclonal preparations. Fluorescent-labelled antigens or haptens are available for a very limited number of analytes with an even smaller variety of labels. The most readily available antigen-conjugated label is fluorescein isothiocyanate, which is probably the most common label used for heterogeneous fluoroimmunoassays. Other commercially available conjugated labels include rhodamine isothiocyanate and its tetramethyl derivative, and Texas Red (Titus et al., 1982), which is a sulforhodamine derivative.

The PRFIA technique was developed using a cross-correlation phase-modulation spectrofluorometer (Spencer and Weber, 1969) equipped with a hardware-based phase-resolution capability (Model 4800S, SLM Instruments, Inc., Mattheis et al., 1983). All of the experiments described in this chapter were performed with a 450 watt xenon arc lamp source, monochromotors to achieve both emission and excitation wavelength selection, Hamamatsu R928 photomultiplier tube detection with a double-beam ratiometric mode, and an on-line APPLE II+ microcomputer for data acquisition and fluorescence lifetime analysis. An independent APPLE IIe was used for post-acquisition PRFIA data analysis. Other details of instrument operation and data collection and analysis can be found in the specific PRFIA references (Bright and McGown, 1985; Tahboub and McGown, 1986).

PRFIA OF PHENOBARBITAL

The first PRFIA was developed for phenobarbital using fluorescein isothiocyanate (FITC)-labelled phenobarbital as Ag* (Bright and McGown, 1985). Fluorescence measurements were made at the approximate excitation and emission maxima of the FITC-phenobarbital (490 nm and 520 nm, respectively), corresponding to those of the unconjugated FITC. A very small intensity difference between the Ag* and Ag*-Ab was first tried as the basis of a steady-state homogeneous immunoassay using the appropriate version of a single Equation 6 and Equation 7 to generate the required two equations. A calibration curve prepared using five standards containing from 0 to 74.8 μg/ml phenobarbital (0-23.7 ng/ml in cuvette) in phosphate buffer (0.010 M, pH 7.5 containing 2 μM human serum albumin) was compared to a curve for the same solutions obtained using a heterogeneous fluoroimmunoassay procedure. A correlation coefficient of 0.6 indicated the poor performance of the steady-state homogeneous approach.

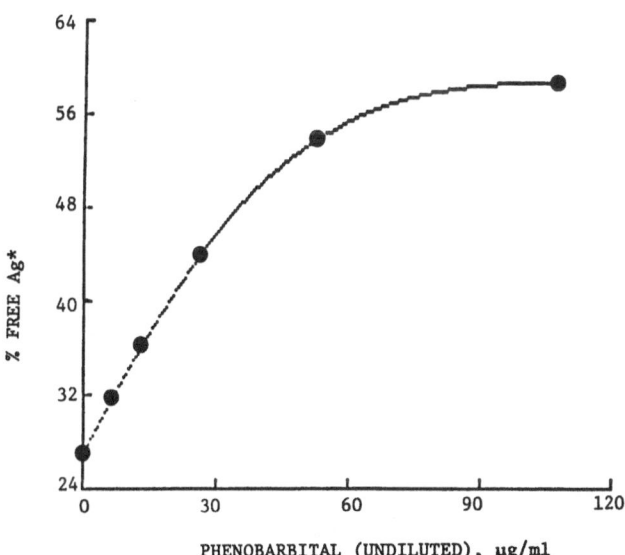

Fig. 3. Calibration curve for phenobarbital obtained
using PRFIA with five detector phase angles
(Bright and McGown, 1985, with permission).

When the small fluorescence lifetime difference between Ag* (4.04 ns) and Ag*-Ab (3.94 ns) of approximately 100 picoseconds was added to the very small intensity difference via the PRFIA technique, a correlation coefficient of 0.97 was obtained between the PRFIA calibration curve generated using six equations (five detector phase angles for Equations 6 plus Equation 7) and the curve for the heterogeneous fluoroimmunoassay. The PRFIA calibration curve is shown in Figure 3. A relative standard deviation of 6.7% was found for a set of four normalized calibration curves obtained one per week over a period of four weeks. The normalization was performed to compensate for reagent degradation and signal loss over the four-week period.

Three artificially prepared "unknown" solutions containing from 7.03 to 28.0 μg/ml phenobarbital in buffer (2.22-8.86 ng/ml in cuvette) were analyzed by the PRFIA technique using from two to six equations (one to five detector phase angles for Equations 6 plus Equation 7). The determination errors for the three solutions are summarized in Table 1. The best overall results (smallest error magnitudes) were obtained for the six equation set. The limit of detection for phenobarbital (defined as the fluorescence signal equivalent to three times the standard deviation of the blank) was estimated to be 3 μg/ml in the original solution, corresponding to 1 ng/ml in the cuvette.

Table 1.   Errors for the PRFIA Determination of Phenobarbital (Bright and McGown, 1985).

	% Relative Error				
Phenobarbital Concentration[a]	Number of Detector Phase Angles				
	1	2	3	4	5
7.03 (2.22)	-5.9	-16.9	-12.5	5.4	1.0
14.0 (4.43)	-11.7	9.6	2.3	1.8	-0.4
28.0 (8.86)	3.1	21.2	11.2	9.5	8.1
Average %Error[b]	-4.8	4.6	0.3	5.6	2.9
Average |%Error| [c]	6.9	15.9	8.7	5.6	3.2

[a] Phenobarbital concentration, μg/ml in the undiluted solutions (ng/ml in cuvette shown in parentheses).

[b] Average relative error for the three solutions.

[c] Average of the absolute values of the relative errors for the three solutions.

ENHANCEMENT OF PRFIA USING MICELLES AND CYCLODEXTRINS

Recent work has focussed upon the use of auxiliary binding reagents such as micellar species and cylodextrins to enhance the fluorescence spectral, intensity and lifetime differences between Ag* and Ag*-Ab by association with the free Ag* (McGown, 1985; Keimig and McGown, 1986). Although the studies described in the previous section indicate that the selectivity provided by the fluorescence lifetime difference in the PRFIA approach can greatly improve the results of the homogeneous fluoroimmunoassay, it is clear that both the intensity and the lifetime differences are very small. Any enhancement of the differences due to the presence of the auxiliary binding reagents can improve the performance of the immunoassay.

The effects of various micellar and cyclodextrin species on the fluorescence properties of FITC-labelled phenobarbital (synthesized on-site, Bright et al., 1986) in unbuffered aqueous solution are summarized in Table 2 (Keimig and McGown, 1986). Each of the micellar species was present at a concentration above the critical micelle

Table 2. Effects of Micelles and Cyclodextrins on the Fluorescence Properties of FITC-Phenobarbital [a] (Keimig and McGown, 1986).

Added Species [b]	$\lambda_{em}(max)$ [c]	τ_F [d]
None	516	4.00
Hexadecyltrimethylammonium chloride (CTAC) (3.0)	526	4.39
Dodecylamine hydrochloride (DAC) (15)	523	4.07
Dodecyltrimethylammonium chloride (DTAC) (22)	526	4.28
N,N-dimethyldodecylamine- N-oxide (LDAO) (4.0)	522	3.68
Tetradecyltrimethylammonium bromide (TTAB) (7.0)	525	4.40
Beta-cyclodextrin (2.0)	516	4.15
Gamma-cyclodextrin (2.0)	517	4.23

[a] FITC-Phenobarbital concentration 0.21 μM in cuvette.
[b] Concentrations in cuvette (mM) shown in parentheses.
[c] Emission maximum, nm. Excitation at 490 nm.
[d] Fluorescence lifetime (ns) calculated from phase-shift.

concentration (CMC) of the species. All of the micelles but neither of the cyclodextrins caused significant spectral shifts. Significant fluorescence lifetime changes were induced by all of the species investigated. In the same work, studies of FITC-phenobarbital in buffered solutions of CTAC, TTAB and LDAO demonstrated the effects of pH and micelle concentration. Fluorescence intensity enhancement increases in all cases as the pH is decreased from 7.8 to 5.8 in 10 mM phosphate buffer. Absorption spectra of FITC-phenobarbital exhibit large increases in molar absorptivity and spectral shifts that parallel the fluorescence intensity and excitation shifts caused by the micelles. The fluorescence excitation and emission spectra of FITC-phenobarbital in the presence of TTAB are shown in Figure 4, and the absorption spectra in Figure 5.

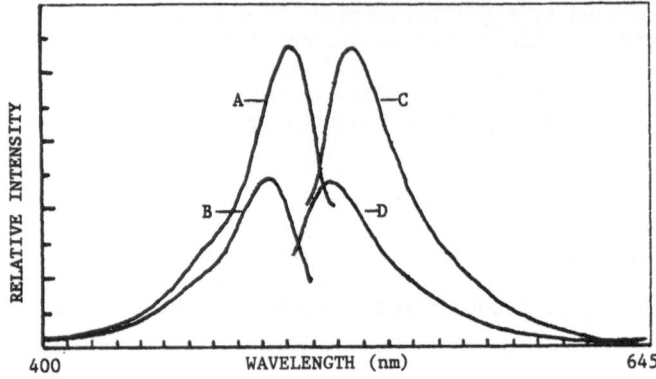

Fig. 4. Excitation (A,B) and emission (C,D) spectra of FITC-
phenobarbital with (A,C) and without (B,D) TTAB in pH
5.8 phosphate buffer. (A) Emission at 525 nm; (B)
emission at 517 nm; (C) excitation at 501 nm; (D)
excitation at 492 nm. Keimig and McGown, 1986, with
permission.

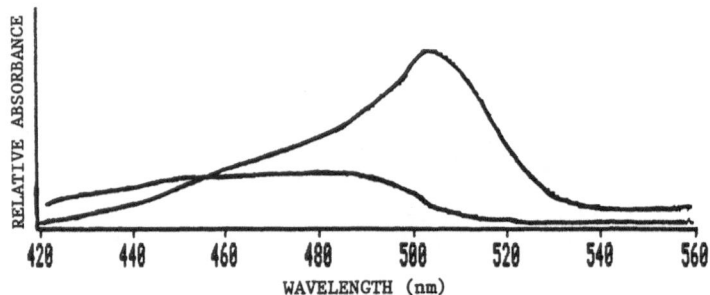

Fig. 5. Absorption spectra of FITC-phenobarbital (A) with and
(B) without TTAB.

Solutions containing FITC-phenobarbital in CTAC above its CMC in 10
mM phosphate buffer were studied in the absence and presence of
anti-phenobarbital antiserum. At pH 6.7, the presence of the antiserum
caused a 10% fluorescence intensity enhancement and a fluorescence
lifetime increase of 0.50 ns. At pH 7.0, a larger intensity enhancement
of 29% and a smaller lifetime increase of 0.31 ns were observed in the
presence of antiserum. At both pHs, excitation maxima occured at 501 nm
both in the absence and presence of antiserum, indicating that the
micelles modify the fluorescence of the antibody-bound Ag* as well as the
free Ag*. Similar studies using LDAO and TTAB did not show such large
fluorescence intensity and lifetime differences between Ag* and Ag*-Ab.
Current research is focussed on the further elucidation of the
interactions between the micellar species and the immunoassay system
components, and on the micellar modification of the Ag* and Ag*-Ab
fluorescence characteristics.

PRFIA OF HUMAN SERUM ALBUMIN

The PRFIA developed for phenobarbital demonstrated the applicability
of the phase-resolved approach for the determination of small molecular
weight hapten analytes. It was also of interest to investigate the use of

the PRFIA for the determination of high molecular weight antigenic analytes. Certain homogeneous fluoroimmunoassay techniques, such as those based on changes in fluorescence polarization properties, are limited to analytes below a certain molecular weight range (Smith et al., 1981). Human serum albumin (HSA) was chosen as a model macromolecular system for the PRFIA (Tahboub and McGown, 1986).

An interesting aspect of the PRFIA for HSA was the use of Texas Red instead of FITC as the HSA-conjugated label. As mentioned above, FITC is a commonly used label in fluoroimmunoassay systems. The fluorescein emission and excitation maxima are longer than those of most common background fluorescence emitters in clinical samples. However, one source of interference, albumin-bound bilirubin, has emission and excitation spectra that overlap extensively with those of fluorescein. Although the quantum yield of the bound bilirubin is very low (10^{-3}) relative to that of fluorescein (0.8-0.9), the high levels of bilirubin in jaundiced patients can present non-negligible interference (Smith et al., 1981). Several approaches to dealing with bilirubin interfence in the determination of fluorescein have been described, including blank subtraction (Smith et al., 1981), a phase-resolved approach (Bright and McGown, 1984) based on the fluorescence lifetime difference between short-lived bilirubin fluorescence (0.4 ns) and the longer-lived fluorescein fluorescence (4 ns), and a first-derivative synchronous excitation technique (Bright and McGown, 1986). However, the simplest solution is to use a different label with emission and excitation spectra removed from those of bilirubin. Texas Red satisfies this requirement, with its excitation maximum at 592 nm and emission maximum at 610 nm.

The PRFIA was applied to the determination of HSA in human serum samples (McGown and Tahboub, 1986). A fluorescence lifetime difference of approximately 90 picoseconds was observed between the Ag* (4.07 ns) and Ag*-Ab (3.98 ns). Matrix effects due to the serum were evaluated by analyzing two sets of six solutions, one set prepared from serum and the other from aqueous standards. Both sets contained the same concentrations, which ranged from 0.10 to 1.0 mg/1 in cuvette. A plot of the concentrations determined for the serum solutions vs. those for the aqueous standards yielded a slope of 1.02 ± 0.03, a y-intercept of 0.04 ± 0.05, and a correlation coefficient of 0.999. Matrix effects were therefore indicated to be negligible, possibly due to the large sample dilution factor of 10^{+6}.

Twenty-four individual serum samples were analyzed by the PRFIA procedure, and results were compared to those obtained using the bromocresol green method (Doumas et al., 1971) on a Technicon RA-1000 autoanalyzer. A correlation plot of the PRFIA results obtained using two detector phase angles for Equations 6 plus Equation 7 vs. the autoanalyzer results yielded a slope of 0.99 ± 0.04, a y-intercept of 0.26 ± 0.17, and a correlation coefficient of 0.996. The sample HSA concentrations ranged from 2.5 to 5.0 g/dl (0.18-0.36 mg/1 in cuvette).

DISCUSSION

The studies performed to date have been aimed towards the demonstration of the feasibility of the phase-resolved approach for homogeneous immunoassay for both small molecular (hapten) and macromolecular analytes. Generation of calibration curves and initial studies of matrix effects for serum samples indicate that PRFIA is a promising technique. The use of auxiliary binding reagents such as micelles will be incorporated into the PRFIA to increase the resolution between the Ag* and Ag*-Ab fluorescence intensity contributions.

The most attractive feature of the PRFIA approach is its simplicity. The basis of the technique is the difference between the fluorescence spectral, intensity and lifetime characterisitics of the Ag* and Ag*-Ab, with one or both of the species suitably modified by the auxiliary binder. All of the fluorescence selectivity parameters are simultaneously exploited in the phase-resolved measurements, which provide phase-resolved fluorescence intensities that are a function of emission and excitation wavelengths and the fluorescence lifetime(s) of the emitting species. The PRFI values are analogous to steady-state (d.c.) fluorescence intensity measurements, requiring the same time per measurement (which is essentially instantaneous) and exhibiting the same direct proportionality to the concentration(s) of emitter(s) and analogous spectral additivity. Therefore, other than the requirement for the specialized phase-resolved fluorescence instrumentation, the PRFIA approach is essentially identical to a conventional homogeneous fluorimmunoassay based only on fluorescence quenching or enhancement (i.e., an intensity difference between Ag* and Ag*-Ab), requiring the same time and effort to generate an equal number of equations (Equations 6 and 7).

Another advantage of the PRFIA approach is the minimal reagent requirements. Only the basic fluoroimmunoassay reagents, Ag*, Ag and Ab, plus inexpensive and readily available auxiliary reagents such as buffers and micelles are necessary. These simple requirements are in contrast to techniques requiring multiple labels, several different antibodies, enzyme markers, or other complex, expensive reagents.

A disadvantage of the PRFIA technique is the loss of incident intensity due to the process of excitation beam modulation, resulting in poorer detection limits (on the order of nanomolar in cuvette for fluorescein) than could be achieved using steady-state fluorescence measurements. Current studies are employing filter rather than monochromator wavelength selection in the emission beam to improve the detection limits.

Another modification to be explored is the use of monoclonal antibodies instead of polyclonal preparations. In addition to the improved selectivity accrued to any immunoassay using monoclonal antibodies, the PRFIA may also benefit from improved fluorescence lifetime homogeneity (i.e., exponential decay). On the other hand, polyclonal preparations may still be advantageous if they exhibit a higher overall affinity for the Ag and Ag*, since competing auxiliary binders will also be present in the solutions.

Future studies will also explore the use of other fluorescent labels besides those for which conjugated analytes are commercially available. The main considerations in the choice of a label for any fluoroimmunoassay include high quantum yield, stability, and emission and/or excitation spectra that do not significantly overlap with those of common fluorescence interferents. An additional requirement for the PRFIA is that the labelled antigen display a sufficient change in fluorescence characteristics upon antibody-binding, possibly aided by the presence of auxiliary binding reagents. On the other hand, a label with a long fluorescence lifetime (greater than 10-15 ns) will be sensitive to dissolved oxygen in solution which will impair precision and detection limits, and may necessitate purging of the solutions. Fluorescein, rhodamine and Texas Red dyes all have short-lived fluorescence (2-5 ns) and are essentially unaffected by dissolved oxygen, while still exhibiting detectible differences between Ag* and Ag*-Ab fluorescence. Appropriate auxiliary binding reagents may help to alleviate the sensitivity of longer-lived labels to oxygen quenching and also may improve detection limits.

254

A future goal for PRFIA is the development of a flow system for rapid sample analysis. If sufficiently large differences between the fluorescence lifetimes and spectral characterisitics of Ag* and those of Ag*-Ab can be achieved, it will be possible to measure sample and standard solutions at the detector phase angle required to null the PRFI of one of the species at the wavelength maxima of the non-nulled species. For example, the PRFI of micelle-associated Ag* could be measured at its emission and excitation maxima, at $\phi_D = \phi_{Ag*-Ab} \pm 90^\circ$ so that the PRFI of Ag*-Ab is zero.

Finally, attention has been directed towards the use of fluorescent-labelled antibodies (Ab*) instead of labelled antigens. This approach is not a competitive immunoassay and the equilibrium involved is non-competitive:

$$Ag + Ab* \rightleftharpoons Ag-Ab*. \tag{9}$$

The phase-resolved approach involves the discrimination between Ab* and Ag-Ab* on the basis of the same fluorescence lifetime changes used in PRFIA. The main advantage of the labelled antibody approach is that the antibody proteins are often easier to label than the analyte species, especially in the cases of hapten analytes. On the other hand, the position of the label may be remote from the antigenic binding site, resulting in less modification of fluorescence properties. A system consisting of HSA and anti-HSA conjugated with Texas Red is under investigation, and a fluorescence lifetime difference of 170 picoseconds was observed between the Ab* and Ag-Ab* (in the absence of auxiliary binders). This is almost twice the difference observed for the HSA PRFIA system. No spectral or intensity changes were observed. The labelled antibody approach should prove to be a reasonable alternative to PRFIA for some analytes and may allow greater flexibility in the selection of label.

ACKNOWLEDGEMENTS

The author is grateful to the National Institute on Drug Abuse and the National Science Foundation for support of the research projects involving phenobarbital and HSA, respectively. Additional support was provided by a Sigma Xi Grant-in-Aid of Scientific Research to Frank V. Bright during his graduate studies at Oklahoma State University.

REFERENCES

Birks, J. B. and Dyson, D. J., 1961, Phase and modulation fluorometer, J. Sci. Instr., 38:282-285.

Bright, F. V., Bunce, R. A., and McGown, L. B., 1986, A selective synthesis of 5-p-aminophenylbarbituric acid, Org. Prep. Proc. Intl., in press.

Bright, F. V. and McGown, L. B., 1984, Elimination of bilirubin interference in fluorimetric determination of fluorescein by phase-resolved fluorescence spectrometry, Anal. Chim. Acta, 162:275-283.

Bright, F. V. and McGown, L. B., 1985, Homogeneous immunoassay of phenobarbital by phase-resolved fluorescence spectroscopy, Talanta, 32:15-18.

Bright, F. V. and McGown, L. B., Minimisation of bilirubin interference in the determination of fluorescein using first derivative synchronous excitation fluorescence spectroscopy, Analyst, in press.

Demas, J. N., 1983, "Excited State Lifetime Measurements", Academic Press, New York.

Doumas, B. T., Watson, W. A., and Biggs, H. G., 1971, Albumin standards and the measurement of serum albumin with bromocresol green, Clin. Chim. Acta, 31:87–96.

Gaviola, E., 1927, Ein Fluorometer. Apparat zur Messung von Fluoreszenzabklingungszeiten, Z. Phys., 42: 853–61.

Hartley, H. O., 1961, The modified Gauss-Newton method for the fitting of non-linear regression functions by least squares, Technometrics, 3:269–80.

Keimig, T. L. and McGown, L. B., 1986, Micellar modification of the fluorescence spectral and lifetime properties of fluorescein-labelled phenobarbital, Talanta, in press.

Lakowicz, J. R. and Cherek, H., 1981, Phase-sensitive fluorescence spectroscopy: a new method to resolved fluorescence lifetimes or emission spectra of components in a mixture of fluorophores, J. Biochem. Biophys. Meth., 5:19–35.

Mattheis, J. R., Mitchell, G. W., and Spencer, R. D., 1983, Phase-resolved nanosecond spectrofluorometry: theory, instrumentation, and new applications of multicomponent analysis by subnanosecond fluorescence lifetimes, in: "New Directions in Molecular Luminescence", D. Eastwood, ed., ASTM Special Technical Publication 822, Baltimore, MD.

McGown, L. B., 1984, Phase-resolved fluorimetric determination of two albumin-bound fluorescein species, Anal. Chim. Acta, 157:327–332.

McGown, L. B., 1985, Phase-resolved fluoroimmunoassay, presented in the Molecular Luminescence Symposium at the National Meeting of the Electrochemical Society, Las Vegas, NV, October 13–18, 1985.

McGown, L. B. and Bright, F. V., 1985, Simultaneous two-component determinations using phase-resolved fluorescence spectroscopy, Anal. Chim. Acta, 169:117–123.

Smith, D. S., Hassan, M., and Nargessi, R. D., 1981, Principles and practice of fluoroimmunoassay procedures, in: "Modern Fluorescence Spectroscopy", E. L. Wehry, ed., Plenum Press, New York.

Spencer, R. D. and Weber, G., 1969, Measurements of subnanosecond fluorescence lifetimes with a cross-correlation phase fluorometer, Ann. N.Y. Acad. Sci., 158:361–76.

Tahboub, Y. and McGown, L. B., 1986, Phase-resolved fluoroimmunoassay of human serum albumin, Anal. Chim. Acta, in press.

Titus, J. A., Haugland, R., Sharrow, S. O., and Segal, D. M., 1982, Texas Red, a hydrophilic red-emitting flurophore for use with fluorescein in dual parameter flow microfluorimetric and fluorescence microscopic studies, J. Immuno. Meth., 50:193–204.

Veselova, T. V., Cherkasov, A. S., Shirokov, V. I., 1970, Fluorometric method for individual recording of spectra in systems containing two types of luminescent centers, Opt. Spectrosc., 29:617–618.

SOLID-PHASE LUMINESCENCE IMMUNOASSAYS USING KINASE AND ARYL

HYDRAZIDE LABELS

William G. Wood, Harald Fricke and
Christian J. Strasburger

Klinische Laboratorien, Klinik für Innere Medizin
Medizinische Universität zu Lübeck, D-2400 Lübeck
Federal Republic of Germany

INTRODUCTION

Although bioluminescence has been known for over 3000 years, and chemiluminescence for over a century, (Radziszewski, 1877), it is only recently that these phenomena have been adapted for routine, in-vitro diagnosis. This has partly been due to the lack of suitable measuring equipment, and partly to the lack of robust immunoassay systems.

The aim of this chapter is to present the reader with practical advice and with examples, so that he or she can set up luminescent immunoassays with a minimum of equipment and time wasting.

All methods described can be performed in a small laboratory, which is equipped for simple organic syntheses and chromatography column separation steps. As one of the authors (WGW) has recently published extensive review article, (Wood, 1985), the topics covered in this chapter will constitute an updating of the state of the art in this laboratory, many of the assays and assay systems being described here for the first time.

The chapter has been effectively divided into two parts; a relatively short one, which deals with bioluminescent immunoassays using pyruvate kinase as marker, and a more extensive review of aryl hydrazide chemiluminescent labels, these being derivatives of isoluminol and naphthalene-1,2-dicarboxylic acid hydrazide.

All assays described use labeled antibodies, as this type of assay has been shown to be both sensitive and robust according to Ekins (1978) and Hunter (1982). Using such "excess-reagent" solid-phase techniques, the sample need not come into contact with the label, thus avoiding possible interference in terms of oxidation or degradation during the incubation. Oxidation of the label is possible, as far as aryl hydrazides are concerned, as the necessary heme compounds, (e.g. cytochrome c, hemoglobin) and enzymes (peroxidases) are often present in serum samples sent for analysis.

MATERIALS

The materials needed to carry out the assays described in this chapter have been listed in terms of either bioluminescent or chemiluminescent systems.

For Bioluminescent Immunoassays

Pyruvate kinase, (E.C. 2.7.1.40) from rabbit muscle - Sigma, Deisenhofen, F.R.G.
Adenosine diphosphate, (potassium salt) and phosphoenolpyruvate - Sigma.
ATP-monitoring reagent - LKB-Wallac, Munich, F.R.G.
Heterobifunctional reagents, (see text for formulas) - Sigma or Boehringer-Mannheim, Mannheim, F.R.G.
5,5'-dithiobis-(2-nitrobenzoic acid), (Ellmann's reagent), Microcrystalline cellulose (20 μm particles), reduced glutathione - Sigma.
Sodium metaperiodate and buffer substances, Merck, Darmstadt, F.R.G.
Schiff's reagent (p-rosaniline reduced with sulfur dioxide) - hospital pharmacy.

Assay buffer 1 - 0.05 mol/L TRIS-HCl containing 0.1 mol/L KCl and 2.5 G/L bovine serum albumin, pH 7.4. Wash solution - 0.1 mol/L KCl.

For Chemiluminescent Immunoassays

ABEI, (6-(4-aminobutyl-N-ethyl) isoluminol - LKB or Sigma. ABEN, (7-(4 -aminobutyl-N-ethyl) naphthaline-1,2-dicarboxylic acid hydrazide was a gift from Dr. Hartmut Schroeder, Miles Laboratories, Elkhart, Indiana, for which the authors are grateful.
Succinic anhydride, N-hydroxysuccinimide and dimethylformamide - Merck.
Dicyclohexylcarbodiimide -Fluka, Neu Ulm, F.R.G.
Antibodies, Cooper Biomedical (Cappel), Frankfurt, F.R.G., Dakopatts, Hamburg, F.R.G., Pelfreez, Heidelberg, F.R.G., Behringwerke, Marburg a.d.Lahn, F.R.G., and Bayer (Miles), Munich, F.R.G.
Other buffer substances -Merck or Sigma.
Assay buffer 2 0.025 mol/L TRIS-HCl, 0.025 mol/L phosphate, 1.25 g/L bovine serum albumin, 0.75 ml/L Tween 20 and 0.15 mol/1 sodium azide, pH 7.6 - 8.0
Pentawash or Proquantum wash system, (Abbott, Wiesbaden-Delkenheim, F.R.G.), with 20 and 60-well trays. Horizontal shaker, (Heidolph / Abbott), 170 - 200 rpm.
Polystyrene balls, 6.4 mm diameter, Spherotech Kugeln, Fulda, F.R.G. or Precision Plastic Ball Co., Chicago, Illinois, or Euromatic, Brentwood, Middlesex, G.B.
Glutaraldehyde, Grade I - Sigma, Catalog No. G-5882.
Polyphenylalanine-Lysine copolymer (Mr 30 - 40 kD) - Miles or Sigma.
Dipotassium hydrogen phosphate and Perhydrit (Urea-hydrogen peroxide) tablets Merck.
Tween 20 (Polyethoxysorbitan monolaurate) - Sigma.
Hemin, cytochrome-c, microperoxidase - MP 11, horseradish peroxidase - Sigma.

Light Initiation and Measurement Systems

The assays using the bioluminescent system were measured on an LKB-1251 luminometer, using a kinetic program. The values were expressed as mV/s. The chemiluminescent-labeled assays were measured on a

Berthold LB-950 300-sample luminometer. The light signal was integrated over the first 10 seconds, the units being arbitrary (cps). All measurements were made at ambient temperature, (18-23°C).

The two-component system for the light initiation of the chemiluminescent systems consisted of:

Reagent a: Hemin-NaCl. Hemin stock solution (1 mmol/L) was made up in distilled water adjusted to pH 8.6 with ammonia to facilitate solution. The working solution was prepared by making a 1:2000 dilution of the stock solution in 0.15 mol/L NaCl, pH 3.

Reagent b: NaOH-hydrogen peroxide. One part of sodium hydroxide (1 mol/L) was mixed with 2 parts of hydrogen peroxide (1 perhydrit tablet in 200 ml distilled water), and left to stand for at least 20 minutes before use.

The hemin working solution is stable for at least a week, if kept in the refrigerator after being used. The hemin stock solution is stable for several months if kept at 4°C. The resistance to microbial attack can be increased by adding benzalkonium chloride (Sigma) to the hemin stock solution. The alkaline peroxide working solution can be used for 3-5 days after being prepared, providing it is kept in a cool dark place, when it is not being used.

BIOLUMINESCENT LABELS

What can be recommended, and what is to be avoided? This question is of importance, as methods have been published, in which different components of the bioluminescent system have been coupled to antigens and antibodies. Examples of such are: NAD, (Schroeder et al., 1976), ATP, (Carrico et al., 1976) and dehydrogenases, (Brolin et al., 1971). In our laboratory, we have used pyruvate kinase as the label, and have published the results in detail elsewhwere, (Fricke et al., 1982, Wood et al., 1982).

The use of an enzyme as label provides for an amplification of the signal by making use of the enzymic activity. Of all kinases tried out in this laboratory, only pyruvate kinase from rabbit muscle, which had a relatively high molecular mass of ca. 225 kD, was found to withstand coupling and retain enzymic activity. Other kinases, which appeared attractive on paper, for example adenylate kinase (myokinase), which is stable to heat and acid pH; proved to be impractical, as the enzymic activity disappeared during the coupling process, when heterobifunctional reagents were used in which the thiol group of the enzyme was used for coupling, (Gadow et al., 1984). Whether the introduction of functional thiol groups, using, for example, SPDP (N-succinimidyl-3-(2-pyridyl dithio) propionate, Pharmacia, Freiburg, F.R.G.) would make it possible to use other kinases, such as adenylate kinase with its low molecular mass, (21 kD) and single substrate (ADP), or acetate kinase, remain interesting speculations.

The main drawback of all bioluminescent systems is their sensitivity to "conventional" antimicrobial agents, such as merthiolate and azide. One advantage of these systems, is that the light-output in some ATP-dependant systems is reported to reach the order of 0.9 Einstein/Mol, compared with less than 0.05 Einstein/Mol for most chemiluminescent systems in aqueous solution.

Labeling of Antigens and Antibodies with Pyruvate Kinase

The labeling of proteins (e.g. antibodies) with another protein, (here pyruvate kinase), presents problems, as far as the production of specific products is concerned. The production of undefined oligomers and polymers can be reduced by using heterobifunctional reagents for coupling. In the case of pyruvate kinase, the first step is the coupling of the heterobifunctional reagent to a thiol group of the enzyme. As pyruvate kinase has over 30 thiol groups, most of which do not form part of the active center of the enzyme, this step can be achieved with minimal loss of enzymic activity. The coupling to the protein takes place after the pH has been adjusted, so that an amino group in the protein can be attached to the second active group of the heterobifunctional reagent. As can be seen, both the enzyme, and the protein to be coupled, contain free amino groups, so that oligomers are still produced which contain enzyme, protein, or a combination of the two. The reaction products must be separated on a suitable column, testing the fractions for enzymic activity, and, in the case where an antibody has been coupled, for antibody-binding capacity. Full details are given elsewhere, (Gadow et al., 1984).

The heterobifunctional reagents used in this laboratory contain a maleimide moiety for coupling to thiol groups, and an N-hydroxysuccinimide group for coupling to amino groups. The thiol group coupling takes place optimally at pH 5-6, the coupling of amino groups betwen pH 8 and 9. It is intersting to note, that after being coupled to transferrin, the pyruvate kinase does not suffer much conformational change, as was seen in the appropriate Lineweaver-Burke plots, (Gadow et al., 1984).

The heterobifunctional reagents used for the coupling, and which gave the best results were: m-maleimidobenzoyl-N-hydroxysuccinimide ester (MBS) and succinimidyl-4-(N-maleimidomethyl)-cyclohexane-1-carboxylate (MCS). The results from these, as well as from other heterobifunctional reagents, have already been published, (Gadow et al., 1984, Fricke, 1986). A typical coupling procedure is described in which anti-IgG can be coupled to pyruvate kinase, using MBS as coupling reagent. The gamma globulin fraction of the antiserum must first be separated from the whole antiserum, for example, using polyethylene glycol or ammonium sulfate precipitation, followed by dialysis and ion-exchange chromatography. (It is often easier, to purchase the ready-purified gamma globulin fraction of the antiserum in question, especially where the labeling takes place at irregular intervals). Ten micromoles of MBS or MCS are dissolved in 1 mL dioxane and 10 µL of the solution given to ca. 100 nmol immunoglobulin dissolved in 1.5 mL 0.15 mol/L sodium chloride solution. The resulting solution is shaken lightly for 45 min at room temperature in the dark, after which it is dialysed twice (2 x 2 h) against 1 L of 0.05 mol/l potassium phosphate buffer containing 0.15 mol/l sodium chloride, pH 7. The dialysate is added to 400 nmol pyruvate kinase in the same buffer and the pH adjusted to 6.0. The reaction with the pyruvate kinase is allowed to continue for 45 min at room temperature, after which, the reaction is stopped by the addition of 100 nmol reduced glutathione. The reaction mixture is separated on a 100 x 1.8 cm Ultrogel A6 column (LKB) using the above buffer as eluent. The elution rate is adjusted to 1.8 ml/h, fractions of 1.8 ml being collected. The antigenicity and enzyme activity is tested for in each fraction, the fractions with both being pooled, portioned and lyophilised. The lyophilised portions are stored at -20°C until required for use.

Full details are given elsewhere (Gadow et al., 1984, Fricke, 1986).

CHEMILUMINESCENT LABELS

Several groups of compounds exhibit the properties of being able to
emit light under relatively simple reaction conditions. These include
derivatives of fluorescein (Shapiro et al., 1984), acridan, (Weeks et
al., 1983), aryl oxalates and oxamides, (Tsuji et al., 1984), and aryl
hydrazides, (Hersh et al., 1979).

Acridinium esters have been widely described as labels in luminescent
immunoassays, (Weeks et al., 1983, 1984) and have been dealt with in
another section of this book. Fluorescein chemiluminescence, although
used in assays, (Shapiro et al., 1984), does not appear to have the sensitivity of either aryl hydrazide or acridinium ester labels. Moreover,
the oxidant commonly used, namely sodium hypochlorite solution, is aggressive towards metal fittings, and can give rise to costly repairs!
Aryl oxalates and oxamides both have a very high efficiency in organic,
water-free solutions. However, the low solubility in aqueous solution,
coupled with a large loss in quantum efficiency, has prevented the widespread use of this group of compounds.

The only practical alternatives to acridinium esters as chemiluminescent labels are the aryl hydrazides, of which luminol is probably the best
known example. All are cyclic 1,2-dicarboxylic acid hydrazides. The two
aryl hydrazides used in the assays described in this chapter are derivatives of isoluminol (ABEI - see materials for full name) and naphthalene
1,2-dicarboxylic acid hydrazide (ABEN). The N-substituted alkylamine
chain allows for coupling to other molecules without steric hindrance of
the aromatic part of the molecule. The use of luminol and isoluminol,
although possible, is not often practised, although the diazotisation
and coupling to proteins is discussed briefly below. Full experimental
details of the procedure, and of the results can be read elsewhere,
(Wood & Gadow, 1983, Wood, 1984).

Luminol and Isoluminol

As stated above, the easiest way of coupling luminol or isoluminol
to proteins is by using the diazo-method. One must bear in mind, that the
protein should contain tyrosine moieties, otherwise the coupling does
not give rise to stable end products, as the reaction of diazoluminol
with primary amines produces labile bonds.

The coupling to phenolic moieties takes place at a pH between 9 and
10, and when antibodies are labeled, there is considerable loss of immunoreactivity and the theoretical sensitivity of the label is never
achieved, (Wood & Gadow, 1983).

In contrast to luminol and isoluminol, the N-alkyl substituted isoluminol derivatives can be easily coupled, providing they have a suitable
terminal function such as a carboxyl or primary amine group. The coupling
reactions are milder and loss of immunoreactivity, as far as antigens
and antibodies are concerned, is much reduced. The light output is also
much higher in these compounds than in those resulting from diazo-coupling.

N-Substituted Derivatives of Isoluminol

These form the most widely used group of aryl hydrazides, and some
are now available commercially (Sigma, LKB). The differences between the

substances lies in the length of the alkyl chain connected to the amino group at position 5 on the benzene ring. The "spacer" arm consists of 2-6 carbon atoms, and terminates in either a carboxyl or amino group. The conversion of a terminal amino group to a carboxyl group can be effected with succinic anhydride in dry, freshly distilled dimethylform-amide, (DMF). In the case of ABEI, the conversion to the hemisuccinamide (ABEI-H) can be followed easily, as the ABEI forms a suspension in DMF, which goes into solution during the formation of ABEI-H. This is the simplest way in seeing that the reaction has taken place! The conversion of the ABEI-H to its N-hydroxysuccinimide ester is carried out in dry DMF with N-hydroxysuccinimide and dicyclohexyl carbodiimide in the molar ratios 1:1.1:1.5. The reactants are allowed to stand at ambient tempera-ture in the dark for 18-24 h to ensure completion of the reaction. Fine needles of dicyclohexyl urea indicate that the reaction has occured. The presence of an amorphous precipitate points to the fact that either the DMF was not freshly distilled and contained free amines and/or water. In such cases, it is probable, that the active ester has not been formed, or has been formed in such a low yield, that the subsequent coupling to a protein or hapten does not give rise to useable products.

Other Aryl Hydrazides

The most important representatives of this group are the derivatives of 7-aminonaphthalene 1,2-dicarboxylic acid hydrazide. An alkyl chain is attached via the 7-position in an analogous way to the derivatives of isoluminol. The light output of the naphthalene hydrazides is higher than that of the isoluminol derivatives. The 4-aminobutyl-N-ethyl deri-vative (ABEN) can be converted to its hemisuccinamide and N-hydroxy-succinimide ester in the same way as described above for ABEI. Recent publications (Caldini et al., 1986, Braun et al. 1986), show that ABEN-labeled antigens and antibodies are usually quantitatively better than the corresponding ABEI-labeled counterparts, as far as light output is concerned.

Hydrazides of other polycyclics have been prepared and their light output measured by Gundermann (1968). Problems occur with their solubility in aqueous solution. As the number of conjugated rings in the system increases, it is often seen, that the light output is unexpectedly low, and it is safe to say that the polycyclic hydrazides have no role to play, as far as labels for immunoassay are concerned.

The chemiluminescent assays, which are used as examples in this chapter, are labeled either with ABEI or ABEN, and can be set up in any laboratory equipped for simple syntheses, and containing personel, who are familiar with immunoassay.

COUPLING PROCEDURES FOR ARYL HYDRAZIDES

The active ester reaction is the method of choice for coupling ABEN and ABEI derivatives to proteins and haptens. The reaction times (coupling times) are different for ABEN and ABEI, the former giving optimal results with a coupling time of 30-60 minutes. The reaction between the ABEI-H active ester can be allowed to proceed for 1 to 6 hours. If the reaction is carried out overnight, then the products of ABEN-labeling are often of poor quality, at least as far as antibody labeling is concerned, and is probably due, at least in part, to overlabeling.

The labeling takes place optimally at a pH between 8.2 and 8.8 in

0.025 mol/L phosphate buffer, when incorporation rates of up to 30% can occur, depending upon the protein to be labeled. The labeling of antibodies takes place with a molar ratio of luminogen-active ester: antibody of around 20:1. As stated above, the reaction times are different for ABEN and ABEI labeling. After coupling, the reaction vessel is centrifuged, (2000g for 5 min), as sometimes a precipitate forms. The clear supernate is transfered to a 30 x 1 cm column of Trisacryl-GF 05 (LKB) using 0.025 mol/L TRIS-HCl buffer, pH 7.5 as eluent. Fractions of ca. 0.5 mL are collected. Three or four colored bands are seen, the labeled protein being present in the first band which elutes from the column.

The fractions containing the labeled antibody are diluted and tested for specific and unspecific binding. The fractions which are suitable, i.e. those with high specific and low unspecific binding, are pooled, aliquoted into portions of 0.25-0.5 mL and stored at -20°C until use.

After thawing, the label is kept at 4-7°C and 15 µl 1 mol/L sodium azide is added as a preservative.

OXIDATION SYSTEMS FOR ARYL HYDRAZIDE LUMINESCENCE

The beauty of fluorescein and acridinium ester chemiluminescence is that the light reaction can be initiated by the addition of a single reagent; in the case of fluorescein this is hypochlorite, for acridinium esters, alkaline peroxide. Many of the systems decribed for aryl hydrazides involve the addition of up to three reagents directly before the light measurement takes place in the luminometer. It is possible to simplify this in such a way, that the injection of an alkaline peroxide solution into the cuvet in the measuring chamber initiates the light reaction. If carefully planned, such a simplification can be accompanied by an enhancement of the light signal, (Wood et al., 1985).

As stated earlier in this chapter, the oxidation of aryl hydrazides requires two components, namely, a heme compound, which can "trap" the active oxygen species, and an alkaline peroxide solution to provide this active OOH^- anion. If the heme compound is added to the solid-phase in the measuring cuvet; in our system, either a polystyrene ball or magnetic microparticles, it serves not only to keep the solid-phase wet, but increases precision and alleviates the need for injection of this solution.

The choice of heme compounds, which are suitable for aryl hydrazide chemiluminescence is relatively large and include microperoxidase - MP 11, (a cytochrome-c nucleus with 11 amino acids still attached), hemin, and peroxidases, but to name those most commonly used. Hemin is the cheapest, and for most assays, the most efficient heme compound, as far as light output is concerned. It is also the cheapest and most stable as far as resistance to microbial attack is concerned. Other compounds which "work" include hemoglobin and myoglobin, as well as cytochrome-c. Other heme nuclei without the iron atom, for example, hematoprotoporphyrin IX, do not function, showing that the iron atom is essential for the light reaction.

The only other metal ion, which approaches the efficiency of the heme compounds is cobalt, which works well when injected as the nitrate in place of, say, hemin. The disadvantage of cobalt salts is the production of brown cobalt hydroxide by the alkaline peroxide, which leads to substantial quenching of the luminescent signal.

The use of other similar substances such as the corrinoids (vitamin B_{12} group of compounds) or chlorophyll as heme-compound replacements, did not give rise to any useable results. The same was true for other metallic salts of the Group VIII elements.

Attempts at "enhancing" the light output in the systems described here, by the addition of d- or 1-luciferin, (Whitehead et al., 1983) succeeded only in increasing the unspecific signal. It is important to note, that the enhancement described by Whitehead is in a system, in which the horse-radish peroxidase still serves as an enzyme, i.e. in a luminescence enhanced enzyme immunoassay, (Amerlite system, Amersham plc). This example has been given, not to dispute that enhancement with luciferin does not take place, but to state, that some phenomena are not transferable to other systems, however similar they may appear on paper!

EXAMPLES OF ASSAYS USING ARYL HYDRAZIDE AND PYRUVATE KINASE LABELED ANTIBODIES

Solid Phase Antigen Luminescent Technique (SPALT)

These assays are analogous to the allergosorbent tests for IgE antibodies to specific allergens. The antigen, or in the case of a hapten, a hapten-protein conjugate, is coupled to the solid phase, which in our case is either a polystyrene ball or acrolein-iron oxide microparticles. Microcrystalline cellulose can also be used in place of the magnetic particles. The microparticulate solid-phases allow an exact dosage of the immobilised antigen, a thing which is not always possible when using polystyrene balls.

For haptens, it is important, that the hapten-protein conjugate used for immunisation is different from the one used in coating the solid phase. This is important, as in SPALT assays, all antibodies raised, i.e. not only those to the hapten, but also those to the "bridge" and carrier protein, play an important role. This is clear, when one considers that the labeled antibody is raised against the IgG of the animal species in which the substance-specific (first) antibody is raised, (Wood, 1984).

Table 1 shows the flow sheet for a SPALT assay for the detection of antibodies to Legionella pneumophila. Specific fractions of the Legionella bacterial coat are coupled covalently to the polystyrene balls, and serve as the immobilised antigen, which "capture" any antibodies in the applied sample. The specificity of the test can be determined by the label, here anti-IgA, anti-IgG or anti-IgM.

TABLE 1. Assay Flow Sheet for the Detection of Legionella Antibodies.

    10 µL serum in dilution steps (e.g. 1:1, 1:10, 1:100)
    200 µL assay buffer 2
    1 Legionella antigen (e.g. Philadelphia 1) coated ball

    Incubate for 120 min on horizontal rotator at 170-190 rpm at
    ambient temperature

    Wash with 2 x 5 mL portions of 0.25 mL/L Tween 20
    (Abbott Pentawash)

    Add 200 µL ABEN-H labeled antibody (anti IgA, IgG or IgM)

Incubate and wash as above, transfer solid-phase to measuring
cuvet, pipet 300 µL hemin working solution, load luminometer,
initiate light reaction and integrate signal for 10 s.

Table 2 shows a SPALT assay for the determination of cortisol levels
in saliva, using a cortisol-ovalbumin conjugate coupled to magnetic
particles. The antiserum used was raised against cortisol-bovine serum
albumin.

TABLE 2.   Flow scheme for a SPALT Assay for Cortisol in Saliva

20 µL sample or standard (the latter in 1g/L ovalbumin solution)
100 µL anti-cortisol from rabbit (1:20,000 working dilution)
100 µL magnetic particle bound cortisol-ovalbumin (= 50 µG
particles/tube)

Incubate for 30 min with occasional shaking to keep particles
suspended.

Wash with 2 x 1 mL wash solution (see Table 1)

Add 200 µL ABEI-H labeled donkey anti-rabbit IgG

Incubate for 60 min as above, wash with 3 x 1 mL wash solution,
add 300 µL hemin to each tube and proceed as in table 1 above.

The sensitivity of the cortisol saliva assay is less than 5 picograms
per tube, and has the advantage that the tubes used for setting up the
assay (55 x 12 mm polystyrene) were directly loaded into the luminometer,
without a tube-transfer being carried out. The magnetic separation
step, carried out in a special rack (Ciba-Corning), allows a
simple and effective washing of 60 tubes in 2 minutes, the latter being
time for the magnetic separation to take place. The rack containing the
tubes can be inverted in the wash basin, there being no need for a
special waste-disposal procedure as in radioimmunoassay. Table 3 shows
a bioluminescent SPALT assay for insulin, using microcrystalline cellulose
as the solid phase. These three tables show the unity in diversity of
the SPALT assays.

TABLE 3. Flow scheme of an insulin SPALT using pyruvate kinase labeled
"second" antibody, and microcrystalline cellulose as solid phase.

100 µL serum / sample
100 µL anti-insulin (guinea pig), (working dilution 1:20,000)
100 µL assay buffer 1

Incubate at 4°C overnight

Add 100 µL cellulose-insulin (= 500 µG solid-phase)

Incubate again overnight as above

Wash with 2 mL 0.1 mol/L KCl, centrifuge the tubes at 2000 g for
5 min and aspirate off the supernate. Repeat the wash step.

Add 200 µL pyruvate-kinase labeled rabbit anti-guinea pig IgG

Incubate 4 h at ambient temperature

Perform wash steps as above and prepare tubes for luminometry
in the following way: Add 400 µL 0.1 mol/L TRIS-HCl buffer
containing 0.01 mol/L EDTA, pH 7.8 Add 100 µL of a 0.01 mol/L
mixture of ADP and phosphoenolpyruvate followed by 100 µL
ATP-monitoring reagent. The luminescent signal was measured
kinetically over 60 s, the slope being used as the parameter
for curve fitting.

Immunoluminometric Assay (ILMA)

This is the analog of the immunoradiometric assay, IRMA, and is the
method of choice where the antigen has at least two independant
antibody binding sites. The use of "excess-reagent" techniques (Hunter,
1982), allow rapid, sensitive and precise assays to be developed. Table 4
shows the assay procedure for a thyrotropin (TSH) ILMA, adapted for
neonatal screening purposes. The assay uses two monoclonal antibodies,
and has a single incubation step. The assay is suitable for a "same-day"
reporting of results, the time for a 200 tube assay, including data
reduction, being under 6 h.

TABLE 4. Flow diagram for a neonatal-TSH one-step ILMA.

200 µL Assay buffer 2
50 µL ABEN-labeled anti TSH
1 anti-TSH coated ball
1 6.5 mm diameter blood spot

Incubate at ambient temperature on a horizontal shaker for 4 h
(see Table 1)

Wash with 3 x 5 mL 0.25 ml/L Tween 20 and prepare for luminometry
as in Table 1.

The assay has a sensitivity of under 2 mU/1 TSH (WHO 80/558).

Chemiluminescent Immunoassay (CELIA)

This assay type has all the disadvantages of labeled antigen assays,
the main ones to note being the effective dilution of the sample antigen
with labeled antigen and the contact between label and sample. The latter
can play a role in chemiluminescent immunoassays using aryl hydrazides,
in which no prior extraction of sample is made, as peroxidases and
heme compounds are present in many serum samples. These could give
rise to oxidation of the label during the incubation step(s), and it is
for this reason, that such assays have not been developed in this laboratory. This does not mean to say that CELIAs do not function. They are
perfectly in order, when a sample extraction step preceeds the assay.

ACTIVATION OF POLYSTYRENE BALLS

The activation of the polystyrene balls has been described elsewhere,
(Wood, 1985), although the methods described here represent the newest
standpoint. The first step is the coating of the balls with a copolymer
of phenylalanine-lysine 1:1, relative molecular mass (Mr) 30-40 kD. This

step is followed by activation of the epsilon-amino groups of the lysine residues with pentan-1,5-dial (better known as glutaraldehyde or glutar-dialdehyde). The resulting activated surface is then utilised for the formation of Schiff's bases, aldol condensation- and Michael addition products with proteins. The final steps include saturation of non-specific binding sites on the ball surface, followed by drying and protection of the surface with polyvinyl alcohol. The procedure is described in detail in the following paragraph.

The polystyrene balls are kept in a dry warm place, for example an incubator, at 35-40°C. 125 mL of a solution containing 2 mg polyphenyl-alaninelysine (poly phe-lys) is poured over 1000 polystyrene balls in a suitable glass container. The container is shaken gently several times during the first hour of coating, in order to remove trapped air bubbles. The coating process is allowed to proceed at ambient temperature for 24-72 h. The balls are then washed twice with distilled water, dried under compressed air or nitrogen, and stored in a stoppered vessel until needed. Balls coated in this way are stable for at least 6 months.

The glutaraldehyde activation follows a similar pattern. 125 mL of a solution containing 2.5 mL of a 25% solution of glutaraldehyde is poured over the poly phe-lys coated balls, and allowed to stand for 30 minutes with occasional shaking to displace gas bubbles from the ball surfaces. The balls are then washed, once with 0.15 mol/L NaCl, and once with dis-tilled water, after which a ball is dropped into a test tube containing Schiff's reagent. A poly phe-lys ball is taken as a control. The develop-ment of a purple shimmer on the surface of the activated ball shows the presence of free aldehyde groups.

The coupling of proteins to the activated balls takes place as follows:
125 mL distilled water containing 2 milligrams of protein is poured over the balls and left to stand for 15 min, after which, the pH is adjusted to between 8.2 and 8.8 with 5 mL 0.5 mol/L dipotassium hydrogen phosphate solution. The coupling reaction is allowed to proceed overnight at ambient temperature, after which, the coating solution is aspirated off, and the balls washed with 0.15 mol/L NaCl. The saturation of the non-specific binding sites is effected with 0.05 mol/L TRIS-HCl containing 2.5 G/L bovine serum albumin, pH 7.6. Depending upon the antibody or protein coupled to the ball surface, the balls are either left in this buffer, with the addition of sodium azide to an end concentration of 0.15 mol/L, or are washed once with saline, and once with water, before being covered with 125 mL 5 G/L polyvinyl alcohol solution and allowed to stand for 5-10 min. The polyvinyl alcohol is aspirated off, and the balls dried as above, before being stored in stoppered vessels at ambient temperature until needed.

The decision to store the balls in a dry form, or under a protein containing buffer depends upon the stability of the protein on the ball surface. As an example, we have a monoclonal antibody against IgE, which is only stable when stored in buffer. The immunoreactivity in the dried form is lost within 8 weeks, in buffer, it is almost constant over 9 months! In contrast, another polyclonal antibody to IgE is stable when stored dry, and gradually loses immunoreactivity when stored in buffer. There appears to be no hard and fast rule, the trial and error method being the only one to suggest!

The activation and preparation of the microparticulate solid phases can be dealt with in a couple of sentences. The magnetic particles have acrolein which allows for direct coupling to proteins by addition of the

267

protein solution at slightly alkaline pH (8.4–8.8). The microcrystalline cellulose is activated with sodium metaperiodate at 50–70°C at pH 5. The resulting aldehyde groups, which result from the cleavage of vicinal hydroxyl groups in the carbohydrate rings are treated in an analogous way to those on the surface of the activated polystyrene balls.

ADAPTION OF COMMERCIALLY AVAILABLE EQUIPMENT AND REAGENTS FOR LIA

The macro solid-phase system described here has been adapted from the Abbott system of beads and trays. The washing is also performed with the Abbott Pentawash or Proquantum. The Proquantum can also pipet reagents automatically, thus saving time, when a lot of assays are performed in the laboratory. The horizontal rotator is also commercially available, and allows assay incubation times to be shortened, without the need for incubation at higher temperatures. The presence of such equipment in laboratories performing hepatitis tests may allow an optimal use of it, when the luminescence immunoassays are carried out with the same equipment.

The use of magnetic particles as solid phase, allows the use of equipment which is already present in the laboratory performing such Serono, Ciba-Corning or Amerlex-M system assays.

FINAL COMMENTS

The systems chosen in this laboratory, allow for full flexibility of choice of measuring equipment. In contrast to coated tubes or microtiter plates, coated balls and micro solid-phases can be measured in any luminometer. It is for this reason alone, that coated tube and coated well techniques have been avoided. The methods and assays described here show the potential advantages and disadvantages of luminescent immunoassays using aryl hydrazides as marker substances.

It remains of interest to see which system(s) will become commercially available, as the main body of luminescence "users" will depend, as is the case for radioimmunoassay, on the commercial kits available. It cannot be avoided, that in this sector, the trend towards idiot-proof black boxes (closed systems) with all their advantages and disadvantages will continue.

ACKNOWLEDGEMENTS

This work was carried out in part with the support of the Deutsche Forschungsgemeinschaft, AZ Wo-351/1, in the laboratories of the klinisch-experimentelle Forschungseinrichtung of the Medical University of Lübeck.

REFERENCES

Braun, J., Schultek, T., Tegtmeier, K.F., Florenz, A., Rohde, C., Wood, W.G. (1986), Luminometric assays of seven acute phase proteins in minimal volumes of serum, sputum, and bronchioalveolar lavage. Clin. Chem. In Print.

Brolin, S.E., Borglund, E., Tegner, L., Wettermark, G. (1971), Photokinetic microassay based on dehydrogenase reactions and bacterial luciferase. Anal. Biochem., 42, 124-135

Caldini, A.L., Orlando, C., Barni, T., Messeri, G., Pazzagli, M.,
    Baldi, E., Serio, M. (1986), Measurement of human seminal plasma
    transferrin by a chemiluminescent method. Clin. Chem. In Print
Carrico, R.J., Yeung, K.K., Schroeder, H.R., Buckler, R.T., Christner,
    J.E. (1976), Specific protein binding reactions with ligand-ATP
    conjugates and firefly luciferase. Anal. Biochem., 76, 95-110
Ekins, R.P. (1978), Quality control and assay design. In:
    Radioimmunoassay and related procedures in medicine 1977, IAEA,
    Vienna, Vol II, pp 39-56
Fricke, H., Strasburger, C.J. Wood, W.G. (1982), Enzyme enhanced
    luminescence immunoassay for the determination of transferrin
    in serum J. Clin. Chem. Clin. Biochem. 20, 91-94
Fricke, H. (1986), Entwicklung biolumineszenzimmunologischer
    Testverfahren als mögliche Alternative zum Radioimmunoassay.
    Inauguraldissertation zur Erlangen der Doktorwurde, Medizinische
    Universität zu Lübeck, 107 pages.
Gadow, A., Fricke, H., Strasburger, C.J., Wood, W.G. (1984),
    Synthesis and evaluation of luminescent tracers and
    hapten-protein conjugates for use in luminescent immunoassays
    with immobilised antibodies and antigens. Part II of a critical
    study of macro solid phases for use in immunoassay systems.
    J. Clin. Chem. Clin. Biochem. 22, 337-347
Gundermann, K.D. (1968), Chemilumineszenz organischer Verbindungen
    - Ergebnisse und Probleme (Hrsg. Broderek, H., Hafner, K.,
    Müller, E.), Springer, Berlin, Heidelberg, New York, 174 pages.
Hersch, L.S., Vann, W.F., Wilhelm, S.A. (1979), A luminol-assisted
    competitive binding immunoassay of human immunoglobulin G.
    Anal. Biochem. 93, 267-271
Hunter, W.M. (1982), Recent advances in radioimmunoassay and related
    procedures. In: Radioimmunoassay and related procedures in
    medicine 1982, IAEA, Vienna, pp 3-21.
Radziszewski, B. (1877), Untersuchungen über Hydrobenzamid, Amarin und
    Lophin. Ber. Chem. Ges. 10, 70-75
Schroeder, H.R., Vogelhut, P.O., Carrico, R.J., Boguslaski, R.C.,
    Buckler, R.T. (1976), Competitive protein binding assay for
    biotin monitored by chemiluminescence. Anal. Biochem. 48,
    1933-1937
Shapiro, R., Chan, J., Pierson, A., Vaccaro, K., Quick, J. (1984),
    Protein-enhanced fluorescein chemiluminescence used in an
    immunoassay for rubella antibody in serum. Clin. Chem. 30,
    889-893
Tsuji, A., Maeda, M., Arakawa, H. (1984), Enzyme immunoassay monitored
    by chemiluminescence reaction using bis(2,4,6-trichlorophenyl)
    oxalate fluorescent dye. In: Procedings of the third
    international symposium on analytical applications of
    bioluminescence and chemiluminescence, Birmingham, U.K.
    Eds. Kricka, L.J., Stanley, P.E., Thorpe, G.H.G., Whitehead,
    T.P., Academic Press, London, Orlando, pp 253-256.
Weeks, I., Beheshti, I., McCapra, F., Campbell, A.K., Woodhead, J.S.
    (1983), Acridinium esters as high-specific-activity labels in
    immunoassay. Clin. Chem. 29, 1474-1479
Weeks, I., Sturgess, M., Siddle, K., Jones, M.K., Woodhead, J.S. (1984),
    A high sensitivity immunochemiluminometric assay for human
    thyrotropin. Clin. Endocrinol. 20, 489-495
Whitehead, T.P., Thorpe, G.H.G., Carter, T.J.N., Groucott, C., Kricka,
    L.J. (1983), Enhanced luminescence procedure for sensitive
    determination of peroxide labelled conjugates in immunoassay.
    Nature 305, 158-159
Wood, W.G. (1984), Luminescence immunoassays - Problems and possibilities
    J. Clin. Chem. Clin. Biochem. 22, 905-914

Wood, W.G. (1985), Luminescence immunoassays in theory and practice –
    The state of the art. In: Immunoassay technology, Vol 1,
    Ed. Pal, S.B., De Gruyter, Berlin, New York, pp. 105–150
Wood, W.G., Fricke, H., von Klitzing, L., Strasburger, C.J., Scriba,
    P.C. (1982), Solid phase antigen luminescence immunoassays
    (SPALT) for the determination of insulin, insulin antibodies
    and gentamicin in serum. J. Clin. Chem. Clin. Biochem,
    20, 825–831
Wood, W.G., Gadow, A. (1983), Immobilisation of antigens and antibodies
    on macro solid phases – a comparison between adsorptive and
    covalent binding – Part I of a critical study of macro solid
    phases for use in immunoassay systems. J. Clin. Chem. Clin.
    Biochem. 21, 789–797
Wood, W.G., Hantke, U., Gross, A.J. (1985), Initiation of luminol-based
    luminescence by the injection of a single reagent coupled with
    an enhancement of light output. J. Clin. Chem. Clin. Biochem.
    23, 47–49

SEPARATION-REQUIRED CHEMILUMINESCENCE IMMUNOASSAYS

FOR STEROIDS

F. Kohen[1], Y. Ausher[1], S. Gilad[1], J. De Boever[2], G.J.R. Barnard[3] and J.B. Kim[4]

[1]Dept. of Hormone Research, Weizmann Institute
of Science, Rehovot, Israel; [2]Akademisch Ziek-
enhuis, Vrouwenkliniek/Poli III, Gent, Belgium; [3] Dept. of Obs./Gyn., King's Coll. Sch. of
Med. and Dent., London U.K.; and [4]Col. of Animal Husb., Kon-Kuk Univ., Seoul, Korea

INTRODUCTION

Recent studies by Schroeder et al. (1978), from our laboratory (Kohen et al., 1985) and elsewhere (Pratt et al.,
1978) have indicated that chemiluminescence immunoassay can
be a feasible alternative to radioimmunoassay (RIA) of haptens. For our studies we chose as a chemiluminescent marker, derivatives of isoluminol possessing alkyl chains of an
optional length of 2-6 atoms and terminating with a primary
amino group (Schroeder and Yeager, 1978). The advantages of
these compounds are that they can be measured at the level
of pmol (Schroeder and Yeager, 1978), are stable (Pazzagli
et al., 1983) and can be conjugated easily to carboxy derivatives of steroid or steroid glucuronides (Kohen et al.,
1983) to yield the corresponding steroid-chemiluminescent
tagged conjugates which can be utilized as markers in immunoassay procedures. Using these markers, we adopted two
types of formats: (a) assays that do not require physical
separation of bound and free hormone, so called "homogeneous", and (b) assays that require a separation step, so
called "heterogeneous".

We shall deal here only with the "heterogeneous" type of
assays, using solid-phase techniques for separation of bound
and free hormone. The "homogeneous" type of assays will be
dealt with separately in this book by Messeri et al. Two
types of solid-phase techniques were adopted: (i) immunoadsorption and (ii) immobilization of antibodies by covalent linkage to solid supports. In the immunoadsorption
technique, IgG fractions of first antibodies or of second
antibodies are adsorbed to the walls of polystyrene tubes,
polystyrene balls or microtiter plates. After the binding
reaction, the reaction mixture is aspirated. The antibody-
bound fraction adsorbed onto the walls of the tubes is then
washed several times to remove factors which may quench

chemiluminescence. This approach is simple and does not require a centrifugation step. Using this approach, chemiluminescence based immunoassays have been developed for plasma steroids such as progesterone (Kohen et al., 1981), estradiol (Kim et al., 1982) and testosterone (Collins et al., 1983) and for urinary steroid glucuronides such as estriol-16α-glucuronide (Barnard et al., 1981a), pregnanediol-3α-glucuronide (Barnard et al., 1981b; Eshhar et al., 1981) and estrone-3-glucuronide (Weerasekera et al., 1982).

In the immobilization of antibodies by the covalent linkage approach, specific polyclonal or monoclonal antibodies are utilized either in the liquid phase or in the solid phase coupled covalently to polymer beads of 5-10 μ. When the first antibody is used in the liquid phase, a solid-phase second antibody (e.g. donkey anti-mouse cellulose suspension) is added after the binding reaction. A centrifugation step is then performed. The label bound to the pellet is dissociated at pH 13, and the light yield of the marker is measured by oxidation with the H_2O_2-microperoxidase system (Kohen et al., 1986). These assays have enabled the development of simple, reliable assays for the measurement of plasma steroids, e.g. estradiol (De Boever et al., 1983), testosterone (Kohen et al., 1983), urinary steroid metabolites e.g. pregnanediol-3α-glucuronide (Kohen et al., 1983) and estrone-3-glucuronide (Kohen et al., 1985), and therapeutic drugs such as digoxin (Kohen et al., 1986). Furthermore, the use of solid-phase techniques enabled the development of a direct immunoassay for progesterone in unextracted serum (De Boever et al., 1984) and in saliva (De Boever et al., 1986), and for estradiol in plasma (De Boever et al., 1985). Results of various types of solid phase chemiluminescence based immunoassays will be reported here.

MATERIALS AND METHODS

Steroids, microperoxidase (MP-11), bovine serum albumin (Cohn Fraction V), and polyoxyethylene sorbitan monolaureate (Tween 20) were obtained from Sigma Chemical Co., St. Louis, MO 63178; diethyl ether, 300 g/L hydrogen peroxide solution, sodium hydroxide pellets, from Merck, Darmstadt, F.R.G.; immunobead matrix, 5-10 u, from Bio-Rad Labs., Richmond, CA 94804; and Sepharose-Protein A from Pharmacia, Uppsala, Sweden.

Reagent Solutions

Assay buffer: Sodium phosphate (50 mmol/L, pH 8.0) containing, per liter, 9 g of NaCl, 100 mg of bovine serum albumin, and 1 g of sodium azide.

Wash solution containing, per liter, 9 g of NaCl, 1 g of sodium azide, and 0.5 ml of Tween 20.

Microperoxidase was dissolved (1 mg/ml) in Tris HCl (10 mmol/L, pH 7.4), and stored at 4°C. The working solution was obtained by diluting the concentrated stock to 10 ug/ml in assay buffer.

Oxidant solution was prepared by mixing 0.1 ml of 300 g/L hydrogen peroxide solution with 15 ml of distilled water.

Steroid stock solutions in ethanol were stored at -20°C and diluted to the desired concentrations in assay buffer when required.

Coating buffer (pH 9.6) containing, per liter, 1.59 g Na_2CO_3, 2.93 g $NaHCO_3$, and 1 gr of sodium azide.

Synthesis of steroid chemiluminescent marker conjugates

Isoluminol derivatives containing alkyl chains of varying lengths and terminating with primary amino groups were synthesized according to Schroeder et al. (1978). These compounds belong to the 6-[N-(α-aminoalkyl)-N-ethyl]-amino-2,3-dihydro-1,4-phtalazine-1,4-dione series and contain a bridging group of two methylene groups (aminoethylethylisoluminol, AEEI; Fig. 1, n = 2); four methylene groups (aminobutylethylisoluminol, ABEI; (Fig. 1, n = 4), five methylene groups (aminopentylethylisoluminol, APEI; Fig. 1, n = 5) or six methylene groups (aminohexylethylisoluminol, AHEI; Fig. 1, n = 6).

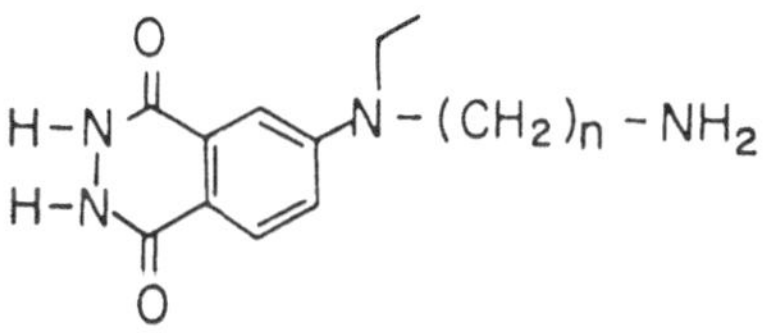

n = 2, AEEI

n = 4, ABEI

n = 5, APEI

n = 6, AHEI

Fig. 1. Structures of amino derivatives of isoluminol.

The steroid-chemiluminescent marker conjugates were prepared in two steps: (i) a carboxy derivative of a steroid or a steroid glucuronide was activated in the presence of carbodiimide and N-hydroxysuccinimide to an activated ester; (ii) the activated steroid ester derivative was reacted with the primary amino group of an isoluminol derivative at basic pH to form a stable peptide bond (Kohen et al., 1986).

The conjugates were purified by thin-layer chromatography (Pazzagli et al., 1981a,b) and characterized by ultraviolet spectroscopy and mass spectrometry (Pazzagli et al., 1983). Fig. 2 shows the structure and fast ion bombardment (FAB) spectrum of a representative conjugate, estradiol-6-[0]-carboxymethyl oxime aminobutylethyl isoluminol.

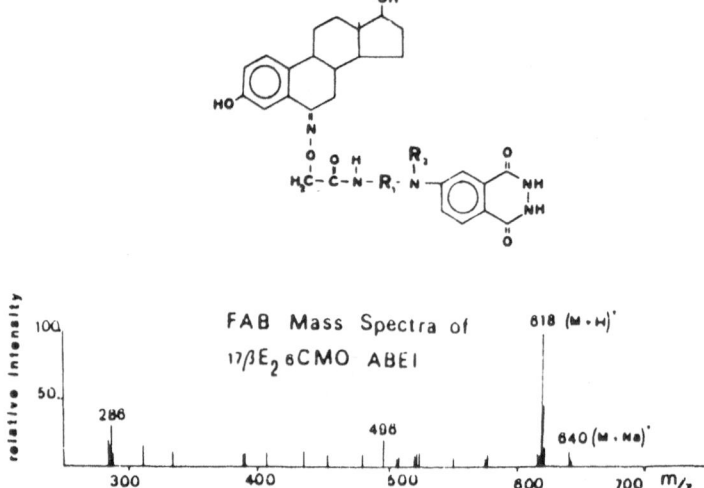

Fig. 2. Structure and fast atom bombardment (FAB) of 17β-estradiol-6-carboxymethyl oxime - ABEI conjugate; R_1 = $(CH_2)_4$; R_2=-CH_2-CH_3 (from Pazzagli et al., 1983, by permission).

Preparation of antibodies to steroids

Specific polyclonal antibodies to steroids conjugated to bovine serum albumin were raised in rabbits and character-ized in terms of titer, affinity and specificity by RIA pro-cedures. The hybridoma technique of Köhler and Milstein (1976) was used to generate monoclonal antibodies to ster-oids (Kohen and Lichter, 1986). Polyclonal and monoclonal antibodies belonging to the IgG class were purified by af-finity chromatography on Sepharose-Protein A (Kohen et al., 1985). The purified antibodies were stored in PBS at -20°C at a concentration of 1 mg protein/ml.

Preparation of the immunoadsorbant:

Immobilization of antibodies by adsorption techniques

Three types of solid phase supports were used for adsorp-tion of antibodies. These include microtiter plates from Nunc (Denmark) polystyrene balls of 0.5 cm in diameter from Northumbria Ltd., Nurthumberland, U.K. and polystyrene test tubes (Lumac Systems, Basel). As a representative example, the antibody-coated tube procedure is described below (Eshhar et al., 1981). IgG fractions of antibodies (300 µl) suitably diluted in coating buffer, pH 9.6) were added to each tube. After an overnight incubation at 4°C, the carbonate buffer was aspirated to waste, and saline containing 0.3% BSA was added to each tube. After an incubation of 2 h at room temperature, the solution was aspirated to waste. The tubes were stored dry at 4°C until required. Tubes prepared in this may retain their activity for at least two weeks if stored at 4°C or longer if stored at -70°C.

Immobilization of antibodies by covalent linkage

Two milligrams of purified IgG fraction of the monoclonal or polyclonal antibody were dialyzed against buffer phosphate (3 mmol/L, pH 6.3, the "coupling buffer") overnight, then mixed with 40 mg of Immunobead matrix in a final volume of 4 mL of coupling buffer. To this suspension was added 15 mg of 1-ethyl-3-[3-dimethylaminopropyl)carbodiimide and the mixture was incubated overnight at 4°C. The reaction mixture was then treated as directed by the supplier (Bio-Rad). The antibodies so prepared were stored at 4°C in phosphate buffer (5 mmol/L, pH 7.2) containing, per liter, 10 g of bovine serum albumin and 10 g of sodium azide, at a concentration of 10 mg of immunoadsorbant per milliliter. Before immunoassay, the immunoadsorbent is diluted 100 to 1000 fold, depending on the titer of the conjugated antibody, with assay buffer containing uncoupled immunobead matrix at a concentration of 1 mg/ml.

Light measurements

Measurements of light emission were made with a Lumac Luminometer Model 2080 (Lumac Systems, Basel) using the automatic injection and integration modes of the instrument, and Lumacuvette P polystyrene test tubes (12 x 50 mm) as reaction vessel. The Luminometer was also connected to a storage oscilloscope (Type 5111, Tektronix Beaverton, Oregon) in order to observe the kinetics of light emission. When the Luminometer was used in the automatic injection mode, readings on the Luminometer started immediately after initiation of the light reaction. The light emission was then integrated for 10 seconds, and the total light production (TLP) was recorded as arbitrary light units.

RESULTS

The measurement of steroids and urinary steroid metabolites using isoluminol-steroid conjugates

The principal aim of this section is to describe current developments in the measurement of steroids from biological fluids. To-date, the most significant advances in steroid methodology for routine clinical application are the determination of steroids or their metabolites in unextracted urine, serum, plasma or saliva. In the following sections we shall describe representative examples of solid-phase chemiluminescence based immunoassay methodology for the direct measurement of steroid hormones from biological fluids.

Urinary Pregnanediol-3 -glucuronide : antibody-coated tube separation

Ovarian function in women can be monitored by measurement of urinary metabolites of estradiol (e.g. estrone-3-glucuronide) and of progesterone (pregnane-diol -3α-glucuronide) (Collins et al., 1970). Conventional assays of urinary steroid metabolites depend on prior hydrolysis of the conjugates, and subsequent extraction and estimation of the free steroids (Klopper et al., 1955; Brown et al., 1958).

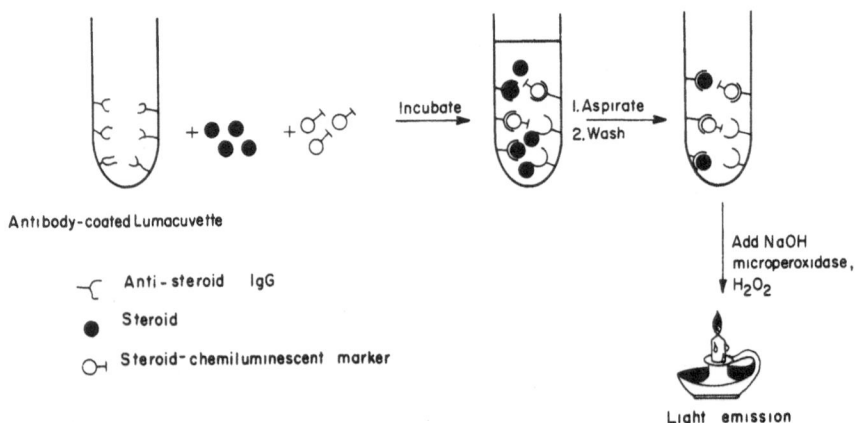

Fig. 3. A schematic diagram of the solid-phase chemiluminescence immunoassay procedure using the antibody-coated tube separation method.

We have developed in pilot studies direct chemiluminescence based immunoassays of steroid conjugates in diluted urine using the antibody-coated method (Lindner et al., 1981). In the method described here, purified polyclonal (Barnard et al. 1981b) or monoclonal antibodies (Eshhar et al., 1981) to pregnanediol-3α-glucuronide are passively adsorbed to the walls of Lumacuvette polystyrene tubes. The labelled antigen is pregnanediol-3α-glucuronide aminohexyl-ethyl isoluminol. The antibody coated tubes are incubated for 2 hr with 100 µl of standards or of samples of unextracted morning urine diluted 1:500 and with the marker conjugate (100 pg in 200 µl of assay buffer). Subsequent rinsing with assay buffer removed the endogenous interfering substances. The light yield of the immunoadsorbed steroid conjugate is measured under conditions that maximize the quantum yield (pH 13) using the oxidation system of microperoxidase and H_2O_2. A schematic diagram of the method is shown in Fig. 3.

Evaluation of the method

The <u>sensitivity</u> of the method was 4 pg/tube and was comparable to results obtained by conventional RIA methods. Fig. 4 shows a dose-response curve as observed in the oscilloscope tracings.

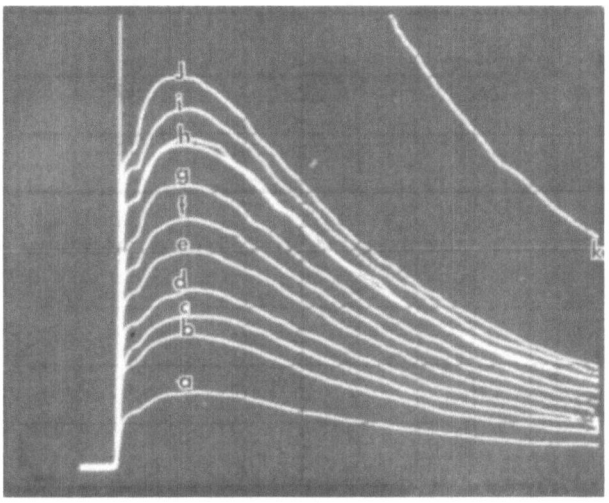

<u>Fig. 4</u>. Oscilloscope tracings of the light signal obtained upon oxidation of pregnanediol-3α-glucuronide AHEI conjugate (100 pg) bound to the antibody-coated tubes in the presence and in the absence of unlabelled pregnanediol-3α-glucuronide. (a) Non-specific binding; (b) through (j): light signal obtained when the bound conjugate was oxidized in the presence of varying amounts of pregnanediol-3α-glucuronide: (b) 1000 pg; (c) 500 pg; (d) 250 pg; (e) 125 pg; (f) 62 pg; (g) 31 pg; (h) 15 pg; (i) 7 pg; (j) 4 pg; (k) signal due to the oxidation of the conjugate in the antibody-coated tube without the addition of pregnanediol-3α-glucuronide; (maximal binding). Scale used on the oscilloscope: Speed: 2 sec/1.22 cm, sensitivity = 200 mV/1.22 cm.

The <u>precision</u> of the interassay-assay and intra-assay variation using the antibody-coated tube method is shown in Table 1.

<u>Table 1</u>. Solid-phase pregnanediol-3α-glucuronide chemiluminescence immunoassay method: within- and between-batch precision.

	No. of estimates (n)	Mean + S.D. (μmol/l)	CV (%)
Intra-assay variation	16	6.79+0.34	7
Inter-assay variation	8	9.95+0.91	9.2

277

Comparison of chemiluminescence immunoassay for urinary pregnanediol-3α-glucuronide with gas liquid chromatography

The concentration of pregnanediol-3α-glucuronide in 24 h urine samples collected by patients (n=30) attending an infertility clinic was measured by the solid phase chemiluminescence immunoassay and by a conventional gas liquid chromatography method. The correlation coefficient "r" between the two methods was 0.96, and the equation was y=1.14X + 0.12 where y corresponds to values obtained by chemiluminescence immunoassay (Eshhar et al., 1981).

Measurement of urinary pregnanediol-3α-glucuronide in early morning urine throughout the normal menstrual cycle

Early morning urine specimens were collected daily by six normal volunteers throughout the entire menstrual cycle. The results are shown in Fig. 5.

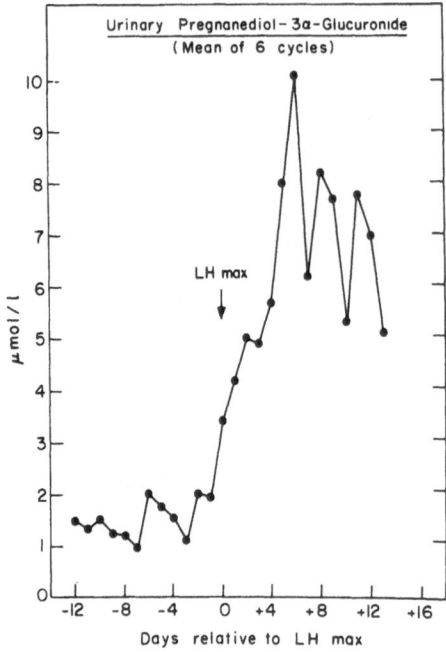

Fig. 5. Mean daily excretion of pregnanediol-3α-glucuronide as determined by chemiluminescence immunoassay.

Remarks

The work reviewed here indicates that it is feasible to determine directly pregnanediol-3α-glucuronide in diluted urine, thus obviating the need of cumbersome hydrolysis and extraction steps. The measurement of pregnanediol-3α-glucuronide can be used to monitor luteal function of the human menstrual cycle and to follow the progressive development of placental function during the early phase of pregnancy. The use of monoclonal antibody preparations in the assay improves specificity and facilitates vigorous standardization of the method.

Plasma estradiol-direct assay: immunobead-separation

One of the most significant achievements in steroid methodology is the direct determination of steroids in unextracted serum or plasma (Ratcliffe, 1983). Extraction with an organic solvent is avoided by the use of reagents which block the binding to, or displace steroids from, serum binding proteins. The measurement of steroids in plasma by a direct assay requires an antiserum with high avidity and specificity and a reliable separation technique. To date, direct chemiluminescent based immunoassays have been reported for cortisol (Lindström et al., 1982), for progesterone (De Boever et al., 1984) and for estradiol (De Boever et al., 1985). We report here the development of a direct solid-phase chemiluminescence immunoassay for estradiol using a homologous monoclonal antibody (clone #2F$_9$; Kohen and Lichter, 1986).

Immunoassay procedure

Stock solution of monoclonal antibody to estradiol (clone # 2F$_9$), coupled to immunobeads is diluted 500-fold with assay buffer containing uncoupled immunobead matrix at a concentration of 1 mg/ml, and 0.1 ml of this solution is added per Lumacuvette. 100 µl of plasma standard (estradiol added to charcoal treated male plasma) or of sample is diluted with 400 µl of assay buffer containing danazol (10 ng) and 5α-dihydrotestosterone (10 ng). The mixture is incubated at 25°C for 30 minutes and 0.1 ml of this solution is transferred in duplicate to Lumacuvettes containing the diluted monoclonal antibody coupled to immunobeads, followed by 100 pg of estradiol-6 CMO-ABEI conjugate in 0.1 ml of assay buffer. The contents are mixed, incubated overnight at 4°C, 1 ml of wash solution is added, and the tubes are centrifuged (10 min, 2000xg, 25°C). The supernatant fluid is decanted, and the washing step is repeated. Sodium hydroxide (2N, 200 µl) is added to each tube. The tubes are incubated at 60°C for 30'. After cooling to room temperature, 0.1 ml of diluted microperoxidase is added to each individual tube which is immediately placed in the luminometer. Chemiluminescence is initiated by the rapid injection of 100 µl of oxidant solution (0.3% of H$_2$O$_2$).

Evaluation of the method

A typical dose-response curve is shown in Fig. 6. The sensitivity of the assay was 1.5 pg of estradiol/tube (220 pmol/l). The intra-assay coefficient of variation was 8.5% and 12% for serum samples containing respectively 1600 and 675 pmol estradiol/l. The specificity of this monoclonal antibody to estradiol (clone 2F$_9$) as determined by chemiluminescence immunoassay is shown in Table 2. Estradiol values obtained by this direct method agreed well with those obtained in a direct RIA using radioiodinated estradiol as a marker (r=0.88, y=1.13x + 16.6, n=23).

Remarks

The results of this study indicate that this direct assay for estradiol can be applied for routine use in clinical

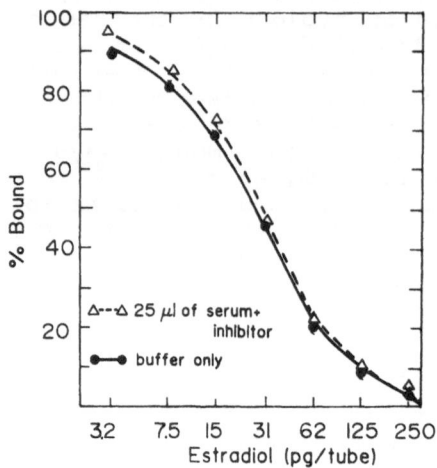

Fig. 6. Dose-response curve for estradiol in buffer ●---●
or in the presence of 25 µl of serum and inhibitors
(◁---▷).

Table 2. Specificity of polyclonal and monoclonal antibodies to estradiol as determined by chemiluminescence immunoassay.

Compound	Cross-Reaction (%)	
	Polyclonal Antiserum	Monoclonal Antibodies (Rat-mouse hybridoma #$2F_9$)
Estradiol-17β	100	100
Estradiol-17α	1	1
Estrone	4.6	<0.05
Estriol	0.4	<0.05
17α-Ethynyl-estradiol-17β	0.2	0.05
Testosterone	<0.2	<0.05
Progesterone	<0.2	<0.05
Heavy chain class		IgG_{2a}

laboratory. In the method described here an overnight incubation is performed. However, due to the high binding affinity of the monoclonal antibody to estradiol, one hour incubation is also adequate. Forty serum samples can directly be assayed in four hours, and the method requires 25 ul of serum (De Boever et al., 1985).

Salivary progesterone-direct assay: Immunobead separation

Saliva is a readily accessible body fluid, and contains free steroid hormones in concentrations that reflect those in the free (biologically active) fraction in serum. Thus, a variety of radioimmunoassay procedures using extraction techniques have been developed for the determination of salivary steroid hormones [e.g. progesterone (Walker et al., 1981), testosterone (Turkes et al., 1979, 1980)] for assessing gonadal and adrenal function.

One limitation of salivary steroid determinations is that the analytes in saliva are in extremely low concentrations. Despite this limitation we have developed a solid-phase chemiluminescence immunoassay for direct measurement of progesterone in 125 μl of saliva using polyclonal anti-progesterone IgG covalently coupled to polyacrylamide beads as the solid-phase reagent (De Boever, 1986).

Immunoassay procedure

Samples of saliva (2 to 4 ml) were collected from seven healthy women, ages 21 to 36 years, every morning throughout their complete menstrual cycles and stored at -20°C until required. 100 μl of 150-fold diluted solid-phase antibody [anti-progesterone-11-BSA IgG covalently coupled to polyacrylamide beads 5-10 μ from Bio-Rad (De Boever et al., 1984)] are added in duplicate to Lumacuvettes followed by 125 μl aliquots of saliva samples or standards freshly prepared in assay buffer. The tubes are mixed briefly. Progesterone-11-ABEI conjugate, 50 pg in 100 μl of assay buffer is then added. The tubes are incubated for 1.5 h at 25°C and 0.9 ml of wash solution is added to each cuvette. The tubes are centrifuged (10 min, 2000 g at 25°C). The supernatant is decanted, and the wash procedure is repeated. The pellet is then processed for luminometry as described in the previous section for the direct assay for estradiol.

Evaluation

A typical dose-response curve is shown in Fig. 7.

The sensitivity was 1.5 pg (equivalent to 38 pmol/L).

The intra-assay variation at 202.1+13.8 pmol/L was 6.8%, and the inter-assay variation at 209.5+14.4 pmol/L was 6.9% in eleven consecutive determinations.

Salivary progesterone concentrations determined by direct solid-phase chemiluminescence were compared with serum progesterone concentrations throughout the menstrual cycles. The salivary progesterone concentration ranged from 38 to 986 pmol/L, and the serum concentrations from 0.3 to 86 nmol/L. The ratio of serum to salivary progesterone was 5.1 in the follicular phase, 16 in the periovulatory period and 56 in the luteal phase of the menstrual cycle. The correlation between serum and salivary progesterone levels was significant (r=0.88, p<0.001, n=96). The concentration of progesterone in samples of saliva throughout a menstrual cycle of a healthy 34 year-old woman is shown in Fig. 8.

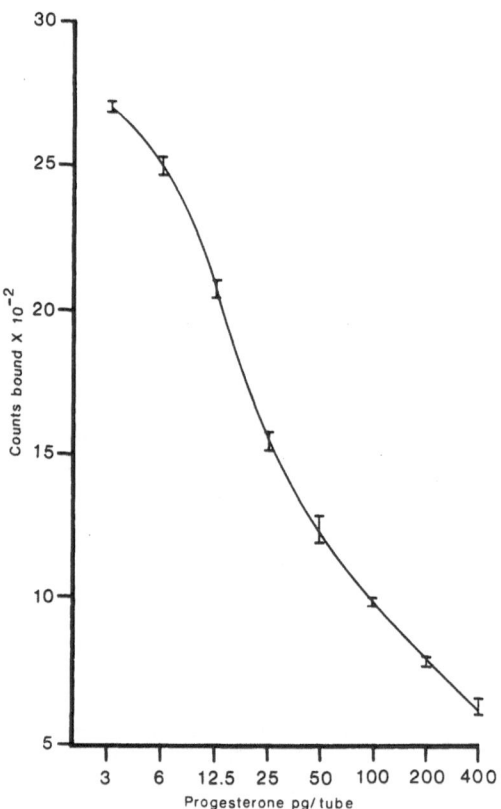

Fig. 7. A dose response-curve for salivary progesterone.

Remarks

The results of this study indicate that salivary progesterone concentrations can be used to monitor luteal function in women. The assay is simple, reliable and relatively fast (4 hr). This direct salivary assay may offer a feasible alternative to progesterone determination in serum.

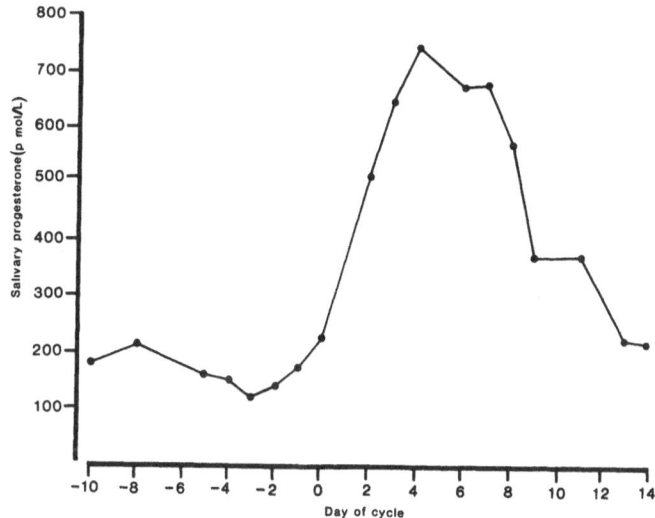

Fig. 8. Salivary progesterone concentration during the menstrual cycle of a 34-year old woman. Day 0=day of
the LH surge.

DISCUSSION

The work reviewed here indicates that it is feasible to
use chemiluminescence based immunoassays for the determination of steroids and their metabolites in biological fluids.
We have shown that the conjugates can be characterized by
physical means and are stable. The procedures are comparable to established RIA in terms of sensitivity, specificity,
speed, accuracy and precision. However, chemiluminescence
immunoassays have certain disadvantages. For instance, the
measurement of the chemiluminescent tracer is more complex
than the measurement of radioactivity. Thus, the precision
of injection of the oxidant prior to the measurement of the
tracer is of utmost importance in obtaining reproducible
measurements. Another limitation is the quenching of light
emission due to the impurities present in the sample. Despite these limitations, chemiluminescence-based immunoassays do provide a viable alternative to radioimmunoassays,
and the procedures avoid the use of a radioactive label.

Acknowledgements

We are grateful to Drs. W.P. Collins, Z. Eshhar and Mrs.
C. Usanachitt, for permission to quote collaborative work;
to Mrs. S. Lichter, Mrs. B. Gayer, Mr. A. Almoznino and
Mrs. D. Leyseale for technical assistance, and to Mrs. M.
Kopelowitz for excellent secretarial assistance.

REFERENCES

Barnard, G., Collins, W.P., Kohen, F., and Lindner, H.R., 1981a, The measurement of urinary estriol-16α- glucuronide by a solid-phase chemiluminescence immunoassay. J. Steroid Biochem., 14:941.

Barnard, G., Collins, W.P., Kohen, F., and Lindner, H.R., 1981b, A preliminary study of the measurement of urinary pregnanediol-3-glucuronide by a solid-phase chemiluminescence immunoassay. In: "Bioluminescence and Chemiluminescence", M.A. DeLuca and W.D. McElroy, eds., Academic Press, New York, pp. 311-317.

Brown, J.B., Klopper, A., and Loraine, J.A., 1958, The urinary excretion of oestrogens, pregnanediol and gonadotrophins during menstrual cycle. J. Endocr., 17:401.

Collins, W.P., Collins, P.O., Kilpatrick, M.J., Manning, P.A., Pike, J.M., and Tyler, J.P., 1979, The concentrations of urinary oestrone-3-glucuronide, LH and pregnanediol-3α-glucuronide as indices of ovarian function. Acta Endocr., 90:336.

Collins, W.P., Barnard, G.J., Kim, J.B., Weerasekera, D.A., Kohen, F., Eshhar, Z., and Lindner, H.R., 1983, Chemiluminescence immunoassays for plasma steroids and urinary steroid metabolites. In: "Immunoassays for Clinical Chemistry: A Workshop Meeting, Edinburgh, March 1982", W.M. Hunter and J.E.T. Corrie, eds., Churchill Livingstone, Edinburgh, pp. 373-397.

De Boever, J., Kohen, F., and Vandekerckhove, D., 1983, A solid phase chemiluminescence immunoassay for plasma estradiol-17β for gonadotropin therapy compared with two different radioimmunoassays. Clin. Chem., 29:2068.

De Boever, J., Kohen, F., Vandekerckhove, D., and Van Maele, G., 1984, Solid-phase chemiluminescence immunoassay for progesterone in unextracted serum. Clin. Chem., 30:1637.

De Boever, J., Kohen, F., and Vandekerckhove, D., 1985, Monoclonal antibody to estradiol suited for a direct solid-phase chemiluminescence immunoassay. In: "Monoclonal Antibodies: Basic Principles, Experimental and Clinical Applications in Endocrinology", Florence, October 1985, Abstract #20.

De Boever, J., Kohen, F., and Vandekerckhove, D., 1986, Direct solid-phase chemiluminescence immunoassay for salivary progesterone. Clin. Chem., in press.

Eshhar, Z., Kim, J.B., Barnard, G., Collins, W.P., Gilad, S., Lindner, H.R., and Kohen, F., 1981, Use of monoclonal antibodies to pregnanediol-3α-glucuronide for the development of a solid-phase chemiluminescence immunoassay. Steroids, 38:89.

Kim, J.B., Barnard, G.J., Collins, W.P., Kohen, F., Lindner, H.R., and Eshhar, Z., 1982, Measurement of plasma estradiol-17β by solid-phase chemiluminescence immunoassay. Clin. Chem., 28:1120.

Kohen, F., and Lichter, S., 1986, Monoclonal antibodies to steroid hormones. In: "Proceedings of the International Symposium on Monoclonal Antibodies: Basic Principles, Experimental and Clinical Applications in Endocrinology", G. Forti and M. Pazzagli, eds., Raven Press, in press.

Kohen, F., Kim, J.B., Lindner, H.R., and Collins, W.P., 1981, Development of a solid-phase chemiluminescence immunoassay for plasma progesterone. Steroids, 38:73.

Kohen, F., Lindner, H.R., and Gilad, S., 1983, Development of chemiluminescence monitored immunoassays for steroid hormones. J. Steroid Biochem., 19:413.

Kohen, F., Pazzagli, M., Serio, M., De Boever, J., and Vandekerckhove, D., 1985, Chemiluminescence and bioluminescence immunoassay. In: "Alternative Immunoassays", W.P. Collins, ed., John Wiley & Sons Ltd., pp. 103-121.

Kohen, F., De Boever, J., and Kim, J.B., 1986, Surface chemiluminescent immunoassays (steroids). In: "Methods in Enzymology on Bioluminescence and Chemiluminescence", M.A. de Luca and W.D. Mc Elroy, eds., Academic Press, Inc., in press.

Köhler, G., and Milstein, C., 1976, Derivation of specific antibody-producing tissue culture and tumor lines by cell fusion. Eur. J. Immunol., 6:511.

Lindner, H.R., Kohen, F., Eshhar, Z., Kim, J.B., Barnard, G., and Collins, W.P., 1981, Novel assay procedure for assessing ovarian function in women. J. Steroid Biochem., 15:131.

Pazzagli, M., Kim, J.B., Messeri, G., Kohen, F., Bolelli, G.F., Tommasi, A., Salerno, R., Monetti, G., and Serio, M., 1981a, Luminescence immunoassay (LIA) or cortisol. 1. Synthesis and evaluation of the chemiluminescent labels of cortisol. J. Steroid Biochem., 14:1005.

Pazzagli, M., Kim, J.B., Messeri, G., Martinazzo, G., Kohen, F., Francheschetti, F., Monetti, G., Salerno, R., Tommasi, A., and Serio, M., 1981b, Evaluation of different progesterone-isoluminol conjugates for chemiluminescence immunoassay. Clin. Chim. Acta, 115:277.

Pazzagli, M., Messeri, G., Caldini, A.L., Monetic, G., Martinazzo, G., and Serio, M., 1983, Preparation and evaluation of steroid chemiluminescent tracers. J. Steroid Biochem., 19:407.

Pratt, J.J., Woldring, M.G., and Villerius, L., 1978, Chemiluminescence-linked immunoassay. J. Immunol. Meth., 21:179.

Ratcliffe, W.A., 1983, Direct (non-extraction) serum assays for steroids. In: "Immunoassays for Clinical Chemistry", W.M. Hunter and J.E.T. Corrie, eds., Churchill Livingstone, Edinburgh, pp. 401-409.

Schroeder, H.R., and Yeager, F.M., 1978, Chemiluminescence yields and detection limits of some isoluminol derivatives in various oxidation systems. Analyt. Chem., 50:1114.

Schroeder, H.R., Boguslaski, R.C., Carrico, R.J., and Buckler, R.T., 1978, Monitoring specific protein-binding reactions with chemiluminescence. Meth. Enzymol., 57:424.

Turkes, A., Turkes, A.O., Joyce, B.G., Read, G.F., and Riad-Fahmy, D., 1979, A sensitive solid-phase enzyme-immunoassay for testosterone in plasma and saliva. Steroids, 33:347.

Turkes, A.O., Turkes, A., Joyce, B.G., and Riad-Fahmy, D., 1980, A sensitive enzyme-immunoassay with a fluorimetric end-point for the determination of testosterone in female plasma and saliva. Steroids, 35:89.

Walker, S., Mustafa, A., Walker, R.F., and Riad-Fahmy, D., 1981, The role of salivary progesterone in studies of infertile women. Br. J. Obstet. Gynaecol., 88:1009.

Weerasekera, D.A., Kim, J.B., Barnard, G.J., Collins, W.P.,
    Kohen, F., and Lindner, H.R., 1982, Monitoring ovarian
    function by a solid-phase chemiluminescence immunoas-
    say. Acta Endocr. Cophenh., 101:254.

SEPARATION FREE CHEMILUMINESCENCE IMMUNOASSAY

Gianni Messeri, Claudio Orlando
Anna L. Caldini and Mario Pazzagli

Endocrinology Unit
Department of Clinical Physiopathology
University of Florence, Italy

INTRODUCTION

As the process of nuclear decay is unaffected by the physico-chemical enviroment, the separation of the antibody-bound and free fraction is an obligatory step in most radioimmunoassay systems. The inclusion of the separation technique has certain disadvantages including increased time of assay, the need for additional equipment, difficulty of automation and finally an increased source of error.
On the other hand, the modification of the tracer output following the binding to the specific antiserum is a potential feature of most non-isotopic labels. Such a behaviour was reported to occur when using antigens or haptens conjugated to different kinds of labels (enzymes, fluorophores, latex particles, chemiluminescent compounds, etc.). As a consequence, it became apparent that simple methods could be devised that omitted the time consuming procedure. Assay requiring no bound/free separation step have been described as "homogeneous" (Rubenstein et al, 1972). Potentially homogeneous assays are simpler, quicker, more precise and easily automated.
In the field of chemiluminescence, very sensitive tracers were prepared by linking isoluminol derivatives (Schroeder et al, 1978a) to steroids, steroid-glucuronides and peptides (Pazzagli et al, 1983; Messeri et al, 1986). Such compounds emit light upon oxidation in the system microperoxidase/ hydrogen peroxide displaying a very fast kinetic emission (Fig.1).
Schroeder et al (1976) first described an "homogeneous" competitive protein binding assay for biotin which was monitored by chemiluminescence. It was found that the light emitted from the oxidation of biotinyl-isoluminol derivative was enhanced upon binding to avidin. This phenomenon could be reversed by increasing amounts of authentic biotyn.
Any modification of the light output following the binding to the specific antibody may be useful for the development of separation-free techniques. The light yield may be enhanced, as described above or it may be quenched or an

alteration of the emission kinetic may occurs (Pazzagli et al, 1980).
If this is the case, "homogeneous" chemiluminescent method can be developed by selecting a suitable choice of the time interval between the oxidant injection and the light measurement. Nevertheless, even if antibody-induced light modifications take often place in the course of the chemiluminescent reactions, only rarely the phenomenon is so evident to allow the elimination of the separation step.
Many steroids and steroid glucuronides were conjugated to several isoluminol derivatives and tested for the ability of changing the light yield upon reaction with the specific antibody. Some attempts were also made in the peptide field by conjugating transferrin and albumin to the same chemiluminogenic compounds.

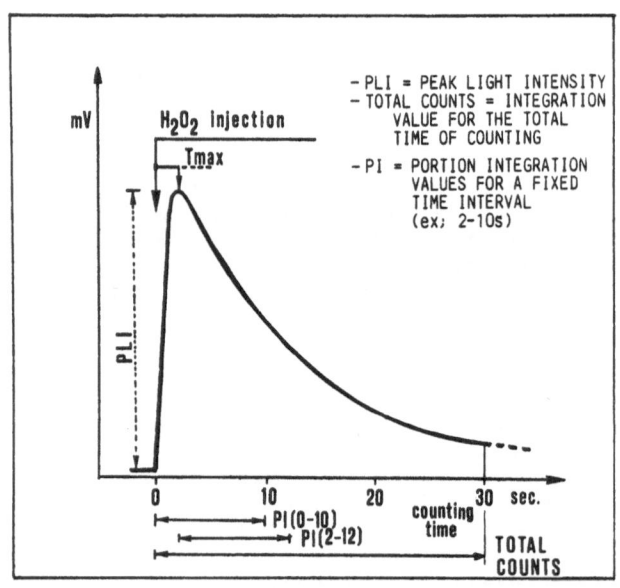

Fig.1 - Typical light emission of isoluminol derivatives and output quantification.

MATERIALS AND METHODS

Reagents

All the steroids, "microperoxidase" (MP-11, sodium salt, from equine heart cytochrome c), bovine serum albumine (BSA), immunoglobulin (HG-II, human Cohn Fraction II), N.N1 -dicyclo-hexylcarbodiimide and N-hydroxysuccinimide were obtained from Sigma Chemical Co., St. Louis, MO.
Hydrogen peroxide (300 mL/L), spectral grade solvents,

Silica gel 60-F254 thin layer chromatography plates and all salts for buffers were obtained from Merck, Darmstadt, FRG. "Helicase" [Suc digestife d'Helix Pomatia, 100,000 Fishman units of beta-glucuronidase (E.C. 3.2.1.31) and 1,500,000 Roy units of sulfatase (E.C. 3.1.6.1.) per milliliter] was purchased from Industrie Biologique Francaise, Villeneuve-La Garenne, France.
All the antisera were raised in rabbits using respectively cortisol-21-hemisuccinate-thyroglobulin, progesterone-11-hemisuccinate-BSA and estradiol-16,17-hemisuccinate-BSA as the immunogens, by the method previously described (Roda et al, 1980).

Oxidation system and light measurement

The oxidation system used was microperoxidase/hydrogen peroxide. Microperoxidase (400 mg/L, 0.2 mmol/L) dissolved in Tris HCl buffer (10 mmol/L, pH 7.4) and stored at 4 C, was stable for at least four months. The working solution was a 40-fold dilution of this stock solution with assay buffer. The starter solution was prepared by diluting the hydrogen peroxide solution 200-fold with doubly distilled water.
We measured light emission with a luminometer Biolumat LB9500 (Berthold, Wilbad, FRG). Using the automatic mode of the instrument, 0.1 ml of hydrogen peroxide are automatically added to initiate the light emission process. The analog output of the luminometer was connected to a storage oscilloscope (Model 5111, Tektronix, Beaverton, OR) to observe the pattern of light emission. In addition, the luminometer was interfaced to a microcomputer (General Processor, Model T/08, dual 13-cm, single-density floppy disk drive, 48 kbite RAM) as previously described (Tommasi et al, 1985).
A typical kinetic of a chemiluminescent reaction and the measuring parameters are shown in Fig. 1.

Chemiluminescent compounds

A number of different isoluminol derivatives have been investigated. The chemiluminescent compounds have been synthesized according to the procedure previously described (Schroeder et al, 1978b) and characterized by electron impact ionization mass spectrometry. The structure and nomenclature of the isoluminol derivatives are reported in Fig. 2.

Synthesis of the steroid-chemiluminescent tracers

The synthesis of steroid-chemiluminescent conjugates was as already published (Pazzagli et al, 1981). Briefly, the steroid-chemiluminescent conjugates were prepared by a two steps reaction which involves the carboxylic group of the steroid derivative (hemisuccinate, carboxymethyloxime or glucuronide) and the terminal amino group of the isoluminol derivative. A dimethylformamide solution of the steroid derivative was first incubated overnight with N-hydroxy-succinimide and dicyclohexyl-carbodiimide to yield the N-succinimide activated ester. The isoluminol derivative was then added and the reaction mixture incubated for an

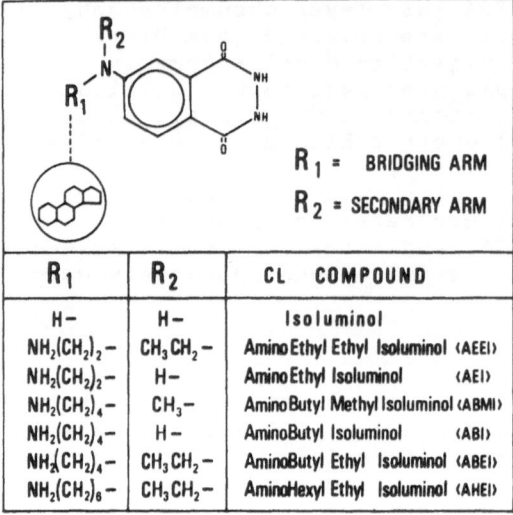

R$_1$	R$_2$	CL COMPOUND
H—	H—	Isoluminol
NH$_2$(CH$_2$)$_2$—	CH$_3$CH$_2$—	AminoEthyl Ethyl Isoluminol (AEEI)
NH$_2$(CH$_2$)$_2$—	H—	AminoEthyl Isoluminol (AEI)
NH$_2$(CH$_2$)$_4$—	CH$_3$—	AminoButyl Methyl Isoluminol (ABMI)
NH$_2$(CH$_2$)$_4$—	H—	AminoButyl Isoluminol (ABI)
NH$_2$(CH$_2$)$_4$—	CH$_3$CH$_2$—	AminoButyl Ethyl Isoluminol (ABEI)
NH$_2$(CH$_2$)$_6$—	CH$_3$CH$_2$—	AminoHexyl Ethyl Isoluminol (AHEI)

Fig.2 - Structure of isoluminol and isoluminol derivatives used for the conjugation to steroid molecules to yield steroid-chemiluminescent tracers.

additional 4-6 hours in the presence of sodium bicarbonate. The resulting steroid-chemiluminescent conjugate was then extracted with ethyl acetate and purified by thin layer chromatography on silica gel plates (methanol-chloroform-ammonia, 30:75:5).

The U.V. absorption spectra of the synthesized compounds were super-imposable on those of equimolar mixtures of the steroid with the isoluminol derivative. All the absorption spectra were recorded in NaOH 0.01 mol/L, 3% ethanol. All the conjugates are listed in Tab. 1.

The identity of the steroid-chemiluminescent tracers has been confirmed by mass spectrometry studies. Because of the thermal instability and low vapor pressure of these molecules, traditional ionization techniques such as electron impact ionization could not be used. Field Desorption and Fast Atom Bombardment techniques were able to indicate the molecular ion peak of the steroid-chemiluminescent tracer and, in the case of Fast Atom Bombardment, the fragmentation spectrum too (Fig.3).

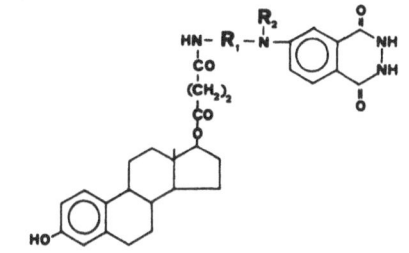

CONJUGATE	R₁	R₂	M.W. a.u.	FD Mass Spectra			
				M⁺˙	(M+H)⁺	(M+H₂)⁺	(M+Na)⁺
17β-E-17HS-ABEI	(CH₂)₄	-CH₂-CH₃	630	630	631		653

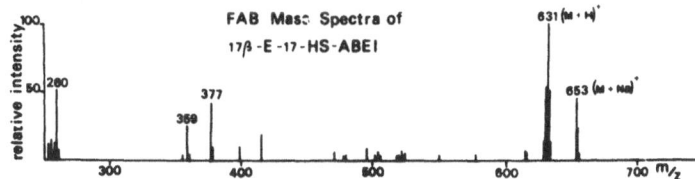

FAB Mass Spectra of
17β-E-17-HS-ABEI

Fig.3 - Example of mass spectra data resulting from a steroid-chemiluminescent tracer (17beta-estradiol-17hemisuccinate-ABEI) as obtained by the Field Desorption (FD) or by the Fast Atom Bombardment (FAB) technique.

All the conjugates, stored at - 20 °C as a 1 g/L solution in ethanol were stable for at least two years. The tracers were diluted to the appropriate concentration in assay buffer at the time of assay.

Assay Procedure for plasma progesterone

Anti-progesterone IgG fraction was prepared as previously described (Kohen et al, 1979).
The assay buffer was 0.02 mol/L phosphate buffer, (pH 8.6), containing 100 mg/L BSA.
Plasma samples (0.1 ml) were extracted with 3 ml of petroleum ether. The organic phase was then dried under a stream of nitrogen and the residue dissolved with 2 ml of assay buffer.
The assay procedure was as following. One hundred microliters of extracted samples or standards (20 - 500 pg/ 0.1 ml) were incubated with 0.1 ml of anti-progesterone IgG (6.2 pmol) for 15 min at 37 C and 30 min at 4 C. Microperoxidase solution (1 μmol/L) and progesterone-11-hemisuccinate-AHEI (200 fmol/0.1 ml) were then added and after a brief incubation (30 sec) at 4 °C, the light emission was measured over a 10 sec interval.

Assay Procedure for urinary "total" estrogens

The assay buffer was 0.1 mol/L borate buffer, pH 8.6, containing 9 g/L NaCl, 80 mg/L BSA and 80 mg/L human IgG. Urine (10 ul) from the first morning voiding or from a 24 h collection were incubated overnight, at 37 °C, with 0.2 ml of 0.1 mol/L acetate buffer, pH 5.2, containing 250 Fishman Units of beta-glucuronidase and 3750 Roy Units of sulfatase. The hydrolyzed samples were diluted to 2 ml with assay buffer.
The standard solution (17-beta-estradiol) was diluted to the appropriate concentration (3 - 500 pg/0.1 ml) with assay buffer containing 5 ml/L of estrogen-free pooled urine from children younger than 3 years.
One hundred microliters of diluted standards or samples were incubated with estradiol-17-hemisuccinate-ABEI (100 fmol/0.1 ml) and 0.1 ml of anti- estriol-16,17-dihemisuccinate-BSA (titer 1:16000) for 1 hour at room temperature.
Light emission was then measured for 10 sec after the injection of the starter reagent.

Assay Procedure for plasma cortisol

The assay buffer was 0.05 mol/L sodium phosphate, pH 8.0, containing 6 g/L NaCl and 20 mg/L BSA.
Anti-cortisol IgG fraction was prepared as previously described (Kohen et al, 1979).
Plasma (0.1 ml) was extracted with 3 ml of methylen chloride. The supernatant was aspirated and the organic phase was washed with 0.1 mol/L NaOH (0.5 ml) and water (0.5 ml). The organic extract was then dried under a stream of nitrogen and assay buffer (5 ml) was added to dried residue. The assay procedure was as following. One hundred microliters of extracted samples or standards (10 - 1250 pg) were incubated with 0.1 ml of specific anti-cortisol IgG (0.23 pmol) for 20 min at room temperature and for 20 min at 4 C. Cortisol-ABEI conjugate (25 pg/0.1 ml) was then added, and the incubation was continued for another 90 min at 4 °C. Light emission was then measured for 10 sec, from the end of the 2nd to the 12th second.

RESULTS

When the light efficiency of all the conjugates, listed in Tab. 1, was recorded before and after the incubation with the specific antibody, only progesterone, cortisol and 17-beta-estradiol appeared to display significant alterations.
As regards progesterone, the lenght of the bridge between the steroid and the isoluminol derivative as well as the nature of the chemiluminescent compound affected the behaviour of the various tracers (Fig. 4).
The light output was quantified by mean of the most common parameters (PLI, TLP and DP), the last one appearing the most effective in revealing the enhancement phenomenum. When testing the same tracer (progesterone-11hs-ABEI) with two different antisera, the enhancement extent varied depending on the antiserum (Fig.5).

Tab.1 - Chemiluminogenic compounds and steroid conjugate
```
-----------------------------------------------------------
CL compound           Optimal       detection limit
                      pH range        (fmol/tube)
-----------------------------------------------------------

Luminol               10 - 11           0.3
Isoluminol            10 - 11           2.3
AEI                   10 - 11           0.7
Prog-11hs-AEI         10 - 11           1.5
AEEI                  12 - 14           0.1
Prog-11hs-AEEI        12 - 14           0.1
T-17hs-AEEI           12 - 14           0.2
T-3cmo-AEEI           12 - 14           0.2
ABI                   10 - 11           1.0
T-17hs-ABI            10 - 11           0.9
T-3cmo-ABI            10 - 11          11.0
C-21hs-ABI            10 - 11           1.0
ABEI                  12 - 14           0.1
Prog-11hs-ABEI        12 - 14           0.1
Prog-6hs-ABEI         12 - 14           0.1
Prog-21hs-ABEI        12 - 14           0.1
T-17hs-ABEI           12 - 14           0.2
T-3cmo-ABEI           12 - 14           0.2
C-21hs-ABEI           12 - 14           0.1
C-3cmo-ABEI           12 - 14           0.2
E2-17hs-ABEI          12 - 14           0.2
E2-6cmo-ABEI          12 - 14           0.1
E1-gluc-ABEI          12 - 14           0.2
Pd-gluc-ABEI          12 - 14           0.4
D-3hs-ABEI            12 - 14           0.2
T-gluc-ABEI           12 - 14           0.2
ABMI                  12 - 14           0.1
Prog-11hs-ABMI        12 - 14           0.1
AHEI                  12 - 14           0.1
Prog-11hs-AHEI        12 - 14           0.1
-----------------------------------------------------------
```

Progesterone (Prog), Testosterone (T), Cortisol (C), 17beta-estradiol (E2), Estrone (E1), Pregnandiol (Pd), Dehydroepiandrosterone (D).
Hemisuccinatenate (hs), carboxymethyloxime (cmo), glucuronide (gluc).

On the bases of these findings the tracer-antiserum system and the recording parameters resulting in the maximum light enhancement were chosen for the assay development.
Similarly to progesterone tracers, 17beta-estradiol-ABEI conjugate is able to increase about fivefold its light efficiency after the incubation with the anti-16,17-dihemisuccinate-BSA serum. Due to the structure of the immunogen, the antiserum is able to bind specifically the main estrogens and their conjugates, allowing the development of a method for the "total estrogens" assay (Figs. 6,7 and Tab.2).

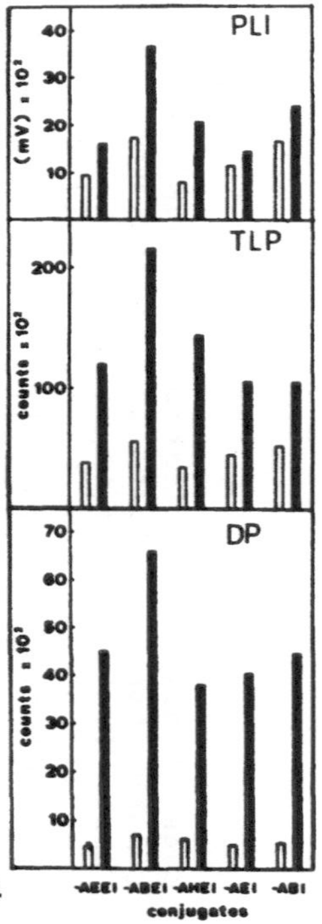

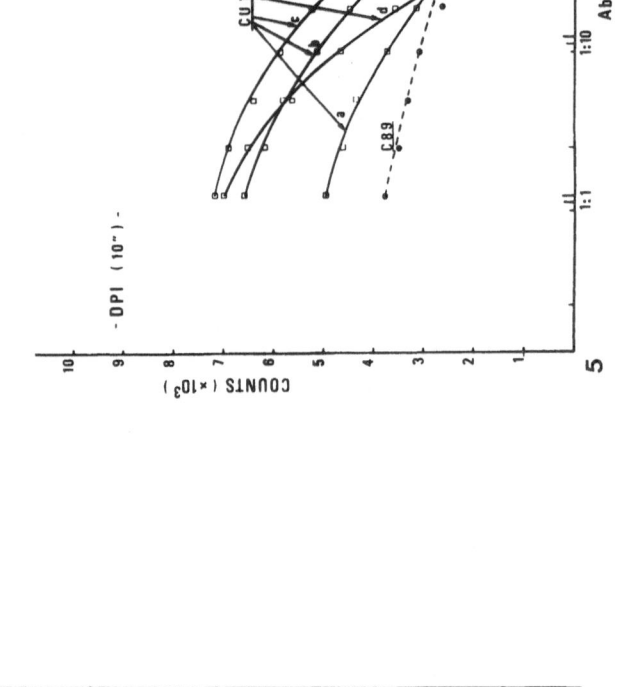

Fig.5 - Titration curves obtained using different antibodies.
1) C 89 - Titre 1:16.000, Kd 0.1 nmol/L
2) CU 113:

Bleeding a:	Titre 1:6.200,	Kd 0.3 nmol/L
" b:	" 1:16.500	" 0.2 "
" c:	" 1:15.000	" 0.2 "
" d:	" 1:7.000	" 0.25 "

Fig.4 - PLI, TLP and DP values produced by different conjugates in the presence (black bars) and in the absence (white bars) of anti-prog-11-hs-BSA-serum. Fifty fmol/tube of each conjugate have been used in all the experiments.

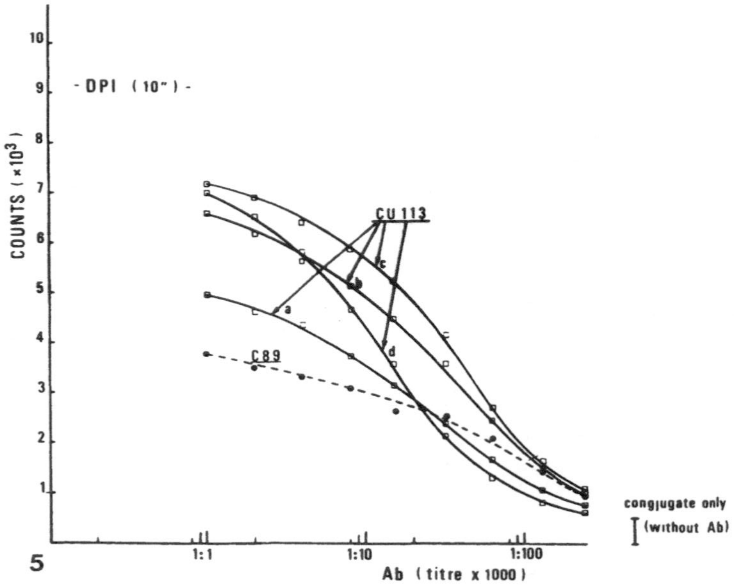

If performing the assay on pregnancy urines, the main component of which is estriol-16alpha-glucuronide (Fig 7), the hydrolytic step can be omitted and a good correlation could be proved with the RIA for total urinary estriol (Fig. 8).
When the cortisol-ABEI conjugate is bound to the anti-cortisol serum, the PLI of the conjugate is decreased but

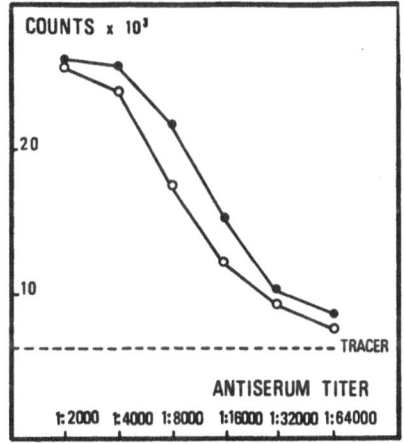

Fig.6 - Curve for the titration of the antiserum to estriol-16,17-dihemisuccinate-BSA with 100 fmol of estradiol-ABEI conjugate as the chemilumine-scent tracer.
The broken line represents the light emission of the tracer alone. The curves represent the light emission of the tracer in the presence of the antiserum (full circles) and in the presence (open circles) of both the antiserum and 50 pg of 17-beta-estradiol.

the total light yield is significantly increased, due to a rise of the light production during the decay part of the reaction. The binding of the tracer and consequent shift of the light emission curve to the right is prevented by the addition of free cortisol in a competitive manner. The calibration curve and the assay performances are reported in Fig. 9 and in Tab. 2.

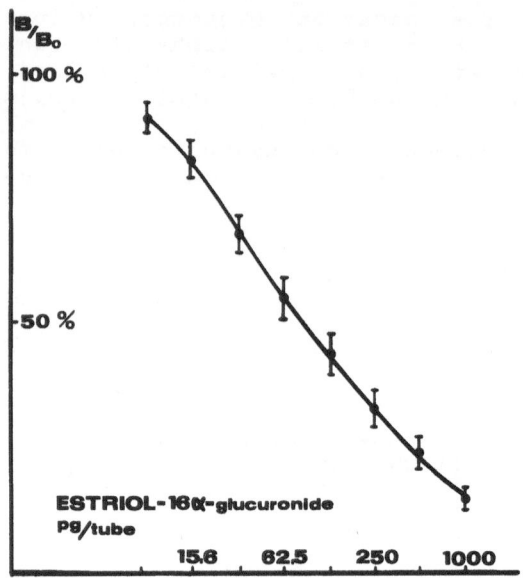

Fig.7 - Calibration curve for estriol-16alpha-glucuronide. Each point represents the Mean +/- S.D. of ten different experiments.

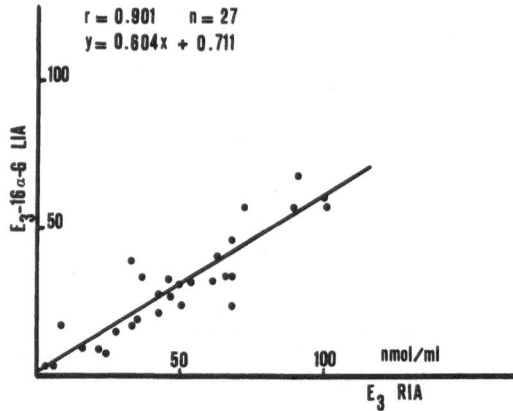

Fig.8 - Correlation between the luminescence immunoassay for estriol-16alpha-glucuronide and a RIA for total urinary estriol.

DISCUSSION

The possibility of developing "homogeneous" methods is one of the most attractive features of non-isotopic tracers. When dealing about chemiluminescence, such a chance appears at the moment limited to few compounds of clinical interest. An even approximate knowledge of the bases of this phenomenon would greatly help in synthesizing tracers for

"homogeneous" luminescent immunoassay. Unfortunately, the
antibody-induced light modification is still a poorly
understood phenomenon and no general rule can be
established about the behaviour of a tracer. It was only
generically suggested that the binding of the ligand of
interest may enhance chemiluminescence by providing a more
hydrofobic enviroment or facilitating the catalytic events.
Any case, apart from these general remarks, no prevision can
be done about the possible behaviour of a new tracer-
antibody system. Our results about the progesterone method
suggest that both the chemical structure of the tracer and
the antiserum can largely affect the enhancement extent.
Modifications of the chemical structure of the bridge as
well as the aminoacid residues of the isoluminol derivatives
can produce significative differences on the reaction
kinetic, on the enhancement and finally on the
dose/response curve slope.
The use of different anti-progesterone sera was proved to
result in a different enhancement of the light yield of the
tracers.

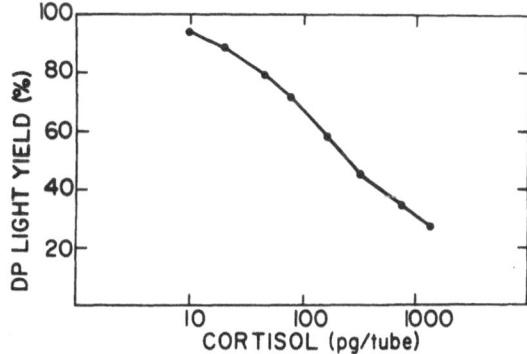

Fig.9 - Calibration curve for cortisol. The
maximum light yield of the tracer is achieved in
the presence of the antiserum and in the absence
of the authentic steroid.

Owing to the peculiar shape of the output (PLI, DPI or
TL),the choice of the mode of light recording may also
largely affect the extent of the enhancement.
In order to prevent the release of the antigen-antibody
binding, the light reaction must be necessarly run at pH 8-9

(Pazzagli et al, 1983). This strongly reduces the quantum efficiency of the isoluminol derivatives (about 50 fold) and, consequently, the sensitivity of the assay.

Tab.2 - Assays performances

| ASSAY | Sensitivity | | Precision (cv) | |
			Intrassay	Interassay
Progesterone	25pg/tube		7.5%	12.4%
Cortisol	4pg/tube		8.3%	11.2%
Total estrogens	2pg/tube	Low level	6.5%	9.8%
		High "	6.7%	8.6%
Estriol-16gluc.	10pg/tube	Low level	7.2%	10.4%
		High "	5.1%	8.5%

Moreover, the chemiluminescent measurements are very sensitive to interferences from biological fluids. Only few microliters of urine or serum can be processed without any previous purification (Tommasi et al, 1984) and this further reduces the sensitivity of homogeneous LIA. This is not a problem in heterogeneous immunoassay because the obligatory bound/free separation step (solid phases, charcoal, etc.) often provides the mean for eliminating the matrix effect. On the other hand, in homogeneous methods the fluid is still present at the light measurement time so that sample tubes can behave differently from the standard ones.
Following these remarks, the use of homogeneous methods seems, at the moment, strictly limited to the determination of compounds which are present in high concentration in the fluid of interest. If this is the case, the sample can be largely diluted before assay in order to minimize the matrix effect. A typical application is the estriol determination in pregnancy urine or serum (Caldini et al, 1984); the assay can be easily automated and it appears useful for meeting the requirements of this kind of test.
As regards the protein-label conjugates (Caldini et al, 1986; Messeri et al, 1986) a few attempts have been made for developing separation-free methods but up to now, no antiserum-induced effect could be proved (data not shown). In the future, a better knowledge of the mechanism underlying the antibody-induced light modification will allow the plan of new properly directed tracer-antibody systems, possibly overcoming the above mentioned drawbacks.

SUMMARY

    The modification of the tracer output following the binding to the specific antiserum is an attractive feature of most non-isotopic labels. Homogeneous methods requiring no bound-free separation step can thus be developed. When using chemiluminogenic labels, the binding to the antiserum may enhance or quenche the efficiency of the light reaction or modify the emission kinetic. Homogeneous immunoassays could be developed for progesterone, cortisol, estriol and "total" estrogens. Separation free chemiluminescence methods are very practicable and can be easily automated, but suffer from poor sensitivity and are very sensible to interferences from sample matrix. Their use seems confined to the measurement of analytes the concentration of which is high enough to allow sufficient pre-dilution of the sample.

REFERENCES

Caldini, A.L., Messeri, G., Buzzoni, P. and Borri, P., 1984, An automated luminescence immunoassay for the measurement of estriol-16alpha-glucuronide in pregnancy urine, in: Analytical Applications of bioluminescence and chemiluminescence, Kricka, L.J., Stanley, P.E., Thorpe, G.H.G. and Whitehead, T.P. (eds), Academic Press, London, 171-174.

Caldini, A.L., Orlando, C., Barni, T., Messeri, G., Pazzagli, M., Baldi, E. and Serio, M., 1986, Measurement of transferrin in human seminal plasma by a chemiluminescence method, Clin.Chem. 32: 153-156.

Kohen, F., Pazzagli, M., Kim, J.B., Lindner, H.R. and Boguslasky, R.C., 1979, An assay procedure for plasma progesterone based on antibody-enhanced chemiluminescence, FEBS Letters 104: 201-205.

Kohen, F., Pazzagli, M., Kim, J.B. and Lindher, H.R., 1980, An immunoassay for plasma cortisol based on chemiluminescence, Steroids 36: 421-437.

Messeri, G., Schroeder, H.R., Caldini, A.L. and Orlando, C., 1986, Synthesis of chemiluminogenic protein conjugates with dimethyladipimidate for sensitive immunoassays, in: Bioluminescence and Chemiluminescence, part B (Methods in Enzymology), DeLuca , M.A., McElroy, W.D. (eds), Academic Press, New York , in press.

Pazzagli, M., Kim, J.B., Messeri, G., Kohen, F., Bolelli, G.F., Tommasi, A., Salerno, R., Moneti, G. and Serio, M., 1981, Luminescent immunoassay (LIA) of cortisol-I. Synthesys and evaluation of two chemiluminescent labels of cortisol, J. Steroid Biochem. 14: 1005-1012.

Pazzagli, M., Messeri, G., Caldini, A.L., Moneti, G., Martinazzo, G. and Serio, M., 1983, Preparation and evaluation of steroid chemiluminescent tracers, J. Steroid Biochem. 19: 407-412.

Roda, A., Bolelli, G.F., 1980, Production of high titre antibody to bile acids, J. Steroid Biochem. 13: 449-454.

Rubenstein, K.E., Schneider ,R.S. and Ullman ,E.F., 1972, Homogeneous enzyme immunoassay. A new immunochemical technique, Biochem. Biophys. Res. Commun. 47: 846-851.

Schroeder, H.R., Vogelhut, P.O., Carrico, R.J., Boguslasky, R.C. and Buckhler, R.T., 1976, Competitive protein binding assay for biotin monitored by chemiluminescence, Anal. Chem. 48: 1933-1937.

Schroeder, H.R. and Yeager, F.M., 1978a, Chemiluminescence and detection limits of some isoluminol derivatives in various oxidation systems, Anal. Chem. 50: 1114-1120.

Schroeder, H.R., Boguslasky, R.C., Carrico, R.J. and Buckhler, R.I., 1978b, Monitoring specific protein-binding reactions with chemiluminescence, in: Methods in Enzymology vol. 57, De Luca M. (ed) Academic Press, New York, p 424-445.

Tommasi, A., Pazzagli, M., Damiani, M., Salerno, R., Messeri, G. and Serio, M., 1984, On-line computer analysis of chemiluminescent reaction with application to a luminescent immunoassay for free cortisol in urine, Clin.Chem. 30: 1597-1602.

SECTION III

IMMUNOASSAY AT LIQUID-SOLID INTERFACE

A PRIMER FOR MULTILAYER IMMUNOASSAY

Carl M. Berke

Hygeia Sciences
330 Nevada Street, Newton, MA   02160

INTRODUCTION

The forces shaping evolution of medical diagnostic technology today have created a growing demand for what is broadly termed "dry chemistry". The success of dipstick products derived from paper impregnated reagents has stimulated the development of a new generation of chemistry delivery systems based on a multilayer format. The principal attraction of multilayer formats is the potential capability for integration of sequential reaction chemistries thereby transforming a complex protocol to one that is "user-friendly", with less reliance on analyst skill or capital-intensive automation. Such a format change may be likened to the development of instant color photographic materials whereby each manual operation of the process along with its specific set of associated reagents, is reduced to a separate thin layer. The reagent layers can be maintained segregated and inactive until initiated by the application of sample fluid. The net effect is to transform operational complexity into media complexity. Just as the photographic laboratory has been elegantly embodied in a sheet of Polaroid self-developing film, so have classical clinical chemistries been translated to a slide format, as magnificently demonstrated by the Ektachem system developed at Eastman Kodak (Curme,1978; Shirey,1983).

The next challenge being faced by practitioners of multilayer chemistry today is the extension from the milli/micromolar range (10^{-3}-10^{-6}M) of conventional blood chemistry profiles and drug monitoring down to the nano/picomolar regime (10^{-9}-10^{-12}M) where RIA and ELISA assays operate (Ebersole, 1981). By current state of the art, the lower limits are only approachable with heterogeneous methods. The requirement for separation in heterogeneous assay mandates the design flexibility and physical configuration of multilayer systems, and as such, will be the main consideration of this article.

GENERAL CONSIDERATIONS

Classical blood analytes, such as glucose or cholesterol, occur at sufficiently high levels to permit direct colorimetric analysis wherein the analyte is stoichiometrically converted to a chromogenic species. In contrast, the difficulties encountered in the development of multilayer immunoassay chiefly arise from the following issues:

1. low analyte concentration
2. high molecular weight analytes and reagents
3. limited sample volume capacity
4. bound/free separation requirements
5. incompatible conditions for immunologic
    versus chromogenic reactions

Predictably, the first dry immunoassay products to have appeared employ homogeneous assay schemes with fluorescent labels for instrumental readout in a single layer paper matrix (Greyson,1981). This is a sophisticated, albeit limited, technology developed and refined by Ames Division of Miles Laboratories, utilizing renaturation of apoenzymes with cofactor-labelled analytes in a competitive mode. Heterogeneous assays are inherently more complex to construct but potentially offer greater sensitivity if background levels can be controlled.

The small dimensions of multilayer systems necessarily constrain sample volume. Consider a 1 mm thick sample deliverylayer, with an 80% void volume and a 1 cm^2 maximum detection area. Even these conservative parameters lead to a maximum sample aliquot size of only 80 uL. Corresponding diffusion distances also scale down such that reactants have to migrate less than 1/100th the radius of a conventional microtitre plate well, to reach an immobilized antibody. To a first order approximation, the Einstein equation for Brownian motion is a reasonable model for estimation (Crank,1968):

$$t=x^2/2D \qquad \text{where } x = \text{distance}$$
$$D = \text{diffusion constant}$$
$$t = \text{time}$$

Thus, for antibody traversal of a 100u thick layer, assuming a relatively high free diffusion coefficient for IgG ($5x10^{-7}$cm^2/sec; Allison,1959), an average of 20 seconds is required. However, in contrast to liquid-phase assays, the reaction kinetics cannot be accelerated by solution mixing. Thus, it is also unlikely that immunometric assays, which rely on removal of large excess of reagent, will ever be realisable in thin films because of the limitations imposed by the format. There is simply not the same opportunity for bulk fluid exchange or washing such as in chromatographic (e.g. Dade Stratus® ; Geigel,1983) flow-through systems (Hybritech ICON®; Valkis, 1985).

SELECTION OF MATRIX MATERIAL

By matrix material, it is meant the host media that

comprises the structure of the layer, akin to the vessel and
solvent in which the reaction chemistry is contained and
dissolved. There are two fundamental types of matrix
material, the choice between will determine both the fluid
transport properties and fabrication process of the assembled structure. Mixed media systems are also possible.

Table I. Summary of Matrix Media Characteristics

media type	examples	fluid transport	fabrication
homogeneous (gel)	gelatin agarose polyacrylamide	diffusion	casting on filmbase
heterogeneous (porous)	Millipore filter paper fabric	capillary action	lamination

GEL MEDIA

In order to maintain layer discreteness under conditions of use, the matrix must be essentially water-
insoluble; hydrogel colloids fulfill this condition. Thin
films are typically prepared by metered application of viscous solution through a slot extrusion orifice to a constant
velocity web of plastic film substrate. Solvent is then
removed by evaporative drying at a subsequent downstream
station. Thermoreversible gels, such as gelatin, are
routinely precision coated, up to 15 layers simultaneously
by photographic manufacturers, but adaptation for immunoassay presents an entire new set of problems. High gel set
temperatures will cause denaturation of reagents such as antibody conjugates. Gelatin (denatured collagen) can be cast
at reasonable temperatures but is not a suitable candidate
for immunoassay due to its lack of inertness as a protein
polyelectrolyte. Agarose qualifies for inertness but its
gel set point is normally above $55^{\circ}C$. Through chemical
substitution on the polymer backbone, the gel point can be
lowered to an acceptable level, for instance, hydrox-
yethylated agarose, $T_g=15^{\circ}C$, available from FMC Marine Colloids as SeaPrep®.

Many other natural and synthetic gellable polymers are
known, but most require chemical curing or in situ
polymerization, such as epichlorohydrin/dextran, calcium al-
ginate, and acrylamide/bisacrylamide (Andrade,1976). All
such systems suffer common flaws:

    a) the crosslinking chemistry is non-specific and can
       attack any other reactable species present
       including the incorporated assay reagents.
    b) in situ cross-linking is slow and difficult
       to control after coating
    c) incompatibility of polymerization or cure conditions
       with assay reagents

Pre-crosslinked polymers, such as the polyhydroxyalkyl-acrylates (used for soft contact lenses) are generally intractable and not amenable to continuous coating processes.

Another essential characteristic for a gel candidate is rehydratability. The cycle of gel set-drydown-reswell is highly susceptible to hysteresis, a physical phenomenon familiar to anyone who has accidently let a gel fractiona-tion column run dry. Gel porosity is directly dependent on water content (or conversely, percent solids). Commercial gel media, prepared and graded for molecular weight exclu-sion limit, are normally supplied at a specified water con-tent because of the irreversible changes which occur upon drying (Kremmer,1970). Likewise, practitioners of gradient electrophoresis are familiar with the dependence of protein permeability on percent gel solids. It is for this reason that agarose, again, is the preferred gel medium for separa-tion of high molecular weight biopolymers, because it main-tains structural coherency at the very low solids level necessary to provide sufficient porosity (Renn, 1984).

It is convenient to quantify gel rehydration in terms of swell ratios, whereby gel dimensions or water weight up-take is determined for a dried film before and after imbibi-tion in the appropriate solvent. To achieve a final gel solids of <5%, the swell ratio must be >19:1. If the gel adheres properly to the filmbase, all of the volume increase is restricted to the thickness dimension. The conversion of dry weight coverage to layer thickness can be expressed as a general rule-of-thumb for gels and most organic materials (approximately unit density):

$$1 \text{ g/m}^2 = 1 \text{ micron thickness}$$

The tertiary and quaternary structure of thermorever-sible gels are principally held together by hydrogen bond-ing, analagous to protein folding. The inclusion of hydrophilic additives into the melt solution can act as a solid diluent once the water is removed, thereby separating the polymer chains and disrupting the conformational proximities that lead to gel formation. In fact, the diluent polymer may best be the major fraction of the cast-ing solution, with the gelling polymer present at the mini-mum concentration necessary to retain cohesiveness. At least two patents refer to means of enhancing the reswell ratio of agarose and other natural gelling polymers (Renn,1970; Boschetti,1977). Liquid phase separation is a frequently encountered problem with such blends, but agarose will combine in all proportions with linear polyacrylamide (no crosslinker) to form stable solutions and high-swelling films. Alternatively, modification of a high T_g polymer, such as alkylation of hydroxyl groups, can also improve res-well ratio, presumably by reducing crystallinity in the solid phase. The rehydration of dry gels is driven by the force of osmotic pressure which is balanced at equilibrium by the elastic restraining forces of the crosslinks. Thus water can be driven very rapidly through a thin film struc-ture while the fate of larger solutes lags by slower Fickian diffusion which begins once the gel sieve is sufficiently swollen.

HETEROGENEOUS MEDIA

One can envision a heterogeneous media multilayer system as a superposition of individual paper dipstick chemistries rather like a chromatographic strip which has been folded over to stack the individual zones thereby translating lateral into transverse development and shortening the elution times. Reagents can also be successively deposited within a single layer by differential solubility techniques (Greenquist,1980). Heterogeneous media are characteristically two-phase systems of mostly void volume interspersed with a random network of open cells. Sample fluid displaces air in the interstices but does not penetrate the cell wall material. The mode of fluid transfer is initially bulk transport with no molecular weight selectivity, rapidly driven by capillary forces throughout the fill volume. High molecular weight analytes and reagents are not directly impeded by matrix materials (which have been properly treated to inhibit surface adsorption) as in the case of gels, aside from some increase in net diffusion path due to the tortuosity of the matrix. Washing with additional applied fluid can be very effective if a sufficient absorptive reservoir is present.

SAMPLE DELIVERY

For the same reasons that thin layer systems possess limited sample capacity, they also exhibit the feature of self-metered fluid uptake. Gel rehydration is generally not rapid enough to perform the task without additional provision for fluid confinement (Wright,1975), so that special elements have been developed for this purpose, all based on porous carriers. The underlying principle is that minimization of surface tension causes an applied drop to imbibe into a wettable capillary matrix. The net effect is to accept a variable size aliquot and meter to a constant volume per unit area, providing a level reservoir from which the underlying gel substrates can draw water. Osmotic pressure of rehydration in the gel layers far exceeds the capillary retention forces in any capillary-type structure.

Pryzbylowicz and Millikan (1976), at Kodak, introduced the basic technical approaches to this problem which has been further refined and extended by others. The current Ektachem products employ the phase inversion casting process to form in situ an integral spreader layer above a gel substrate. This method, also called "blush-coating", is accomplished by dissolving a non-swelling, water-insoluble polymer, such as cellulose acetate, in a mixture of low-boiling solvent, such as acetone, and a higher boiling non-solvent such as toluene. During evaporation, the low-boiling solvent volatalizes first, causing the polymer to precipitate from the mixture enriched in non-solvent. When finally dried, an open structure remains that is typically >80% void (Kesting,1985). Property modifiers such as TiO_2, for reflectance, carbon black dispersion, for opacity, plasticizers to prevent cracking, and surfactants to promote wetting, can be added to the formulation. Direct inclusion of immunoreagents in a non-aqueous formula is problematic due to insolubility, denaturation and polymer entrapment.

Pre-formed porous web also functions well as a spreader layer when laminated to an underlying gel support. Almost any type of filter stock will suffice with higher void volumes being preferred for capacity and imbibition rate. An attractive feature is that assay reagents can readily be impregnated into the web prior to lamination with minimal activity loss. Fuji has published extensively on the use of fabric spreaders, both textiles and non-wovens, which are treated to ensure surface wetting (Kitajama,1981).

Highly porous particulate layers may be fabricated <u>in situ</u> by coating and drying a metered coverage of aqueous slurry (Pierce, 1981). The slurry is composed of inert, non-swelling particles of sufficient size and proper shape to ensure adequate interstitial voids in a randomly packed dry matrix. The layer can be bonded together by inclusion of a small amount of adhesive binder in the slurry, that coats the particles to ultimately provide point-to-point bonding in a three-dimensional rigid lattice. However, in such a structure, the amount of binder must be held to a minimum to ensure sufficient adhesive strength but not decrease void volume to the point where flow is impeded. The adhesive is preferably rendered insoluble after the initial drydown in order to preserve the structural integrity of the array upon rewetting with sample fluid. A low T_g colloidal latex, such as Rhoplex (Rohm & Haas), is an adhesive materials of this type. Larger beads, in the 50-100u diameter range, are optimal for rapid flow, especially for whole blood which tends to clog pores of <10u with erythrocytes. Stabilization of the slurry for coating purposes is best arranged by choosing beads of neutral buoyancy such as polystyrene. Suspension thickeners will improve coatability but care must be exercised to avoid additives that impede flow of sample by space-filling or viscosity. To avoid the need for adhesive additives, Konishoroku (another Japanese photographic manufacturer) has proposed the use of beads that include a minor fraction of glycidyl methacrylate, which is alleged to render the beads self-adhering (Koyama,1984).

The simplest concept for self-metered fluid uptake is a mechanical arrangement of a wettable plane surface at a fixed capillary gap above another wettable surface such as gel, with access for fluid introduction and vent for displaced air (Lewis,1981).

IMMOBILIZATION AND REACTION SEQUENCING

The ability to maintain segregation of reagents during fabrication and storage and to localize reactions during operation plays a key role in the design of multilayer systems. Gel layers must be cast, set and dried quickly to prevent premature mixing. Porous layers are less susceptible to this problem because they may be prefabricated then laminated in the dry state by mechanical means such as a retaining frame or adhesive. Unlike chromatographic systems, where elution is unidirectional, multilayer systems are subject to so-called "cross-talk", undesirable exchange of reagents between functional zones. It is of particular concern with catalytic labels (e.g. enzymes) which may undergo premature reaction as opposed to directly stimulated fluorescent labels. Chromagenic substrate may be rendered immobile to minimize high background elevation due to "shuttling" between the immunosorbent (reaction) layer and the detection layer, somewhat analagous to "carryover" due to incomplete washing on an ELISA plate. Examples of immobilized substrates are cellulose-coupled dianisidine (Scharf, 1984) and reaction-precipitated indoxyl derivatives. The immobilizing carrier can be the matrix material itself, a derivitizable latex, or a soluble substituent that will sufficiently ballast the substrate in gels.

Reaction sequencing is controlled to some extent in gel systems by the limited rate of diffusion and lack of bulk mixing. Non-immobilized species comingle almost instantaneously in porous system and consequently there is less opportunity for controlling sequence. To further extend residence times in individual zones (i.e. incubation) the concept of timing layers has been proposed but not yet substantially reduced to practice.

RADIATION-BLOCKING LAYERS

The function of the radiation-blocking element is to provide discrimination between spatially-separated (e.g. bound and free) label in heterogeneous assays. In other words, since multilayer systems do not have the option of removing unwanted material, and the zones are viewed superimposed, the background must be concealed instead. Normally the structure is viewed from the opposite side of sample through the supporting transparent filmbase. Opacification may be built-in to any layer by inclusion of an appropriate light-absorbing dye (soluble) or pigment (insoluble). Discontinuous media, such as microporous sheet or bead layers, are to some degree naturally opaque by reflective scattering, which is enhancable by loading the opacifying agent in the precursor material used to make the media. Gel films may be similarly modified but particular care must be exercised to passivate pigments so as to avoid adsorption. Soluble dyes are to be avoided unless they are immobilized in some fashion to prevent wrong-way transfer to the detection layer. Titania is commonly employed for its covering power and because it also serves as a reflective surface to enhance detection from an adjacent layer. Carbon black is the most effective absorptive pigment on a per weight basis. Fuji (Kitajima,1981) has developed a method for increasing the opacity of membrane layers, with minimal additional material by vacuum depositing a thin layer of reflective metal on the irregular surface of a porous sheet.

EXAMPLES

Immunoprecipitation Assays

Fuji has published a number of patents which describe prototype elements for multilayer immunoassay, with some commentary on the inherent difficulties encountered relative to conventional blood chemistries (Hiratsuka,1982). One example is cited to illustrate the concept of a bound/free separation layer, including a novel detection system. Un-crosslinked polyacrylamide coated over photographic emul-sion, was spotted directly with a reaction mixture of in-sulin, labelled insulin, and anti-insulin that had been pre-viously incubated then subsequently precipitated with second antibody. In this case, the label is a carbocyanine dye which behaves as a spectral sensitizer for silver halide ex-posure. Purportedly, the polymer overcoat acts to retard the migration of label so that only free species can reach the emulsion layer. The film is subsequently light-exposed and processed in a conventional manner to yield an optical density proportional to the original concentration of analyte. A dose-response curve for insulin is claimed but no data are included nor is the control experiment, i.e. deletion of the separation layer.

In a patent issued to Polaroid Corporation (Berke,1984), a similar concept is revealed in reference to a proposed integral self-processing unit. Agarose solution containing sorbitol as permeator and fluorescein-labelled anti-albumin was coated on transparent polystyrene base. The film was imbibed in varying concentrations of albumin solution and subsequently incubated in wash solution. Ac-cording to the principle asserted, the gel acts to differen-tiate bound and free antibody by virtue of the molecular weight increase that occurs upon immunocomplexation. In the presence of higher antigen levels, an increase in residual fluorescence was observed in the film after washing. Struc-tural schematics were outlined wherein the excess labelled antibody would migrate from the reaction zone through an opaque separator layer (protein-blocked TiO_2 in agarose) to a detection zone. An immunosorbent, such as anti-species antibody immobilized to latex, may be included in the detec-tion layer to drive the equilibrium of antibody to con-centrate in the detection zone instead of partitioning to the higher volume spreader layer.

Heterogeneous Competitive Assays

Pierce and Frank (1981), at Kodak, have disclosed an immunoassay structure comprised of two successively coated layers of 6-8u beads with an adhesive latex, to form a reac-tion zone and a detection zone. The bead material of the reaction zone was "loaded", at the polymerization stage, with an opacifying dye. Anti-bovine IgG was passively ad-sorbed on this solid-phase before coating. The detection layer, prepared from beads that were blocked but not dyed, comprises the major volume fraction of the final structure. For demonstration of operation, serum samples were prepared with varying levels of bovine IgG (BGG) and then spiked with fluorescent-labelled BGG. After spotting and incubation at 37° for 15 min, the fluorescent label was distributed be-tween the reaction and detection zones in proportion to the level of analyte and the volume ratio of the layers. Only the unbound fraction of label in the bottom layer is detec-table through the base, because the fluorophore captured in the reaction zone is screened by the absorption pigment. The dose-response range spans [BGG] from $0-10^{-5}M$, with ~5 ug/mL sensitivity cutoff and maximum signal-to-noise ratio, S/N <2.

In a later patent, a close analogue of the above example has been reported by Konishoroku (Koyama,1984). They used self-adhering beads of larger diameter that contained a different opacifying agent. The standard curve for alpha-fetoprotein extended from 10^{-5}M down to 10^{-8}M cutoff also with an S/N <2.

Affinity-Mediated Immunoassay

Liotta (1984) has described the closest example, as of this writing, to a true self-working multilayer immunoassay. The principle is similar to other immunoaffinity assays, such as the duPont AcA test for digoxin (Freytag,1984), except that it is reconfigured from the original column geometry to the flat equivalent in a porous multilayer sheet assembly. The affinity or capture layer consists of antigen passively adsorbed to nitrocellulose membrane such as that used for protein and nucleic acid blotting. The reagent layer is cellulose paper impregnated with antibody against the specific analyte and conjugated to peroxidase enzyme. When sample is applied, fluid permeates the matrix and resolubilizes the conjugate which goes on to form competitive immunocomplexes between sample antigen and the immobilized affinity antigen. That fraction which binds in the affinity zone becomes unavailable for reaction with substrate. The remainder is free to migrate to the detection sheet where it can react with chromogenic substrate to form visually detectable color in proportion to the original analyte concentration. The peroxidase label requires a source of peroxide which, in the example, is supplied exogenously, but he suggests it may be generated in situ by including glucose oxidase in the detection layer and glucose elsewhere.

REFERENCES

Allison, A.C., Humphrey, J.H., 1959, Estimation of the Size of Antigens by Gel Diffusion Methods, Nature, **183**, 1591-1592.
Andrade, J.D., ed., 1976, "Hydrogels for Medical and Related Applications", ACS Symposium Series #31, Washington, DC.
Berke, C.M., 1984, Multilayer Element for Analysis, US Pat 4,459,358.
Boschetti, E., Moroux, P., and Tixier, R., 1977, Dried Rehydratable Film Containing Agarose or Gelose and Process for Preparing Same, US Pat. 4,048,397.
Crank, J., ed., 1968, "Diffusion in Polymers", Academic Press, New York.
Curme, H.G., Columbus, R.L., Dappen, G.M., Eder, T.W., Fellows, W.D., Figueras, J., Glover, C.P., Goffe, C.A., Hill, D.E., Lawton, W.H., Muka, E.J., Pinney, J.E., Rand, R.N., Sanford, K.J., and Wu, T.W., 1978, Multilayer Film Elements for Clinical Analysis, Clin Chem, **24**, 1335-1342.
Dudman, W.F., 1964, Unsuitability of Gelatine for Immune Diffusion in Plates, Nature, **201**, 995-996.
Ebersole, R.C., and Chait, E.M., 1981, Clinical Analysis: A Perspective on Chromatographic and Immunoassay

Technology, Anal Chem, 53, 682A-692A.

Freytag, T.W., Dickinson, J.C., Tseng, S.Y., 1984, A Highly Sensitive Affinity-Column-Mediated Immunometric Assay as Exemplified by Digoxin, Clin Chem, 30, 417-420.

Geigel, J.L., Brotherton, M.M., Cronin, P., D'Aquino, M., Evans, S., Heller, Z.H., Knight, W.S., Krishnan, K., and Sheiman, M., 1982, Radial Partition Immunoassay, Clin. Chem., 28, 1894-1898.

Greenquist, A.C., Rupchock, P.A., Tyhach, R.J., and Walter, B.,1984, Preparing Homogeneous Specific Binding Assay Element to Avoid Premature Reaction, US Pat 4,447,529.

Greyson, J., 1981, Problems and Possibilities of Chemistry on Dry Carriers, J Auto Chem, 3(2), 67.

Hiratsuka, N., Mihara, Y., Masuda, N., and Miyazako, T., 1982, Method for Immunological Assay Using Multilayer Analysis Sheet, US Pat 4,337,065.

Kesting, R., 1985, "Synthetic Polymeric Membranes", 2nd edition, John-Wiley, New York.

Kitajima, M., Arai, F., and Kondo, A., 1981, Multilayered Integral Element for the Chemical Analysis of Blood, US Pat 4,255,384.

Kitajima, M., Arai, F., and Kondo, A., 1981, Multilayer Analysis Sheet for Analyzing Liquid Samples, US Pat 4,292,272.

Koyama, M., Kikugawa, S., Okaniwa, K., and Tamaki, K., 1984, Analytical Element and Method of Use", US Pat 4,430,436.

Kremmer, T., and Boross, L., 1979, "Gel Chromatography: Theory, Methodology, Application", Wiley-Interscience, New York.

Lewis, R.L., 1981, Test Device for Measuring a Plurality of Analytes, EPA 0340491.

Liotta, L.A., 1984, Enzyme Immunoassay with Two-Zoned Device Having Bound Antigens, US Pat 4,446,232.

Pierce, Z.R., and Frank, E.S., 1981, Element Structure and Method for the Analysis of Liquids, US Pat 4,258,001.

Przyblowicz, E.P., and Millikan, A.G., 1976, Integral Analytical Element, US Pat 3,992,158.

Renn, D.W., and Mueller, G.P., 1970, Dried Agarose Capable of Rehydration for Use as Porous Chromatographic Medium, US Pat. 3,527,712.

Renn, D.W., 1970, Agar and Agarose: Indispensable Partners in Biotechnology, Ind Chem Prod Res Dev, 23, 17-21.

Scharf, G., 1984, Carrier Bonded Analytical Indicators, US Pat 4,446,085.

Shirey, T.L., 1983, Development of a Layered-Coating Technology for Clinical Chemistry, Clin Biochem, 16, 147-155.

Valkers, G.E., and Barton, R., 1985, Immunoconcentration: a New Format for Solid-Phase Immunoassays, Clin Chem, 31(9), 1427-1431.

Walter, B., 1983, Dry Reagent Chemistries in Clinical Analysis, Anal Chem, 55(4), 498A-514A.

Walter, B., Greenquist, A.C., and Howard, W.E., 1983, Solid-Phase Reagent Strips for Detection of Therapeutic Drugs in Serum by Substrate-Labelled Fluorescent Immunoassay, Anal Chem, 55, 873-878.

Wright, R.F., 1975, Device for Determining the Concentration of a Substance in a Fluid, US Pat 3,915,647.

REFLECTANCE METHOD FOR IMMUNOASSAY ON SOLID SURFACES

Hans Arwin, Stefan Welin, and Ingemar Lundström

Laboratory of Applied Physics
Department of Physics and Measurement Technology
Linköping Institute of Technology
S-581 83 Linköping, Sweden

INTRODUCTION

Background

In a solid phase immunoassay, one specifically wants to measure antibody adsorption from a liquid phase onto an antigen covered surface or vice
versa. The objective is therefore to characterize and quantify adsorbed
organic material on a solid substrate surface. Quantification is the most
important task from a clinical point of view. We can either use the rate
of adsorption or the total adsorbed amount of material as data. Most methods for quantification of binding reactions on solid surfaces, e.g. radioimmunoassay (RIA) or enzyme-linked immunosorbent assay (ELISA) (Engwall
and Pesce, 1978), do not permit continuous (kinetic) measurement of binding reactions. In clinical studies one is therefore confined to the type
of measurements in which the reaction under study either is terminated after a certain time or alternatively allowed to run to an equilibrium. In
this communication we will refer to this type of measurements as steady-
state measurements, although it is not always an equilibrium situation we
study. We will here describe a simple optical technique by means of which
it is possible to do either kinetic or steady-state measurements.

Several surface oriented optical methods is currently used or suggested for biomedical analysis (Ivarsson and Lundström, 1986). Some are indirect like staining with Coomassie brilliant blue (Adams et al., 1973) or
detection of changes in wettability by water condensation (Elwing, 1980;
Wikström et al., 1982). Other methods can be direct but involve often spe-
cially prepared solid surfaces. Giaever (1973), for example, made use of
the changes in light scattering from surfaces coated with thin indium films
which occur due to antigen-antibody complex formation on the surfaces.
Sandström et al. (1983) have shown that changes in interference colors of
thin layers can be used to assay immunoreactions occuring on the surface
of the layers. Immunoassays based on evanescent waves are described by
Sutherland in this volume. It has also been shown that surface plasmon resonance can be used to monitor immunological reactions (Liedberg et al.,
1983). A review of opto-electronic immunosensors has recently been given
by Place et al. (1985) and the use of surface oriented optical methods in
biomedical analysis has been discussed by Arwin and Lundström (1986).
Ellipsometry (Azzam and Bashara, 1977) has been widely used for studies of thin (monomolecular) organic films. Porter and Pappenheimer (1939)

and Rothen (1973) have demonstrated its usefulness for studying antigen-antibody reactions and Trurnit (1953) for enzymatic reactions. However, to the authors' knowledge there are no routine applications of ellipsometry in clinical laboratories. The limited clinical use of ellipsometry is probably due to cost and handling difficulties. The isoscope (Stenberg et al., 1980) is a simplified ellipsometer based on the use of a reference sample onto which light is reflected before it hits the sample to be studied. The isoscope has potentials for clinical use due to its simplicity and ease of operation. Direct optical methods are generally more attractive because they are easier to implement in routine work and often comprise fewer steps in the laboratory procedure. This facilitates understanding and thereby increases the confidence in the methods.

The method we describe here is a simple reflectance technique based on a special choice of experimental conditions. It can be designed to work at either air/solid or liquid/solid interfaces, and can thus be used to study both steady-state and kinetic measurements of binding reactions of immunological interest. We will shortly describe the principle of the technique in this section. In Appendix I we present a reflectance theory as a physical background to the method. A simple instrument and sample preparation are described under material and methods. Results obtained so far are presented next and finally we discuss limitations, possible improvements of the instrument, and potential applications.

Principle of the reflectance technique

The method is based on that there is a minimum in reflectance at a certain angle of incidence when light polarized in the plane of incidence (p-polarized) is reflected at the interface between two dielectric media. This angle of incidence is called the (pseudo-) Brewster angle. The underlying theory is presented in Appendix I, and in Fig. 1 we show, as an example, the reflectance for p-polarized light, R_p, versus angle of incidence for a silicon surface in air.

Reflectometry is generally rather insensitive to surface conditions except in certain cases. One such case is reflection of p-polarized light at or near the pseudo-Brewster angle, $\emptyset_B$, of a near-dielectric substrate. Around this angle of incidence, the value of R_p is very sensitive to the presence of overlayers as shown in Fig. 2 (Arwin and Lundström, 1985). The absolute magnitude of the reflectance is low, but the relative change due to an overlayer is large, especially if a substrate with high refractive index is used.

High resolution, sensitivity, and reproducibility of surface-oriented physical methods for immunological reactions require surfaces of high quality. Polished silicon, used in the semiconductor industry, is a very attractive candidate. Its surface is optically flat and it can be chemically treated to become hydrophobic or hydrophilic and to provide chemical groups for covalent coupling of molecules to its surface. Silicon also has a high refractive index ($n=4$ for visible light), which provides a high optical contrast to organic layers which normally have n around 1.5.

We have combined the use of silicon surfaces with reflectance measurements at the pseudo-Brewster angle to develop a simple optical method to quantify the amount of adsorbed organic material on a surface. The method

314

is one of the simplest continuous optical methods described and it can be implemented with simple equipment.

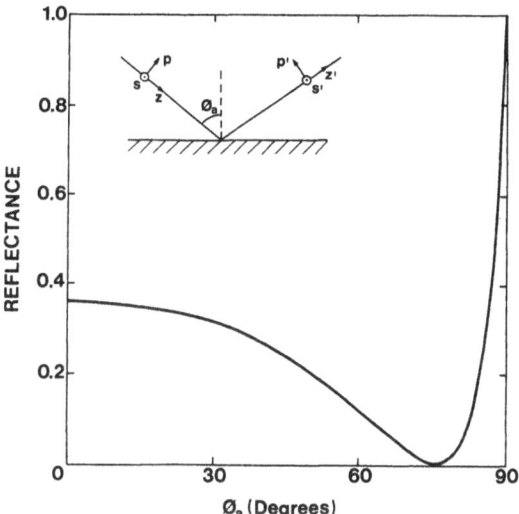

Fig. 1 Reflectance in air versus angle of incidence for a silicon surface for light polarized in the plane of incidence. The insert shows the definition of the coordinate systems used in Appendix I.

Some calculations

The refractive index for silicon is N_s=4.05+i0.03 at the wavelength 546.1 nm. In first approximation we may neglect the small imaginary part of N_s. More exact calculations show that a small imaginary part do not change the pseudo-Brewster angle significantly but causes the minimum reflectance to be finite (but still very small) instead of zero. With N_a=1, we now obtain $\emptyset_B$=76.1^o for the air/silicon interface at λ=546.1 nm by using Eq. (A6). Similarly we have $\emptyset_B$=71.2^o for the water/silicon interface at a wavelength of 632.8 nm assuming N_a=1.334 and N_s=3.90+i0.02.

At the pseudo-Brewster angle, any layer adsorbed on, transferred to, or grown on a silicon surface will change the reflectance. In Fig. 2 we exemplify this with the calculated reflectance around the pseudo-Brewster angle for the silicon/silicon dioxide/air system for the oxide thicknesses 0 nm, 5 nm and 10 nm.

In Fig. 3, the solid line shows the change in reflectance versus oxide thickness close to the pseudo-Brewster angle. The increase in slope in Fig. 3 indicates that it would be advantageous to use preoxidized silicon substrates. This would increase resolution and give a more linear relation between the amount of adsorbed material (layer thickness) and the reflectance. There is a maximum in the slope around 40 nm. An optimal choice of preoxidation is therefore around 30 nm. A subsequent formation of a second layer on top of the oxide can then be detected with optimal resolution and linearity up to a total layer thickness of around 70 nm. The dynamical range for the second (organic) layer is therefore 40 nm.

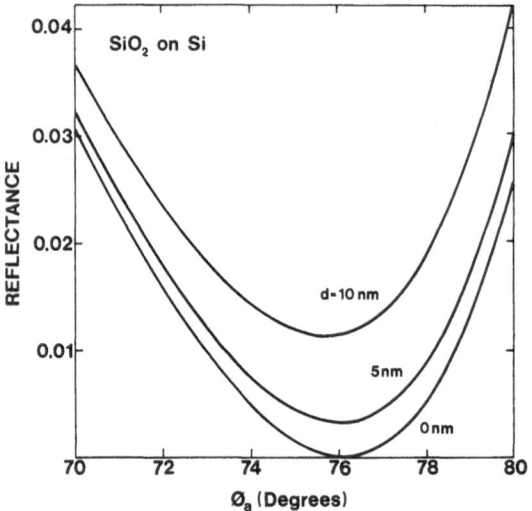

Fig. 2   Reflectance versus angle of incidence for light polarized in the plane of incidence at a silicon-air interface for different oxide thicknesses.

We end this section by showing in Fig. 4, the calculated reflectance versus thickness of a second layer on top of a 30 nm silicon-dioxide film on a silicon substrate. The second layer is assumed to have a refractive index of 1.5 which represents an estimated average of refractive indices for organic films of interest here (Arwin, 1986). In Fig. 4 we see that the reflectance change due to an overlayer is much larger for a silicon/air interface than for a silicon/water interface.

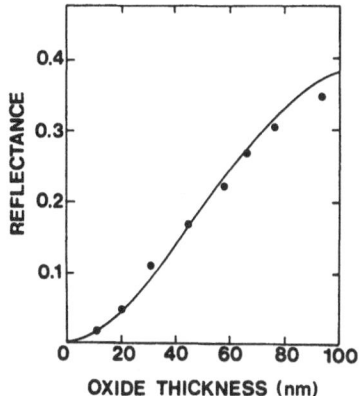

Fig. 3   Measured (dots) and calculated (solid line) reflectance at an angle
of incidence equal to 76.1° for a series of thermally oxidized silicon samples with oxide thicknesses in the range 10-100 nm.

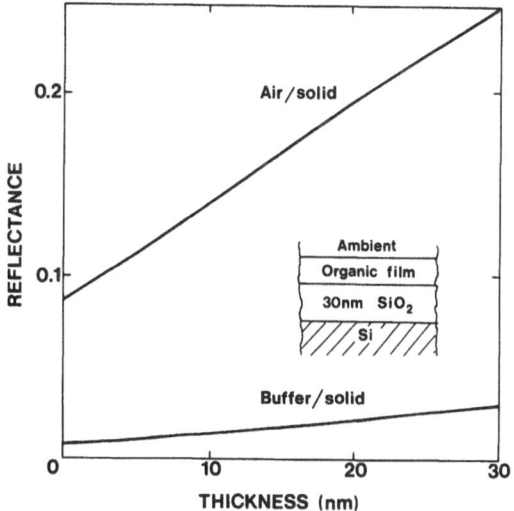

Fig. 4   Calculated reflectance versus thickness of a second layer on top
of a 30 nm silicon-dioxide film on a silicon substrate in air (top
curve) and in buffer (bottom curve). Assumed refractive indices
for the buffer and the adsorbed protein layer were 1.3338 and
1.500, respectively.

MATERIALS AND METHODS

Silicon substrates

Silicon substrates were obtained by cutting 0.3 mm thick silicon wafers
(Wacker Chemie, GMBH) into pieces with (typical) dimensions 10x5 mm. Such
silicon wafers are polished close to atomic flatness but are nevertheless
rather inexpensive due to their extensive use in semiconductor industry.
Preoxidized substrates were fabricated by thermal oxidation of silicon substrates at 900°C in dry oxygen.

In most experiments we have used silanized surfaces obtained by immer-
sion of silicon substrates for 10 min in dichlorodimethylsilane (Merck)
dissolved in trichloroethylene (1+9) (Merck, p.a.). After this silanization,
the silicon substrates were washed with ethanol, rinsed in trichloroethylene,
and stored in ethanol until used. The silanization results in silicon surfaces which are extremely hydrophobic which was wanted in our applications.
There are other silanization procedures which may be more favourable in certain applications (Jönsson et al., 1986).

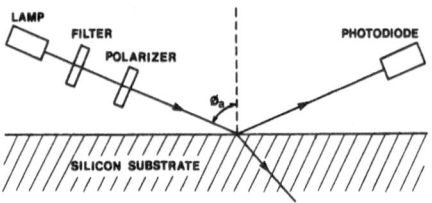

Fig. 5a   Experimental setup. The light beam is polarized in the plane of
incidence and the intensity of the reflected light is measured
by a detector. The lamp, filter, and polarizer can be replaced
by a laser.

Apparatus

A possible experimental setup is schematically shown in Fig. 5a. A low-
cost instrument is realized with a microscope lamp, a blue filter, a polaroid filter, a photodiode, and a simple electronic unit containing an amplifier and a display showing the measured light intensity. These components
are mounted on a machined aluminium housing which also contains the sample
holder and suitable apertures. The angle of incidence is 76.1°. For normalization, a second photodiode (not shown in Fig. 5a) measures the light intensity of the lamp. The angular position of the polarizer can be adjusted
to obtain light polarized in the plane of incidence.

The lamp, filter and polarizer can be replaced by a low-power polarized laser (e.g. Hughes 3221H-PC). The whole laser is then rotated to
achieve p-polarized light. A laser also gives a collimated beam which simp-
lifies alignment in designs for kinetic studies on liquid/solid interfaces.
However, we have sometimes observed long-term (drift) and short-term (noise)
instabilities when using lasers.

In kinetic studies it is also necessary to have a compartment (cell)
with entrance and exit windows for the light beam. The cell can be of the
flow-through type, whereby the liquid sample under test is pumped into the
cell and comes into contact with the test surface which is pressed over an
opening in the cell. Cell volumes down to 30 µl can be used without loss
of accuracy. The instrument can be made very compact by mounting the cell
directly on the laser and the detector directly on the cell as shown in
Fig. 5b. The instrument is under development and different designs will
soon be commercially available (Sensistor AB, Sweden).

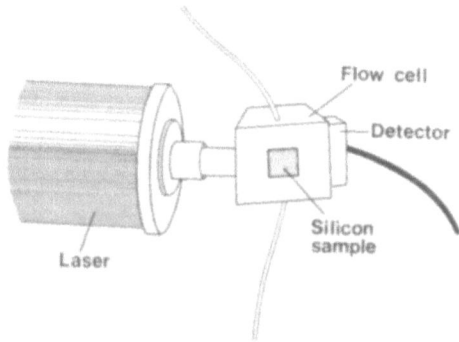

Fig. 5b   A flow-through cell mounted directly on a laser. The sample surface is mounted over a hole in the cell and held in position with
a small spring (not shown).

Reagents

Human serum albumin (HSA) and 16.5% human gammaglobulin (γ-glob) was
obtained from KabiVitrum, Stockholm, Sweden. Antiserum containing anti-
human albumin (a-HSA) was obtained from Calbiochem Behring Corporation. The
anti-human IgG was gamma chain specific and produced in swine (Orion Diagnostica, Finland; lot no IE 38). Gelatin (Sigma Chemical Company) was used
at a concentration of 1 mg/ml. All reagents were dissolved in 0.15 M NaCl
(Merck, p.a.).

TYPICAL RESULTS

Calibration: silicon dioxide on silicon

An absolute calibration of the reflectometer is generally not necessary to do. In an immunoassay one compares the responses of unknown samples to those of known reference samples. What counts is rather resolution
and sensitivity than absolute accuracy. A calibration on surfaces with
"standard" reflection properties therefore only play the role of veryfying
that 'an instrument is in working order.

Such an instrumental check can be made by taking data on a set of oxidized silicon samples. The silicon-silicon dioxide system is optically well characterized and provides a near-ideal model system. In Fig. 3 we present data on thermally oxidized samples for which the oxide thicknesses were independently measured with null ellipsometry. The agreement between theory and experiment is good. We found as standard an instrumental resolution on the order of ± 0.02 nm. This resolution is generally much better than the reproducibility in organic layer thickness observed when proteins are adsorbed from a given solution.

<u>Antigen-antibody reactions: steady-state measurements in air</u>

The adsorption of antibodies (a-HSA) from antiserum against human serum albumin (HSA) to HSA-coated silicon surfaces was used as a test of the applicability of the method. Hydrophobic silicon pieces (10x5x0.3 mm) were incubated for 30 min in 1% HSA in 0.15 M NaCl followed by incubation in diluted serum containing anti-HSA. An incubation time of 1 h was used for serum with different dilutions. The reflectance for each surface was measured and the reflectance change due to adsorption of antibodies was calculated for each sample by subtraction of the reflectance from a surface incubated at infinite dilution ("blank" sample). The result is shown in Fig. 6.

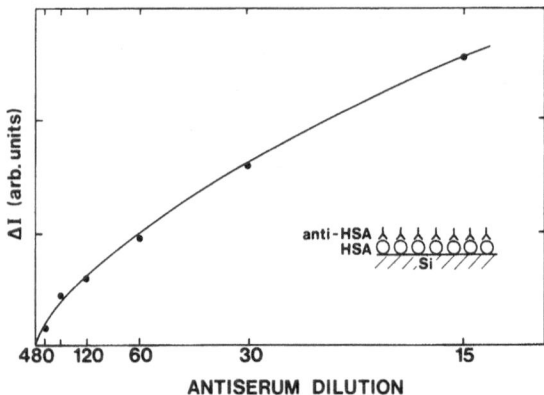

Fig. 6   Reflectance change due to adsorption of antibodies from antiserum against HSA on HSA-covered silicon substrates.

A saturation occurs at low dilutions due to the fact that the surface is filled with a complete monolayer of antibodies which prevents further adsorption. When the serum is more dilute the coverage gets lower and the amount of adsorbed antibodies depends on incubation time, specificity and affinity between the antigen and the antibody, experimental conditions such as temperature and incubation volume, and, of course, on the antibody concentration in the diluted serum. Quantification in relation to normal level of antibody concentration may be made by comparing results like those in Fig. 6 with those obtained on a reference serum. An efficient way of doing this is to use end-point titration; that is, a series of experiments is carried out to determine the dilution of serum which corresponds to the detection limit of the method.

As discussed above, there are some improvements if oxidized silicon surfaces are used. We found, for example, that a 12 nm thick oxide reduced the detection threshold five times. With preoxidized surfaces, we also obtain a more linear relation between the amount of adsorbed material and detector output (see also Fig. 4). Further linearization can be made electronically if desired. However, it is the antibody concentration in the serum which is of primary interest and an appropriate linearization for a dilution series should therefore include adsorption kinetics and isotherms. These are complicated and not well understood, and a full linearization is therefore meaningless at present.

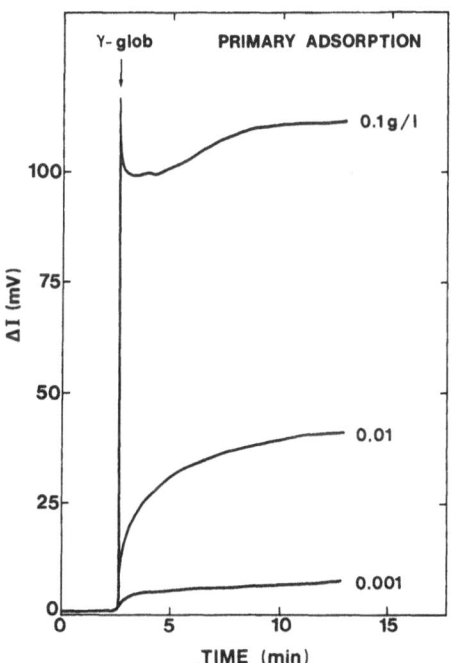

Fig. 7a   Primary adsorption of γ-glob at the concentrations indicated on the curves. ΔI is the change in detector output upon adsorption.

Although, the primary goal is to quantitate the antibody concentration in the sample under investigation, it can also be of interest to know the amount of surface-bound antibodies. Therefore we made parallel ellipsometric measurements to determine the absolute thickness of the surface films. We found that the saturation thickness at high antibody concentration (low dilution of antiserum) was around 7 nm, corresponding to a surface concentration of $0.8 \ \mu g/cm^2$. At large dilutions the accuracy is not limited by instrumental resolution but depends on the reaction under study. Nonspecific binding gives rise to a background, and in practice we observe a detection limit in the range 0.2-0.4 nm (0.02-0.05 µg/cm^2). In a flow-through cell we can do kinetic adsorption measurements and the detection threshold is improved considerably.

Antigen-antibody reactions: kinetic measurements in buffer

Primary adsorption on a hydrophobic silicon surface. γ-globulin was added to a measurement cell with a mounted hydrophobic silicon substrate surface, and the adsorption kinetics was recorded. Typical adsorption curves with an initially rapid, and later, a slow adsorption were obtained as shown in Fig. 7a. At high protein concentrations, the initial signal was frequently unstable because of refractive index variations in the solution. In an immunoassay, the primary adsorption step is the means by which a substrate surface is made specific for a class of antibodies. It is absolutely necessary to maintain accurate control over this precoating procedure.

Antibody binding to an antigen precoated surface. A typical stepwise experiment demonstrating antigen-antibody binding is shown in Fig. 7b. γ-globulin was first adsorbed on a silicon surface. Incubation with gelatin gave a further adsorption, indicating the presence of uncovered binding sites on the surface. After rinse with buffer, antiserum containing antibodies to IgG was added and the specific binding was followed. In Fig. 7c, this specific binding is shown at two antiserum dilutions. Binding from (1/1000) dilution was at the detection limit, and (1/100) dilution showed a slow rise in signal over 10 min.

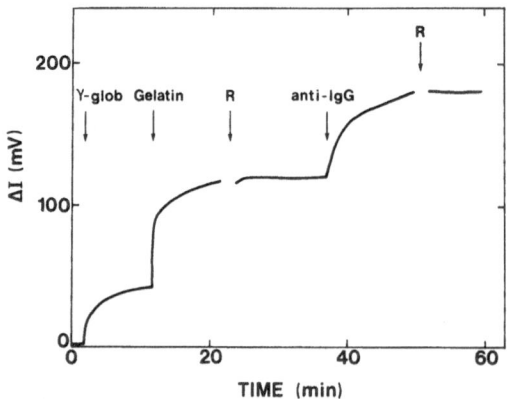

Fig. 7b Stepwise experiment showing an antigen-antibody binding reaction. Arrows indicate addition of reagents: 0.01 gl^{-1} γ-glob, 1 gl^{-1} gelatin and (1+99) anti-IgG; R indicates rinsing 5 times with 75 µl of 0.15 M NaCl.

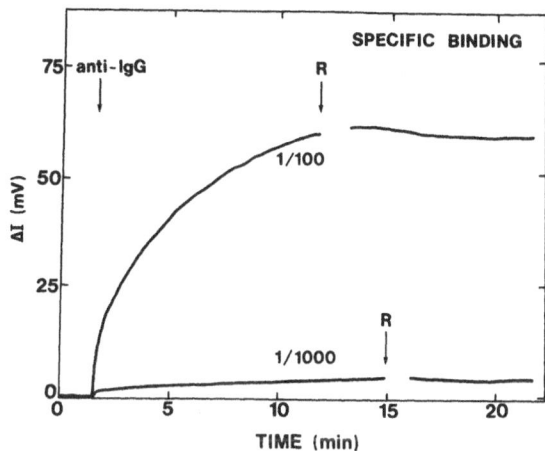

Fig. 7c Specific binding of Anti-IgG to a silicon surface coated with γ-glob and gelatin. Adsorption from control serum without anti-IgG was not significant (R as in Fig. 7b).

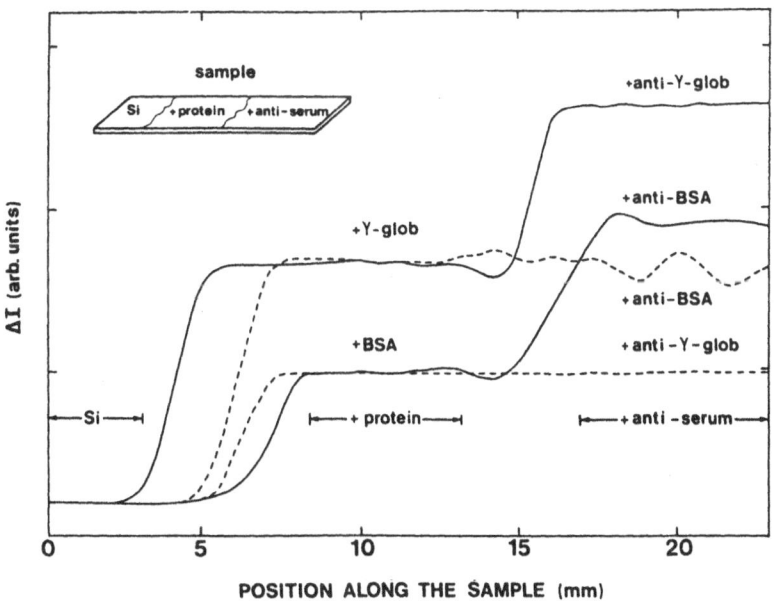

Fig. 8 Thickness profiles measured along an antigen-covered silicon sample incubated in antisera. The solid lines show samples partly inserted in sera containing antibodies specific to the surface-bound antigen, while the dashed lines show samples exposed to nonspecific sera. Further details are given in the text.

Thickness profiles - lateral scanning

Figure 8 illustrates the interesting possibility of measuring thickness profiles along a silicon surface. In these experiments we incubated two hydrophobic silicon substrates for 30 min on about 2/3 of their total length (30 mm) in a solution containing bovine serum albumin (BSA). Two other identical substrates were incubated in a γ-globulin solution in the same way. One of each type of sample was then dipped to about 1/3 in serum containing antibodies to γ-globulin. After an incubation time of 60 min, we analyzed the four samples in an experimental setup which permitted us to determine the variation in reflectance along the surface. Figure 8 clearly demonstrates the biochemical specificity characteristic for antigen-antibody reactions as well as the lateral and thickness resolution obtainable with the method.

DISCUSSION

Two important features of optical methods are their non-destructive character and that high surface sensitivity can be achieved without the necessity of vacuum. These properties facilitate application of optical methods in biomedical analysis, because most molecules and reactions of interest in this area are confined to the liquid phase. Thus optical techniques offer possibilities to study biochemical reactions either afterwards ex situ in air, or in situ during the reactions. It should also be observed that one do not need any marker for identification of the macromolecules.

The drawback with optical methods using photon energies in the visible is the poor identification capability, basically due to the absence of sharp and well separated optical structures in the optical response of the organic material under study. Therefore, in this context we use light probes mainly for detection and only rarely for identification. The implication of this is that we must know what we are measuring on.

It has been stated that "at their present state, surface oriented optical techniques are semiquantitative" (Genshaw, 1984). This statement is based on the observed semilogarithmic relation between number of adsorbed molecules and their concentration in bulk. The accuracy in solid phase determinations of bulk concentration of e.g. antibodies, will therefore be reduced considerably even though the amount of adsorbed antibodies can be determined very precisely and reproducibly. However, this is not a principal limitation and can be circumvented by using other assay procedures, like end-point titration, where instead the detection limit of the method determines the ultimate accuracy.

The drawback discussed above applies only to the classical type of measurement where the bulk concentration of a substrate is of primary interest. However, in many problems in modern biochemical analysis, other aspects of macromolecules are more important and parameters like conformation, activity, binding capacity, specificity, and general function are of more relevance. New biochemical analytic methods must therefore be developed. Surface oriented optical methods will here play an important role.

It is obvious that certain requirements have to be put on the solid substrate used in an immunoassay based on the reflectance method. A metal is generally not good since its reflection minimum is very shallow. The most important factor is a large real part of the refractive index. Silicon with $n = 4$ for visible light is therefore very suitable. Furthermore, silicon can be oxidized and the oxide surface can be treated in ways similar to those for normal glass beads. It also turns out that an oxide on the silicon surface increases the sensitivity and linearity towards adsorption of organic molecules as illustrated in Fig. 3. We conclude that silicon is an excellent substrate to use for optical studies in general and, thus, also for the simple optical method described in this paper.

The increase in differential sensitivity with oxide thickness is due to the fact that R_p is proportional to the square of the thickness for small thicknesses. There is an optimum in the rate of change of R_p for an oxide thickness of 40 nm. In practice, the optimal oxide thickness is lower (around 20-30 nm) because the reference reflectance, that is, the reflectance with no organic layer, increases rapidly with oxide thickness. Thus, the relative reflectance change decreases with increasing oxide thickness. Ultimately, the instrument performance is limited by noise, stability and resolution determined by the mechanical, optical, and electronic design of the instrument.

The use of preoxidized substrates makes an assay somewhat more expensive. If linearity is not a primary requiest, it may therefore be more cost-effective to use natural silicon substrate.

A successful application of the reflection method stands and falls with the possibility to prepare reproducible surfaces. No matter how accurate the instrument is, the overall assay accuracy is relying on that the different test surfaces have identical properties. In most cases, surface specificity is required, and surface preparation therefore has at least two critical steps. The first step is to produce a reproducible substrate surface. There are two principally different ways of preparing reproducible surfaces: either one cleans a natural or intentionally preoxidized surface to obtain an intrinsically clean substrate, or one modifies the surface in a controlled way. The latter is probably the best approach from stability aspects. A commonly used method is silanization, as described under "Materials and Methods" (see also Jönsson (1986)). Silicon dioxide ("glass") also provides a type of surface known to many biochemists and can be used for immobilization of various organic molecules.

The second step in sample surface preparation is the coupling or adsorption of specific macromolecules, in our case antigens or antibodies. Special precautions have to be taken in order to retain the biological function of the macromolecules.

As the reflection method not is analytic it is important to consider the effect of interference. We may distinguish between specific and non-specific interference. The specific interference is due to the inability of the surface bound haptenes to differentiate between similar determinants on related antibodies. This is strictly a biochemical problem. Non-specific interference, on the other hand, is a technical problem. As proteins do not in general, bind to each other without specificity, we mainly have to worry about non-specific interference caused by adsorption onto surface sites where no antigen molecules have adsorbed. The preparation of the test surfaces is therefore very important, and the coverage of antigen should be high. It is also possible to "neutralize" free adsorption sites by adsorption of an inert molecule like gelatin as described in Fig. 7.

In our case we have used serum, and, in control experiments with serum without antibodies specific to the surface-bound antigen, we typically observed residual adsorption of less than 0.3 nm. If more crude biological material is used, the residual adsorption may be overwhelming and the practical use of the method will be limited.

We have presented and discussed data in form of a "thickness of a layer" on the substrate surface. This may be misleading as we are dealing with discrete macromolecules adsorbed in the form of a discontinuous "film" on a surface. The way out of this dilemma is to use conversion formulas to calculate surface concentration, Γ, in units of $\mu g/cm^2$ instead. As long as the size of the macromolecules is much smaller than the wavelength of the light, the surface "film" constitutes an effective medium, which in first approximation can be described by its physical thickness and porosity. In this limit ($d \ll \lambda$), the light probe cannot resolve finer microstructural details.

The measured optical thickness of a surface layer is therefore an average thickness of a fictous dense film. The simplest conversion formula to surface concentration Γ is

$$\Gamma = \rho d, \tag{7}$$

where ρ is the density of the material in the layer and d the experimentally determined thickness. If we use the density of a typical protein, 1.3 g/cm^3 we find that d = 1 nm corresponds to Γ = 0.13 $\mu g/cm^2$.

If necessary, an approximate calibration may be carried out as follows. A series of substrates with different oxide thicknesses is prepared and a calibration curve like that in Fig. 3 is measured. This requires that the oxide thicknesses can be independently measured with ellipsometry. For each sample we calculate Γ using ρ = 2.65 for silicon dioxide (CRC , Handbook of Chemistry and Physics, 1972/73) and then plot Γ versus the reflectance measured with the reflectometer. The resulting curve can now be used to convert a measured reflectance value to a value independent of the properties of the surface film. A small correction factor needs to be incorporated due to difference in refractive-index increment, dn/dc, between silicon dioxide and proteins (Welin and Arwin, 1986). Different proteins have very similar refractive-index increments (De Feijter et al., 1978) and thus this approximative calibration procedure has general validity for proteins. Notice that we do not need to know the density or the refractive index of the material in the film. The reason why this simple calibration procedure works is that light reflected at a silicon surface is only sensitive to the total number of dipoles in the surface film if the film is dielectric.

In practice, a typical instrument for air/silicon operation has a resolution of about 0.1 nm in average thickness change (corresponding approximately to 13 ng/cm^2) of an organic layer for an optimally chosen oxide thickness. For liquid/silicon designs, the principal resolution is less due to the smaller optical contrast between water and silicon than between air and silicon. However, as one in the liquid/silicon case during the whole experiment measures on the same spot in situ, the uncertainty due to reference reflectance is reduced considerably. The overall resolution is therefore even in in situ measurements around 13 ng/cm^2.

It seems that the reflectance technique has approximately the same sensitivity and resolution as ellipsometry for quantification of the mass of organic molecules on solid/liquid interfaces. However, in contrast to ellipsometry, the reflectometer is simple and inexpensive and has no moving parts. These attractive features greatly simplify quantitative investigations of binding reactions at solid/liquid interfaces and should expand interest in such applications. One interesting application is the use of the reflectometry principle in connection with small-volume flow cuvettes.

In many cases it is of interest to measure the amount of antigen in a sample. In this case one should first adsorb antibodies on the silicon surface. However, due to the small number of determinants on the antibody molecules it is important to have ways of orienting them on the surface. One would also like to have monoclonal antibody preparations for this purpose.

The reflectometer has also been used in other biochemical applications. One of these is the characterization of precipitates adsorbed on surfaces, where the measurements of surface-bound lateral adsorption profiles are of importance (Elwing et al., 1984). Preliminary studies of the complement system has also been made (Elwing et al., 1986). Enzymatic degradation of protein layers has been studied (Welin et al., 1985) and Mandenius et al., (1986) have made in situ studies of the interaction of dehydrogenases with a coenzyme coated surface.

There are still a lot of details that can be improved in both the instrument, sample surface preparation, and the assay procedures. By replacing the laser with a light-emitting diode, a very small, compact and inexpensive instrument may be realized. In connection with modern electronics, one can design a complete microcomputer controlled instrument not larger than a pocket calculator. For specific applications it is possible to develop and prefabricate substrate surfaces, which then can be used directly in clinical investigations. The whole assay procedure is thereby reduced to two steps: (I) incubation in sample solution: and (II) reflectance measurements.

An at present visionary development is the immunochip on which several microscopic (say 100 μm^2) areas are independently precoated with different reagents (antigens). With a proper design of the detection system, the reflectance properties of all areas may be analyzed in a single measurement. What is gained is not only speed due to the fact that several components are measured simultaneously, but also increased resolution and accuracy due to what is called matrix detection. With the rapid development seen in the area of micromachining in silicon, the realization of such immunochips may come sooner than expected.

ACKNOWLEDGEMENTS

We would like to thank H. Elwing for valuable discussions, and A. Askendal for technical assistance. We also acknowledge the National Defence Research Institute, Umeå, Sweden, for construction of the flow cuvette. Our research on development of methods in the field of biomedical analysis is supported by grants from the National Swedish Board for Technical Development.

SUMMARY

A reflectance technique suitable for quantification of the amount of organic material adsorbed on a solid surface is described. The method is based on reflection of light at the interface between a transparent ambient, which can be air or a liquid, and a solid substrate with high refractive index. The light is polarized in the plane of incidence. At an angle of incidence equal to the so-called pseudo-Brewster angle, a well defined minimum in reflectance occurs. The reflectance at this minimum is very sensitive to the presence of a surface layer, and by using silicon substrates it is possible under optimal conditions to resolve changes in surface concentrations down to 10 ng/cm^2. Applications of the reflectance method for studying antigen-antibody reactions are presented. The importance of having well characterized substrate surfaces and coating procedures are stressed.

REFERENCES

Adams, A.L., Klings, M., Fisher, G.C., and Vroman, l., 1973, Three Simple Ways to Detect Antibody-Antigen Complex on Flat Surfaces., J. Immunol. Methods, 3:227.

Arwin, H., and Lundström, I., 1985, A Reflectance Method for Quantification of Immunological Reactions on Surfaces, Anal. Biochem., 145:106.

Arwin, H., 1986, Optical Properties of Thin Layers of Bovine Serum Albumin, γ-Globulin, and Hemoglobin, Appl. Spectr. 40:313.

Arwin, H., and Lundström, I., 1986, Surface Oriented Optical Methods for Biomedical Analysis, in: "Meth. Enzymol.," in press.

Azzam, R.M.A., and Bashara, N.M., 1977, "Ellipsometry and Polarized Light", North-Holland, Amsterdam.

De Feijter, J.A., Benjamins, J., and Veer, F.A., 1978, Ellipsometry as a Tool to Study the Adsorption Behavior of Synthetic and Biopolymers at the Air-Water Interface, Biopol., 17:1759.

Elwing, H., 1980, "Thin Layer Immunoassay", Thesis, Gothenburg.

Elwing, H., Welin, S., and Arwin, H., 1984, Precipitate Adsorption on Surface: A Simple Method for Analysis of Macromolecules Adsorbed on a Solid Surface, National Symposium on Measurements Techniques in Biotechnology, Gothenburg.

Elwing, H., 1986, personal communication.

Engvall, E., and Pesce, A.J., 1978, Quantitative Enzyme Immunoassay, Scand. J. Immunol., 8:7.

Genshaw, M.A., 1984, Immunoassays Monitored by Ellipsometry and Related Techniques, in: "Clinical Immunochemistry: Principles of Methods and Applications", R.C. Boguslaski, E.T. Maggio, and R.M. Nakamura, eds., Little, Brown and Co., Boston/Toronto.

Giaver, I., 1973, The Antibody-Antigen Reaction: A Visual Observation, J. Immunol., 110:1424.

Ivarsson, B., and Lundström, I., 1986, Physical Characterization of Protein Adsorption on Metal and Metaloxide Surfaces, to appear in: "Critical Reviews in Biocompatibility", D.S. Williams, ed., CRC Press Inc., Boca Raton.

Jönsson, U., Malmqvist, M., Olofsson, G., and Rönnberg, I., 1986, Surface Immobilization Techniques in Combination with Ellipsometry, in: "Meth. Enzymol.", in press.

Liedberg, B., Nylander, C., and Lundström, I., 1983, Surface Plasmon Resonance for Gas Detection and Biosensing, Sensors and Actuators, 4:299.

Mandenius, C.F., Mosbach, K., Welin, S., and Lundström, I., 1986, Reversible and Specific Interaction of Dehydrogenases with a Coenzyme Coated Surface Continuously Monitored with a Reflectometer, submitted for publication.

Place, J.F., Sutherland, R.M., and Dähne, C., 1985, Opto-Electronic Immunoassay at Continuous Surfaces, Biosensors, 1:321.

Porter, E.F. and Pappenheimer, A.M., 1939, Antigen-Antibody Reactions between Layers Adsorbed on Built-up Stearate Layers, J. Exp. Med., 69:755.

Rothen, A., 1973, Immunologic and Enzymatic Reactions Carried out at a Solid-Liquid Interface, Physiol. Chem. Phys., 5:243.

Sandström, T., Stenberg, M., and Nygren, H., 1985, Visual Detection of Organic Monomolecular Films by Interference Colors, Appl. Opt., 24:472.

Stenberg, M., Sandström, T., and Stiblert, L., 1980, A New Ellipsometric Method for Measurements on Surfaces and Surface Layers, Mater. Sci. Eng., 42:65.

Trurnit, H.J., 1953, Studies on Enzyme Systems at a Solid-liquid Interface. II. The Kinetics of Adsorption and Reaction, Arch. Biochem. Biophys., 48:176.

Weast, R.C. ed., "Handbook of Chemistry and Physics", CRC Press, Cleveland (1972/73) p. F1.

Welin, S. and Arwin, H., 1986, to be published.

Welin, S., Elwing, H., Arwin, H., Lundström, I., Wikström, M., 1984, Reflectometry in Kinetic Studies of Immunological and Enzymatic Reaction on Solid Surfaces, Anal. Chim. Acta, 163:263.

Wikström, M.B., Elwing, H., and Möller, A.J.R., 1982, Proteins Adsorbed to Hydrophobic Surface Used for Determination of Proteolytic Activity, Enzyme Microb. Techn., 4:265.

APPENDIX I

Reflectance theory

In this appendix we give some background in electromagnetic theory to aid in understanding the reflectance technique. An incident plane wave (the light probe) can be represented by its complex electric field vector E_i (see e.g. Jackson, 1975). In the z-direction of a local orthogonal coordinate system $\hat{p}, \hat{s}, \hat{z}$) as shown in the insert in Fig. 1, E_i can be expressed as

$$E_i = \text{Re}\{(E_p\hat{p} + E_s\hat{s})\exp(ikz - i\omega t)\}, \tag{A1}$$

where Re means real part, and E_p and E_s are the complex electric field vectors describing the amplitude and phase of the projections of E_i on the p and s axes. The factor $\exp(ikz - i\omega t)$ shows the propagation characteristics along the z axis; k is the propagation constant; ω is the angular frequency of the light; t is the time; and i denotes $\sqrt{-1}$. If the electromagnetic wave is reflected by a smooth surface, the outgoing wave can be represented in another local coordinate system $(\hat{p}', \hat{s}', \hat{z}')$ as

$$E_o = \text{Re}\{(r_p E_p\hat{p}' + r_s E_s\hat{s}')\exp(ikz - i\omega t)\}, \tag{A2}$$

where the effect of the surface is described by two complex reflectances, r_p and r_s, corresponding to the action of the sample on the field components parallel (p) and perpendicular (s) to the plane of incidence. Thus we need four parameters, amplitude and phase of the two complex numbers E_p and E_s, to completely describe the incoming wave and four more to describe the sample through r_p and r_s. The properties of the sample is therefore obtainable if the properties of both incident and reflected waves are known.

In reflectometry one measures the change in intensity due to reflection. We distinguish between p- and s-polarized light, and by Eqs. (A1) and (A2), the independent intensity ratios are

$$\left[\frac{I_r}{I_i}\right]_p = \frac{|r_p E_p|^2}{|E_p|^2} = |r_p|^2 = R_p, \tag{A3a}$$

$$\left[\frac{I_r}{I_i}\right]_s = \frac{|r_s E_s|^2}{|E_s|^2} = |r_s|^2 = R_s, \tag{A3b}$$

where the real quantities R_p and R_s are called the reflectances.

Except for normal incidence, R_p and R_s are generally different and are functions of the angle of incidence $\emptyset_a$, the wavelength λ, and the optical properties of the ambient and the sample.

In the simple case of a film-free substrate, r_p and r_s are given by

$$r_p = \frac{N_s\cos\emptyset_a - N_a\cos\emptyset_s}{N_s\cos\emptyset_a + N_a\cos\emptyset_s} \tag{A4a}$$

$$r_s = \frac{N_a \cos\emptyset_n - N_s \cos\emptyset_s}{N_a \cos\emptyset_a + N_s \cos\emptyset_s} \qquad\qquad (A4b)$$

respectively, where N_a and N_s are the refractive indices of the ambient and the substrate, respectively, and $\emptyset_s$ is the angle of refraction obtainable from Snell's law

$$N_a \sin\emptyset_a = N_s \sin\emptyset_s . \qquad\qquad (A5)$$

Of special interest for us is the condition for obtaining zero reflectance for p-polarized light, i.e. $R_p = 0$. This occurs at the angle of incidence

$$\emptyset_B = \arctan(N_s/N_a) , \qquad\qquad (A6)$$

which can be verified from Eq. (A4a). $\emptyset_B$ is called the Brewster angle of the substrate. However, in the more general case of a transparent ambient and an absorbing substrate the substrate refractive index is a complex number $N_s = n_s + ik_s$. If k is small compared to n we call the material near-dielectric. In this case the reflectance for parallel polarized light, R_p, never becomes zero but has a minimum at the so-called pseudo-Brewster angle. For highly polarizable materials with low absorption (n_s high and k_s low), this minimum is very pronounced. An example of this is shown in Fig. 1, where the dependence in R_p on angle of incidence is given for silicon in air. The pseudo-Brewster angle is given approximately by Eq. (A6) with $N_s = n_s$.

References

Jackson, J.D., 1975, "Classical Electrodynamics", Wiley, New York.

INTERFACE IMMUNOASSAYS USING THE

EVANESCENT WAVE

Ranald Sutherland and   Rudolf Slovacek and
Claus Dähne   Barry Bluestein

Biomedical Group   Biomedical Research Laboratories
Battelle Institute   Ciba Corning Diagnostics Corp.
Geneva, Switzerland   Cambridge, M.A., U.S.A.

SUMMARY

The use of the evanescent wave to detect fluoro-immunoassay reactions
at an interface is reviewed. Two systems are discussed in detail as
practical examples of applying the evanescent wave detection of immunoassays.
These systems are based on light guides of different geometry, a planar
quartz slide and a quartz optical fibre. Both systems are based on
attaching antibodies covalently to the waveguide surface and using
fluoresceine as the label. Examples are given of measuring human IgG
and human albumin by two-site immunofluorometric assay with the slide
waveguide, and ferritin using the fibre optic. An example is also given
of measuring digoxin by a limited reagent assay using the optical fibre.
The future application of evanescent wave immunoassay is analyzed in the
discussion.

INTRODUCTION

Modern health care in Western society is subject to rapid changes
due to economic, social and technical pressures. One of the major and
continuing changes in recent times is the (re)introduction of in-vitro
diagnostic testing to the non-laboratory environment (e.g. physician
offices, clinics, out-patient departments, intensive care units). This
trend to decentralized testing has been the subject of considerable
discussion (Rinsler, 1977; Mitchell, 1979; Weiner, 1980; Anderson,
Lindsell and Mitchell, 1980; Watson, 1980; Free and Free, 1984; Hicks,
1985; Marks and Alberti, 1985) and carries its own set of problems (e.g.
quality control) and advantages (e.g. speed of response) in efficient
patient care. One of the more challenging aspects of this trend to
instrument developers is to carry out immunoassays in the non-laboratory
environment. As many biochemical compounds of significant clinical
interest are measured solely or most efficiently by immunoassays (e.g.
hormones, drugs, microbial particles) there is a demand to carry out
these tests in a decentralized environment. However, most of the current
approaches to immunoassays are designed for the hospital laboratory, and

are inappropriate to decentralized testing, due mainly to technical and technological factors. The requirements for a non-laboratory immunoassay system are very demanding in terms of system performance and design due mainly to the lack of technical expertise of the operator and the relatively uncontrolled environment in which the system will operate. A list of some of the requirements of an ideal immunoassay system for such an environment is given (Table 1). Many of these requirements can be built into the hardware and software of an instrumental system, others are dependent on the analytical system itself. The ideal analytical system should be a non-isotopic, manipulatively simple (homogeneous), rapid, sensitive immunoassay.

Table 1. Requirements for Decentralized Clinical
          Chemistry Analytical System

Minimum sample pre-treatment (sample separation,
                                dilution, etc..)
Foolproof and simple in operation.
Not operator dependent.
Stable, reproducible.
Pre- or self-calibrated.
Small, relatively inexpensive.

A number of approaches are currently being investigated as a means to introducing simple immunoassays to the decentralized testing environment. One of the more promising techniques is based on the use of the evanescent wave to monitor immunological reactions at the interface between an optically transparent support and the sample solution.

When a light beam is totally internally reflected within the optical support, an electromagnetic waveform is generated in the sample solution close to the reflecting surface. This evanescent wave is part of the internally reflected light beam and penetrates a fraction of a light wavelength into the solution. The evanescent wave is the "sensing" component and can optically interact with compounds close to or at the surface. This optical interaction can be detected in at least three ways; as a reduction in the transmitted light (attenuated total reflection (Harrick, 1967)); as an increase in light scattered from the surface (Sutherland, Dähne and Place 1984); or by an increase in fluorescence (Total Internal Reflection Fluorescence; TIRF (Harrick and Loeb, 1973)).

We have previously reviewed these techniques along with other approaches to optically monitoring interface immunoassays (Place, Sutherland and Dähne, 1985; Sutherland and Dähne, 1986). For the purposes of this article we have chosen to review what is probably one of the most sensitive of these techniques, TIRF, and illustrate its utility with examples of two different instrumental approaches which have been independently developed at the laboratories of Battelle-Geneva and Ciba Corning Biomedical Research Laboratories.

THEORY

Principles of Internal Reflection Spectroscopy (IRS)

When a light beam irradiates the interface between two transparent media (Figure 1A), striking from the medium of higher refractive index ($n1 > n2$), total internal reflection occurs (Harrick, 1967) when the angle of reflection θ is larger than the critical angle θc.

$$\theta c = \sin^{-1}(n2/n1) \qquad (1)$$

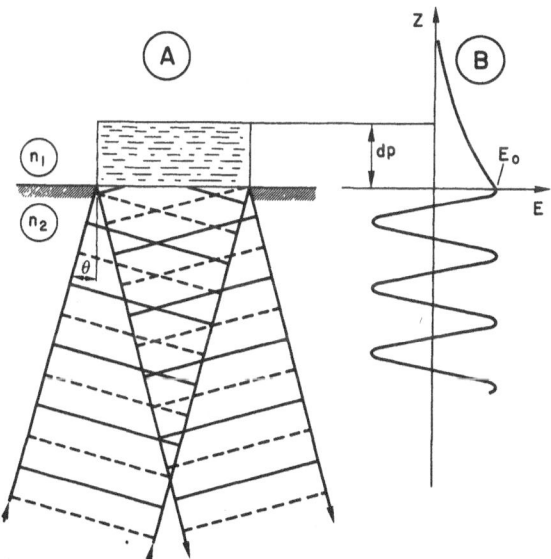

Fig. 1. Generation of the evanescent wave at an interface between two optical media; (A) where n1 > n2 and $\theta > \theta c$, θc is the critical angle at which refraction occurs; the evanescent wave is generated at the reflecting surface. (B) same as (A), but representing the electric field amplitude E, on both sides of the reflecting surface (Z = distance into the rarer medium, dp = the characteristic penetration depth of the evanescent wave).

In this case the evanescent wave penetrates a distance (dp), of the order of a fraction of a wavelength, beyond the reflecting surface into the rarer medium (n2). According to Maxwell's equations, a standing sinusoidal wave, perpendicular to the reflecting surface, is established in the denser medium (Figure 1B). Although there is no net flow of energy into a non-absorbing, rarer medium, there is an evanescent field in that medium. Because of continuity conditions of the field vectors the electric field amplitude (E) is largest at the surface interface (Eo) and decays exponentially with distance (Z) from the surface :

$$E = Eo \cdot \exp(-Z/dp) \qquad (2)$$

The depth of penetration (dp), defined as the distance required for the electric field amplitude to fall to exp(-1) of its value at the surface, is given by :

$$dp = \frac{\lambda/n1}{2\Pi\left\{\sin^2\theta - (n2/n1)^2\right\}^{1/2}} \qquad (3)$$

This quantity dp decreases with increasing θ and increases with closer index matching (i.e. as n2/n1 → 1). Also, because dp is proportional to wavelength, it is greater at longer wavelengths.

Thus, by an appropriate choice of the refractive index n1 of the IRE, of the incident angle, and of the wavelength, one can select a dp to promote optical interaction mainly with compounds close to or affixed at the interface and minimally with bulk solution.

As an example, if the waveguide is made of quartz (n1 = 1.46), and the rarer medium is an aqueous sample (n2 = 1.34), θc is 66° (equation 1). If θ is selected as 70°, λ as 500 nm, the resultant dp is approximately 150 nm into the solution (see equation 3). The estimated size of an IgG molecule (i.e. an antibody) is approximately 10 nm by 6 nm. Thus a "sandwich-type" immunological complex at the surface, consisting of three layers of IgG may have an average diameter of around 25 nm. At 25 nm the field strength is still 91 % of Eo (see equation 2). However, at double or treble this distance, the field strength falls off to 83 % and 76 % respectively, due to the exponential decay characteristics.

The depth of penetration is one of four factors which determine the attenuation caused by absorbing films in internal reflection. The other factors are, the polarization dependent electric field intensity at the reflecting interface, the sampling area which increases with increasing θ, and matching of the refractive index of the denser medium to that of the rarer medium which in turn controls the strength of the optical coupling. The appropriate quantity which takes account of all these factors is the effective thickness, de. It represents the actual thickness of film that would be required to obtain the same absorption in a transmission experiment.

In order to enhance sensitivity, multiple reflection elements are often used. The number of reflections (N) is a function of the length (L), thickness (T) of the waveguide and angle of incidence (θ).

$$N = L/T \cdot \cot\theta \qquad (4)$$

The longer and thinner the waveguide, the larger is N and the more frequently the evanescent wave interacts with the surface layer of antibody-antigen complexes. If for one reflection the reflectivity (R) is :

$$R = 1 - \alpha . de \qquad (5)$$

where α is the absorption coefficient for a weakly absorbing material and de is the effective thickness of the absorbing layer, after N reflections, the reflection loss is :

$$R = 1 - \alpha . Nde \qquad (6)$$

i.e. it is increased by a factor of N.

Total Internal Reflection Fluorescence (TIRF)

When an absorbing material is placed in contact with the reflecting of an IRE, the resultant internally reflected light beam is said to be attenuated (Harrick, 1967). In the case of ATR techniques, what is measured is the attenuated intensity as a function of incident wavelength. In TIRF, fluorescent materials are used, and thus the absorbed energy is partly re-emitted as fluorescent light which is in turn detected.

To collect the fluorescence at a waveguide/liquid interface different signal collection techniques can be employed. Fluorescence emitted at an interface can be detected either conventionally where the detector is placed at right angles to the interface (Figure 2A) or in-line with the primary light beam (Figure 2B) either at the exciting light input or output port. Considering the very small solid angle of emission in in-line detection compared with the emission angle of right-angle geometry detection, the former would not seem to be very efficient. However, there is an enhancement effect, and theory predicts that for a fused silica waveguide with water as the n2 medium, the in-line fluorescence intensity can be 50 times higher than the fluorescence emitted at right-angles to the waveguide. This effect, that fluorescence is tunneled back into the waveguide is verified both theoretically and experimentally (e.g. Lee, et al., 1979; Carniglia, Mandel and Drexhage, 1972).

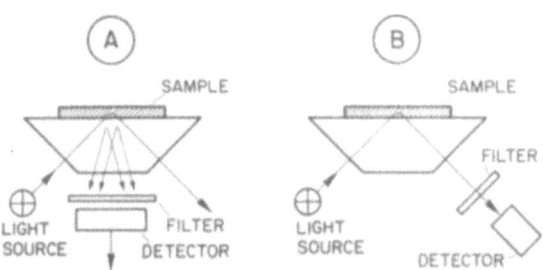

Fig. 2. Schematic representations of a single reflection prism; with detection of (A) right-angle fluorescence; (B) detection of in-line fluorescence.

Theory shows that the fluorescent intensity emission peaks at the critical angle of total internal reflection so that it can be internally reflected. To increase sensitivity, the IRE can be structured to collect fluorescent light from several reflections and guide it to the detector.

This is especially advantageous when an optical fibre is used as an IRE, as fluorescent light emitted at right-angles from an elongated fibre cannot be easily collected onto a detector. In-line detection also avoids measurement of fluorescence through the bulk of the sample solution surrounding the fibre, which otherwise can give significant interference, depending of the fluorescent dye used (Soini and Hemmilä, 1979).

METHODS AND INSTRUMENTATION

There are three major factors involved in the development of a TIRF immunoassay system, the waveguide/IRE, the method of immobilizing antibody to the IRE surface, and the associated instrumentation. The work of the two laboratories summarized herein has a number of common elements. Firstly, the method used for covalent attachment of antibodies is the same, and has been previously described (Lynn M., 1975). Secondly, the IRE material is the same (i.e. quartz) and finally both laboratories choose to measure fluorescence (i.e. TIRF). However, the instrumental approaches and IRE formats reviewed here are different. Two different IRE formats are described, the planar fused quartz microscope slide and the fused quartz optical fibre. Two different optical approaches were also followed. With the microscope slide, fluorescence was measured both perpendicular to the slide (Figure 2A) and at the light output (Figure 2B). With the fibre, fluorescence was measured at the light input (Figure 2B). Furthermore, one instrument is based on conventional fluorescence intensity measurement and the second instrument based on photon counting. Both these systems will be described in the following paragraphs.

Microscope Slide Waveguide System

Selection of the Multiple Reflection Plate. Out of the large variety of different geometries for multiple internal reflection elements we have chosen the simplest and most commonly used. This is a plate having parallel reflecting surfaces. Here the light is propagated down the length of the plate via multiple internal reflections from the opposing reflecting surfaces.

In our experiments we used microscopic slides made from fused silica (Suprasil 1, Heraeus Quarzschmelze GmbH, Germany) as available for fluorescence microscopy. Their size was 75 x 25 cm, thickness 1 mm.

Fused silica is ideal for these experiments because of the following reasons : (i) silica has an excellent optical transmission from the near ultraviolet to the near infrared allowing coverage of the whole spectrum, (ii) the internal reflection angle is maintained in the plate because of its parallel reflecting surfaces, (iii) the microscope slides are available, and polished with high optical quality as required for total internal reflection spectroscopy, (iv) the fluorescence of high quality silica is extremely low. This is necessary for fluorescence experiments, in particular when ultraviolet light is used for excitation, (v) fused silica is outstandingly resistant to solutions used in immunoassay, such as water, salt solutions, and acids, (vi) immobilization techniques on silica (covalent and non-covalent) are known from literature.

The Optical System. The layout of the system is shown in Figure 3. Its basic subunits are discussed in the following sections.

The Monochromator - Light Source. The light source was a Xenon flash lamp (E.G. & G., Salem, Mass., USA) which generates light in the spectral region of 200 to 1100 nm. The in-house developed power supply allows to emit up to 500 flashes per second at 0.2 Joule per flash. The monochromator is equipped with a concave holographic grating. After diffraction, part of the monochromatic light is directed onto a reference detector to measure flash intensity. A large band-pass filter is selected to block off stray light and light from higher diffraction orders. Additional filters can be inserted such as filters with transmission bands for FITC excitation.

The Light Injection System. In order to study the interaction of the evanescent wave with an absorbing/fluorescing thin layer, e.g. the signal as a function of the penetration depth of the evanescent wave, an optical system was constructed which allows varying the angle of incidence. As shown in Figure 3 the waveguide plate is contacted with index matching oil on two quarter round silica prisms which couple the light in and out of the plate. The parallel light beam coming from the monochromator is directed vertically down by mirror M1 and towards the left quarter round coupling prism by mirror M2. A cylindrical lens focuses the light into the centre of the plate close to the sharp prism edge. When the mirror M2 is displaced vertically the reflected beam is displaced parallel, focused into the same plate centre but at a different angle of incidence. The linear position of the mirror M2 was calibrated to allow selection of angles of incidence of 64 degrees (close to the critical angle of 66°) down to 86 degrees.

This design has the following advantages : (i) continuous variation of the angle of incidence is obtained by a simple linear mirror movement without readjustment of the optics, (ii) as the light pass is limited to the central part of the waveguide gaskets can be applied on its periphery to seal the flow cell without any optical adverse effects.

The Fluorescent Light Detection System. Two different light detection systems were installed for alternative use to detect fluorescent light which is emitted from the fluorochromes on or very close to the top surface of the waveguide.

For in-line fluorescence detection a photomultiplier (Hamamatsu R928) measures fluorescence emitted in-line with the exciting radiation. This is the fluorescence which is tunneled back into the waveguide preferably near and larger than the critical angle. It is therefore totally internally reflected and exits at the waveguide ends. A cylindrical lens system L2 images the output light (see Figure 3) onto a ground glass in front of the photomultiplier. The position of the image is independent of the reflection angle. The ground glass distributes the light more uniformly onto the photomultiplier window to minimize signal variations caused by variations of the angle. Light of the exciting wavelength was blocked by a steep edge band-pass filter (Schott) exhibiting low fluorescence, placed in the parallel beam path between the lenses of the lens system L2.

337

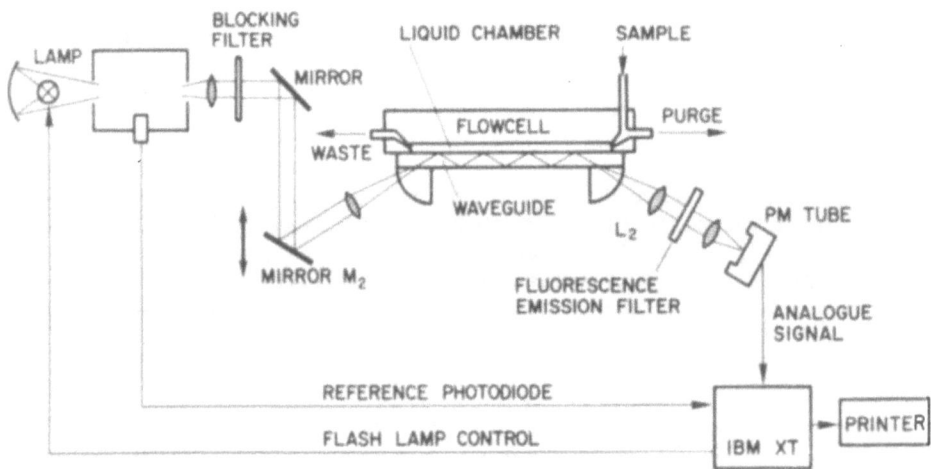

Fig. 3. Diagram of instrumental layout used for measuring immunoassays with a multiple-internal reflection plate.

For right-angle fluorescence detection a photomultiplier was placed below the waveguide to detect fluorescent light emission emitted at right angles out of the waveguide. The fluorescent light emitted close to the upper surface of the waveguide passes through the transparent waveguide and through the free air space onto the detector. A band-pass filter in front of the photomultiplier blocks scattered radiation of the excitation wavelength. This detection geometry is preferably used with a plate waveguide and not with a fibre waveguide as in the latter case the fluorescent radiation would have to pass the analyte solution with the risk of interfering absorption effects by substances in the bulk of the solution.

The Data Acquisition and Processing System. By an in-house preprocessing system the photomultiplier signals from either the in-line or the right-angle position were amplified, integrated during flashing time and converted by a 12-bit A/D converter to a digital format. The entire measurement cycle and the fast signal averaging was performed by an IBM-AT microcomputer. Among others the operator can select from a menu the lamp flash power, the number of flashes per single measurement, the number of measurements and the delay between measurements. Results during each measurement are monitored on the CRT and later on, stored on a disk for print-out and plotting.

Fibre Optic Waveguide System

Selection of the Fibre Optic Material. In the fibre case, light is also propagated down the length via multiple internal reflections which produce a series of meridianal rays passing through the fibre axis and skew rays which do not. As both contribute to the formation of an evanescent wave, the mathematical treatment of fibres is similar to the planar system though geometrically more complex.

The fibres themselves were drawn from optical quality fused quartz (General Electric, Type 214) to a diameter of 500 μm and cut to 7 cm lengths. Optical quality end-faces were produced by diamond blade scribing and a simple snap break technique. When mounted in a flow cell (Figure 4), the active sensing length was 5 cm long while the surface area was 78.5 mm^2, considerably less than the planar configuration. The advantages using such glass fibres are similar to those described for slides.

The Optical Instrumentation. The optical illumination and detection system was assembled from commercial components in the configuration outlined in Figure 5. Illumination was provided by a 100 W tungsten halogen lamp powered by a constant 24 V DC supply. Excitation at 492 nm was obtained by passage of the light through an Instruments SA (Model H-10) monochrometer with 2 nm slits and a 590 nm interference filter with a 10 nm half band width. The pinhole image was refocused on the fibre face with a microscope objective lens to produce a spot size of 240 μm and containing a maximum launch angle corresponding to an input numerical aperture (NA) Of 0.6. Fluorescent light at 525 nm was also collected from the fibre face by the objective and passed through an Instruments SA (Model DH-10) double monochrometer with 1 mm slits before impinging on a cooled Hamamatsu R-928 PMT. Photon counting was performed with an EG & G (Model 1120) amplifier discriminator and (Model 1112) photon counter. Unless otherwise stated, the integration or actual counting interval was

10 seconds whereas all the reported data was reduced to counts per second (cps). The purpose of the above design was to produce a flexible research oriented instrument for fibre optic sensors.

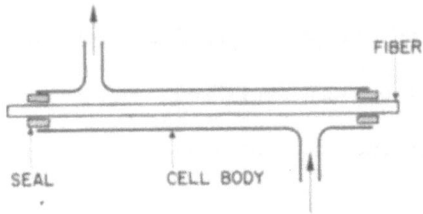

Fig. 4. Fibre optic flow cell sensor. The flow cell was composed of a length of glass tubing (4 mm ID) with sample inlet and outlet ports 180° apart and perpendicular to the tube axis. Fibres of 500 μm diameter were inserted through the watertight seal material.

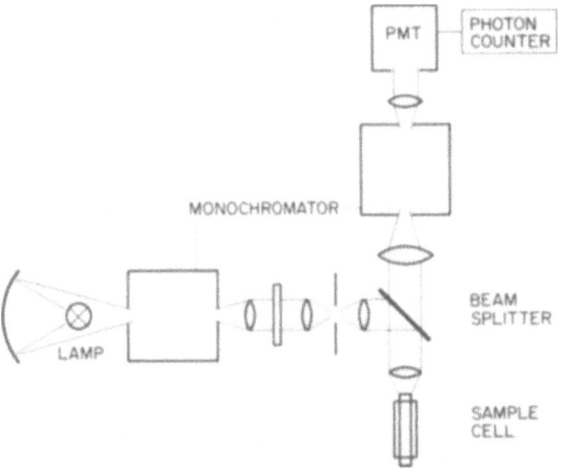

Fig. 5. Optical system for fibre sensors. For a description of components see text.

RESULTS

Planar Waveguide

A standardized two-site immunofluorometric assay procedure was used where the antibody coupled waveguide was first incubated with a standard antigen solution for 15 minutes. After the antigen was washed out of the system using assay buffer an alternative specific antibody labelled with FITC was allowed to incubate with the immunologically immobilized antigen and the reaction monitored. In some instances unbound FITC-protein was then removed by washing with assay buffer and the bound signal measured. Similarly, antibody-coupled waveguides could be reused by washing the surface with dilute (0.01 N) HCL which disrupted the immunological bond.

A schematic of the typical evolution of signal is given (Figure 6) along with an explanation of the various components.

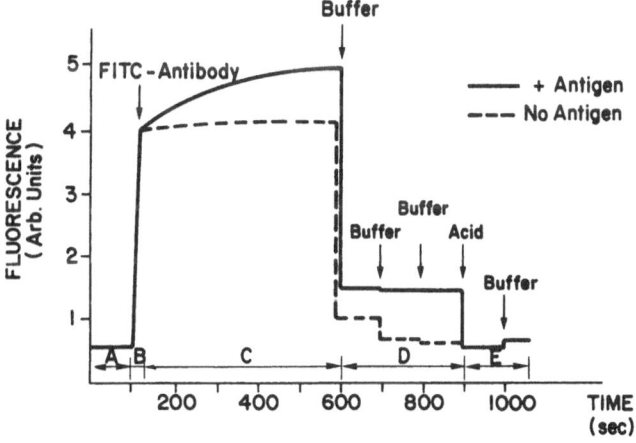

Fig. 6. Binding curves and a schematic representation of a two-site immunofluorometric assay monitored by TIRF. Here the IRE surface is coated with antibody raised against human IgG. Following reaction of the coated waveguide with a standard IgG solution (10 µg/ml) a second FITC-labelled anti-IgG is allowed to react with the immobilized antigen and the reaction monitored by measuring fluorescence with an in-line geometry. (A) The prepared waveguide; (B) Signal generated by unbound FITC-anti-IgG within the dp of the evanescent wave; (C) Binding curves of FITC-anti-IgG in the presence (solid line) and absence (dotted line) of antigen; (D) Washing away unbound materials leaving the specific (solid line) and non-specifically (dotted line) bound signals; (E) Disrupting the immunologically bound materials with dilute acid.

With the planar waveguide two methods of detecting fluorescence were used; detection at right-angles to the long axis of the waveguide (perpendicular fluorescence Figure 2A) and detection of fluorescent light which after being tunneled back into the waveguide exits at the excitation light output/input (in-line fluorescence Figure 2B).

Detection at right-angles was originally advocated by Kronick and Little (1973), and we have reported data using this system elsewhere (Sutherland et al., 1984[a]). The following data is based on in-line detection solely.

Both biochemical and optical factors were investigated. The biochemical factors included optimization of the assay buffer conditions to minimize the non-specific component of signal generation yet maximize specific signal (Sutherland et al., 1984[a]). Optical factors included choice of optimal filters particularly important for in-line detection system, especially with labels where the Stokes shift between excitation and emission is small (e.g. FITC 30 nm) with a considerable overlap of the excitation and emission spectra. This will be discussed in a future publication. A second major factor from the optical standpoint is the choice of nominal incident angle of excitation light. This will be discussed in the following section.

Choice of Incident Angle

In the theoretical section it was described how the penetration depth of the evanescent wave is a function of the three factors; refractive index ratio; incident angle θ and wavelength of incident light. With the planar waveguide system the refractive index ratio was fixed by the choice of waveguide material and the use of an aqueous sample solution. The choice of wavelength was fixed by the choice of fluorochrome (FITC, $\lambda ex = 490$ nm). The third factor, incident angle was a variable which could be investigated. It is also a simpler task to have a variable angle device using the planar waveguide than with an optical fibre, purely due to the geometric factors.

For these experiments the nominal angle of incident light (θn) was varied by using the apparatus illustrated in Figure 2. Here by simply adjusting the height of the mirror (M2) θn could be varied from 64° to 86°. In order to achieve a higher light throughput, the light beam was convergent with an angular aperture of $\pm 2^o$ around the nominal angle of incidence focused to the centre of the waveguide. With the above system the critical angle θc was calculated as 66°.

To test the effect of θn on a two-site immunofluorometric system a standard procedure was used with human IgG and human serum albumin as antigens. At various points in the reaction illustrated in Figure 6, effects of changing θn from 64° to 86° were investigated. Measurements were carried out of A) background, the thin film of bound fluorescent materials before C) and after D) washing away non-bound fluorescence. The results for IgG (Figure 7A) and HSA (Figure 7B) are illustrated. The upper part of the figures represent the raw data and the lower figure is the data normalized for background (i.e. divided by background). It can be seen in both cases that the most sensitive measurement (i.e. higher ratio of signal to background) of the thin film is when θ is close to θc. This is probably due to two factors. Firstly, at smaller angles there is an increased number of reflections and this can increase opportunity to detect the surface reaction. Secondly, at the angle close to θc the dp

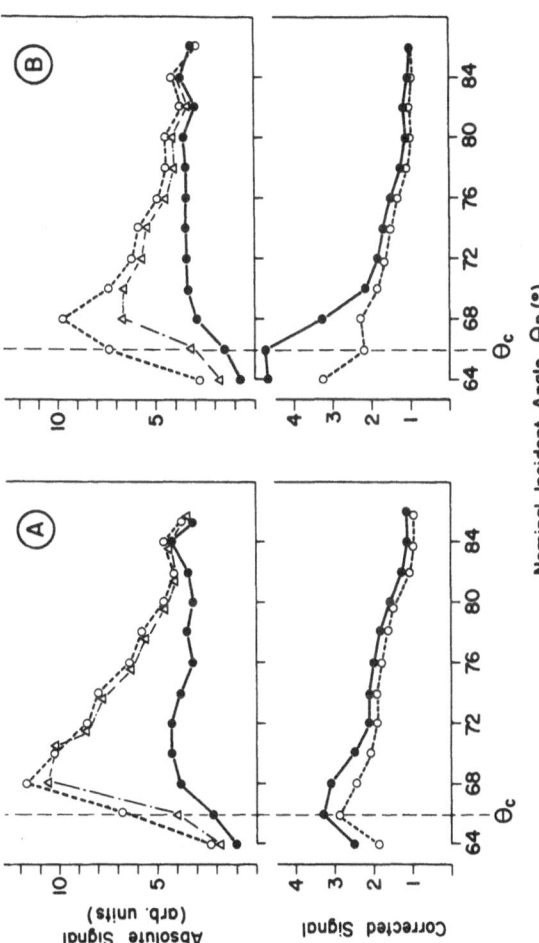

Fig. 7. Effect of nominal incident angle on TIRF signal generated by FITC-antibody immunologically immobilized to a planar waveguide surface for two antigens A) IgG and B) human serum albumin. Data is expressed as absolute signal (upper figures) and corrected for background (lower figures). Upper curves are; background (●), after immunological reaction before (○) and after (△) washing. Lower curves are; after immunological reaction before (●) and after (○) washing.

is greater and also the field strength more intense at the surface, increasing the possibility of sensing the surface reaction. However, although the thin film is monitored with greater sensitivity, there is also a concomitant increase in monitoring of the non-surface component, due purely to the increased dp. One possibility to reduce this effect is to use more dilute concentration of FITC-Ab. If the kinetics can be maintained this should not result in a loss of sensitivity. This is shown in the section of fibre optic detection.

To illustrate the effect of θ on a kinetic reaction, the reaction of FITC-anti-IgG with immunologically immobilized IgG was detected using θn at 67°, 68°, 70°, 72° and 74°. Each experiment was run with its concomitant zero antigen curve. The zero antigen curve was subtracted from the 10 μmg/mL IgG standard curve and the final data plots compared (Figure 8). It can be seen that the previous results are confirmed where operation close to, but not at, θc gives the most sensitive system. In this fashion, it was decided to use $\theta = 68^{\circ}$ for all further experiments.

Dose-Response Curves

Dose-response curves for the two antigens HSA and hIgG were constructed in duplicate by following the above protocol. One set of binding curves for the hIgG assay are illustrated in Figure 9 and plots of the absolute change in fluorescent signal, following a 10 min incubation are given in Figure 10. The S.D. at 0 concentrations was $\pm$ 4 units and $\pm$ 2 units for the hIgG and HSA assays respectively. An estimate of minimum detection limit can be made by taking three times the S.D. at zero antigen and dividing them by the slope of the curve at zero antigen (Ekins, 1981). The minimum detection limits were estimated as 3 mg/L Ig and 5 mg/L HSA.

Fibre Optic Waveguide

Examples of the data recorded with a fibre optic system are presented in Table 2 for a two-site immunofluorimetric assay involving the clinical analyte ferritin (a high molecular weight iron storage protein important in diagnosing certain anemias). Similar to the description of Figure 6, the addition of fluoresceinated antibody brought about an immediate signal increase. Comparison of the ferritin treated and control fibres shows that this was, in part, due to the high concentration of fluoresceinated antibody which resided near the surface and sensed in the unbound state. The immunologically active ferritin fibre displayed an even larger initial signal and a rapid rate of increase which virtually ceased after 30 minutes. In each case removal of label by a washing step caused a signal drop roughly equivalent to the amount of fluoresceinated material originally added. The remaining signal thus constituted a measure of the specifically bound and non-specifically adsorbed antibody in the sample and control fibres respectively.

When the observed fluorescent signals were corrected for background interference from unbound label, the kinetic pattern illustrated in Figure 11 emerged. As expected, the accumulation rate of labelled antibody appeared to increase in proportion to the amount of ferritin initially present during the pre-incubation period. Although one could utilize these differential rate measurements for constructing a standard curve, the curve in Figure 12 was alternatively derived by subtraction of the non-specific binding component from the ferritin sample values after 30 minute label incubation and wash steps. Note, there is a

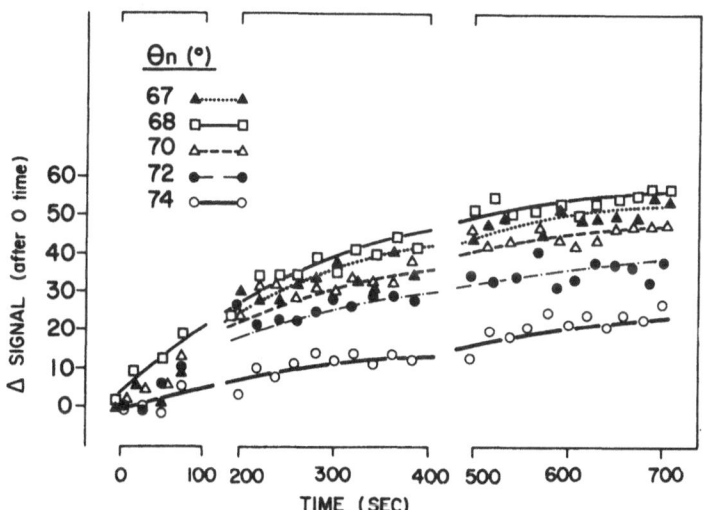

Fig. 8. Effect on nominal incident angle (θn) of excitation light on TIRF
signal obtained from the immunological binding of FITC-anti-IgG
to a planar waveguide surface.

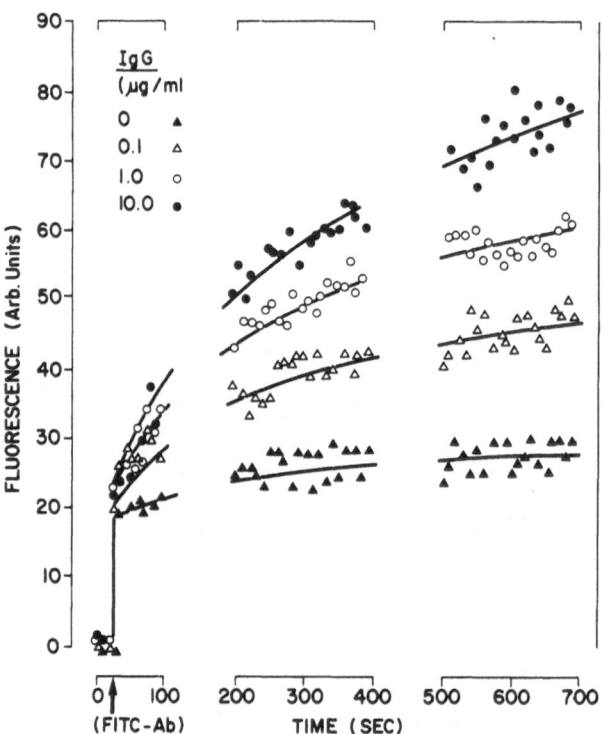

Fig. 9. Two-Site Immunofluoremetric Assay for Human IgG. Kinetic
binding curves of FITC-anti-IgG monitored by TIRF at a
planar waveguide/liquid interface.

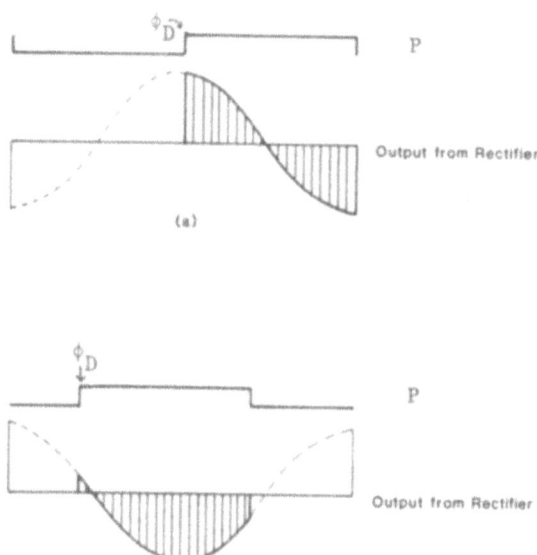

Fig. 10. Dose-Response Curves for the Estimation of Human IgG and Human Serum Albumin, by Two-Site Immunofluorometric Assay Monitored by TIRF, Using a Planar Waveguide. Response is expressed as the absolute change in signal following a 10 minute incubation with FITC-antibody.

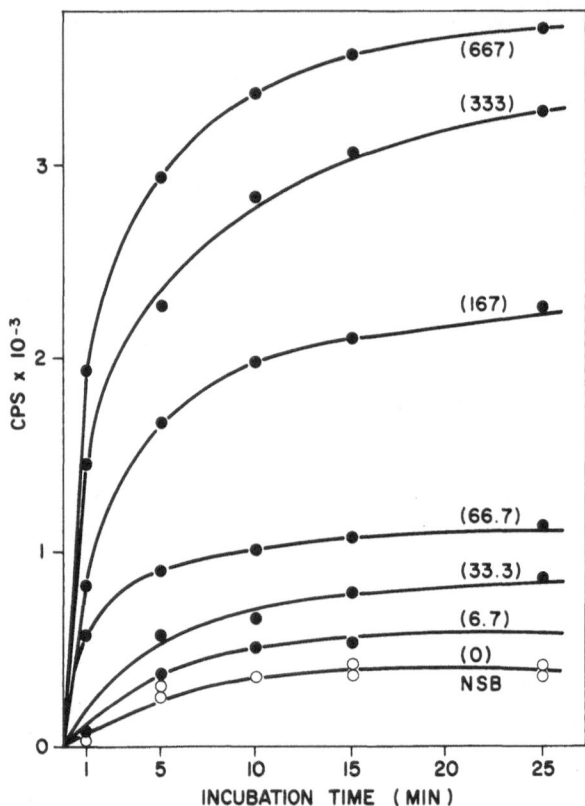

Fig. 11. Kinetics of Labelled Secondary Antibody Accumulation on
Ferritin Treated Fibres. Numbers in parenthesis give
the concentration of ferritin present in ng/ml in the
initial pre-incubation mixture. All reactions were
conducted in a 0.2 % BSA-PBS buffer at pH 7.4. Values
for plots were calculated by a subtraction of the
background due to fluoresceinated antibody (i.e.
fluorescein equivalent to 2.4×10^{-6}M or 1950 cps).

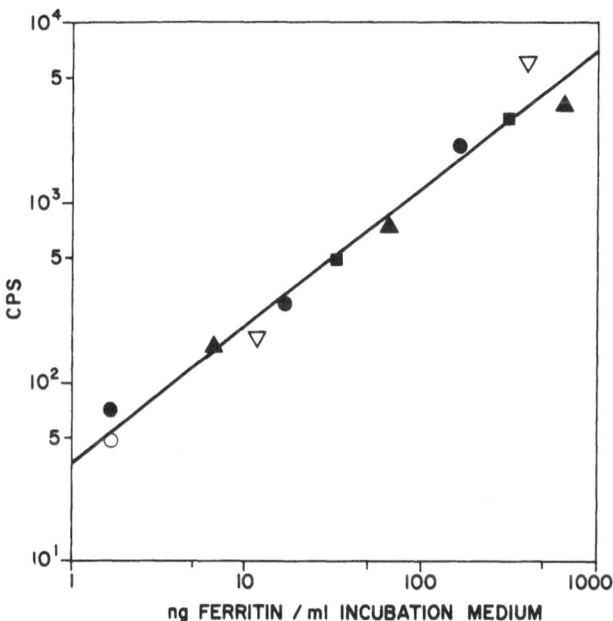

Fig. 12. Ferritin Standard Curve. Values were derived from a two-
site immunofluorimetric assay as described in Table 2.
Note, all reactions were performed in 0.2 % BSA-PBS buffer.
Fibres were incubated with ferritin samples for 15 minutes,
washed, then incubated with fluorescein labelled
antiferritin antibody for 30 minutes. The label was removed
by a wash step and the specific signal determined after
correction for non-specific binding (NSB) on a control
fibre. Different experiments are represented by the
symbols.

reasonably good correlation between the ferritin level and signal over the clinically relevant range between 1 and 1000 ng/ml when assayed in a 0.2 % BSA-PBS buffer.

Table 2. Experimental Data for Ferritin Immunoassay

Event	Specific and Non-Specific	Non-Specific
Pre-incubated with	Ferritin (0.666 µg/mL)	0.2 % BSA
Dark current	8*	9
(A) Illuminated	273	305
Incubation with FITC-Ab		
1 min	4470	1953
10 min	5905	2350
15 min	6120	2480
30 min	6245	2490
Washed with 0.2 % BSA	4000	727
Correction for (A)	3727	422

*all data expressed as cps.

To demonstrate concordance with present radioimmunological practice, both RIA and TIRF techniques were performed under similar assay conditions using fibres as the solid phase support material. For comparison, the amounts of bound secondary labelled antibody have been calculated and plotted in Figure 13, for each of the two methods. The values obtained by RIA were computed from the specific activity of the tracer and the counts per minute bound. In the TIRF case, the values were computed by comparison of the fluorescent signals with those of free fluorescein solutions and division by the fluor to Ab ratio. This was followed by a conversion to mole equivalents present within the small cylindrical space interrogated by the evanescent wave. Given some uncertainty in comparisons between free fluorescein values and those associated with an antibody in the bound state, there is a reasonable equivalence between the methods. This should not be surprising since both techniques rely upon the same fundamental diffusion mechanism for concentrating secondary labelled reagent on the fibre surface.

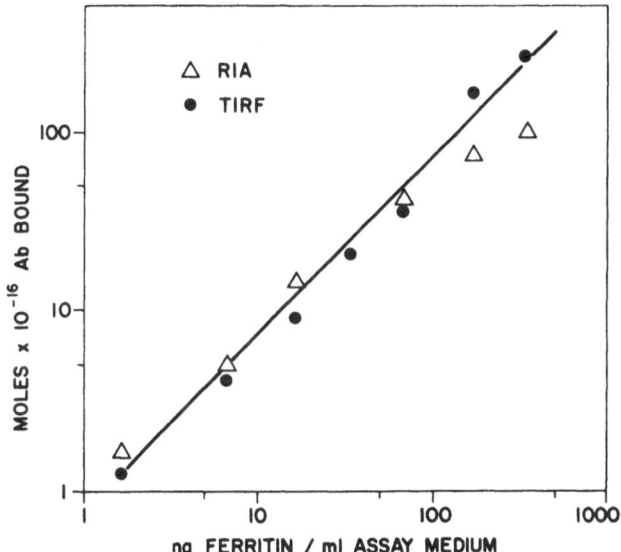

Fig. 13. Comparison of RIA and TIRF Assays or Fibres. Radioimmunoassays
(RIA) for ferritin were conducted in a manner analogours to those
in Figure D except that I^{125} labelled secondary antibody was used
in place of fluoresceinated antibody. Total Internal Reflected
Fluorescence (TIRF) assays were performed as in Figure 12.

Ideally one would wish to drive the specific immunological reaction toward equilibrium as rapidly as possible to achieve complete analyte or label loading for maximum signal detection. However, this may pose a distinct problem in a single step assay when unbound fluorescent tracer is not removed prior to making a measurement. The advantage of a direct readout is lost if the concentration of fluoresceinated material is itself sufficiently high enough to be largely sensed within the evanescent wave zone but without a specific immunological reaction as in Figure 6 and Table 2. Low-end detection limits for antigen are compromised when attempting to measure small signals on a large background. Consequently, the following experiments were conducted in the presence of dilute concentrations (i.e. 1×10^{-9}M) of fluoresceinated conjugate to insure that unbound material was not, or at most only weakly, detectable at the fibre surface. Specifically, this level of fluorescein was observed to produce only 5 – 10 cps in the system described here.

A second noteworthy difference is that the assay architecture was changed to a competitive format to illustrate the applicability to a low molecular weight analyte such as digoxin (a cardiac glycoside used to manage arrhythmias). For demonstration purposes, antidigoxin antibody coupled fibres were chosen for incubation in human serum samples doped with the indicated digoxin levels (in ng/ml) and premixed with a small volume of a fluorescein-digoxin conjugate. Figure 14, thus displays the kinetic responses obtained directly when an assay was conducted in essentially pure serum. A series of discrete binding curves was observed to bear an inverse relationship to the amount of digoxin present.

Plots of the relative proportion of bound label are presented in Figure 15 for the indicated incubation times. Clearly, a standard digoxin curve was achievable within short time intervals. More importantly, the assay could be performed as a single step using a real time readout without the requirement for an additional washing step as evidenced by the dashed curve in Figure 15.

DISCUSSION

The original work using TIRF to measure immunological reactions was carried out by Kronick and Little (1973, 1975, 1976). Haptens such as phenylarsonic acid or morphine were immobilized to the surface of a quartz microscope slide via a hapten-albumin conjugate. FITC-labelled antibody which bound to the immobilized hapten could be detected by exciting fluorescence at the surface using the evanescent wave. On addition of free hapten to the bulk of solution the binding rate of the FITC-antibody to the surface was reduced in a concentration-dependent fashion. Monitoring the fluorescence at right-angles, a minimum detection limit of 0.2 µmol/L of morphine could be measured.

Thompson, Burghardt and Axelrod (1981) used a similar approach combined with fluorescence correlation spectroscopy. Here, dinitrophenol was immobilized to a quartz slide via adsorbed albumin. Both monovalent and bivalent rhodamine-labelled antibodies were reacted with the surface-bound hapten. One of the main findings associated with these experiments was the large proportion of non-specifically bound antibody (up to 60 %

of the signal). One conclusion by these authors was that the high level
of non-specific binding prevented continuous measurement of the kinetics
of binding using this system. Intrinsic fluorescence was also suggested
as a method to detect immunoassays at interfaces (Andrade and Van Wagenen,
1983; Andrade, Van Wagenen et al., 1985). Here the intrinsic fluorescence
of tryptophan and tyrosamine moieties is measured when the IgG molecules
are immunologically bound to the surface (λex = 284 nm, λem = 335 nm).

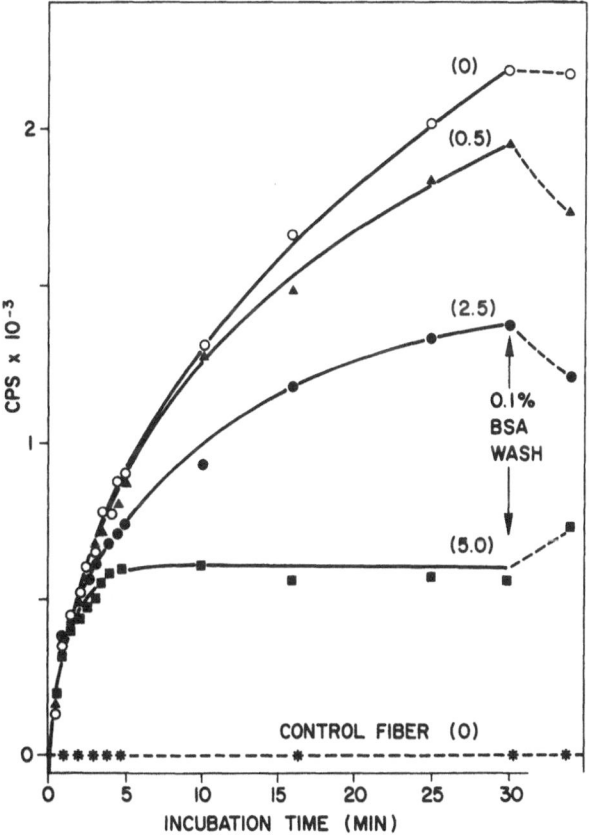

Fig. 14. Kinetics of a Competitive Assay for Digoxin in Serum.
Serum samples contained the values indicated in ng/ml
digoxin in parenthesis and 1 x 10^{-9}M fluorescein-
digoxin conjugate. The samples were introduced to the
anti-digoxin antibody coupled fibres at time t = 0 min.
The control fibre was coupled with antiferritin
antibodies.

These authors all used right-angle detection of fluorescence. Systems to measure in-line fluorescence for immunoassays have been described previously by the Battelle Group (e.g. Sutherland et al., 1984[b]) and others (e.g. Block and Hirschfeld, 1984; Hirschfeld and Block, 1984). What is presented here is an extension of this work giving a more detailed description of the possible approaches and an interesting comparison between two systems developed in different laboratories.

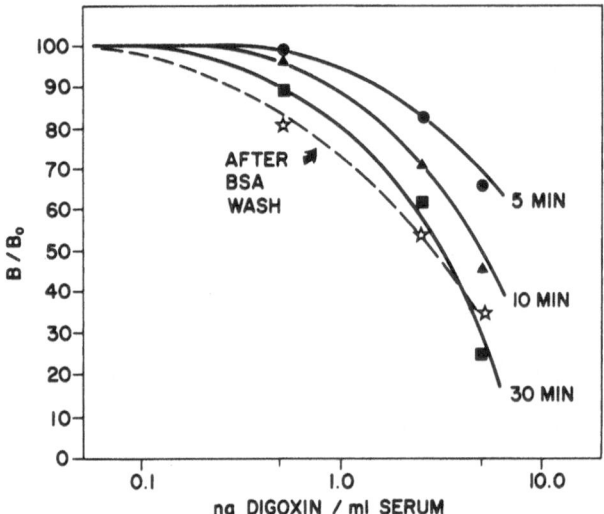

Fig. 15. Digoxin Standard Curve. Values were computed as the fraction of label bound relative to that obtained with zero free digoxin present. Curves were derived from the measurements performed after 5, 10 and 30 minutes as in Figure 14. The dashed curve represents the results obtained after a 0.1 % BSA wash to remove both unbound label and the serum sample.

There are certain inherent advantages and disadvantages in both systems. The fibre optic form of waveguide has a number of advantages in terms of sensitivity due to the thinness of the fibre (i.e. large number of reflections). Used with the appropriate optics, nearly the entire waveguide surface is used for sensing. Because of the cylindrical symmetry of the fibre, evanescent fluorescent light is guided and collected down the waveguide in two dimensions, whereas in the case of the planar waveguide some fluorescent light is lost as it guides and collects light in one dimension only. However, small pieces of thin fibre may present handling and manufacturing problems. There is probably also a significant optical problem to be addressed in accurately positioning and handling the fibre reproducibly.

The planar waveguide can be readily handled and can be manufactured as a single piece, including all the optical and support structures using injection moulding techniques. This allows large numbers of waveguides to be produced at low cost without difficult assembly steps. With the planar waveguide there is also a large surface area for sensing, although only one surface is currently used limiting the number of possible interactions (reflections) of the light beam, unlike the fibre.

With quartz as the waveguide material the chemical covalent attachment of antibodies is a relatively straightforward procedure using current well proven techniques. However, changing to a plastic, as suggested above, may introduce a new variable in finding a covalent attachment procedure which works well on plastics (e.g. as polystyrene or polymethylmethacrylate).

A further difference between the two systems reviewed above is the optical layout and detection means. With both systems in-line detection of the tunneled fluorescence was carried out. However, with the planar waveguide fluorescent light was detected at the output (i.e. where the excitation beam was coupled out of the waveguide (Figure 3)) while with the fibre the fluorescent light was detected at the waveguide input (Figure 5). The latter system involves a more complicated optical beam splitter and monochrometer arrangement to separate excitation light from the fluorescent signal. However, in principle, a dichroic beam splitter with appropriate blocking filters would serve the same function.

A final difference between the two systems was the signal processing arrangement. With the fibre optic, photon counting was used. This system is less susceptible to drifts than the analogue signal treatment, but requires a higher degree of light segragation since the detector is intrinsically sensitive.

Overall, the key factor on which the evanescent wave immunoassay technique should be judged is the ability to monitor surface reactions without major interferences from the bulk of solution components i.e. a homogeneous, optical immunoassay. This factor and the potential simplicity of a developed commercialized test should mean that these EVIA systems could be operated in a relatively non-skilled environment. Also to achieve this potential requires design of a simple photometric system in which to carry out measurements. As yet such a system is not readily available but is being developed at both laboratories.

ACKNOWLEDGEMENTS

We wish to acknowledge all the scientific and technical staff and both laboratories who donated their expertise which give rise to the above data. We also wish to thank Ms. J. Boëque for typing the manuscript, and Prutec Ltd., who supported financially some of the work at the Geneva Laboratories.

REFERENCES

Anderson, J. R., Lindsell, W. D. and Mitchell, F. M. 1980, Chemical pathology on the ward, Lancet, 1:487.

Andrade, J. D. and Van Wagenen, R. A., 1983. U.S. patent No. 4,368,047. Process for conducting fluorescence immunoassays without added labels and employing attenuated internal reflection.

Andrade, J. D., Van Wagenen, R. A., Gregonis, D. E., Newby, K. and Lin, J. N., 1985, Remote fibre-optic biosensors based on evanescent-excited fluoro-immunoassay : concept and progress, IEEE Trans., ED-32:1175 - 1179.

Block, M. and Hirschfeld, T. B., 1984, World patent application No. WO 84/00817. Immunoassay apparatus and method.

Carniglia, C. K., Mandel, L. and Drexhage, H., 1972, Absorption and emission of evanescent photons, J. Opt. Soc. Amer., 62:479 -486.

Ekins, R., 1981, Misunderstanding about assay sensitivity and precision, Ligand Q., 4:60 - 65.

Free, A. H. and Free, H. M., 1984, Self-testing, an emerging component of clinical chemistry, Clin. Chem., 30:829 - 838.

Harrick, N. J., 1967, in: "Internal reflection spectroscopy", Interscience, New York, 327 pp.

Harrick, N. J. and Loeb, G. I., 1973, Multiple internal reflection spectrometry, Anal. Chem., 45: 687 - 691.

Hicks, J. M., 1985, In-situ monitoring, Clin. Chem., 31:1931 - 1935.

Hirschfeld, T. B. and Block, M., 1984. U.S. patent No. 4,447,546. Fluorescent immunoassay employing optical fibre in capillary tube.

Kronick, M. N. and Little, W. A., 1973, A new fluorescent immunoassay, Bull. Amer. Phys. Soc., 18:782.

Kronick, M. N. and Little, W. A., 1975, A new immunoassay based on fluorescence excitation by internal reflection spectroscopy. J. Immunol. Methods, 8:235 - 242.

Kronick, M. N. and Little, W. A., 1976. U.S. patent No. 3,939,350.

Lee, E. H., Benner, R. E., Fenn, J. B. and Chang, R. K., 1979, Angular distribution of fluorescence from liquids and monodispersed spheres by evanescent wave excitation, App. Opt., 18:862 - 870.

Lynn, M., 1975, Inorganic support intermediates : covalent coupling of enzymes on inorganic supports, in: "Immobilized enzyme, antigens, antibodies and peptides", H. H. Weetall, ed., Marcel Decker Inc., New York, pp 1 - 48.

Marks, V. and Alberti, K. G. M. M., 1985, in: "Clinical biochemistry nearer the patient", Churchill Livingstone, Edinburgh, 240 pp.

Mitchell, F. L., 1979, The trend towards devolution in clinical chemistry. J. Auto. Chem., 1:179 - 181.

Place, J. F., Sutherland, R. M. and Dähne, C., 1985, Opto-electronic immunosensors; a review of optical immunoassays at continuous surfaces, Biosensors, 1:321 - 353.

Rinsler, M. G., 1977, Chemical pathology services, Lancet, 1:946 - 947.

Soini, E. and Hemmilä, 1, 1979, Fluoroimmunoassay : present status and key problems, Clin. Chem., 25:353 - 361.

Sutherland, R. M., Dähne, C. and Place, J. F., 1984, Preliminary results obtained with a no-label, homogeneous, optical immunoassay for human immunoglobulin G, Anal. Lett., 17(B1):43 - 55.

Sutherland, R.M., Dähne, C., Place, J. F. and Ringrose, A. R., 1984[a], Immunoassays at a quartz-liquid interface : Theory, instrumentation and preliminary application to the fluorescent immunoassay of human immunoglobulin G, J. Immunol. Methods., 74:253 - 265.

Sutherland, R. M., Dähne, C., Place, J. F. and Ringrose, A. R., 1984[b], Optical detection of antibody-antigen reactions at a glass liquid interface, Clin. Chem., 30:1533 - 1538.

Sutherland, R. M. and Dähne, C., 1986, IRS devices for optical immunoassays, in: "Biosensors : Fundamentals and Applications", A.P.F. Turner, I. Karube and G. S. Wilson, eds., Oxford University Press, Oxford, in press.

Thompson, N. L., Burghardt, T. P. and Axelrod, D., 1981, Measuring surface dynamics of biomolecules by total internal reflection fluorescence with photobleaching recovery and correlation spectroscopy, Biophys. J., 33:435 - 454.

Watson, D., 1980, Analytical investigations closer to the patient, Brit. Med. J., 281:31 - 35.

Weiner, K., 1980, Pathology measurements closer to the patient, J. Clin. Path., 33:857 - 863.

SECTION IV

MEMBRANE IMMUNOASSAY

SPIN MEMBRANE IMMUNO ASSAY IN SEROLOGY

A. I. Vistnes

Department of Physics
University of Oslo
Blindern, 0316 Oslo 3
Norway

In 1964 artificial membranes were prepared and described for the very first time by the English scientists Bangham and Horne (1964). These membranes formed spontaneously when phospholipids were put into an aqueous solution. The model membranes were onion-like structures where the different layers of the "onions" consist of lipid bilayers. Between the layers aqueous spaces were entrapped. These model membranes are called multilammellar liposomes, or in short liposomes.

Since Bangham's discovery model membranes have been utilized extensively for everything from general studies of membrane characteristics to attempts to use liposomes as a carrier e.g. for drugs in cancer treatment. A large number of papers have dealt with new methods for preparing these model membranes to get more homogenous or single layered systems compared to Bangham's original liposomes.

In this chapter we deal with model membranes used in immuno assays. In particular we will discuss systems where complement, in conjunction with antigens and antibodies, is utilized to damage the membrane. Since membranes are an important constituent in this kind of test, it is often called a membrane immuno assay (MIA).

The basic principle in membrane immuno assay is illustrated in Figure 1 and can be explained as follows. Model membranes are made in the presence of a membrane impermeable, water soluble, marker. The marker will be present in the aqueous space inside as well as outside the liposomes. The marker on the outside is washed away. An antigen is incorporated into the model membrane. This is most often done by mixing antigen molecules with the lipid prior to the preparation of the liposomes. The antigen molecules are amphiphatic, that are either antigens or haptens themselves or have an antigen covalently bound to the hydrophobic part of the molecules. The resulting system is often called sensitized liposomes. If antibodies directed toward the antigen are added, they become bound to the antigens. When complement is added, incubation of the structures leads to complement lysis via the classical pathway in which trapped marker is released.

The presence of released marker can be detected in various ways depending on which marker is used. For spin membrane immuno assay release of marker can easily be observed by ESR. Trapped marker in high local concentration gives a spectrum with three broad, low-amplitude lines because of the exchange broadening effect. Spin labels that have escaped the vesicles experience a low local concentration, and their spectrum consist of three intense, narrow lines. Simple amplitude measurements yield information of the degree of lysis in "real time" (as it happens). The whole procedure is carried out without any separation process.

Marker will only be released if all reagents : antigens, antibodies and complement are present; so in principle the test can be utilized for detection of each of these components. The test is quite sensitive since in

THE SMIA TECHNIQUE

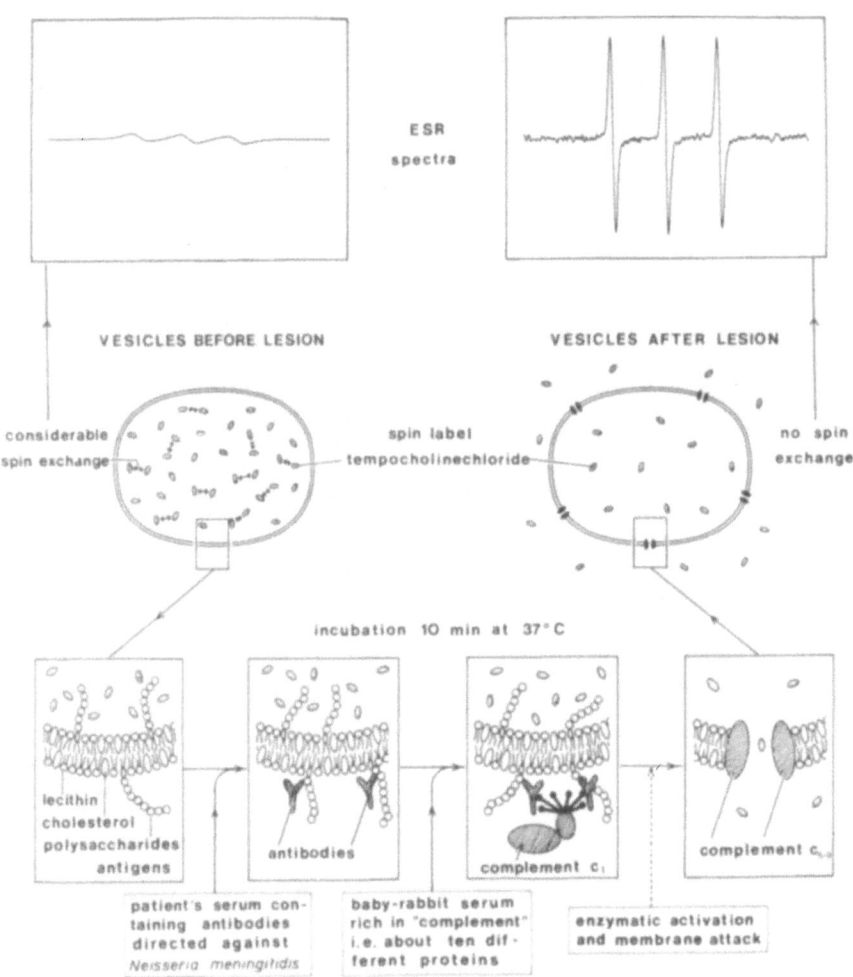

Fig. 1.   Basic principle for the membrane immuno assay is the utilization of complement mediated lysis of vesicles filled with marker molecules. If the marker is a spin label, the method is called spin membrane immuno assay (SMIA), and lysis can be detected without separation or any time-delay by measuring increase in amplitude in the electron spin resonance (ESR) spectrum.

principle, one single hole in a liposomes can result in thousands of released marker molecules. We can talk about a multiplicative effect.

Membrane immuno assays were first performed a few years after Bangham's discovery. S. C. Kinsky and his coworkers worked out the basics in this kind of immuno-assay (Kinsky et al., 1969). Later, other groups have further exploited the system by introducing new marker molecules, testing more antigens, and working out new ways to treat the data in order to obtain as much information as possible. Different aspects of the membrane immuno assay has been reviewed in papers by Kinsky (1972), Kinsky and Nicolotti (1977), Hsia and Tan (1978), Blaedel and Boguslaski (1978), Esser (1980), and Schall and Tenoso (1981).

In spite of hard work in more than a decade, no membrane immuno assay has been evaluated promising enough to be introduced in large scale immuno assays for routine serological measurements. This may indicate that MIA have intrinsic problems that make it less popular than other immuno assays. Later in this chapter reasons for the lack of popularity in routine measurements will be discussed, and speculations of the method's future will be put forward.

Even if MIA is not used in routine immuno assays so far, the technique has been widely used for research purposes. The reaction involved in membrane immuno assay mimics the process that takes place in real systems, probably more than any other immuno assay does. Thus, the MIA has been used in studies of basic immunological questions as :

- What kind of requirement do we have for the antigen concentration in the membrane in order to get complement fixation/lysis ?

- How will the membrane fluidity (lateral diffusion) influence the complement lysis ?

- How will antigens' exposure to antibody binding be influenced by changes in membrane conditions ?

- What kind of time dependence is involved in antibody binding/complement lysis ?

Furthermore, since many antigens normally are confined to membranous environments, MIA is a most natural way to perform an immuno assay for some systems. As an example, test for anti-lipid antibodies is probably studied more directly by MIA than by any other technique.

So far we have talked about membrane immuno assay as if only one class exists. This is done on purpose since the present author really thinks that this is the case. However, depending on the type of marker, it is possible to divide the class into subclasses. If the marker is a spin label (nitroxide radical) and the measurements are carried out using an Electron Spin Resonance (ESR) spectrometer, the subclass is called Spin Membrane Immuno Assay (SMIA). The real details in this chapter will deal with this subclass, and Hsia's chapter is also devoted to this technique. Yasuda will discuss the subclass that results when fluorescent markers are used. However, in spite of various properties of the different markers the key properties for all subclasses are all the same.

In the following one particular SMIA application will be described: the assay for anti meningitides polysaccharide antibodies. The full set of information is given by Vistnes et al. (1983), but brief versions of materials and methods and results are included here for completeness. It is

meant to be background information for a more extensive examination of the
various ingredients and procedures used in MIA in general. This "discussion" section makes up the main bulk of this chapter. It is also supposed
to be the most useful part since it introduces the reader to the kind of
judgments that have to be carried out in order to optimize membrane immuno
assays for a particular situation. It is also the kind of information that
normally is left out from ordinary papers. Suggestions for improvements in
the future is also included in that part.

MATERIALS AND METHODS

Lipids. Egg yolk phospatidylcholine was obtained from Avanti Biochemi-
cals Inc., Birmingham, Ala., and cholesterol and L-α-dipalmitoyl phosphatidic acid were from Sigma Chemical Co.,St. Louis, Mo.

Complement. As a source for complement serum from 4-week-old rabbits
was used. The serum was shown to be nonbactericidal in bactericidal activity test using meningitides meningococci. It was stored in small aliquots
at $-25^{\circ}C$ for maximum 3 months and was thawed just prior to use.

Antisera. Human serum, taken three weeks after immunization with a Men
A+C capsular polysaccharide vaccine, was used as reference serum. This
serum contained 2 µg of anti-A antibodies per ml and 21 µg of anti-C antibodies per ml, as determined by Helena Käyhty, Central Public Health Laboratory, Helsinki, Finland, who kindly supported us with this serum.
    As a reference for the group B system, mouse ascites fluid containing
monoclonal antibodies in the IgM class directed against the B capsular
polysaccharide from meningococci was used. This serum was a gift from W. D.
Zollinger, Walter Reed Army Institute of Research, Washington, D.C., and
contained about 5 mg/ml IgM as estimated from agarose gel electrophoresis.
    Meningococcal agglutinating sera purchased from Wellcome Reagents
Ltd., Beckenham, England, were used as group-specific antisera for this
work. The sera contained rabbit antibodies against N. meningitides group A,
B, or C, which had been absorbed as necessary to make them group specific
in agglutination tests.
    "Clinical samples" were obtained from two sets of vaccination pro-
grams. In the first one sera were taken from five adults 0, 4, and 20 weeks
after vaccination with Merieux A+C vaccine. In the other program sera from
five volunteers were taken at 0 and 2 weeks after vaccination with a
combined capsular B polysaccharide and type 2 outer membrane protein
vaccine. Further details on the vaccinations are given in the original work
(Vistnes et al., 1983).
    All vaccination sera were heat inactivated at $56^{\circ}C$ for 30 min before
use.

Polysaccharides / antigens. As antigens capsular polysaccharides from
N. meningitides were used. Group A polysaccharides were taken from the
corresponding vaccine from Institut Merieux, Lyon, France, and a mixture of
group A and C polysaccharide were similarly from a Men A+C vaccine from
Connaught Laboratories Inc., Cherry Hill, N.J. Polysaccharides of group B
and C (separated) were gifts from W. A. Hankins, Connaught Laboratories,
Swiftwater, Pa. The first two sets of polysaccharides were stabilized with
lactose, while the two last ones were stored at $-25^{\circ}C$ but not stabilized in
any other way.

Preparation of vesicles. Large oligo-lamellar vesicles were made by
reverse-phase evaporation essentially as described by Szoka and Papahadjopoulos (1978). The composition of lipids was phosphatidylcholine-choles-
terol-dipalmitoylphosphatidic acid in the mole fractions 60:35:5. Typically
10 µmole lipid and 1 ml aqueous phase were used. The aqueous phase consisted

of 400 µl of 150 mM spin label tempocholine chloride, prepared as described
by Kornberg and McConnell (1971), 400 µl of iso-osmolar veronal buffer, and
200 µl of water. The iso-osmolar buffer was made out of 145 mM NaCl, 3.1 mM
diemal, 0.97 mM sodium diemal, 0.83 mM $MgCl_2$ and 0.19 mM $CaCl_2$, and the pH
was 7.2. A hyperosmolar veronal buffer was made the same way but all
concentrations were 1.5 times the values given above. The vesicles were
washed for untrapped marker either by dialysis or passing the preparation
through a Sepharose CL-2B column equilibrated with iso-osmolar veronal
buffer. The vesicles were stored at room temperature and diluted extensively with iso-osmolar veronal buffer before use. In the final assay the
lipid concentration was typically 20 nmol of lipid per ml.

SMIA. Four-fold dilution series of antisera were made using iso-
osmolar veronal buffer as dilutant. 30 µl of this antiserum dilution was
mixed with 10 µl of the hyperosmolar veronal buffer, followed by 10 µl of
vesicle solution and 10 µl of complement solution. The samples were incubated under shaking at $37^{\circ}C$ for 10 min, drawn up in 50-µl capillary tubes,
sealed with Sigillum wax (Modulohm, Denmark), and incubated 10 more min at
room temperature before ESR recording.

ESR measurements. The ESR signal intensity was recorded on a Bruker
ER-200-D ESR spectrometer equipped with standard rectangular cavity. The
modulation amplitude and frequency were 1.25 G and 100 kHz, respectively,
and the microwave power was 20 mW. The capillary tubes were placed in a 3
mm outer diameter quartz tube, and the solution completely filled the part
of the capillary inside the cavity. The amplitude of the midfield line was
determined by a g-value controlled unit described previously (Vistnes and
Wormald, 1982).

TYPICAL RESULTS

Background release. The absolute amount of spin label released from
the vesicles was determined as follows : A sample in which buffer was
exchanged for antiserum and complement gave, by definition, 0% release. The
value obtained by adding Triton X-100 (final concentration 2%) was defined
as 100% release. A linear dose-response between the two limits was assumed.
The background release of spin label due to serum or complement (no antibodies) was typically a few percent.

The phospholipid composition of the vesicles influenced the background
release. Vesicles with phosphatidlylserine or dicetyl-phosphate as the
negatively charged lipid instead of phosphatidic acid often possessed an
enhanced nonspecific lysis. As also observed by Allen and Cleland (1980),
the nonspecific leakage of marker induced by serum or complement was higher
if the hyperosmotic buffer was not included in the reaction mixture.

In spite of efforts in choosing the optimal phospholipid composition
and osmotic conditions, the nonspecific lysis and the "natural leakage"
from vesicles with polysaccharides in the membrane was higher than for pure
lipid antigen systems (Rosenqvist and Vistnes, 1977). Thus, vesicles
typically had to be used within 8 h after removal from the dialysis bag,
and they were generally unsuitable 3 days after the first washing. Because
of leakage and nonspecific lysis, the ratio between ESR amplitudes for a
positive serum response and negative serum response was only a factor 2 to
10.

Immune lysis. When capsular polysaccharides were incorporated in the
vesicles, the addition of group-specific anti-polysaccharide antibodies and
rabbit complement resulted in lysis of the vesicles, as monitored by a
rapid increase in the ESR amplitude. The reaction was complete after 30 min

at room temperature or 10 min at 37°C. No significant lysis beyond the nonspecific one discussed above was observed unless all of the following three components were present: antiserum (antibodies), active complement, and polysaccharide incorporated in the membrane.

With a constant amount of complement per sample, the maximum immune lysis when titrating an antiserum was apparently independent of the actual antiserum. This maximum immune lysis was defined as 100% immune lysis and corresponded to 40 to 80% of the Triton X-100 release of spin label. The 0% immune lysis was defined as the degree of lysis after incubation with only complement and buffer. Again, a linear relationship between ESR amplitudes and the corresponding percent immune lysis was adopted between the 0 and 100% limits.

The shapes of the antibody dilution curves for the A, B, and C test systems appeared to be similar when antisera from the same species were compared. Strong and weak antisera differed mainly by horizontal displacements of the curves (See e.g. Fig.4). To indicate the relative strength of the antisera, the serum dilution factor corresponding to 20% immune lysis was defined as the titer of that serum.

In some cases high antibody concentration gave less immune lysis than somewhat less antibodies, thus giving a bell-shaped dose-response curve instead of a sigmoidal curve. This so-called prozone effect could be overcome by reducing the amount of antibody or by increasing the amount of complement. Speculation on the cause of the prozone is given in the original paper.

At the polysaccharide concentrations tested (6 to 25 µg of polysaccharide per µmole of lipid), the degree of lysis and the sensitivity of the test were fairly independent of the antigen concentration.

Group-specific antisera (meningococcal agglutinating sera) were tested for cross-reactivity. For most combinations of antisera and vesicles, the cross-reactivities between the A, B, and C systems were quite low. Typically, the titer for one group-specific antiserum dropped by a factor of 100 or more when its own group polysaccharide was replaced by that of another serogroup. The only exception from this was the anti-C antiserum that showed a titer for the B test system of about 1/10 of the titer for the C test system. On the other hand, with the monoclonal anti-B antibodies and the C test liposomes, the titer was reduced to about 1/100 of the titer with the B liposomes. Some speculations on the implications of the cross-reactivity between the C and B systems is given in the original paper.

Liposomes prepared with A+C polysaccharides behaved very much like the pure A system, but responded now to both anti-A and anti-C antibodies. The pure group C polysaccharide coated liposomes also responded very similarly to the above-mentioned systems. The B polysaccharide system, however, tended to give less reproducible results. The maximum lysis was somewhat lower (40 to 60% of the Triton value) and varied significantly for different preparations of liposomes and different antisera. With the group B polysaccharides, the test system was also degraded by time. About 30 to 40 h after the vesicles were made, the percent immune lysis suddenly decreased from the maximum value to nearly zero within a few hours. This effect was also observed to some degree for the liposomes made from group C polysaccharides but not from those prepared from the lactose-stabilized polysaccharides.

Sensitivity. By using the quantitation procedure discussed above, the reference/calibration serum had a titer of about 2000 in the group A polysaccharide system and about 7000 in the C system. With the concen-

tration of antibodies given in materials and methods above, the present
tests apparently have a sensitivity limit 1 to 3 ng of antibodies per ml.
The sensitivity of the B system was analyzed in a similar manner, and the
limit was about 100 ng of IgM per ml.

   Vaccination sera. The sera from five people vaccinated with Men A+C
vaccine were tested both with A+C liposomes and with A and C liposomes
separately. Fourfold titration series in veronal buffer were made, and the
titers were determined as described above. The complete results are given
in Vistnes et al. (1983), but an extract is as follows :

   In the prevaccination sera, three vaccinees had relatively low titer
in all systems, and two of them had significantly higher values. These high
values were mainly caused by anti-A antibodies in one case and anti-C antibodies in another case. In the postvaccination sera, we found that all had
responded significantly to the vaccine, both with anti-A and anti-C antibodies. The data from the three people with low pre-vaccination titer is
summarized in table 1 in form of their mean values. Anti-A antibodies
increased five-fold during the 4 first weeks after immunization, but
dropped thereafter by about 30% during the next 16 weeks (based on values
before the round-off used to get the numbers in the table). For the C
system antibody concentration increased 180 times during the first four
weeks for this particular group of people, and was reduced by 20% during
the next 16 weeks. The data were somewhat different for the two cases with
high prevaccination titers.

   When the paired sera from the five vaccinees who had received the
combined group B polysaccharide and type 2 outer membrane protein vaccine
were analyzed with the B system, we could not find significant (more than
two-fold) rise in titers against B polysaccharides for any of the vaccinees. However, the instabilities of the B system discussed above made it
difficult to be sure about such small changes in the titer. Using an
enzyme-linked immunosorbent assay technique, significant increase in
antibodies against outer membrane antigens from a noncapsular variant of a
type 2, group B meningococcus was found in all of the five postvaccination
sera. That technique, however, do not explicitly measure antibodies against
group specific polysaccharides, so in principle the two methods measure
different items.

DISCUSSION

   One particular application of MIA was described above. From a single

Table 1.   Relative antibody concentrations
            (titer values) for humans vaccinated
            with Men A+C vaccine. The numbers
            given are mean values for three people

Test system	0 wk	4 wk	20 wk
A+C	11	160	130
A	22	120	80
C	1	240	200

example like that the reader will know very little about actual design of such an assay. In this section we will therefore try to discuss the different ingredients and procedures by surveying several MIA works, and also suggest some possible future developments. It is the hope that this section can be useful for membrane immuno assays in general even if some parts of it, like instrumentation, only handle SMIA. Let us start with a discussion of the membrane system.

Liposomes. In most of the membrane immuno assays the bags used to encapsulate the marker have been of the multilamellar liposome type (MLL). Multilamellar liposomes were first studied extensively by Bangham and Horne (1964). MLL have typically 10-20 bilayers in an onion-like structure. They are very stabile, but rupture of the outer membrane alone will only lead to about 10% release of the marker. In membrane immuno assays this has not proven to be a serious problem so far since the complement lysis can rupture many successive layers of lipid membrane. A possible effect of many-layered structure as compared with single layered should be considered, though, since in other membranous lysis system the difference between these systems can be rather dramatic (Rosenqvist et al., 1980). MLL can be prepared in few hours (depending on how well the solvent for the lipid is removed by vacuum). The size distribution is very wide, - the diameters are in the 0.1 to 10 μm range. The amount of trapped marker per unit lipid is moderate (≈ 2 ml per mole lipid). Thus a rather large amount antigen is used in order to get sufficient trapped marker for the assay.

Since Szoka and Papahadjopoulos started to prepare their reverse-evaporation-vesicles (REV) (1978), such membrane structure has also been utilized in some spin membrane immuno assays (Vistnes et al., 1983, Vistnes, 1984, Boggs et al., 1984). REV have a narrower size distribution than MLL (0.1 to 2.0 μm diameter, typically), they have fewer layers (1-5 typically), and the trapping efficiency is better than for MLL (≈ 20 ml per mole lipid). Reverse evaporation vesicles are somewhat more tricky to prepare than MLL, and takes about 1-2 hours longer. REV are not as stable as MLL. However, if high sensitivity is an important requirement and the antigen is limited, REV might be the membrane system that should be used.

Several groups have tried to use small unilammellar liposomes (SUV) in membrane immuno assays (unpublished results and private communication). SUV have a very narrow size distribution (Brunner at al., 1976) of about 0.03 μm diameter and they are truly single layered. Thus, the homogeneity seems attractive. However, so far it has not been demonstrated that complement is able to lyse these structures to a considerable degree. The reason for this is unclear, but it might be that stearic requirements for antibody orientation necessary for complement fixation are not fulfilled because of SUV's small size.

There are a variety of other methods to prepare membranes than the above mentioned, and Szoka and Papahadjopoulos (1980) review some of them. When designing a particular membrane immuno assay one should keep this in mind and experiment to find a system that fits the requirements as well as possible.

Some people discriminate between vesicles and liposomes as if they were two different systems. In this chapter the two expressions will be used as synonyms.

The vesicles can be washed for the outside marker by dialysis, Sephadex or Sepharose gel column chromatography, or centrifugation. Neither separation procedure is perfect; dialysis takes a long time, gel columns tends to be clothed by the largest vesicles in a preparation and phospholipids adsorb to the gel. Finally, if centrifugation is used, only the

largest vesicles spin down at the speeds used, while small vesicles remain
in the supernatant. Personally, I prefer dialysis since liposome concentration do not change much during the washing, and vesicle samples can be
withdrawn repeatedly from the dialysis bag just before you need them.

Lipid composition of vesicles. A variety of lipid composition have
been used for MIA but some common trends can be mentioned. The main bulk of
lipid (40 - 50%) is often neutral lipids like phosphatidylcholine or
sphingomyelin. Cholesterol is frequently included in 30 - 50 mole %, - of
two reasons. First of all cholesterol tends to stabilize the membranes and
make them less leaky. Secondly, efficient immune lysis seems to require
lateral diffusion of the antigen in the membrane. Cholesterol modulate
membrane fluidity in a favorable direction and has also been shown to
enhance antigen exposure (Humphries and McConnell, 1975, 1977, and Brulet
and McConnell, 1976, 1977).

Another common feature used in choosing lipids for MIA is to include
about 5 - 10 mole % negatively charged lipids. It is known that the trapping efficiency is higher when charged lipids are included in the formation
of liposomes. Thus, negatively charged lipids might increase the sensitivity of the assay. Beside of this, the negative charge is believed to
decrease unwanted interaction between serum proteins and the membranes,
even if positively charged liposomes give better stability for some systems
(Alving, 1984).

Stability of the vesicles. A perfect vesicle system for MIA would not
release marker in any other situation than immune lysis. Such perfect
vesicles do not exist, and the leakage of marker certainly is one of the
most serious problem involved in MIA. The natural leakage of marker will be
discussed under "Markers" below. In "The complement source..." we will
concentrate on factors in serum that will destroy the vesicles, and here
emphasis is put on how the vesicles can be modified to make them most stable. One possibility is to use special phospholipid compositions that can
make the vesicles extra stable in serum (Kirkby and Gregoriadis, 1981,
Finkelstein and Weissmann, 1979, Gupta et al., 1981). Several groups have
explored this possibility since it has important implications in pharmaceutical research with regard to liposome encapsulated drug transport
within an organism. Several research groups claim to have found lipid
composition that are particularly stable (e.g. Finkelstein and Weissmann,
1979, Gupta et al., 1981), but the results so far look not too optimistic
from a MIA point of view. It should also be pointed out that experiments to
test how lipid composition influences the stability are often ambiguous
since different lipid composition will normally lead to difference in the
vesicles (size distribution and number of layers). Thus, the observed
effects might be more related to morphological changes in the vesicles than
to the lipids used.

Two other approaches appear quite interesting. Cuppoletti and coworkers (1981) introduced a procedure for preparing vesicles; the so-called
"erythrosomes". The cytoskeleton from red blood cells is cross-linked and
then covered with a phospholipid membrane. These structures might be more
resistant for leakage and disruption than ordinary vesicles. For erythrosomes to be useful for membrane immunoassays, they should not expose any
antigens from the cytoskeleton, and the complement must be able to lyse the
membrane. So far, however, no information is available on these points.

Another interesting approach is the use of cross linked lipids for
the structure of the membranes (see e.g. Dorn and Ringsdorf, 1982, and
references therein). The corresponding vesicles seem to be much more stable
and far less permeable than ordinary vesicles. The crucial question is
whether antigens can be incorporated in these membranes and if the vesicles

can be lysed by an antigen-antibody-complement reaction. The fact that lateral diffusion of the lipids seems to be necessary for complement lysis (Parce and McConnell, 1980) probably indicates that the "whole" lipid structure can not be cross-linked. Further work is necessary in order to elucidate these possibilities.

If the vesicles used in membrane immunoassays can be made more stable, it will have important implications for the method. The number of false positive samples will be reduced because of less serum induced leakage. Furthermore, the vesicles can be stored for long periods of time without leakage of too many marker molecules. It will be easier to wash the vesicles. For SMIA this implies that the signal from unlyzed vesicles will be a clean exchange broadened one. In this way it will be possible to increase the so-called amplification ratio (or factor), which will increase the accuracy of the method (see below).

Antigens. Antigens that are used for MIA must be amphiphatic, e.g. one portion of the molecule is hydrophilic and a well separated other portion is hydrophobic. The hydrophobic part is responsible for the anchoring to the lipid bilayer. The antigenic site must be located to the hydrophilic part, which is exposed to the aqueous space. Many antigens (or haptens) are amphiphatic to begin with, e. g. cardiolipin. Others have to be covalently bound to e.g. a lipid in order to make it amphiphatic artificially (Hsia and Tan, 1978, Budker et al., 1982, Ishimori et al., 1984).

There are numerous antigens that are used up to date, and Table 2 list some of them.

The antigen can be incorporated into the membrane when the vesicles are formed ("active sensitization"). Alternatively, in some cases antigen can be incorporated into already formed liposomes just by simple incubation of liposomes and antigens ("passive sensitization") (See e.g. Kataoka et al., 1971b, Uemura and Kinsky, 1972). Passive sensitization may in general work less well for large antigens like proteins than for small lipid-like antigens. Thus, only 10 - 15% release of marker was observed in a couple such systems (Budker et al., 1982, Boggs et al., 1983), and Boggs suggest that this comes from the limitation that only one layer is ruptured in the multilammellar liposomes.

The concentration of antigens in the membrane must be above some critical limit. The limit seems to be somewhat dependent on the immuno-globulin class (Six et al., 1973), but above the limit the immune lysis is remarkable independent of the concentration. In most cases the limit is in the order of one percent antigen relative to phospholipids in the membrane. The matter of how antigen concentration will influence the sensitivity of the test is touched in some publications. However, a careful study in light of the possibilities to dilute liposome concentration to improve sensitivity (Vistnes, 1984) has not been carried out so far but would be of great interest.

Antigens, when incorporated into the membrane, usually are at least as stable as the vesicles. Group B capsular polysaccharides from *N. meningitides* happened to be unstable as discussed in "results" above. This should be kept as a reminder since similar effects might also show up in future systems.

Buffer. Almost any type of buffer at about pH 7 will do the job in MIA, and various PBS and veronal buffers have successfully been applied. To avoid deficiency in divalent cations at low serum concentration (antiserum and serum used for complement source combined) the buffer should contain Ca and Mg at about 0.1 and 1.0 mM, respectively. Calcium tends to destabilize

370

Table 2   Some of the antigens used in membrane immuno assays

Antigen	Test for disease	Selected references
Glycolipids, mainly Forssman hapten		Wei et al., 1975 Uemura et al., 1980,1982 Yasuda et al., 1981 Wood and Kabat, 1981 Mori et al., 1982 Boggs et al, 1984 Crook et al., 1986a,b
	Hemolytic anemia	Yasuda et al., 1982
Cardiolipin	Syphilis	Rosenqvist and Vistnes, 1977 Takashi et al., 1980 Umezawa et al., 1984
Gangliosides		Geiger and Smolarsky, 1977 Zemmour et al., 1984
	Multiple sclerosis	Feix et al., 1984 Boggs et al., 1984
Phospholipids		Alving, 1984
Dnp-capro-PE deriv.		Uemura and Kinsky, 1972 Six et al., 1973 Hsia and Tan, 1978 Chan et al., 1978
Thyroxine-PE		Tan et al., 1981a,b Hsia an Tan, 1978 Braman et al., 1984
Lipopolysaccharides		Kataoka et al., 1971 Humphries and McConnell, 1974
Polysaccharides	Meningitides	Vistnes et al., 1983
Egg albumin		Humphries and McConnell, 1974
Basic protein	Multiple sclerosis	Boggs et al., 1983 Boggs et al., 1984
Myoglobin		Budker et al., 1982
IgG		Ishimori et al.,1984

some model membranes, so that it is in general wise not to include more Ca than necessary.

Vesicles acts as osmometers, that is they swell and shrink according to osmotic pressure (Vistnes and Puskin, 1981), and they can easily burst in hypo-osmolar medium. Because of this the osmolarity of the buffer should be close to what is found in serum. In order to avoid that variations in osmolarity plays a role in MIA it is wise to prepare and store the vesicles in a solution that has 10 - 15 % less osmolarity than serum or the final reaction mixture (Vistnes et al., 1983). It might be worthwhile to consider osmotic pressure conditions even when choosing marker concentration for the preparation of the vesicles. For SMIA work the spin label concentration is often 0.075 - 0.1 M (e.g. for tempocholine chloride), which means that other salts have to be reduced accordingly.

Marker. A membrane immuno assay can be carried out with a variety of marker/detector system combinations. In the classical MIA works from Kinsky's group, glucose was used as a marker and assayed spectrophotometrically after an enzymatic process. This system is simple and use instrumentation found in almost every chemical/clinical laboratory. However, the drawbacks are quite evident. First of all, glucose is found in serum which makes it impossible to use MIA in its most sensitive region where small liposome-to-serum ratios are common. Furthermore, kinetics of the marker release is difficult to determine since the color changes does not follow immediately after the glucose is released from the vesicles. Also, serum is often slightly turbid, which will influence the spectrophotometric readings.

Glucose can be exchanged with fluorescent probes, and the next chapter discuss the corresponding technique. Fluorescent probes are superb in one respect : very small concentration of the probe is necessary for its detection. However, natural fluorescent molecules in the serum and turbidity again makes it impossible to use the full potential of the fluorescent technique itself.

This chapter is concerned with spin membrane immuno assay where the marker is a spin label. In common with fluorescent probes, the release of marker can be followed immediately as it takes place, which make kinetics measurements very easy and straightforward. In contrast to most other markers, serum gives no influencing background signal to the spin label spectrum, - not even turbidity plays any role for the detection. This means that in spin membrane immuno assay one can utilize the full sensitivity that the ESR technique can offer. Even if ESR as a general technique is less sensitive than fluorescence, preliminary comparison between MIA carried out with spin labels and with fluorescent label reveals that the spin membrane assay seems to be the most sensitive of the two (personal communication). However, a more thorough study using various fluorescent dyes and a large number of different sera is necessary for a more strict comparison.

Glucose, fluorescent dye and spin labels are the most common markers for MIA up to date. However, quite a few other markers have been utilized, and table 3 give a survey of markers and some representative papers. The list is included to show the variety of possibilities, and it is not supposed to be complete in any way. Various chapters in this book will treat a few of the more exotic versions in some detail.

Before we leave the survey of markers, a couple more aspects to be considered in the choice of marker should be discussed. For a marker to be suitable for MIA it should have low permeability through the membrane. Charged probes generally leak less than neutral ones. Thus, the spin label tempocholine penetrates membranes slower than glucose (unpublished observation). A thorough study comparing many different labels in the same membrane system is not carried out so far but would be of considerable interest. In general, the more leakage of marker between washing and detection, the less sensitive the method will be. It should be mentioned that some antigens when incorporated into the membrane (e.g. polysaccharides) tend to increase the leakage of marker (Vistnes et al., 1983). In extreme cases it might be necessary to use high molecular weight marker like enzymes to overcome such problems.

Table 3. Some markers used in membrane immuno assays

Marker	Detected by	Some references
Glucose	Spectrophotometry	Kinsky et al., 1969 Six et al., 1973 Kataoka et al., 1971a,b Uemura and Kinsky, 1972 Alving et al., 1977 Alving, 1984
Dichromate	Spectrophotometry	Wood and Kabat, 1981
Fluorescent dye	Fluorescence spectroscopy	Geiger and Smolarsky, 1977 Smolarsky et al., 1977 Uemura et al., 1980, 1982 Yasuda et al., 1981, 1982 Okada et al., 1982a,b, 1983a,b, 1985 Mori et al., 1982 Ishimori et al., 1984 Zemmour et al., 1984
ATP	Luminescence	Budker et al., 1982
Spin labels	Electron spin resonance spectroscopy	Humphries and McConnell, 1974 Wei et al., 1975 Rosenqvist and Vistnes, 1977 Chan et al., 1978 Hsia and Tan, 1978 Tan et al., 1981a,b Vistnes et al., 1983 Boggs et al., 1983, 1984 Feix et al., 1984 Crook et al., 1986a,b
$^{86}Rb^+$	Scintillation counter	Shin et al., 1978
Ions	Ion selective electrodes	D'Orazio and Rechnitz, 1977 Shiba et al., 1980 Umezawa et al., 1984
Enzymes	Ion selective electrodes	Haga et al., 1980, 1981
Enzymes	Spectrophotometry	Braman et al., 1984

Another important aspect of markers is stability. Most of the markers used in MIA are more than sufficient stable up to the point where serum is added. Serum is a complicated mixture of molecules, however, and the actual mixture varies considerably from person to person. In our laboratory we have observed that some sera reduces the spin labels with time, and similar effects probably exist also for other markers. Even if reduction in marker signal is not a serious problem in most cases, the possible effect should be considered when choosing which controls to include. The effect is most evident at high serum concentrations, and may vary rather dramatically from serum to serum.

A final comment on markers: Two or more test systems could be used simultaneously if the marker from the different test systems were distinguishable. This can e.g. be done in SMIA if both ^{14}N and ^{15}N spin labels were used, as pointed out by Hsia and Tan (1978).

Amplification ratio and use of ESR quenchers. The amplification ratio in SMIA is defined as the ESR signal amplitude at maximum immune lysis divided by the amplitude at no lysis. Amplification ratios can probably be defined similarly for any other marker. When the spin membrane immuno assay is based on *one* calibration curve and only one sample per serum (no dilution series), the amplification ratio is an important parameter. This is due to the limited accuracy by which the ESR amplitude can be determined (typically standard deviation is 1 - 10% under normal SMIA conditions). If standard deviation is given by d and the amplification ratio by R, the number of significant different antibody-levels, N, that can be determined is approximately given by (Vistnes, 1983) :

$$(1 + d)^{N-1} = R$$

or

$$N = 1 + (\log R)/\log(1 + d)$$

If we start out with a standard deviation of 5 % (d=0.05) for the ESR amplitude determination, and an amplification ratio of 40, the number antibodylevels that can be determined by one calibration curve will be approximately 77. The accuracy in the antibody-concentration determination will of course also depend on the steepness of the "dose-response" curve ("calibration curve") (Vistnes, 1983), but as seen from the argument above the amplification ratio is a key parameter.

Hsia and coworkers (Chan et al., 1978, Tan et al., 1981a,b) obtained a ratio of 40 while in experiments where the antigen makes the membrane more leaky a ratio of 2 - 10 was the best obtainable. That small a ratio might look like an impossible situation, but it works actually very satisfactorily. The reason is that for low amplification ratio and for very large variation in antibody concentrations in the different sera, dilution series of sera are introduced as discussed above. With that approach the number of significant different measurable antibody-levels will depend more on the choice of the dilution series than on the amplification ratio.

There are different ways to increase the amplification ratio beyond the level used today. So far only "self-broadening" (that is exchange between the spin label molecules themselves) has been used. The ESR signal can be substantially more reduced by introducing ESR quenchers. One effective quencher is chromium oxalate (See Vistnes and Puskin (1981) and references therein). Chromium oxalate exhibits a considerable quenching effect on ESR signals from the spin label used in SMIA (tempo-choline-chloride) (unpublished observations). Thus, an amplification ratio of about 1000 may be attained if the vesicles can be made stable and nonpermeable. For the vesicles used today, however, experiments have shown that use of special quenchers only give marginal improvements compared to the self-quenching case (unpublished results).

The amplification ratio and the sensitivity of SMIA can be improved somewhat by introducing ^{15}N and perdeutrated spin labels (see e.g. Yost et al., 1980). The ESR amplitude will increase by 50% since the ESR spectrum will consist of two lines instead of normally three. In addition the linewidth will be narrower than for non-perdeutrated labels, making the increase even more pronounced. The anticipated increase in the amplification ratio is due to the reduced linewidth for the free spin label.

The complement source in membrane immunoassays. The complement can not be synthesized, implying that we have to use natural sources. In general human or some particular animal sera are used. It is nontrivial to isolate the complement components from the serum, which implies that in most cases the serum is used per se. It is evident that the serum contains other factors like lipoproteins and immunoglobulins that may influence the membrane immunoassay.

In most MIA work guinea pig serum is used as complement source with good results. The complement activity is normally rather high per unit volume, which makes it possible to work with rather low concentration of the guinea pig serum. The lipoprotein concentration is therefore low, with the result that non-immunological disruption of the membranes is almost negligeable.

When dealing with the meningococcus polysaccharide antigens (Vistnes et al., 1983) several problems became evident. First, guinea pig serum often contains antibodies against polysaccharides. These undesired antibodies may probably be removed by absorption techniques, but this possibility was not pursued. Instead, baby rabbit serum was used which most often is free from these particular antibodies. The complement activity in this serum was, however, lower than for guinea pig serum. Furthermore, the prozone effects which are present in this system (see above), require a rather high rabbit serum concentration in the assay reaction mixture. This leads to some instability because of serum induced non-immunological lysis of the membranes.

Naturally occurring antibodies can also create problems if they appear in the complement source serum. Alving (1984) claims that about half out of 43 sera have some anti-liposome antibodies and about 6% have quite high activity. The anti-phospholipid antibodies could be removed by adsorption. Various aspects of antibodies against phospholipids have been studied by MIA in several laboratories (see e.g. Uemura et al, 1980, Wood and Kabat, 1981, and Yasuda et al., 1982).

Lipoprotein in the serum is believed to be the main reason for serum induced lysis (Scherphof et al., 1981, Kirkby and Gregoriadis, 1981). In several experiments, using different techniques, the lipoprotein was removed completely or partially from the rabbit serum. Unfortunately, whenever the lipoprotein was removed, the complement activity also dropped below detection limits (unpublished observations). Golden hamster serum was tried as a source for complement since it was reported to be free from lipoproteins (Gujral et al., 1981). However, the complement activity of this serum was too low to be useful (unpublished observations). More research should probably be aimed toward the problem of maximizing complement activity simultaneously with keeping lipoprotein concentration low.

The reaction utilized in the membrane immunoassay is the so-called "classical pathway", which is trigged by the antibody-antigen complexes. The complement can also react through an alternative pathway (see e.g. Hobart and McConnell, 1975), which can be trigged without any antibody--antigen complexes. It is known that certain substances, like yeast cell--wall polysaccharides, can activate this alternative reaction pathway. Okada and coworkers describe a variety of compounds that can trigger the alternative pathway for complement activation in MIA systems (Okada et al., 1982a,b, 1983a,b,c, 1985). Consequently, uncertainties exist whether an observed lysis is due to the antibody-antigen complex or not. This uncertainty can be large when the serum concentration is high, since the alternative pathway outlined above works most satisfactorily at high serum concentrations. The classical pathway also works in dilute solutions, implying that the experiments should be aimed at such conditions.

The complement activity influences the rate of release of marker in most MIA systems published so far. In some cases even the maximum lysis depend on the complement activity and the source of complement. It is therefore a drawback for the majority of published MIA work that the complement activity is not well characterized. For guinea pig serum even slight differences in handling can result in significant changes in complement activity. It would be very valuable if a method could be developed that could give complement activity in "absolute values" in a fast, easy and accurate way. A method based on the MIA technique would be preferable. Plant (1986) has worked out such a method and she claims that her assay is qualitatively similar to the erythrocyte-based hemolytic assay used in clinical laboratories, but is faster, more accurate and more reproducible. The work is interesting and a good step in the right direction, but the method will vary with liposome preparation from one laboratory to another, and the antibody coating step is not standardized enough either. We hope someone will take the extra efforts needed to make the test so good that it could be the reference test for everyone in the future.

Freytag and Lichfield (1984) and Litchfield et al. (1984) have described MIA systems that do not use complement for disruption of the membrane. This approach might be useful for a variety of antigens, but it is probable not as general as the more classical approach. Feix et al. (1984) do not use external complement, only the complement activity normally present in human serum (antiserum is not heat-inactivated). This approach seems to work fine for their multiple sclerosis case, but it would be impossible to do extensive dilution series of antiserum without external complement.

Antisera. In most MIA work the source of antisera is immunized animals, most often rabbits. In a few cases antibodies from humans not subject to a particular immunization are measured. This was the case in the test for anti-cardiolipin antibodies in syphilis serology (Rosenqvist and Vistnes, 1977), for anti-glycolipids in several diseases like hemolytic anemia (Yasuda et al., 1982), and in the test for anti-ganglioside antibodies in multiple sclerosis patients (Feix et al, 1984). With a few exceptions complement in the test sera is heat inactivated so that variations in patients sera complement activity shall not influence the antibody measurements.

Mixing ratio. The ingredients of the membrane immuno assay can be mixed in a variety of concentration ratios, and the final choice will depend highly on which marker/detection system that is used. Optical methods like fluorescence often require dilute solutions to avoid turbidity etc., while ESR is free from that problem. SMIA in its most sensitive form often use a high serum-to-liposome ratio. Typically, antiserum (or dilution thereof), complement (moderately diluted), and vesicles (highly diluted) are mixed in roughly equal volumes (a combination that also minimize pipetting errors). Thus, in the final reaction mixture as little as 40 nmole of lipid per ml antiserum has been used (Vistnes et al., 1983). The different solutions can be mixed in any order, but the kinetics of the reaction do depend on the order even if the final result (plateau value) seems to be rather independent of the order (Kinsky et al., 1969).

Membrane immuno assays often require very small volumes for the reaction and the measurement. For example, SMIA as it has been carried out so far only need sample sizes of 50 µl, and with a new type sample-holder (loop-gaps, see below) the size could even be reduced to 5 µl or less. Small volume like this are nice in one respect since we need very little of sera

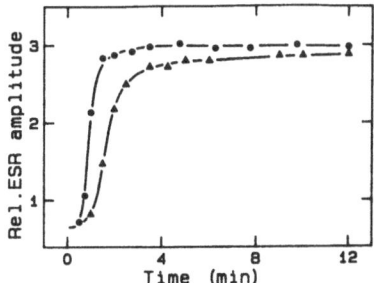

Fig.2.   Typical time response for complement
lysis of liposomes. The incubation
temperature was 37^{0}C (filled circles)
and 22^{0}C (filled squares). The antigen
in this experiment was cardiolipin.

and liposomes to carry out a test. On the other hand experiments gets more
difficult when the volumes get so small, since pipetting and mixing of the
different solutions is rather tricky for tiny samples. The vial used is
also of importance. We have found tiny round-bottomed glass tubes much
better with respect to mixing capabilities than the plastic dishes used for
many other types of immunoassays. It should be stressed that bad mixing
leads to poor reproducibility, so this matter should be taken seriously.

    Incubation time and temperature. The immune lysis is a rather fast
reaction, but it still takes some time. In general the reaction rate depend
on temperature (see e.g. Fig. 2), on antigen concentration (Kinsky et al.,
1969), on complement activity (Geiger and Smolarsky, 1977), and antibody
concentration (Rosenqvist and Vistnes, 1977). However, even if release rate
depend on temperature the limiting value for immune lysis seems to be
rather independent of this parameter (Smolarsky et al., 1977). Fig. 2 show
a typical response for the lysis as function of time for incubation temperatures of 22 and 37^{0}C. For excess complement and high antibody concentration 10 min is sufficient for the reaction to have reached it limiting
value. Longer incubation time will not change the result in this case. For
less complement or less antibodies the reaction might take somewhat longer.

    In a way it is a matter of taste how long incubation time that should
be used. It should not be much shorter than 10 min, and it should not be
much longer than 30 min because of natural leakage from the liposomes. In
our experiments we usually use 10 or 20 min incubation, but keep this time
the same for every sample within rather narrow tolerances (15 sec). This
require strict planning of how the samples should be mixed, withdrawn for
measurement, and measured, but it works very well.

    The incubation temperature is usually chosen to be 37^{0}C or room
temperature, and there is no major differences between them. The incubation
time should normally be longer for room temperature than for 37^{0}C, and it
is certainly worthwhile to carry out a time study for the temperature and
mixing ratio chosen to be sure that the system is working properly.

    "Calibration and the calibration curve." The MIA technique can in

377

principle not give antibody concentration in absolute measure, only in relative. Absolute units can only be obtained by comparison between a test sample and a calibration sample where the antibody concentration is determined by other methods, e.g. by spectrophotometry. This often requires isolation of specific antibodies in the calibration sample, e.g. by absorbtion technique.

The inability to give absolute values without further calibration has led to an unfortunate situation for MIA in serology. Antibody concentration in test sera has been given in ill-defined "titer" values, which is satisfactory only if several sera are compared using exactly the same test system. Sometimes comparison between different test systems would be interesting, and for those cases more work should be put into the calibration procedure than what has been common in the past.

Let us take the data in Table 1 as an example. Here variations in anti-A and anti-C antibodies can be followed correctly separately. However, the sensitivity, or calibration, of the A, C, and A+C test systems are not identical. Thus, a serum with titer 22 in the A test system can still have a titer of only 11 in the A+C system. If the three systems had been better calibrated using the calibration sample with known absolute antibody concentration, comparisons could be done to some degree between the different systems. The procedure can never be without ambiguity, however, because of the complexity involved in sensitivity questions (see below), but at least it would give more useful information than table 1 can give today.

There are several ways to make calibration curves for MIA. Here we will discuss three approaches, two of which have already been applied while the third one, suggested by Vistnes (1983), has never been tried out in practice. Several of the published papers report that the immunological release of marker vs antibody concentration most often follows a sigmoidal curve (see fig. 3). For low concentration or no antibodies, a background release level is present. For very high antibody concentration, the immuno

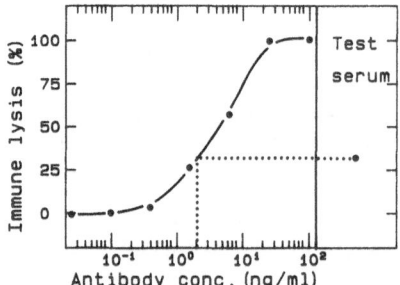

Fig. 3.   Standard curve (dose-response curve) based on a calibration sample can be used to quantitate antibodies in a test sample. Only one sample (dilution) is now necessary for the determination of antibodies in the test serum.

-lysis is constant and corresponds to 40 to 90% release of the marker. Some authors define the no-antibody level of release as "0% immune lysis" and the plateau at high antibody "100% immune lysis". In an about 10 fold concentration range, the immune lysis is smoothly going from the low to the high level. There are now two ways the immuno assay has been performed:

Using dilution series of the standard sample, a calibration curve like the one in fig. 3 is obtained. The sera to be assayed is then studied with a fixed ratio of reagents and serum (one serum concentration). From the measured immune response one can determine the antibody concentration (rel. to the standard). The limitation is as follows: If the immune response is in the range of about 10-90%, the antibody concentration can be determined relatively accurately. If it is outside this response, one can say nothing except that the antibody-concentrations must be above or below some corresponding values. If antibody concentration from patients varies only about 10-fold, this method is quite good for determining antibody levels provided the test is arranged properly. The method works less satisfactorily for low amplification ratio (see above) that results from leaky membranes in some antigen systems. Another limitation with the method is its implicit assumption that the shape of the "dose-response" curve ("calibration curve") as well as the antibody affinities are identical for all the test sera. Working with a large number of sera reveals that this is far from correct.

The second method is as follows: From every serum to be analyzed as well as for the calibration sample, a dilution series is made (e.g. 2-fold dilution, about 8-10 tubes). The immune lysis is determined from all these samples. The immune response for the dilution series are plotted as in Fig. 4, and from the displacement of the corresponding sigmoidal curve for the test serum compared with the curve from the standard, the antibody concentration in the serum of interest can be determined. By using the 20 or

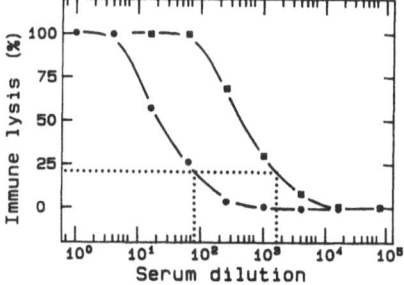

Fig. 4.   Dilution series of the calibration
          sample as well as for the test sample
          often leads to two almost identical
          sigmoidal curves shifted relative to
          each other. (Note: Logarithmic axis)
          The antibody concentration in the
          test serum can be determined from this
          shift provided the antibody concentra-
          tion in the calibration sample is known
          by some other technique.

50% lysis levels for the two curves, the displacement can be measured accurately. Note that a logarithmic serum concentration axis is necessary in order to keep the sigmoidal curve shape constant even if it is shifted along the axis. Many investigators use linear axis for antiserum or antibody concentration, which makes it difficult to recognize the similarities in the shape of the curves.

The second method has a clear advantage compared to the first one since a much wider range of antibody concentration can be determined accurately. It also works for low amplification ratio as discussed above. However, the extra cost by making dilution series for each serum assayed is quite considerable.

In our first SMIA experiments we used method one and pointed out that one single sample (no dilution series for every serum) could give the antibody concentration, which was a real advantage compared to the complement fixation test. In later work we switched to the second method since the latter could better handle the increased range for antibody concentration variation as discussed above, as well as the prozone effect that was observed. The prozone is characterized by a reduced lysis for high serum concentrations compared to some lower concentration. The effect could be due to degradation of released label due to some particular compound in serum. The result would be that the apparent immune lysis (based on ESR amplitude only) could be lower at high antibody concentration than at a somewhat lower antibody level. In the cases reported by Vistnes at al. (1983) degradation of the marker was not the reason for the prozone. Prozone effect has also been observed by Ishimori and coworkers (1984).

In systems where prozone happens for some sera, method 1 for antibody concentration determination is not useful, while method 2 has proven to give quite acceptable results. The problem is to define a good plateau value, but due to the steepness in the interesting part of the dose-response curve, the uncertainty due to an ill-defined plateau becomes almost negligible.

It has been known for some years that the antibody range corresponding to 10-90% immune lysis can be shifted by varying the relative amount of reagents involved in the assay. In short, the result is as follows: If the antigen concentration (vesicle concentration) is decreased, less antibody is necessary for 50% immune lysis. Thus decreasing the vesicle concentration often leads to a more sensitive test, but only up to some limit that depend on the binding constant. Vistnes (1984) offers a way to determine this limit for sensitivity and also a way to analyze data so that relative binding constants can be obtained. Boggs utilized this method in her studies on modulation of antigen-exposure in vesicles as a function of various physico-chemical parameters (Crook et al., 1986a,b). As a pure research tool, this method is expected to be of considerable value.

The possibility to shift the steep region in the dose-response curve by varying the vesicle concentration opens up a third method for analyzing serum in SMIA (Vistnes, 1983). It combines the advantages of both method 1 and 2, and can be explained as follows: Several dose response curves of the calibration sample are determined for different vesicle concentrations. The vesicle concentration should be chosen so that the antibody concentration of the sera under study will fall into the variable part of one of the curves, and the variable part of the different dose response curves should overlap. This procedure would have the advantage that one test serum dilution would be sufficient to determine the antibody concentration. Thus, the errorprone and time consuming dilution procedure for every serum tested could be avoided. In practice the number of samples involved per test serum would probably be 2 to 4, which is considerably smaller than the number

used in the dilution series experiments (usually 8 in Vistnes et al.,
1983). Care should be taken for a correct use of this method if working
close to the sensitivity limit. In extreme cases, very close to this limit,
the method will not work.

Sensitivity. The first decade of the MIA technique most people worked
with rather high liposome (antigen) concentration in the reaction mixture.
This resulted in a rather insensitive system and several tenths of a μg of
antibodies were necessary for significant immune lysis.

Later on the possibility to increase sensitivity by decreasing liposome concentration (Vistnes, 1984) has been utilized. This opens up new
sensitivity limits in MIA, especially in SMIA. Earlier, strong antisera
gave no immune lysis after 10-100 dilutions while with today's more sensitive approach 1000 - 10000 dilutions are often necessary. For one particular monoclonal anti-B polysaccharide antiserum even a 100,000-times
dilution gave significant immune lysis. In absolute values, 1 - 3 ng
antibodies per ml (approximately 10^{-11} M) was detected by SMIA for the
meningitides system (Vistnes et al., 1983). If loop-gap resonators are used
in SMIA (see "Instrumentation" below), this implies that less than 10 pg
(picogram) antibodies (or in the order of 10^7 immunoglobulin molecules) may
be sufficient for significant immune lysis. Thus, the sensitivity of the
MIA (SMIA) technique is quite good. It should be noted that the high
sensitivity can only be obtained for high affinity binding of antibodies to
the antigen.

The sensitivity of the MIA technique is claimed to depend on the class
of immunoglobulins in the antiserum e.g. in the papers by Six et al.,
(1973) and Vistnes et al., (1983). In the lastmentioned work high affinity
IgG antibodies seems to give higher sensitivity than IgM if measured in ng
per ml, while the difference between IgG and IgM probably is less pronounced on a molar basis. The differences in the immunoglobulin classes might
be real, but it is also possible that the binding constants are more
important. The immune response in humans is so complex, with several
classes of immunoglobulins involved, and several clones of antibodies
within each class with various affinities, that the matter of sensitivity
is an intricate matter.

Sensitivity limits are given in many papers, but it is very confusing
to try to compare them. The two important parameters for such comparison
are 1) minimum *concentration* (e.g. given in nM or picomol per ml), and 2)
minimum absolute amount (in e.g. picomol), which is proportional with
minimum volume multiplied by minimum concentration for detection. Sensitivity in weight units are less useful unless compounds of same molecular
weight are compared. An example can illustrate the problem : One published
paper claim to have a sensitivity in the order of "one picomolar", while it
should have been one picomol, – which make quite a difference !

Specificity and false responses. MIA is based on a specific antigen-
antibody reaction, and in the few examples where specificity is tested by
these techniques, it is found to be very satisfactory. Quite a few MIA
papers deal with specificity and only a few will be mentioned here. Haga
and coworkers (1981) find that caffeine and theobromine cross-react in
systems with theophylline as antigen, while cross-reactivity with other
xanthine derivatives were negligible. Vistnes et al. (1983) tested specificity for the different groups of the polysaccharide system. It was found
that anti-A, anti-B and anti-C antibodies did cross-react to some degree
since one antiserum could react with an antigen from another group, but
only if the concentration was about 100 times higher than for the similar
reaction for its own group. Ishimori and coworkers (1984) find no cross-
reactivity between anti-IgA, anti-IgG and anti-IgM antibodies in their test

for anti-IgG antibodies. Zemmour et al. (1984) find that the cross-reactivity as measured by the MIA and the hemagglutination-inhibition test for anti-ganglioside antibodies are different, and argue that the MIA test is the most reliable one of the two because of difference in ganglioside exposure.

It is not a surprise that the specificity is high in MIA. The antigens used are in general equal to the natural ones. Furthermore, their exposure to antibodies in the MIA membrane is very much identical to that in the microorganism.

The method is not foolproof, however. Immune lysis can take place when antibodies against some of the lipids in the MIA membrane are present, even if no antibodies exist for the antigen under study (see above). This false response can be revealed if vesicles without antigen are included as a control for the assay. False response can also occur when the complement is activated through the alternative pathway. A control sample without anti-serum should tell if that's the case, unless factors in the antiserum itself can trigger the alternative complement reaction. In that case vesicles without antigen could again be useful.

Which controls to include. Because of the finite leakage of marker from the liposomes even in absence of antibodies, some appropriate controls must be included in every MIA. The controls are treated as closely the same as the true test samples as possible. Thus, dilution of vesicles, mixing, incubation etc are carried out the same way for controls as for the other samples. The controls we most often use are :

- A sample with only liposomes (that is: the volumes for antisera and complement exchanged with buffer only).

- A sample with liposomes and complement (not antiserum).

- A sample with liposomes and antiserum (not complement).

- A sample with 100% lyzed liposomes, e.g. liposomes added 1-2% Triton X-100 or any other convenient detergent.

The samples with only liposomes and liposomes plus complement gives normally about the same reading. These samples are used to define the zero immune lysis level. The sample with 100% lyzed vesicles define the 100% lysis level that is used to measure the maximum lysis found in the assay. The sample with liposomes and antiserum only is used to detect any non-immune lysis activity in some sera, e.g. in cases of high lipoprotein concentrations in test sera. If immune lysis measured from this sample is above 0%, the actual measured value for this serum might be far from true immune lysis. Other control samples are discussed in the specificity section above.

Instrumentation. In all SMIA works reported to date, elaborate general purpose ESR spectrometers (both manufactured by Varian or Bruker/IBM) have been used for detection of the responses. The samples have been placed in glass capillaries (50 ml) which were sealed by wax or by melting the ends. Normally, the samples were changed by hand, but an automatic sample exchanger has been considered by Hsia (private communication). The samples have been studied by ESR either at room-temperature, at $37^{o}C$ or at $4^{o}C$. Normally only one line in the ESR spectrum has been recorded on graph paper, and the amplitude measured manually. This is a procedure that is poorly suited for a large scale immuno assay situation with hundreds of samples. Therefore, an automatic amplitude measuring unit was constructed (Vistnes and Wormald, 1982). With this unit the marker concentration in the final mixture can be reduced significantly, which implies a more sensitive assay. Time consumption is also brought down so that the automatic amplitude measurement unit has proven to be very useful in our SMIA experiments.

When a computer is interfaced to the spectrometer, spectral subtraction (removal of exchange broadened part of the spectrum) followed by digital filtering and automatic amplitude measurements also has turned out to be a very useful procedure (unpublished observations). In this way an accurate measure of the percent release of marker can be obtained. The numbers would, however, not be much different from the approximate values used in all published work so far, values calculated from the expression:

$$\text{percent release} = 100 \times (A_x - A_0) / (A_{100} - A_0)$$

where A_x is the amplitude measured for the test sample, A_0 is the amplitude for the unlysed vesicles, and A_{100} is the amplitude for 100% release.

If SMIA should be used for large scale immuno assays, the ESR instrument had to be less expensive and have a larger capacity (throughput) than instruments today. New microwave equipment like loop gap resonators are developed (Fronsisz and Hyde, 1982) which can be made cheaper than ordinary ESR cavities. With loop gaps the sample size can be reduced to 5 µl and below, so that the amount samples used in the assay could be almost negligible. Furthermore, the small sample size gives opportunities to use smaller and cheaper magnets in the instrument.

In the last years other microwave equipments also open up the possibility to produce strip-line microwave bridges that could bring down the prices of ESR instruments even more (J. S. Hyde, private communication). Consequently, within few years it should be possible to produce ESR instruments in the 5-10 kilo-dollar range that are fully suitable for SMIA work. The limitation for this development will probably not be of technical character but rather a too small demand for it.

Test for antigens. Most of this chapter discussed MIA as used for detection of specific antibodies. However, as pointed out before, the method can be used also to determine complement activity or antigen concentration. The presence of antigens is detected by inhibition of lysis due to competition between antibody binding to antigens on the vesicles and antigens in the serum under study. The method is somewhat less sensitive than ordinary MIA since it is less favorable to detect a slight reduction in a large signal (no external antigens give often 100% immune lysis) than to see an equal but positive change in amplitude in a small signal. In addition the competition process itself tends to reduce sensitivity compared to a pure binding situation. In spite of this, the possibility to use MIA for antigen detection is a useful extension of the method, and table 4 give a list of some published applications.

Patents. Several research groups have applied for patents on various aspects of MIA. Hsia and Tan have applied for patent on a MIA method to determine antigenic material by inhibition of immune lysis as discussed above (U.S.Patent, application 4,235,792, Nov.25, 1980). Cole has applied for patent on an MIA test for detecting antigens or antibodies by enzyme loaded liposomes (UK Patent application GB 2,069,133A, Feb.3, 1981 and International Patent Classification, application PCT/US81/00040, Jan.13, 1981). Weiner has invented a method for long time storage of loaded lipid vesicles by rapid freezing, and has applied for a U.S.Patent for it (application 4,397,846, Aug.9, 1983).

Table 4. Some antigens that can be detected by MIA.

Antigen	Selected references
Glycolipids	Wei et al., 1975 Geiger and Smolarsky, 1977 Uemura et al., 1982
Dinitrophenyl deriv.	Chan et al., 1978 Shiba et al., 1980
Thyroxine	Hsia and Tan, 1978 Tan et al., 1981a,b Braman et al., 1984
Digoxin	Freytag and Litchfield, 1984 Litchfield et el., 1984
IgG	Braman et al., 1984
Theophylline	Haga et al., 1980, 1981

FINAL COMMENTS

In this chapter we have discussed membrane immuno assay and emphasized details useful for designing new assays. It should be evident for the reader that a lot of possible improvements of the method can be carried out, and some suggestions for changes leading to better systems have been given in this chapter. What we have not discussed so far is the method's potential in the future. Will membrane immuno assays survive in the competition between a variety of good immunoassays ?

Membrane immunoassays are sensitive, relatively rapid, specific and general methods. One of the main drawbacks for the method is its lack in well-defined and homogenous vesicles with long-time stability after washing of external marker. Future experiments might reveal some tricks in liposome preparation so that stable, non-leaky liposomes with high sensitivity and great flexibility can be made. However, so far no real good systems have been published.

In order to develop the SMIA technique to a powerful assay in the future hard work is needed. Another question is the situation on the personnel side? In the authors opinion it is far from good. For the time being only 3-4 laboratories in the world are working with SMIA, and probably none of these are involved in the detailed developments necessary for large-scale assays. In short, it is a rather pessimistic picture for routine use of spin membrane immuno assay. However, as a research tool SMIA might still be attractive. Wether another subgroup of MIA will be utilized for large scale immunoassay in the future is an open question, but at least the method has so many nice features that it deserves to be taken seriously.

ACKNOWLEDGEMENTS

This work was supported by grants from The Norwegian Research Council for Science and the Humanities.

REFERENCES

Allen, T. M. and Cleland, T. M., 1980, Serum induced leakage of liposome contents, Biochim. Biophys. Acta, 597:418-426.

Alving, C. R., Richards, R. L., and Guirguis, A. A., 1977, Cholesterol-dependent human complement activation resulting in damage to liposomal model membranes, J. Immunol., 118:342-347.

Alving, C. R., 1984, Natural antibodies against phospholipids and liposomes in humans, Biochem. Soc. Trans., 12:342-344.

Bangham, A. D. and Horne, R. W., 1964, Negative staining of phospholipids and their structural modification by surface-active agens as observed in the electron microscope, J. Mol. Biol., 8:660-668.

Blaedel, W. J. and Boguslaski, R. C., 1978, Chemical amplification in analysis: A review, Amer. Chem. Soc., 50:1026-1032.

Boggs, J. M., Samji, N., Moscarello, M. A., Hashim, G. A., and Day, E. D., 1983, Immune lysis of reconstituted myelin basic protein-lipid vesicles and myelin vesicles, J. Immunol., 130:1687-1694.

Boggs, J. M., Samji, N., and Adamo, S. A., 1984, Immune lysis of lipid vesicles containing myelin basic protein or glycolipid antigens by multiple sclerosis and normal sera, J. Neurolog. Sci., 66:339-348.

Braman, J. C., Broeze, R. J., Bowden, D. W., Myles, A., Fulton, T. R., Rising, M., Thurston, J., Cole, F. X., and Vovis, G. F., 1984, Enzyme membrane immunoassay (EMIA), Biotech., 2:349-355.

Brulet, P. and McConnell, H. M., 1976, Lateral hapten mobility and immunochemistry of model membranes, Proc. Natl. Acad. Sci. USA, 73:2977-2981.

Brulet, P. and McConnell, H. M., 1977, Structural and dynamical aspects of membrane immunochemistry using model membranes, Biochem., 16:1209-1217.

Brunner, J., Skrabal, P., and Hauser, H., 1976, Single bilayer vesicles prepared without sonication; Physico-chemical properties, Biochim. Biophys. Acta., 455:322-331.

Budker, V. G., Mustaev, A. A., Pressman, E. K., Roschke, V. V., and Vakhrusheva, T. E., 1982, Adsorption of non-membrane proteins on the surface of model phospholipid membranes, Biochim. Biophys. Acta., 688:541-546.

Chan, S. W., Tan, C. T., and Hsia, J. C., 1978, Spin membrane immunoassay: Simplicity and specificity, J. Immunol. Methods, 21:185-195.

Crook, S. J., Boggs, J. M., Vistnes, A. I., and Koshy, K. M., 1986a, Factors affecting surface of glycolipids: Influence of lipid environment and ceramide composition on antibody recognition of cerebroside sulfate in liposomes, Biochem., in press.

Crook, S. J., Boggs, J. M., Vistnes, A. I., and Zalc, B., 1986b, Characterization of anti-cerebroside sulfate antisera from liposome immune lysis data, submitted for publication.

Cuppoletti, J., Mayhew, E., Zobel, C. R., and Jung, C. Y., 1981, Erythrosomes: Large proteoliposomes derived from crosslinked human erythrocyte cytoskeletons and exogenous lipid, Proc. Natl. Acad. Sci. USA, 78:2786-2790.

Dorn, K. and Ringsdorf, H., 1982, Polymeric monolayers and liposomes as model for biomembranes and cells, in: "Transport in Biomembranes: Model systems and reconstitution," R. Antolini, A. Gliozzi, and A. Gorio, eds., Raven Press, New York, p13-25.

Esser, A. F., 1980, Principles of electron spin resonance assays and immunologic applications, in: "Immunoassays, clinical laboratory techniques for the 1980s," R. M. Nakamura, W. R. Dito, and E.S.Tucker, Alan R. Liss, New York. p213-233.

Feix, J. B., Khatri, B., McQuillen, M. P., and Koethe, S. M., 1984, Immune reactivity against membranes containing ganglioside GM_1 in chronic-progressive multiple sclerosis : Observation by spin-membrane immunoassay, Immunol. Com., 13:465-.

Finkelstein, M. C. and Weissmann, G., 1979, Enzyme replacement via liposomes. Variations in lipid composition determine liposomal integrity in biological fluids, _Biochim. Biophys. Acta_, 587:202-216.

Freytag, J. W. and Litchfield, W. J., 1984, Liposome-mediated immunoassays for small haptens (digoxin) independent of complement, _J. Immunol. Methods_, 70:133-140.

Froncisz, W. and Hyde, J. S., 1982, The loop-gap resonator: A new microwave lumped circuit ESR sample structure, _J. Magn. Reson._, 47:515-521.

Geiger, B. and Smolarsky, M., 1977, Immunochemical determination of ganglioside GM_2, by inhibition of complement-dependent liposome lysis, _J. Immunol. Methods_, 17:7-19.

Gujral, S., Patel, N., Lovekar, C. D., and Seth, D., 1981, Absence of lipoproteins in serum of golden hamster _Mesocricetus Auratus_, _Curr. Sci. India_, 50:363-364.

Gupta, C. M., Bali, A., and Dhawan, S., 1981, Modification of phospholipid structure results in greater stability of liposomes in serum, _Biochim. Biophys. Acta_, 648:192-198.

Haga, M., Itagaki, H., Sugawara, S., and Okano, T., 1980, Liposome immunosensor for theophylline, _Biochem. Biophys. Res. Com._, 95:187-192.

Haga, M., Sugawara, S., and Itagaki, H., 1981, Drug sensor: Liposome immunosensor for theophylline, _Analyt. Biochem._, 118:286-293.

Hsia, J. C., and Tan, C. T., 1978, Membrane immunoassay : Principle and applications of spin membrane immunoassay, _Ann. N. Y. Acad. Sci._, 308:139-148.

Hobart, M. J. and McConnell, I., 1975, "The immune system, a source on the molecular and cellular basis of immunity," Blackwell, Oxford, 357pp.

Humphries, G. K., and McConnell, H. M., 1974, Immune lysis of liposomes and erythrocyte ghosts loaded with spin label, _Proc. Natl. Acad. Sci. US._, 71:1691-1694.

Humphries, G. M. K. and McConnell, H. M., 1975, Antigen mobility in membranes and complement-mediated immune attack, _Proc. Natl. Acad. Sci. USA_, 72:2483-2487.

Humphries, G. M. K. and McConnell, H. M., 1977, Membrane-controlled depletion of complement activity by spin-label-specific IgM, _Proc. Natl. Acad. Sci. USA_, 74:3537-3541.

Ishimori, Y., Yasuda, T., Tsumita, T., Notsuki, M., Koyama, M., and Tadakuma, T., 1984, Liposome immune lysis assay (LILA): A simple method to measure anti-protein antibody using protein antigen-bearing liposomes, _J. Immunol. Methods_, 75:351-360.

Kataoka, T., Inoue, K., Galanos, C., and Kinsky, S. C., 1971a, Detection and specificity of lipid A antibodies using liposomes sensitized with lipid A and bacterial lipopolysaccharides, _Eur. J. Biochem._, 24:123-127.

Kataoka, T., Inoue, K., Luderitz, O., and Kinsky, S. C., 1971b, Antibody- and complement-dependent damage to liposomes prepared with bacterial lipopolysaccharides, _Eur. J. Biochem._, 21:80-85.

Kinsky, S. C., Haxby, J. A., Zopf, D. A., Alving, C. R., and Kinsky, C. B., 1969, Complement-dependent damage to liposomes prepared from pure lipids and Forssman hapten, _Biochem._, 8:4149-4158.

Kinsky, S. C., 1972, Antibody-complement interaction with lipid model membranes, _Biochim. Biophys. Acta_, 265:1-23.

Kinsky, S. C. and Nicolotti, R. A., 1977, Immunological properties of model membranes, _Ann. Rev. Biochem._, 46:49-67.

Kirby, C. and Gregoriadis, G., 1981, Plasma-induced release of solutes from small unilammellar liposomes is associated with pore formation in the bilayers, _Biochem. J._, 199:251-254.

Kornberg, R. D. and McConnell, H. M., 1971, Inside-outside transitions of phospholipids in vesicle membranes, _Biochem._, 10:1111-1120.

Litchfield, W. J., Freytag, J. W., and Adamich, M., 1984, Highly sensitive immunoassays based on use of liposomes without complement, _Clin. Chem._, 30:1441-1445.

Mori, T., Fujii, G., Kawamura, A. Jr., Yasuda, T., Naito, Y., and Tsumita,

T., 1982, Forssman antibody levels in sera of cancer patients, Immunol. Com., 11:217-225.

Okada, N., Yasuda, T., Tsumita, T., and Okada, H., 1982a, Activation of the
alternative complement pathway of guinea-pig by liposomes incorporated
with trinitrophenylated phosphatidylethanolamine, Immunol., 45:115-
124.

Okada, N., Yasuda, T., Tsumita, T., Shinomiya, H., Utsumi, S., and Okada,
H., 1982b, Regulation by glycophorin of complement activation via the
alternative pathway, Biochem. Biophys. Res. Com., 108:770-775.

Okada, H., Okada, N., and Yasuda, T., 1983a, Activation of the alternative
complement pathway by IgM antibody reacted on paragloboside incorporated into liposome membrane, Molec. Immunol., 20:499-500.

Okada, N., Yasuda, T., Tsumita, T., and Okada, H., 1983b, Membrane sialoglycolipids regulate the activation of the alternative complement
pathway by liposomes containing trinitrophenylamino-caproyldipalmito-
ylphosphatidylethanolamine, Immunol., 48:129-140.

Okada, N., Yasuda, T., Tsumita, T., and Okada, H., 1983c, Activation of the
alternative complement pathway by natural antibody to glycolipids in
guinea-pig serum, Immunol., 50:75-84.

Okada, N., Yasuda, T., and Okada, H., 1985, Antibody-mediated alternative
complement pathway activation resists inhibition by sialoglycolipids,
J. Immunol., 134:3316-3319.

D'Orazio, P. and Rechnitz, G. A., 1977, Ion electrode measurements of
complement and antibody levels using marker-loaded sheep red blood
cell ghosts, Anal. Chem., 49:2083-2086.

Parce, J. W. and McConnell, H. M., 1980, Kinetics of antibody-dependent
activation of the first component of complement on lipid bilayer
membranes, Biochem. Biophys. Res. Comm., 93:235-242.

Plant, A. L., 1986, A serum complement assay system based on lysis of lipid
vesicles, (Submitted for publication).

Rosenqvist, E. and Vistnes, A. I., 1977, Immune lysis of spin label loaded
liposomes incorporating cardiolipin; a new sensitive method for detecting anticardiolipin antibodies in syphilis serology, J. Immunol.
Methods, 15:147-155.

Rosenqvist, E., Michaelsen, T. E., and Vistnes, A. I., 1980, Effect of
streptolysin O and digitonin on egg lecithin/cholesterol vesicles,
Biochim. Biophys. Acta, 600:91-102.

Schall, R. F. Jr. and Tenoso, H. J., 1981, Alternatives to radioimmunoassay: Labels and methods, Clin. Chem., 27:1157-1164.

Scherphof, G., Damen, J., and Hoekstra, D., 1981, Interactions of liposomes
with plasma proteins and components of the immune system, in: "Liposomes : From physical structure to therapeutic applications," C. G.
Knight, ed., Elsevier/North-Holland biomedical Press, Amsterdam, p299-
322.

Shiba, K., Umezawa, Y., Watanabe, T., Ogawa, S., and Fujiwara, S., 1980,
Thin-layer potentiometric analysis of lipid antigen-antibody reaction
by tetrapentylammonium (TPA[+]) ion loaded liposomes and TPA[+] ion
selective electrode, Anal. Chem., 52:1610-1613.

Shin, M. L., Paznekas, W. A., and Mayer, M. M., 1978, On the mechanism of
membrane damage by complement: The effect of length and unsaturation
of the acyl chains in liposomal bilayers and the effect of cholesterol
concentration in sheep erythrocyte and liposomal membranes, J. Immunol., 120:1996-2002.

Six, H. R., Uemura, K-i., and Kinsky, S. C., 1973, Effect of immunoglobulin
class and affinity on the initiation of complement-dependent damage to
liposomal model membranes sensitized with dinitrophenylated phospholipids, Biochem., 12:4003-4011.

Smolarsky, M., Teitelbaum, D., Sela, M., and Gitler, C., 1977, A simple
fluorescent method to determine complement-mediated liposome immune
lysis, J. Immunol. Methods, 15:255-265.

Szoka, F. Jr. and Papahadjopoulos, D., 1978, Procedure for preparation of

liposomes with large internal aqueous space and high capture by
reverse-phase evaporation, <u>Proc. Natl. Acad. Sci. USA</u>, 75:4194-4198.

Szoka, F. Jr. and Papahadjopoulos, D., 1980, Comparative properties and
methods of preparation of lipid vesicles (liposomes), <u>Ann. Rev.
Biophys. Bioeng.</u>, 9:467-508.

Takashi, T., Inoue, K., and Nojima, S., 1980, Immune reactions of liposomes
containing cardiolipin and their relation to membrane fluidity, <u>J.
Biochem.</u>, 87:679-685.

Tan, C. T., Chan, S. W., and Hsia, J. C., 1981a, Specific anti-thyroxine
antisera induced by thyroxine sensitized liposomes, <u>Immunol. Com.</u>,
10:27-34.

Tan, C. T., Chan, S. W., and Hsia, J. C., 1981b, Membrane immunoassay : A
spin membrane immunoassay for thyroxine, <u>in:</u> "Methods in Enzymology,
Vol 74 : Immunochemical techniques, Part C," J. J. Langone and H.
VanVunakis, eds., Academic Press, New York, p152-161.

Uemura, K-i., and Kinsky, S. C., 1972, Active vs. passive sensitization of
liposomes toward antibody and complement by dinitrophenylated derivatives of phosphatidylethanolamine, <u>Biochem.</u>, 11:4085-4094.

Uemura, K-i., Yuzawa-Watanabe, M., Kitazawa, N., and Taketomi, T., 1980,
Liposome agglutination and liposomal membrane immune-damage assay for
the characterization of antibodies to glycosphingolipids, <u>J. Biochem.</u>,
87:1641-1648.

Uemura, K-i., Hattori, H., Kitazawa, N., and Taketomi, T., 1982, Immunochemical determination of Forssman and blood group A-active glycolipids
in human gastric mucosa by inhibition assay of liposome lysis, <u>J.
Immunol. Methods</u>, 53:221-232.

Umezawa, Y., Sofue S., and Takamoto, Y., 1984, Thin-layer ion-selective
electrode detection of anticardiolipin antibodies in syphilis serology, <u>Talanta.</u>, 31:375-378.

Vistnes, A. I. and Puskin, J. S., 1981, A spin label method for measuring
internal volumes in liposomes or cells, applied to Ca-dependent fusion
of negatively charged vesicles, <u>Biochim. Biophys. Acta</u>, 644:244-250.

Vistnes, A. I. and Wormald, D. I., 1982, A g-value controlled device for
amplitude measurements in electron paramagnetic resonance, <u>J. Magnetic
Reson.</u>, 46:125-128.

Vistnes, A. I., 1983, Spin label methodology in membrane research with
applications to immunology, Dr. Philos. dissertation, University of
Oslo, Norway.

Vistnes, A. I., Rosenqvist, E., and Froholm, L. O., 1983, Spin membrane
immunoassay for use in meningococcal serology, <u>J. Clin. Microbiol.</u>,
18:905-911.

Vistnes, A. I., 1984, A new method of evaluating complement mediated lysis
of liposomes, <u>J. Immunol. Methods</u>, 68:251-261.

Wei, R., Alving, C. R., Richards, R. L., and Copeland, E. S., 1975, Liposome spin immunoassay: A new sensitive method for detecting lipid
substances in aqueous media, <u>J. Immunol. Methods</u>, 9:165-170.

Wood, C. and Kabat, E. A., 1981, Immunochemical studies of conjugates of
isomaltosyl oligosaccharides to lipid, <u>J. Exp. Med.</u>, 154:432-449.

Yasuda, T., Naito, Y., Tsumita, T., and Tadakuma, T., 1981, A simple method
to measure anti-glycolipid antibody by using complement-mediated
immune lysis of fluorescent dye-trapped liposomes, <u>J. Immunol. -
Methods</u>, 44:153-158.

Yasuda, T., Ueno, J., Naito, Y., and Tsumita, T., 1982, Antiglycolipid
antibodies in human sera, <u>Adv. Exp. Med. Biol.</u>, 152:457-465.

Yost, Y., Polnaszek, C. F., Mason, R. P., and Holtzman, J. L., 1981, Application of spin labeling to drug assays, <u>J. Label. Compoun. Radio-
pharm.</u>, 18:1089-1097.

Zemmour, J., Portoukalian, J., and Dore, J. F., 1984, Serological specificity of the liposome lysis test for measurement of anti-ganglioside
antibodies. A comparison with hemagglutination inhibition, <u>J. Immunol.
Methods</u>, 66:331-340.

IMMUNOASSAY USING FLUORESCENT DYE-TRAPPED LIPOSOMES

LIPOSOME IMMUNE LYSIS ASSAY (LILA)

Tatsuji Yasuda, Yoshio Ishimori and Mamoru Umeda

Laboratory of Biological Products
The Institute of Medical Science
The University of Tokyo, Minato-ku, Tokyo 108

INTRODUCTION

The liposome immune lysis assay (LILA) system was developed by S. C. Kinsky and coworkers in 1968 (Haxby, 1968). They used glucose as an internal liposome marker, and measured the glucose release in an enzymatic reaction. They also introduced umbelliferone phosphate and alkaline phosphatase as the first fluorescent marker system (Six, 1974). The enzyme reaction resulted in both the generation and the amplification of fluorescence. In 1977, Smolarsky et al. developed a novel LILA system using a complex of a fluorescent molecule (1-aminonaphthalene-3,6,8-trisulfonic acid (ANTS)) and a quencher (dipyridinium-p-xylene) as a marker trapped in liposomes. The complex, when entrapped in liposomes, shows little fluorescence. Upon lysis of the liposomes, dilution of the quencher in the external volume leads to a high fluorescence signal. One of the problems of the use of ANTS as a fluorophore is, however, that its excitation and emission wavelengths coincide with those of a fluorescent component existing in complement sources, and precise measurements may not be possible because of an increase in the background fluorescence.

In the same year, Weinstein et al. used carboxyfluorescein (CF) as an internal marker to detect the fusion between cells and liposomes (1977). CF is a highly water soluble fluorophore and emits high fluorescence. Moreover, CF is commercially available and inexpensive. Since we presumed that this fluorophore is suitable for simplifying the LILA using glucose as an internal marker, we applied this fluorophore to determine the antibody titers against various glycolipid antigens (Yasuda, 1981). Recently we used this method to measure the antibody against protein antigen (Ishimori, 1984) and to determine the serum protein concentrations (Umeda, 1986a; Umeda, 1986b). This method, which is also based on the quenching phenomenon of the fluorophore, is not affected by fluorescent components in sera. The trapping of CF at high concentrations within liposomes leads to the quenching of the fluorescent signal. Upon lysis, dilution of the released substances markedly reduces the quenching, and thus the fluorescent signal for the titrating antibody or antigen can be detected. Furthermore, this LILA system has the following advantages; 1) CF is commercially available, and 2) the liposomes can be stored for as long as over one year without substantial leakage of the fluorophore.

MATERIALS AND METHODS

Buffers

A 5 times concentrated stock solution of veronal-buffered saline (VB) is prepared by dissolving 10.19 g sodium veronal and 83.0 g NaCl in 1500 ml H_2O, then the pH is adjusted to pH 7.4 with 1N HCl and then volume made up to 2000 ml with H_2O. For experimental use, the stock solution is diluted and supplemented with 2% gelatin to a final concentration of 0.1% (GVB$^-$), or, in addition, with 3.5 mM $MgCl_2$ and 0.15 mM $CaCl_2$ (GVB^{++}). The diluted stock solution containing 10 mM disodium ethylenediaminetetraacetate (EDTA-VB; pH7.4) is prepared as a stopping reagent for the reaction. The GVB$^-$ and GVB^{++} are kept at 4°C until use.

The components of HEPES-buffered salt solution (HBS) are as follows: NaCl, 8.00 g; KCl, 0.40 g; $Na_2HPO_4 \cdot 12H_2O$, 0.25 g; glucose, 1.00 g; 4-(2-hydroxyethyl)-1-piperazineethanesulfonic acid (HEPES), 2.38 g; H_2O, 1000 ml. The pH is adjusted to 7.40 with 1 N NaOH.

The acetate-buffered saline contains 8.20 g sodium acetate and 8.50 g NaCl in 1000 ml of H_2O. The pH is adjusted to 4.50 with 0.1 N acetic acid. These two buffer solutions are stored at room temperature.

Lipids

L-α-Dipalmitoylphosphatidylcholine (DPPC), L-α-dimyristoylphosphatidyl-choline (DMPC), DL-α-dipalmitoylphosphatic acid (DPPA) and dipalmitoyl-phosphatidylethanolamine (DPPE) are available from Sigma (St. Louis, MO) and variety of other commercial suppliers, including Calbiochem-Behring (La Jolla, CA), Avanti (Birmingham, AL), Nihon Shoji Kaisha Ltd. (Osaka, Japan) and Nippon Fine Chemical Co. Ltd. (Osaka, Japan). Cholesterol (Chol) is obtained from Sigma. We usually use these commercial lipids without further purification. DPPC (5 mM), DMPC (5 mM), Chol (10 mM) and DPPA (1 mM) in chloroform or chloroform/methanol (2/1, v/v) are stored at -20°C after substitution of air with nitrogen gas in a test tube with a Teflon liner cap (Pyrex, TST-SCR 13-100).

Preparation of dithiopyridyldipalmitoylphosphatidylethanolamine (DTP-DPPE)

DTP-DPPE is prepared by the method of Leserman et al. (1981) with minor modifications.

(1) Seventy mg of DPPE is suspended in 30 ml of chloroform/methanol (5/1) and then dissolved with the aid of a bath type sonicator (Bransonic-12). Sixty μl of triethylamine (Wako Pure Chemicals Co., Osaka, Japan) and 50 mg of N-succinimidyl-3-(2-pyridyldithio)propionate (SPDP) (Pharmacia Fine Chemicals, Uppsala, Sweden) are added to the DPPE solution.

(2) The reaction mixture is stood for 60 min at room temperature after thorough bubbling with nitrogen gas and shaking. Almost all of the DPPE used is converted to DTP-DPPE because a ninhydrin spot due to free NH_2 groups in DPPE completely disappears on silica gel thin layer chromatography with chloroform/methanol/H_2O (65/35/8) as the developing solvent during the reaction.

(3) The reaction mixture is evaporated to dryness with a rotary evaporator at 40°C. Sixty ml of chloroform/methanol (2/1) is added to redissolve the product, followed by the addition of 15 ml H_2O and shaking in a partition flask. The clear organic layer is collected and evaporated.

(4) The product is dissolved again in 5.5 ml of chloroform/methanol (10/1) and then charged on to a silica gel (Kiesel gel 60) column (8 mm X 190 mm) activated by overnight heating at 120°C in an oven and pre-equilibrated with ca. 200 ml of chloroform. The column is washed with chloroform and then 50 ml of chloroform/methanol (10/1) successively. Then, 80 ml of chloroform/ methanol (5/1) is applied and collected. Thin layer chromatography of this elute shows one spot (phosphorus positive and ninhydrin negative). The purified product is concentrated to about 10 ml and then stored at -20°C under nitrogen. The yield of DTP-DPPE is usually between 80 and 90%. Total phosphorus is determined

according to the method of Gerlach and Deuticke (1964).

Complement

Young guinea pig (body weight around 250 g) sera (Shizuoka Agricultural Cooperative Association for Lab. Animals, Hamamatsu, Japan) are used as a complement source. The sera should be obtained from unanesthetized animals. An individual guinea pig may have natural antibodies against several molecules, especially glycolipid antigens, or the Fc portion of the antibody. We routinely test natural antibody titer of an individual serum against liposomes for each assay before selecting a suitable group as a complement source. Specific pathogen free guinea pigs are the best source of complement, if available, because their natural antibody titer is very low. The sera can be frozen indefinitely at -75°C, but they should be aliquoted and should not be refrozen. One unit of complement (1 CH_{50}) is defined as the amount of serum needed for 50 % hemolysis of sensitized sheep red blood cells (5 X 10^8/7.5 ml) after incubation for 60 min at 37°C (Kabat, 1961).

Preparation of carboxyfluorescein (CF) stock solution

CF is available from Eastman Kodak (Rochester, NY). Purification of CF is necessary, since commercial CF usually contains some impurities which interfere with a LILA. Several purification methods have been reported (Ralston, 1981; Weinstein, 1984; Lelkes, 1984). Our purification method is as follows:
(1) Dissolve CF in 1N NaOH. Add slowly in a bath type sonicator. When insoluble particles are observed, centrifuge and discard the precipitate.
(2) Add 1N HCl slowly until the yellow precipitates do not increase. Centrifuge and discard the supernatant.
(3) Repeat the above process at least twice, and then wash with distilled water.
(4) Dry the precipitate in a desiccator in vacuo.
(5) Make up a 100 mM CF solution in distilled water through titration to pH 7.2 with 1N NaOH, monitoring the pH with a pH meter.
(6) The CF solution, protected from light, can be stored in a freezer for months or years with little deterioration.
For CF purification Sephadex LH-20 column chromatography is performed. Our CF purification method is very simple, but enough for LILA with our system. The concentration of CF solution is changed from 200 mM to 100 mM, since the backgroud fluorecence is always low in liposomes prepared from 100 mM CF solution.

Preparation of liposomes containing lipid antigen(s)

Preparation of liposomes according to Alving et al. (1984) is a good way. Multilameller liposomes are sensitized by the addition of a lipid antigen (glycolipid or haptenated PE) to a lipid mixture containing DMPC (0.5 μmoles; 100 μl of the stock solution), Chol (0.5 μmoles; 50 μl of the stock solution), DPPA (0.05 μmoles; 50 μl of the stock solution) in a pear shaped flask of 10 ml volume. The solvent is removed by rotary evaporation at 40°C under reduced pressure and subsequent vacuum desiccation for 1 hr. The dried lipid film is then dispersed with a Vortex mixer in 0.1 ml of 0.1 M CF solution. Warming the flask up to 40°C facilitates smooth dispersion. The liposome preparation is washed by centrifugation at 4°C (20,000 g) for 20 min in 10 ml of sterilized GVB⁻ until no fluorescence is observed in the supernatant (usually repeat 3 to 4 times). The packed liposomes are finally resuspended in 2 ml of GVB⁻ containing 0.1% sodium azide and then stored in a refrigerator.

Preparation of protein antigen-bearing liposomes

Human IgG antigen is used here as a typical case. Five mg of human IgG is dissolved in 2 ml of HBS and then treated with 0.1 μmoles of SPDP under nitrogen gas. The reaction mixture, after being kept for 30 min at room temperature, is charged on to a Sephadex G-25 fine column (10 mm x 170 mm) pre-equilibrated with saline. The protein peak fraction (2 ml) elute with acetate-buffered saline is reduced with 30 mg of dithiothreitol (DTT) under nitrogen gas. After 20 min incubation at room temperature, the reaction mixture is applied to a Sephadex G-25 fine column (15 mm x 150 mm) pre-equilibrated with HBS. The first main peak fraction (2 ml) is stored at 4°C under nitrogen gas.

The liposome suspension (DPPC:Chol:DTP-PE in molar ratio of 1:1:0.1, respectively) (2 ml) is added to the modified human IgG (2 ml), prepared as above, followed by reaction overnight at room temperature under nitrogen gas with slow shaking. The liposomes are collected by centrifugation under the same conditions as mentioned above. The liposomes are resuspended in 2 ml of GVB⁻ containing 0.1% sodium azide, and then stocked at 4°C under nitrogen gas.

Preparation of antibody-bearing liposomes

For the sandwich assay, liposomes conjugated with antibody are used. In the case of serum C-reactive protein (CRP) measurement, anti-CRP antibody to human CRP is obtained from sera of both goats and rabbits immunized with an emulsion of the purified CRP solution and Freund's complete adjuvant (Difco Lab., Detroit, MI). Monospecific antibody is prepared by means of absorption using CRP-negative human sera. The IgG fraction is separeted from the antisera by precipitation with 33 % ammonium sulfate, followed by ion-exchange chromatography on a column of DE-52 cellulose (Whatman Ltd., Maidstone, Kent, England) with 0.05 M Tris-HCl (pH 8.0). Furthermore, the IgG fraction is passed through an affinity chromatography column containing CRP antigen immobilized gel according to a conventional procedure. The purity of IgG is examined by the reversed single radial immunodiffusion method.

Multilamellar liposomes encapsulating CF are prepared from a lipid mixture solution containing DPPC (0.5 µmol), Chol (0.5 µmol) and DTP-DPPE (0.05 µmol). The packed liposomes are suspended in one ml of GVB⁻. The preparation of liposomes with covalently coupled anti-CRP antibody (IgG fraction) is performed by the same method as that described for the protein antigen-bearing liposomes. Five hundred ul of the freshly prepared liposome suspension (about 0.5 µmol of DPPC/ml) is added to 500 µl of the anti-CRP antibody (IgG fraction, 0.25 mg/ml) treated with SPDP and DTT, and then the mixture is stood overnight at room temperature with gentle shaking. Unbound anti-CRP antibody is removed by centrifugation (20,000 g) three times for 20 min in GVB⁻ at 4°C. After the final centrifugation, the pellet of liposomes is resuspended in GVB⁻ (1 ml) and stored as above.

Antibody titer assay on a microtiter plate

The diluent is always GVB⁺⁺. Sera are diluted serially 10-fold. To each well of a microtiter plate (Nunc, Roskilde, Denmark), 25 µl of a diluted serum, 5 µl of the liposome suspension diluted 1/100 and 25 µl of fresh guinea pig serum diluted 200 times (or 1 CH_{50}) are added in duplicate. The hard type of microtiter plate can be used, since flexible one contains fluorescent materials. The total releasable CF from liposomes is determined by the addition of 10% Triton X-100 instead of complement. For a background control, 25 µl of GVB⁺⁺ is added instead of the serum sample. A PB-600 dispenser with a 250 µl microsyringe (Hamilton, Reno, NY) is convenient for dispensing liposomes. After incubation for 60 min at 37°C, 100 µl of EDTA-VB is added using a Jet Pipet (York Instruments, Berkeley, CA) to stop the reaction.

Inhibition assay and sandwich assay

The procedures for the inhibition and sandwich assays are similar to that for the antibody titer assay. The standard assay procedures in our laboratory are as follows:

One step inhibition assay

```
Microplate
        |  <-  Sample (25 µl)
        |  <-  Antibody (25 µl)
        |  <-  Liposomes (5 µl)
        |  <-  Complement (25 µl)
Mix with a microtiter plate mixer
        |
Incubate for 60 min at 37°C
        |  <-  EDTA-VB (100 µl)
Measure fluorescence
```

Two step inhibition assay

```
Microtiter plate
        |  <-  Sample (25 µl)
        |  <-  Antibody (25 µl)
Incubate for 90 min at room temperature or overnight at 4°C
        |  <-  Liposomes (5 µl)
        |  <-  Complement (25 µl)
Mix with a microtiter plate mixer
        |
Incubate for 60 min at 37°C
        |  <-  EDTA-VB (100 µl)
Measure fluorescence
```

Sandwich assay

```
Microplate
        |  <-  Sample (25 µl)
        |  <-  Liposomes (5 µl)
        |  <-  Antibody (25 µl)
        |  <-  Complement (25 µl)
Mix with a microplate mixer
        |
Incubate for 60 min at 37°C
        |  <-  EDTA-VB (100 µl)
Measure fluorescence
```

Instruments

A fluorescent microplate fluorometer MTP-12F (Corona Electric Co.,
Katsuta, Japan) is used for the measurement of CF released from liposomes in
our laboratory. But any type of microplate fluorometer equipped with
suitable filters (excitation at 490 nm and emission at 520 nm) can be used.
When a regular type spectrofluorophotometer is used instead of a microtiter
fluorometer, the reaction can be performed in a 12 mm x 75 mm Pyrex test
tube (Corning Glass Works, Corning, NY). In this case, 2 ml of EDTA-VB is
added to stop the reaction and the intensity of the fluorophore is directly
measured in a glass tube with a spectrofluorophotometer equipped with a test
tube adapter. The total releasable CF in liposomes is determined by lysing
the liposomes with 25 µl of 10% Triton X-100 instead of complement. Spontaneous release is estimated from the fluorescence in the wells of complement and GVB^{++}.
Specific marker release is calculated as follows:

$$\% \text{ specific marker release} = \frac{\text{experimental release} - \text{spontaneous release}}{\text{total release} - \text{spontaneous release}} \times 100$$

A microcomputer is connected with a MTP-12F through an RS232C interface for
data processing. The calculation is performed automatically according to
the assay format.

TYPICAL RESULTS

Antibody titer assay

There are several factors that influence the CF release from liposomes. One is the composition of the liposomes. To obtain stable liposomes, the Chol content is very important. Our experiments have shown that a liposome preparation containing PC and Chol in a molar ratio of 1:1 is better than one with a ratio of 1:0.75, which was used originally (Kinsky, 1974). The addition of a negatively charged lipid (DPPA in our case) in a molar ratio of 0.1 over PC increases the stability and easiness of handling during centrifugation.

The effect of the molecular species of PC is shown in Fig. 1. As can be seen in the figure, DMPC or DPPC is the best PC for the antibody titer assay. These PCs are very stable and the peroxidation which is often seen for PC containing polyene fatty acids need not be considered. Another important factor which influences the assay sensitivity and the specificity is the epitope density of the antigen on lipid bilayers. The marker release increases in parallel with the epitope density of the antigen. This factor should be taken into consideration when one wants to compare data.

When assaying glycoconjugate antigen(s) one should be aware that all sources of sera contain naturally occurring antibodies against numerous glycolipids (Alving, 1979; Okada, 1983b; Kaise, 1985), even a complement source. The experimenter should carefully select the complement and choose controls for the assay system.

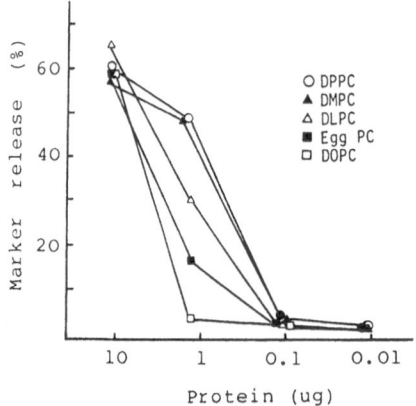

Fig. 1: Influence of the acyl chain composition of phosphatidylcholine on liposomes sensitized with trinitrophenylaminocaproylphosphtidyl ethanolamine (TNP-Cap-PE). Liposomes containing various PCs/Chol/ PA/TNP-Cap-PE (1:1:0.1:0.1) were reacted with anti-TNP monoclonal antibody (6F3) purified by affinity chromatography. DPPC, dipalmitoylphosphatidylcholine; DMPC, dimyristoylphosphatidyl-choline; DLPC, dilauroylphosphatidylcholine; Egg PC, egg yolk phosphatidylcholine; DOPC, dioleoylphosphatidylcholine.

Inhibition assay

It is important to check the specificity of the antibody which is to be used for the inhibition assay. As mentioned by numerous investigators, animal antisera contain antibodies against not-immunized antigen(s). Before using an antibody for an inhibition assay one should check the specificity of the antibody.

Fig. 2 shows a standard curve for Forssman glycolipid. The IgG fraction of Rabbit anti-Forssman antibody is obtained by a DEAE-cellulose column chromatography. Whole sera contains antibodies against a variety of glycolipid antigens, but the IgG fraction is specific to the Forssman antigen. A standard sample of Forssman glycolipid is trapped in sonicated liposomes which do not contain CF. A standard sample and the antibody are incubated overnight at 4°C and then CF containing Forssman liposomes (the Forssman concentration is 0.5% compared to PC) and complement are added followed by further incubation.

The same method can be applied to protein antigens (Ishimori, 1984; Umeda, 1986a). Preincubation of the antibody and a sample (two step method) is not an absolute condition. The simultaneous addition of a sample, antibody, liposomes and complement (one step method) can be performed. In this case the standard curve becomes wider than that in the case of the preincubation method, and relatively high CF release occures. Although for practical use the one step method is better than the two step one, the two step method can use to determine a lower amount of antigen and shows better reproducibility.

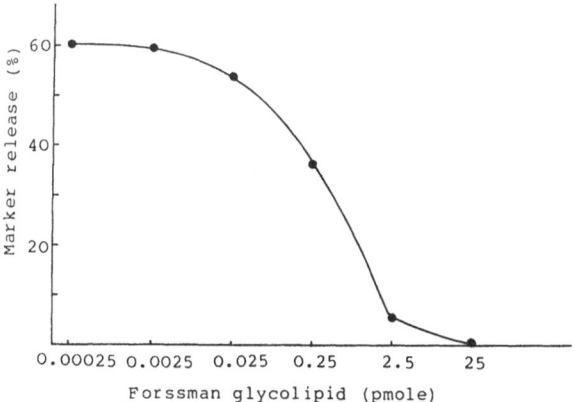

Fig. 2: Standard curve for Forssman glycolipid antigen determined by the inhibition assay.

Sandwich assay

Liposomes containing equal molar amounts of DPPC and Chol (1:1) show the least spontaneous release of CF in the sandwich assay. We choose DPPC and Chol (1:1) as base liposomes and examined the effect of the amount of DTP-DPPE on the binding of an antibody to liposomes. The amount of antibody bound to liposomes is proportional to the antibody concentration in the coupling reaction and the epitope density of DTP-DPPE. Although the sensitivity is proportional to the amount of the antibody bound to liposomes, the liposomes become fragile and their spontaneous release of CF during storage increases when more than 400 μg of anti-CRP antibody per μmol of DPPC is bound to them, irrespective of their composition. Liposomes composed of DPPC, Chol and DTP-DPPE, in a molar ratio of 1:1:0.1, are very stable and show a favorable sensitivity when 160 μg of IgG/μmol of DPPC is bound to the liposomes at an antibody concentration of 250 μg/ml in coupling reaction in the the case of CRP determination.

The concentration of the second antibody and the titer of complement also affect both the sensitivity and the dynamic range in the sandwich assay. As a rule, the higher the concentrations of the second antibody and complement are, the higher is the sensitivity and the narrow is the dynamic range. Fig. 3 shows a typical standard curve for different second antibody concentrations.

For the sensitization of liposomes, the IgG fraction of the rabbit antibody is not suitable because the liposomes are lysed using guinea pig complement. However, the Fab' fragment of rabbit IgG can be sensitized on the liposomes as well as goat IgG antibody. Both antibodies cannot activate guinea pig complement.

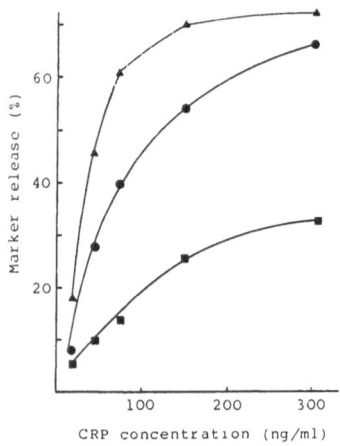

Fig. 3:   Standard curve for human serum CRP measurement by the sandwich assay with different concentrations of the second antibody. The concentration of the second antibody was 0.7 μg/ml (—■—), 7 μg/ml (—●—) or 70 μg/ml (—▲—). The concentration of complement was 2 CH_{50}.

DISCUSSION

The unique features of LILA using CF as an internal liposome marker are as follows:

(1) The assay system is completely homogeneous. This means that the separation of free antigens from bound ones is not required and the washing procedure is eliminated. The assay procedure consists of only four steps of the addition of reagents (the tested serum, antigen-loaded liposomes, complement and the solution for stopping the reaction) to detect the antibody. In the case of the inhibition or sandwich assay one more step (the addition of antibody) is required. These simplicity may facilitate the development of a fully automated determination technique. We are now developing a new machine, which is combination of a diluter and an incubation system, for a semiautomatic assay.

(2) The sensitivity of LILA is almost the same as that of a regular enzyme immunoassay (EIA) or radio immunoassay (RIA). Moreover, for the detection of antiglycolipid antibody, LILA is superior to EIA (Endo, 1984).

(3) Using a microtiter plate fluorophotometer equipped with microcomputer, the assay can be performed on a microtiter plate and a large number of serum samples can be analyzed at one time.

(4) The liposomes with CF entrapped as an internal marker fluorophore are very stable. In some cases, liposomes for LILA can be stored for over a year at 4°C in a storage solution containing 0.1 % sodium azide under nitrogen gas.

(5) The total running cost is very low, since no expensive reagents like isotopes and enzymes are needed. Liposomes which contain antigen(s) associated with one μmol of PC could suffice for 80,000 samples.

Several unexpected results were obtained in the course of establishing the standard assay system. The release of the trapped marker from the liposomes sensitized with haptenated PE was observed with a high concentration of complement (in the range of 2-32 times dilution of guinea pig fresh serum) alone. This marker release was induced through the alternative complement pathway (ACP), since CF was released as in GVB^{++} even in a Mg^{++}-EGTA solution . However, in the presence of antisera, the release of the marker was still maximally observed up to 2^{10} times dilution of the complement. On analysis of the ACP activation mechanism, it was found that liposomes containing haptenic determinants (such as trinitrophenyl or fluoresceinisothiocarbamyl residues) induced the ACP activation in guinea pig sera, whereas SH reactive haptens (dithiopyridyl or maleimidobenzoyl residues) induced the ACP activity in not only guinea pig but also human sera (Okada, 1982; Okada, 1983a). These results indicate that human complement activity can be determined by using these hapten conjugated phosphatidylethanolamine containing liposomes. Liposomes containing haptenic determinants sensitized with anti-hapten antibody can be used for determination of the classical complement pathway activation activity (CH_{50}) and SH reactive hapten liposomes in Mg^{++}-EGTA for the alternative complement pathway activation activity (ACP_{50}). We are investigating a complement activity assay system using liposomes based on these phenomena.

LILA has several potential problems. The most serious one is that a serum sample contains nonspecific factor(s) that damage liposomes and thus the marker dye is released. Preliminary experiments showed that the factor(s) can be absorbed with liposomes not containing an antigen. Another point is that a biological active material (complement) is still used in this system. As mentioned under MATERIALS AND METHODS careful attention is required for preparation and storage.

The applicability of LILA has become wider than in an early investigation, in which it was limited to lipid antigens. The development of the sandwich assay may facilitate the determination of minute amounts of protein antigens in sera in a homogenous way. We are now working on the application of LILA to alpha-fetoprotein, carcinoembryonic antigen, streptolysin O and so on.

We are studying the application of monoclonal antibodies and the oriented attachment of antibodies to liposomes, because it is evident that they will be useful for improving the sensitivity of this assay system. The authors are also working on the application of LILA to other serum components.

SUMMARY

A simple, reproducible, and micro quantitative method has been described to measure antibody or antigen. The multilamellar liposomes containing carboxyfluorescein (CF), which is self-quenched at a high concentration, are prepared by vortexing the dried lipid films consisting of lecithin, cholesterol, phosphatidic acid and lipid antigen(s) or dithiopyridyl phosphatidylethanolamine (DTP-PE). In the case of protein antigens they are conjugated to the surface of liposomes containing DTP-PE through disulfide bonds. The addition of antiserum plus active complement induces liposome lysis and the trapped CF is released. The dilution of CF in the external volume abolishes quenching, resulting in a high fluorescence signal. An inhibition assay and sandwich assay system based on this reaction to measure antigens in serum are developed.

REFERENCES

Alving, C. R., Shichijo, S. and Mattsby-Baltzer, I. (1984) Preparation and use of liposomes in immunological studies. in "Liposome Technology", G. Gregoriadis ed., CRC Press, Boca Raton, FL, 2:157-175.

Endo, T., Scott, D. D., Stewart, S., Kundou, S. K. and Marcus, D. M. (1984) Antibodies to glycosphingolipids in patients with multiple sclerosis and SLE. J. Immunol., 132:1793-1797.

Gerlach, E. and Deuticke, B. (1964) Eine einfache Methode zur Mikrobestimmung von Phosphat in der Papierchromatographie. Biochem. Z., 337: 477-479.

Haxby, J. A., Kinsky, C. B. and Kinsky, S. C. (1968) Immune response of a liposomal model membrane. Proc. Natl. Acad. Sci. USA, 61:300-307.

Ishimori, Y., Yasuda, T., Tsumita, T., Notsuki, M., Koyama, M. and Tadakuma, T. (1984) Liposome immune lysis assay (LILA): a simple method to measure Anti-protein antibody using protein antigen-bearing liposomes. J. Immunol. Methods, 75:351-360.

Kabat, E.A. and Mayer, M.M. (1961) "Kabat and Mayer's Experimental Immunochemistry", Charles C. Thomas Pub., Springfield, IL, p. 133

Kaise, S., Yasuda, T., Kasukawa, R., Nichimaki, T., Watarai, S. and Tsumita, T. (1985) Antiglycolipid antibodies in normal and pathogenic sera and synovial fluids. Vox Sang., 49:292-300.

Kinsky, S. C. (1974) Preparation of liposomes and a spectrophotometric assay for release of trapped glucose marker. Methods in Enzymology, 32:501-513.

Lelkes, P. I. (1984) Methodological aspects dealing with stability measurements of liposomes in vitro using the carboxyfluorescein. in "Liposome Technology", G. Gregoriadis ed., CRC Press, Boca Raton, 3:225-246.

Leserman, L. D., Mache, P. and Barbet, J. (1981) Targeting to cells of fluorescent liposomes covalently coupled with monoclonal antibody or protein A. Nature, 288:602-604.

Okada, N., Yasuda, T., Tsumita, T. and Okada, H. (1982) Activation of the alternative complement pathway of guinea-pig by liposomes incorporated with trinitrophenylated phosphatidylethanolamine. Immunology, 45:115-124.

Okada, N., Yasuda, T., Tsumita, T. and Okada, H. (1983a) Differing reactivity of human and guinea pig complement on haptenized liposomes via the alternative pathway. Mol. Immunol., 20:857-864.

Okada, N., Yasuda, T., Tsumita, T. and Okada, H. (1983b) Activation on the alternative complement pathway by natural antibody to glycollipids in guinea-pig serum. Immunology, 50:75-84.

Ralston, E., Hjelmeland, L. M., Klausner, R. D., Weinstein, J. N. and Blumenthal, R. (1981) Carboxyfluorescein as a probe for liposome-cell interactions: effect of impurities, and purification of the dye. Biochim. Biophys. Acta, 694:133-137.

Six, H. R., Young, W. W., Jr., Uemura, K. and Kinsky, S. C. (1974) Effect of antibody-complement on multiple vs. single compartment liposomes. Application of a fluorometric assay for following changes in liposomal permeability. Biochemistry, 13:4050-4058.

Smolarsky, M., Teitelbaum, D., Sela, M. and Gitler, C. (1977) A simple fluorescent method to determine complement- mediated liposome immune lysis. J. Immunol. Methods, 15:255-265.

Umeda, M., Ishimori, Y., Yoshikawa, K.,Takada, M. and Yasuda, T. (1986a) Homogeneous determination of C-reactive protein in serum using liposome immune lysis assay (LILA). Jpn. J. Exp. Med., 56:35-42.

Umeda, M., Ishimori, Y., Yoshikawa, K., Takada, M. and Yasuda, T. (1986b) Liposome immune lysis assay (LILA) : Application of sandwich method to determine a serum protein component with antibody bearing liposomes. J. Immunol. Methods, submitted.

Weinstein, J. N., Yoshikami, S., Henkart, P., Blumenthal, R. and Higgins, W. A. (1977) Liposome-cell interaction: transfer and intracellular release of a trapped fluorescent marker. Science, 195:489-492.

Weinstein, J. N., Ralston, E., Leserman, L. D., Klausner, R. D., Dragsten, P., Henkart, P. and Blumenthal, R. (1984) Self-quenching of carboxy-fluorescein fluorescence: uses in studying liposomes stability and liposome-cell interaction. in "Liposome Technology", G. Gregoriadis ed., CRC Press, Boca Ralton, FL, 3:183-204.

Yasuda, T., Naito, Y.., Tsumita, T. and Tadakuma, T. (1981) A simple method to measure anti-glycolipid antibody by using complement-mediated immune lysis of fluorescent dye-trapped liposomes. J. Immunol. Methods, 44:153-158.

SECTION V

"PARTICLE" - MEDIATED IMMUNOASSAY

SHELL-CORE PARTICLES FOR TURBIDIMETRIC IMMUNOASSAYS

William J. Litchfield

E. I. du Pont de Nemours & Company, Inc.
Biomedical Products Department
Glasgow Research Laboratory, Wilmington, DE 19898

INTRODUCTION

In 1980, we began to research and develop a number of homogeneous immunoassays for the Du Pont aca® discrete clinical analyzer. In doing so, we desired a technology that could provide rapid and accurate results, be easily automated on commercial equipment, and be sufficiently versatile to use with many different analytes of clinical importance. After evaluating a variety of immunoassay technologies, we selected latex agglutination which was measured turbidimetrically and which used novel shell-core particles synthesized by emulsion polymerization. These submicron particles contained cores of high refractive index polymers and were optimized in their size distribution to give maximal signal upon agglutination. They were coated with thin chemically reactive shells that could covalently bind antigens or antibodies and that were necessary for long term stability.

In developing this technology, assays were investigated by attaching low molecular weight haptens such as theophylline, gentamicin and tobramycin to the polyglycidyl methacrylate shells of particles with polystyrene cores. In some cases haptens were attached directly to reactive epoxide groups, while in others they were attached to epoxides using either albumin or polyethylene ether polyamine (PEPA) as a linker. Since specific antibodies caused agglutination of these particles and free analyte inhibited agglutination, we referred to these methods as PETINIA or particle enhanced turbidimetric inhibition immunoassays. Using particles with polyvinyl-naphthalene cores, other PETINIA methods were investigated for haptens such as cortisol and for proteins such as fibrin degradation products (FDP).

With antibodies directly attached to the shells, we demonstrated assays for C-reactive protein, IgG and ferritin, which we called PETIA for direct particle enhanced turbidimetric immunoassays. Also, a dual particle PETINIA method for digoxin was researched using both ouabain-albumin coated particles and antibody coated particles.

To date, PETINIA methods for theophylline and FDP have been commercialized for use on the Du Pont aca, while PETINIA kits for gentamicin and tobramycin are available from Du Pont for use on other clinical chemistry systems. This article will review some of our experiences with these assays and others that may be forthcoming.

403

MATERIALS AND METHODS

Shell-core particle syntheses. To produce research quantities of submicron shell-core particles that could be used in a number of methods,
we first synthesized an emulsion of polystyrene core particles as described
by Litchfield et al. (1984). These were prepared at 70°C under a nitrogen
atmosphere by adding 50 mL of styrene to 1.0 L of water containing 6.0 g
of Gefac RE-610 surfactant and 2.0 g of potassium per sulfate initiator.
After 30 min. of incubation, we slowly added to this a mixture of 400 mL
of styrene and 1.5 g of Aerosol OT-100 surfactant at a rate of 4 mL/min.
The emulsion was then held at 70°C for 1 hour to assure complete polymerization. Polyvinylnaphthalene core particles were made in a similar manner
by slowly adding 2-vinylnaphthalene into a seed emulsion of polystyrene
particles prepared with sodium dodecyl sulfate. For syntheses of polyglycidyl methacrylate shells at 80°C, we added 200 mL of the core particles
to 50 mL of water containing 0.2 g each of anhydrous potassium carbonate
and potassium persulfate. Glycidyl methacrylate was then pumped in at a
rate of 0.1 mL/min. until 3.9 mL of monomer was added. After 45 min., the
suspensions were cooled and stored refrigerated.

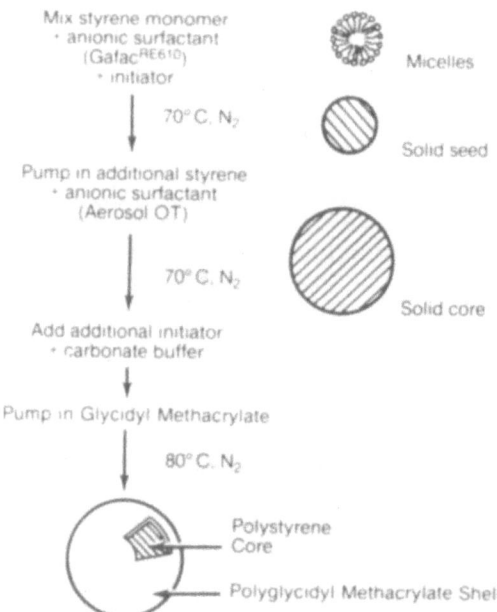

Fig. 1.   Schematic representation of novel shell-core
               particle synthesis.

Conjugate syntheses. For each low molecular weight analyte studied, we
investigated a number of ways to attach immunochemicals to the shell-core
particles. For the PETINIA theophylline method, the best performance was
obtained using a theophylline derivative, 8-(3'-carboxypropyl)-1,3-
dimethylxanthine, attached to a PEPA linker. The derivative was synthesized according to Cook et al. (1976) and activated with N-hydroxysuccinimide before adding the polyamine. PEPA was also used with
cortisol-3-carboxymethyloxime. With methods for gentamicin and tobramycin, we found that human serum albumin was the preferred linker, so these
drugs were routinely attached to this protein using glutaraldehyde. In
the case of the dual particle digoxin method, an analog of digoxin known
as ouabain was also covalently bound to albumin using periodate oxidation

404

to activate the ouabain and using cyanoborohydride to reduce the albumin-ouabain complex. To obtain optimal performance in each of the assays, many separate batches of conjugates with differing ratios of hapten to linker had to be synthesized and tested.

Particle reagent syntheses. Particle reagents containing antigens or antibodies coupled to the shell-core particles were prepared by reacting immunochemicals with the epoxide groups of glycidyl methacrylate. With hapten coated particles, the reaction was generally performed by incubating hapten-linker conjugates ($\sim$1 to 2 mg/mL in phosphate buffer) with dilute suspensions of particles ($\sim$0.1 to 2% solids) at 50 to 70°C for 1 to 2 hours. To obtain optimal performance in each assay, the exact conditions of temperature, timing, and pH (between 7.8 and 10) differed for each particular particle reagent. For proteins such as FDP, the incubation was carried out for 24 hours at 4°C, while for specific antibodies, the incubations were usually 1 to 3 hours at 33 to 37°C. After completing the coupling reactions, particles were washed repeatedly by centrifugation and resuspension in buffers until all significant amounts of unattached immunochemical were removed. All centrifugations were performed using a Du Pont Sorvall® Model RC-58 centrifuge, and most resuspension buffers were 15 mM phosphate pH 7.8 containing 0.1% Gafac RE-610. In the dual particle digoxin assay, the antibody coated particles were prepared with affinity purified rabbit anti-digoxin (10 µl/mL mixed with 2 mg/mL human serum albumin), and the particles were washed repeatedly with 200 mM glycine pH 7.5. After thorough washing, all particle reagents were stored refrigerated in the presence of preservative such as 0.1% sodium azide or 0.01% thimerosal.

In our early work with these particle reagents, most preparations required some sonication after the centrifugation steps since the particles were packed too tightly to resuspend with gentle mixing. As the coupling and washing conditions were adjusted for each assay, however, the need for sonication (a difficult to reproduce variable) was overcome.

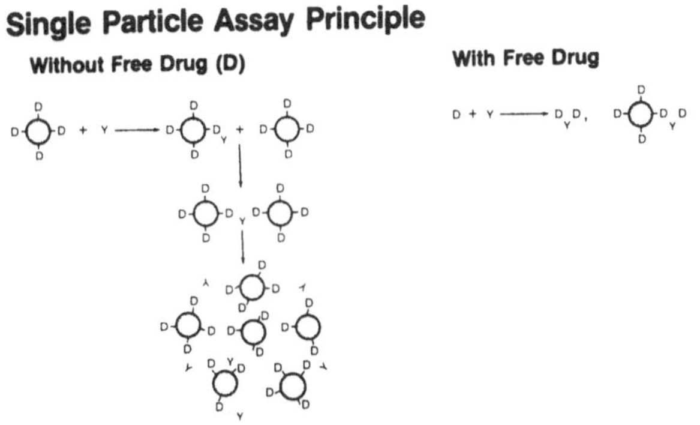

Fig. 2 The single particle PETINIA requires specific antibody (Y) which aggregates latex particles coated with antigens or haptens. In this figure the hapten is represented by a drug (D). Free drug from a patient's sample inhibits both the rate and extent of aggregation.

405

Immunoassay conditions. Although all immunoassay conditions were first
determined at 37°C using small test tubes and conventional spectrophotometers, the final conditions for most assays were established using
analytical test packs as described by Babcock et al. (1983) for the
Du Pont aca. Samples of 20 to 60 μL for most assays and 500 μL for the
digoxin assay were pipetted into the test packs along with sufficient
phosphate buffer (150 mM, pH7.8) to bring the volume to 5.0 mL. For the
PETINIA theophylline method, packs with 40 μL samples were incubated at
37°C for 84 seconds before adding the particle reagent along with 11.1 g/L
polyethylene glycol 8000(PEG), 7.5 mM dithioerythritol, 1.2 g/L sorbitol
and 1.83 g/L Gafac RE-610 (all final concentrations in the reaction). The
agglutination was begun 3.5 minutes later by adding monoclonal antibody to
theophylline, after which the rate of agglutination was measured by
following the absorbance change at 340 nm. The order of addition of
antibody and particle reagent was reversed for the PETINIA FDP method;
however, PEG, dithioerythritol, sorbitol and sodium dodecyl sulfate (instead
of Gafac) were added at the same time. Dithioerythritol was present to
remove potential interferences from serum rheumatoid factors but was not
included in some of the earlier assay designs (particularly digoxin and
cortisol). Before making large lots of analytical test packs for any
of these immunoassays, the amounts of antibody and particle reagent used
were varied until the best clinical performance was obtained.

    For the dual particle PETINIA digoxin assay, the antibody coated
particles were added to the diluted sample at 84 seconds, and the antigen
coated particles were added later to initiate agglutination. No other
additives to the reaction such as PEG or surfactant were necessary.

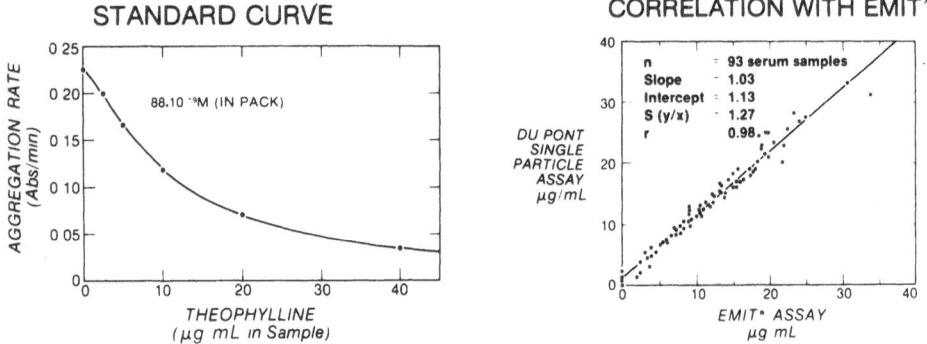

Fig. 3   Performance of the single particle PETINIA method for theophylline
          on the Du Pont aca.

RESULTS AND DISCUSSION

    Synthesis of submicron shell-core particles, illustrated in Figure 1,
was critical to the successful demonstration of the PETINIA and PETIA

immunoassays diagrammed in figures 2, 5, and 7. Seed emulsions of polystyrene were first prepared by initiating the polymerization of styrene
in surfactant micelles with persulfate. The seed emulsions were then
further reacted in a controlled manner with either styrene or 2-vinyl-
naphthalene monomers to increase the particle diameters to between 70
and 100 nm. In these stages, it was important to add the proper amounts
of each material and to carefully control the time and temperature conditions to assure the desired particle size and complete conversion of the
monomer to polymer. Using gas chromatography, the percent conversion of
monomer to polymer was found to be greater than 99%. In the final stages
of synthesis, glycidyl methacrylate was pumped into the emulsions of core
particles at a slow rate to assure that the cores would be covered with a
reactive shell. Analyses of these particles by IR spectroscopy indicated
that between 95 and 98% of the shell monomer was polymerized and that
only 5% or less of the reactive epoxide groups had hydrolysed. From both
calculations and electron microscopy of the particles, we determined that
the shells were on the order of only 1 to 2 nm thick and that essentially
all of the core particles were coated with the shells.

Table 1. Refractive Indices of Polymers at 569 nm.

	Refractive Index
Polytetrafluoroethylene	1.34
Polymethylmethacrylate	1.49
Polybenzyl methacrylate	1.57
Polystyrene	1.59
Polychlorostyrene	1.61
Polyvinylidene chloride	1.63
Polyvinylnaphthalene	1.68
Polyvinylcarbazole	1.68

Polystyrene and polyvinylnaphthalene were chosen for the core materials
because of their high refractive indices relative to water - which has a
refractive index of 1.33 at the sodium D line (569 nm). As the refractive
index increases so does the signal generated by light scattering, so more
turbidity is obtained from polyvinylnaphthalene particles than from polystyrene particles of a comparable size. Core particles of polyvinyl-
carbazole were tried in our studies; however, the strong absorbance of
this polymer at 340 nm was too high for practical use. Core particles of
polytetrafluoroethylene were not tried since they show no significant
turbidity in the submicron size range. The desired size of polystyrene
core particles for the Du Pont aca was determined experimentally to be a
little less than 0.1 micron (Looney, 1984).

Results with a single particle PETINIA for theophylline are shown in
figure 3. This method using polystyrene cores detected as low as 2 µg/mL
theophylline in the sample (or 88 nanomolar drug in the reaction) with good
precision of 2.6% CV at the 10 and 1.7% CV at the 20 µg/mL levels. Very
good correlation between this method and an EMIT assay for theophylline
was observed, and additional correlations to RIA and HPLC verified the
acceptability of this approach. Reproducibility and the shape of the
standard curve were significantly improved by using shell-core particles
with the PEPA linker rather than a human serum albumin linker.

Shell-core particles with cores of polyvinylnaphthalene were first prepared for use in the digoxin assay but significantly improved the signals for other methods such as FDP and cortisol. As shown in figure 4, the immunoassay for FDP not only displayed better signal (turbidity/min) using polyvinylnaphthalene cores but also the sample size was reduced from 200 to 60 μL per test. In single particle immunoassays for cortisol, similar results were found when comparing polyvinylnaphthalene and polystyrene cores.

In the case of digoxin, immunoassays using the single particle PETINIA, even with the polyvinylnaphthalene cores, could only measure down to 10 ng/mL digoxin in the sample (or about 1.3 nanomolar digoxin in the reaction). The single particle assay was beneficial for screening antibodies; however, it did not have sufficient sensitivity for clinical

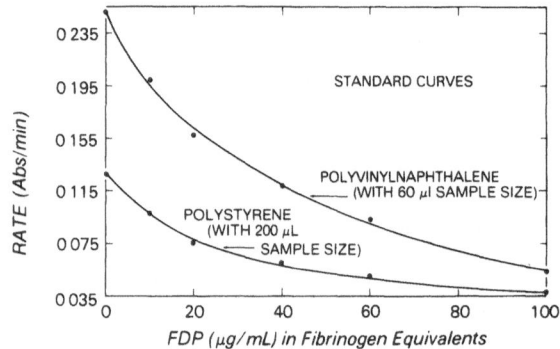

Fig. 4.   Comparison of polystyrene and polyvinylnaphthalene core particles in an immunoassay for fibrin degradation products

use. To improve the sensitivity another 10 to 20 fold, we developed the dual particle PETINIA approach (Figure 5) where both antigen coated and antibody coated particles were employed. In this approach, adequate clinical sensitivity down to 0.5 ng/mL digoxin in the sample was obtained by carefully limiting the number of antibodies placed on the polyvinylnaphthalene particles ($\sim$2 to 6 active antibodies per particle as determined by equilibrium dialysis) and by using high affinity polyclonal antibodies ($<10^9$ L/mol affinity constants). Since agglutination in the dual particle approach did not require the formation of a "transient antibody particle" species, as did single particle assays, the initial rates of agglutination were considerably more sensitive to the levels of free digoxin present. To perform this assay on the Du Pont aca, we typically measured the rates of absorbance change as soon as possible ($\sim$5 sec) after mixing the two particles.

Dual Particle Assay Principle

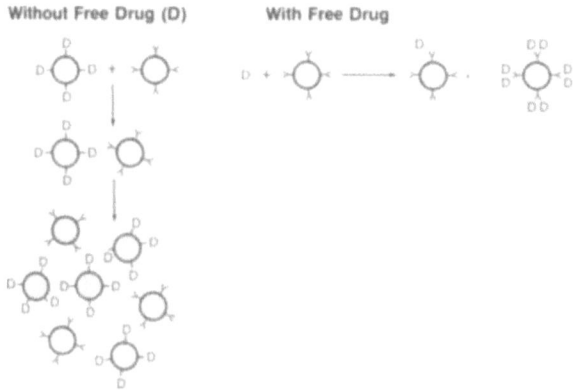

Fig. 5.   The dual particle PETINIA requires particles coated with
a small number of specific antibodies (Y) which aggregate
other particles coated with antigen or haptens, such as
drugs (D). Free drug from the patient's sample inhibits
the rate of aggregation.

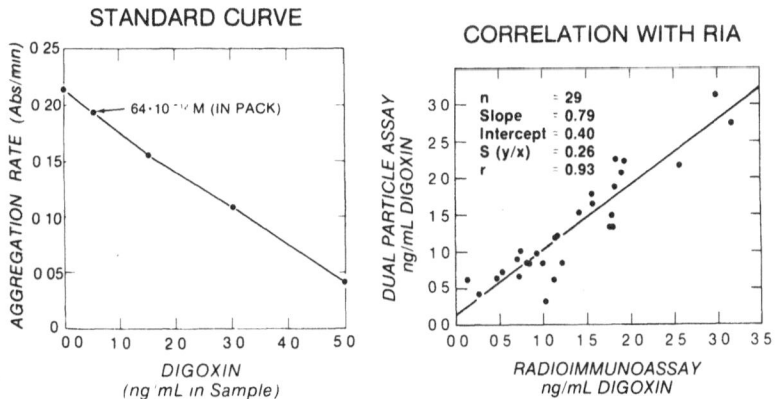

Fig. 6   Performance of the dual particle PETINIA approach for
digoxin on the Du Pont aca.

Results of the dual particle PETINIA for digoxin, shown in figure 6, demonstrate that the standard curve was nearly linear (as expected from theory and the pseudo-first order kinetics) and that the assay correlated fairly well to a RIA method for digoxin. Two RIA methods from Corning and Nuclear Medical Laboratories were commonly run as references, and these showed no better correlation with each other than the results in figure 6. The digoxin PETINIA displayed comparable within-run precision to the RIA methods with 15% CV at the 0.5 ng/mL level, 6.9% CV at 1.5 ng/mL, and 3.6% CV at 3.0 ng/mL. Cross reactivity was also assessed in the absence of digoxin by adding deslanoside at 1 ng/mL, digitoxin at 100 ng/mL, and ouabain at 20 ng/mL. Digoxin values of 0.5, 1.5, and 0 ng/mL were observed, respectively.

Although ouabain instead of digoxin was covalently linked to the antigen particles, the affinity of the antibody to ouabain was four orders of magnitude less than that for digoxin. This did not affect rates of agglutination (comparing freshly prepared ouabain to digoxin coated particles) since each antigen particle was coated with hundreds of ouabain molecules and since an excess of antigen particles was used in the assay. Ouabain instead of digoxin was actually beneficial since any desorption from the particle surface would not be detected as free drug by the antibody. Particle reagents for the digoxin assay were found to be stable, with no detectable change in activity, over 80 days of storage at 4°C.

Direct Antibody Particle Assay Principle

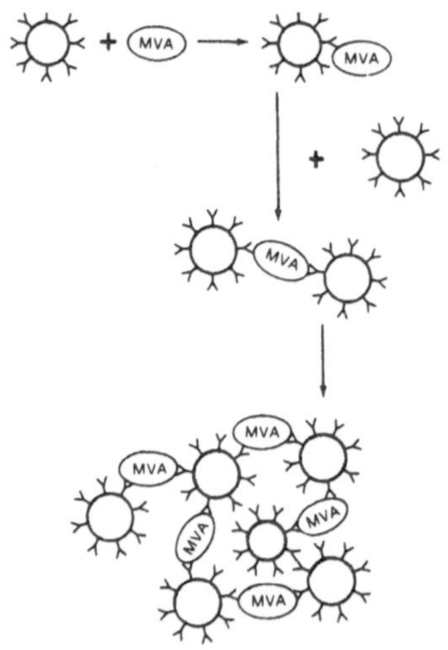

Fig. 7   The direct PETIA approach requires specific antibody (Y) coated particles which aggregate in the presence of multivalent antigens (MVA)

One of the first particle based immunoassays studied on the Du Pont aca was a PETIA approach for antibody to streptolysin-O using reagents from an agglutination kit of Calbiochem-Behring. This approach used antigen coated polystyrene particles that were directly agglutinated by the presence of antibody in patient samples. A direct antibody particle assay principle (figure 7) was later demonstrated using specific antibodies bound to the shell-core particles which were agglutinated by the presence of analytes with multiple antigenic determinants. Unlike the dual particle PETINIA, a large number of antibodies per particle were desired in this case to optimize rates of reaction and overcome restrictions caused by the intrinsic antibody affinity. Antibodies that normally were too low in affinity to work in other types of immunoassays gave acceptable results in the direct PETIA approach. This was expected because analytes diffusing toward the particles had a lower probability of escaping the high concentrations of antibody on the shell.

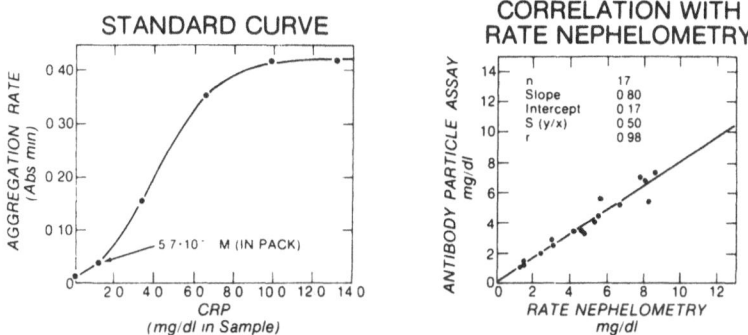

Fig. 8.   Performance of the direct PETIA approach for C-reactive protein on the Du Pont aca.

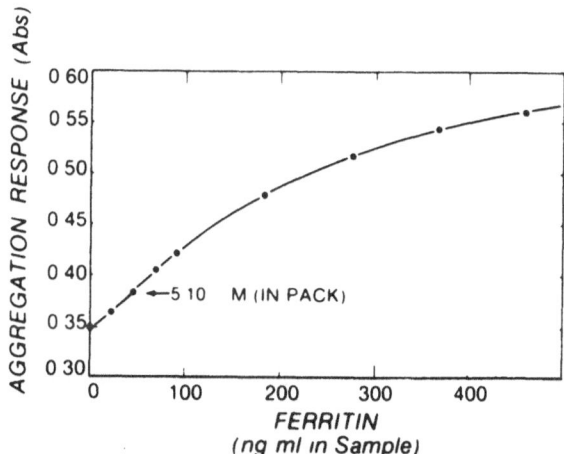

Fig. 9   Performance of the direct PETIA approach for ferritin

Results with direct antibody particle assays for C-reactive protein and ferritin are shown in figures 8 and 9. Using shell-core particles with polyvinylnaphthalene cores, the CRP assay gave good correlation to rate nephelometry, and the lowest detectable levels of CRP and ferritin were 0.57 nanomolar and 5 picomolar, respectively, in the reactions. The increased sensitivity observed with the ferritin assay may be partially attributed to the large number of epitopes present on the ferritin molecule (containing 24 subunits). In addition, agglutination caused by ferritin was measured as an endpoint after 3.5 minutes instead of following initial rates. Further details of the ferritin assay may be found in the preceding article by Peter Horsewood.

SUMMARY

Few advances in latex particle design have been published since Singer and Plotz (1956) first described an agglutination test for rheumatoid factor. Until recently, most immunoassays based on latex agglutination used polystyrene particles in the micron size range and most were semiquantitative in nature, requiring frequent calibration. These tests were not commonly used for clinical analyses mainly because of their poor reliability and because they were not designed to give optimal performance on commercial instruments.

In our work at Du Pont, we decided to build the assays from the ground up, starting with the basic particle design. High refractive index cores were used to get maximal turbidimetric signals with the optics of the Du Pont aca, and particle sizes were chosen specifically to complement the desired signal per particle and the rates of desired agglutination. To develop assays which would require infrequent calibration and be highly reproducible, polyglycidyl methacrylate shells containing reactive epoxides were chosen over other polymer containing carboxyl, aldehyde, and amino groups. We investigated ways to covalently attach a number of different analytes to the shell-core particles via albumin and other linkers, and then we developed the specific assays using these particle reagents. Along the route, we found a number of surprises which led to differences in how each assay and each particle was constructed, and this ultimately led to superior performance. Some methods, like theophylline, have been very successful on the Du Pont aca; some methods have been very successful as kits; some, like digoxin, were not commercialized because other approaches were available; and some PETINIA and PETIA methods are still forthcoming. The decision to develop our own particles starting with emulsion polymerization was definitely the correct one for us at the time.

ACKNOWLEDGMENTS

I wish to extend my sincere appreciation to A. R. Craig, W. A. Frey, C. C. Leflar, C. E. Looney, M. A. Luddy and J. P. Galvin who were pioneers of this work at Du Pont. I also wish to thank D. K. Vickery, R. R. Charlton, E. G. Gorman and R. S. Schifreen who were responsible for developing many applications of this technology as well as J. Gauldie and P. Horsewood at McMasters University who helped develop the ferritin assay.

REFERENCES

Babcock, D. A., Berger, J., Dautlick, J. X., et al. (1983) The Du Pont
    aca® III discrete clinical analyzer. In Laboratory medicine,
    G. J. Race, Ed., Harper & Row, Philadelphia, pp. 1-34.
Cook, C. E., Twine, M. E., Myers, M. et al. (1976) Theophylline radio-
    immunoassay: Synthesis of antigen and characterization of antiserum.
    Res. Commun. Chem. Pathol. Pharamacol. 13, 497-505.
Litchfield, W. J., Craig, A. R., Frey, W. A., Leflar, C. C., Looney, C. E.,
    and Luddy, M. A. (1984) Novel shell/core particles for automated
    turbidimetric immunoassays. Clinical Chemistry 30, 1489-1493.
Looney, C. E. (1984) High-sensitivity light scattering immunoassays.
    J. Clinical Immunoassay 7, 90-95.
Singer, J. M. and Plotz, C. M. (1956) The latex fixation test, 1. Applica-
    tion to the serologic diagnosis of rheumatoid arthritis. Am. J. Med. 21,
    888-892.

413

METALLOIMMUNOASSAY

Michael Cais

Department of Chemistry
Technion Israel Institute of Technology
Haifa 32000, Israel

INTRODUCTION

The role of metal ions in biological processes has received increasing attention in recent years both among inorganic chemists and biochemists. The emergence of "bioinorganic chemistry" as a discipline in its own rights is attested to by the appearence of new scientific journals and many symposia devoted entirely or largely to the area of bioinorganic chemistry research in general and to metallobiochemistry in particular.

Our own interest in metallo-biochemical interactions stemmed from the report by Rosenberg et al., (1969) that platinum coordination complexes exhibited potent antitumor activity. To date, the antineoplastic activity of platinum complexes has been demonstrated in many animal tumor-screening systems (Leh and Wolf, 1976). Furthermore, cis-dichlorodiammineplatinum(II), in combination with other chemotherapeutic agents, is a Food and Drug Administration approved drug for the treatment of testicular cancers and ovarian carcinoma. However, a major problem in the application of platinum compounds to antitumor therapy is the apparent lack of target specificity when the platinum compounds are introduced into living systems. It is well known that both biological activity and side effects of drugs can be correlated to their distribution, retention, and excretion in the various organs of animals and humans. The distribution of cis-dichlorodiammineplatinum(II) labeled with ^{193m}Pt and ^{195m}Pt has been evaluated in mice, rabbits, and humans (Lange et al., 1972, 1973). Although the rate of clearance in tumor-bearing animals was significantly higher than control animals, the organ distribution in mice, rats, and rabbits with tumors was similar to that in the control animals. After 45 hr, organs containing most of the radioactivity were the kidney, liver, and the intestine. In all the work carried out so far, no clear distinction has emerged between effects on tumor cells and on normal cells or preferred concentration of platinum in tumor tissue versus normal tissue. This lack of selective toxicity of platinum compounds in tumor therapy became in 1973 the subject of a research project in our laboratory, where we had already established some expertise in the field of organometallic and coordination chemistry. We approached this problem by directing attention to the three major biological systems of high specificity of interaction: antibody-antigen; receptor-hormone; and enzyme-substrate. The concept of targeting a reagent by attaching it to a carrier, which will be recognized selectively by the target site, has formed the basis of much research, in particular with the receptor-hormone system (Katzenellenbogen et al., 1980). The studies initiated with the antibody-antigen system resulted

in the development of a novel concept for the use of metal atoms in the form
of their organometallic and/or coordination complexes as nonradioisotopic
labeling agents in biological systems in general and in immunochemical
applications in particular. The new method (Cais et al., 1977, 1978; Cais and
Josephi, 1977; Cais, 1979; Cais and Adler, 1979; Cais, 1980; Cais and
Shimoni, 1980; Cais and Tirosh, 1981; Gandolfi et al., 1981; Cais, 1983) was
designated metalloimmunoassay (MIA) by analogy to the well-known procedure of
radioimmunoassay (RIA).

This chapter will review the general principles of the MIA concept and
will summarize the work which has emanated from our laboratory, including
some research results which have not yet been published elsewhere.

GENERAL PRINCIPLE

The general principles of MIA and RIA (Abraham, 1977) are the same. In
both systems, the basic immunological reaction is that of antibodies (ab)
specifically recognizing and binding antigens (ag) to form a strongly bound
antibody-antigen complex (ab.ag). This reaction follows the law of mass
action and is described by the equilibrium in Eq. (1)

$$ab + ag \rightleftharpoons ab.ag \qquad\qquad (1)$$

$$ab + ag\text{-}M \rightleftharpoons ab.ag\text{-}M \qquad\qquad (2)$$

If the antigen, ag, is replaced by a metal-labeled antigen (metalloantigen),
ag-M, it is possible that a similar equilibrium, Eq. (2), will be established
provided the metal labeling is carried out in such fashion that there will be
as little interference as possible with the antibody-antigen recognition
process. In other words, the binding constant (or affinity constant) $\underline{K}$ for
the complex ab.ag should be the same (or very nearly so) as the binding
constant K' for the complex ab.ag-M. Under such conditions, if mixtures of
variable amounts of antigen [ag] and a constant amount of metalloantigen
[ag-M] are allowed to compete for a limited and constant concentration of
antibody-binding site, [ab], the reaction mixture, upon equilibrium, will
consist of a "free" antigen fraction (unbound ag and ag-M) and a "bound"
fraction of antibody-antigen complexes (ab.ag and ab.ag-M). After separation
of the two fractions, the amount of metal present in either the "bound" or
"free" fraction (or both) can be determined by suitable methods for the
detection and quantitative analysis of trace metals. Preparation of a
calibration curve plotted for standardized amount of metalloantigen and unlabeled antigen provides the means for determining the concentration of the
analyzed substance in unknown samples, by interpolation from the standard
curve.

The four major steps in the development of a desired metalloimmunoassay
system are (a) production of specific antibodies; (b) synthesis of metallo-
antigens; (c) selection of techniques for separating the "free" unbound
antigens from the antibody-antigen complexes; and (d) choice of analytical
method to determine the metal concentration.

The most important component of MIA, from the novelty aspect of the
concept, is the metal-labeling of analyzable haptens and antigens. (The
terms metallohapten and metalloantigen will be used interchangeably in the
remaining text.)

In contrast to RIA, where only a rather limited number of radioisotopes
are available for radiolabeling the antibodies or/and the antigens, the MIA
concept incorporates, in principle, a very high degree of flexibility in the
choice of labeling agent. It is now an established fact that every metal
element in the periodic table can be made to react with suitable organic
ligands to form either coordination complexes or organometallic derivatives.
Both types of compounds, organometallic and coordination complexes, are
suitable for metal-labeling. The organic moiety of the complex can be any

organic compound provided it incorporates a functional group with which one can form a bond to the metal atom. The main requirements for an ideal metal-lolabeling agent are (a) ready availability in highly purified and stable form at reasonable cost; (b) solubility in aqueous media; (c) that the metal in the label should not occur in significant concentrations in biological fluids; (d) prolonged shelf life, (e) no health hazards; (f) amenability to detection by suitable low-cost, easy-to-operate analytical instruments, preferably of existing technology; (g) high potential sensitivity in assay.

Two general strategies for the synthesis of metal-labeled analytes can be envisaged. One approach would be to introduce a metal atom (or atoms) directly into the analyte, if the structure of the latter is suitable for reaction with metals. In the other approach, one would plan the synthesis of a functionalized metal-containing reagent that could react with the analyte to produce the desired metal-labeled species. We have employed both these approaches as illustrated by several selected examples in the following section.

METAL-LABELING SYSTEMS FOR MIA

In the more general approach to metal-labeling the scheme requires the synthesis of a metal-bearing reagent, which also incorporates a suitable functional group by means of which one can attach the metallo-reagent to the analyte antigen and/or antibody. For example, if the analyte has, or it can be modified to have, a carboxyl group, one would try to synthesize a metallo-reagent in which one of the ligands, complexed to the metal atom, incorporates an amino group available for subsequent reaction with the carboxyl function of the analyte to form an amide bond. In this way the metallo-reagent would become the label for that analyte and the resulting compound would be the metalloantigen or the metalloantibody for use in the metallo-immunoassay of that particular system. Clearly, the same functionalized metallo-reagent could be used for the metal-labeling of other antigens or antibodies containing the appropriate functional group for amide-bond formation.

A readily-available, stable, non-toxic and versatile metallo-reagent for labeling antigens and/or antibodies is ferrocene (and its many derivatives). This iron-containing organometallic complex can be functionalized with practically any desired functional group by classical organic chemistry methods. The reaction scheme shown in Figure 1 indicates one of the pathways used in our laboratory (Cais et al., 1978) whereby the N-hydroxysuccinimide active ester (III) of estrogens was used as starting material for synthesizing 3-O-ferrocenyl labeled estrogens (V) as well as bovine serum albumin (BSA)-estrogen conjugates (VI) for animal immunization in the. production of anti-estrogen antisera.

Additional examples of Fe-labeled derivatives of estrogens, a cannabinoid metabolite of marijuana and of BSA are shown in Figure 2.

Similar chemistry to that mentioned above was applied (Cais, 1983) to the synthesis of cobalt-labeled estrogens using carboxycobalticenium as labeling reagent (Figure 3) and to the preparation of manganese-labeled derivatives of estrogens (Cais and Tirosh, 1981) and barbiturates (Cais and Adler, 1979) using cyclopentadienylmanganese tricarbonyl (cymantrene) (Figure 4) as starting material.

Suitably substituted acetylacetonate metal complexes provide useful metal-labeling reagents, examples of which are shown in Figure 5 for rhodium and platinum and in Figure 6 for chromium.

An additional system investigated by us (Gandolfi et al., 1980, 1981) for anchoring transition metals to organic molecules of biological interest has been that of Pd(II), Pt(II)-functionalized-o-catecholato complexes, prepared by the general scheme shown in Figure 7. We prepared and character-ized twenty four such complexes (Gandolfi et al., 1980, 1981) and we utilized some of them for the synthesis and characterization of steroid-Pt(II)-o-catecholato complexes from functionalized derivatives of estrone, estradiol and testoterone, as summarized in Figures 8 and 9. These compounds possess high chemical stability and they are potentially suitable substrates for investigating Pt(II) effects in bioinorganic studies. One important limitation to be considered in the use of these compounds is their very low solubility in aqueous media.

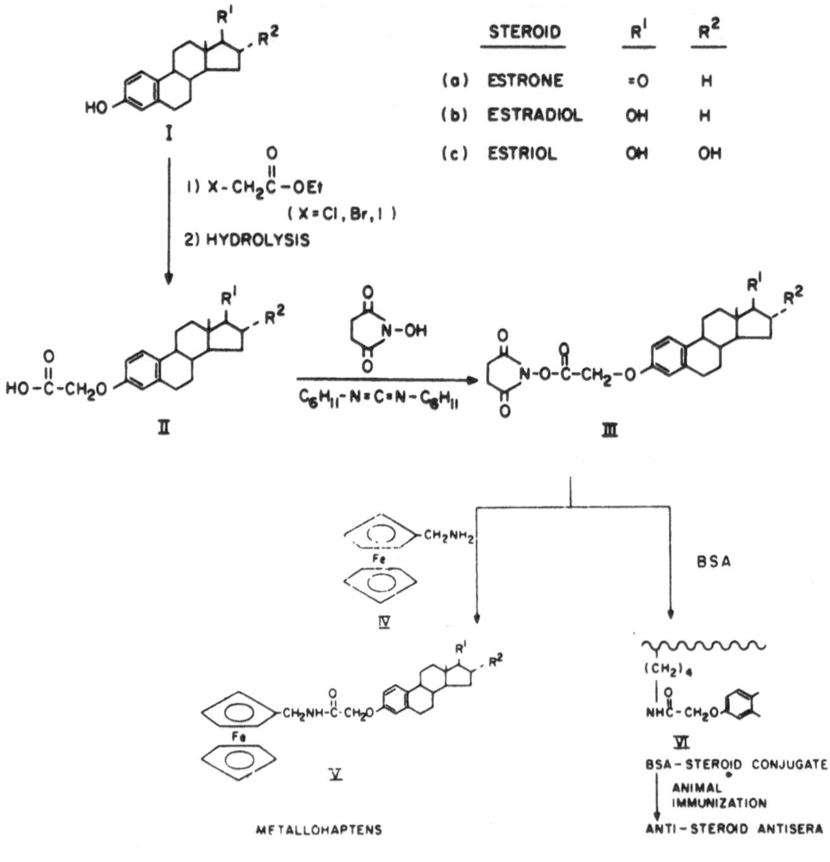

Fig. 1. Synthesis of 3-0-ferrocenyl-containing estrogens and bovine serum albumin (BSA)-estrogen conjugates for production of antisera.

Recently we have reported the synthesis and characterization of Pt(II)-dichloro-diamino acid complexes which could be convenient intermediates in the synthesis of potentially selective targeted-antitumor drugs as well as markers in metalloimmunoassay reagents. It was shown by X-ray analysis that in the complexes (L-ornithine)PtCl$_2$; (DL-diaminobutyric acid)PtCl$_2$; and (DL-2,3-diamino propionic acid) PtCl$_2$ obtained as major products under our experimental conditions, complexation to the Pt(II)Cl$_2$ moiety had occurred through the carboxylate group and one of the two amino groups. However in all cases we also isolated from the mother liquors minor products which, by

I.R. spectroscopy, were shown to have a free carboxyl group, complexation having occurred via the two amino groups. The presence of this type of bonding involving the two amino groups was demonstrated by X-ray analysis for one of the two products obtained from the reaction between DL-2,3-diamino-propionic acid and K_2PtCl_4 (Cais et al., 1986).

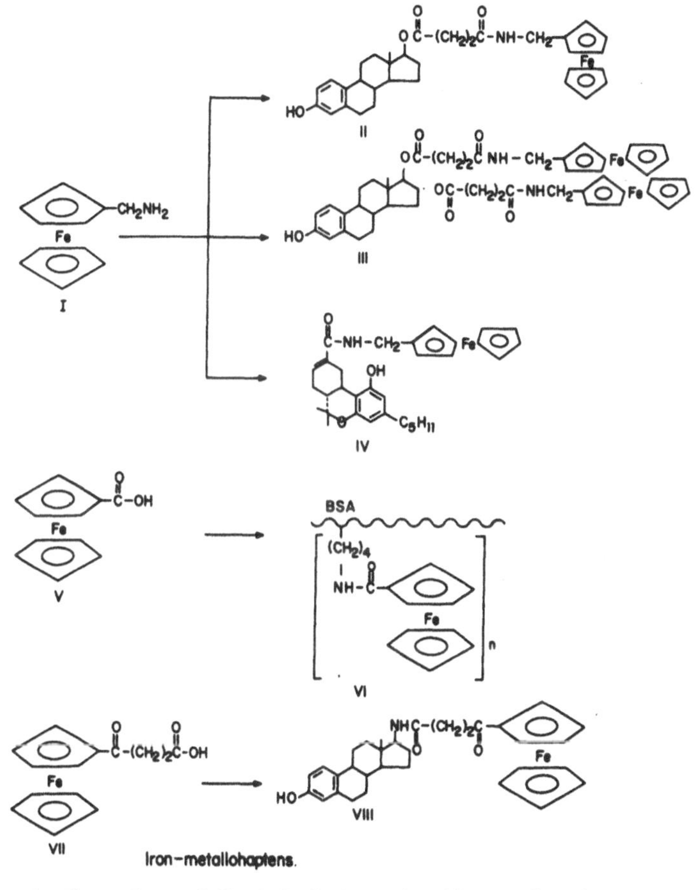

Fig. 2. Examples of Fe-labeled derivatives of estrogens, Δ^6-tetrahydrocannabinol and bovine serum albumin (BSA).

As already mentioned, one approach for coupling transition metals to organic molecules of immunological importance can be achieved by anchoring the metals directly into the biological molecule. An example can be provided by the important and extensive work on palladium(II) and platinum(II) complexes with nucleobases and nucleosides in the frame of metal coordination complexes in cancer chemotherapy (Beck et al., 1979; Pneumatikakis et al., 1977, 1978; Hadjiliadis and Pneumatikakis, 1978; Chu et al., 1977; Marzilli and Kistenmacher, 1977; Dehand and Jordanov, 1977; Hadjiladis and Theophanides, 1976; Hadjiliadis et al., 1973). Recently, the synthesis of

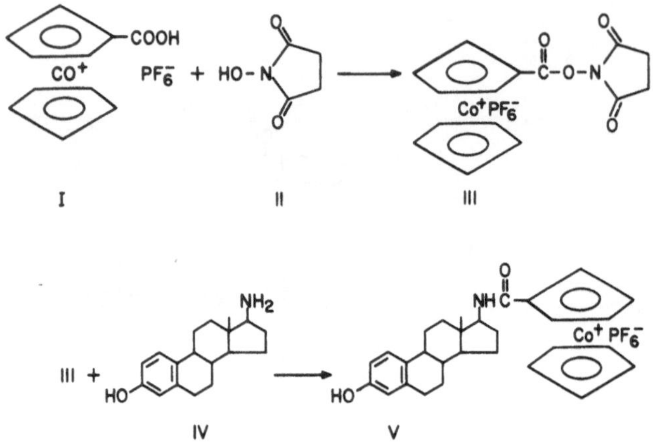

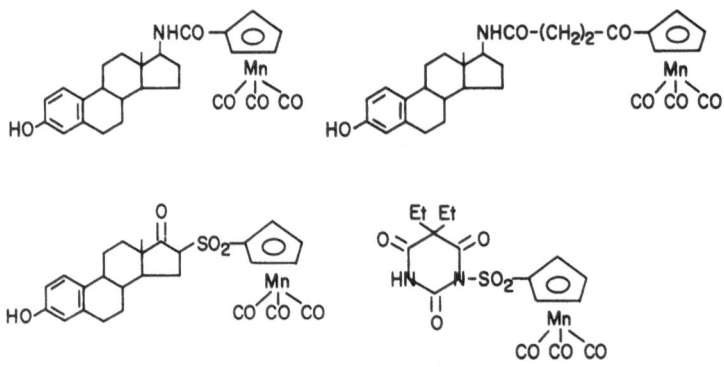

Synthesis of a cobalt–labelled estrogen

Fig. 3. Reaction scheme for Co-labeling of an estrogen derivative.

Manganese – labelled Metallohaptens.

Fig. 4. Examples of Mn-labeling of estrogens and
of 5,5-diethylbarbituric acid.

several complexes with steroids was accomplished by using (π-allyl)palladium chemistry (Trost and Verhoeven, 1976, 1978; Harvie and McQuillin, 1978; Henderson and McQuillin, 1978; Barton and Patin, 1977). Nevertheless, since biological molecules anchored to metal complexes have to undergo immunological reactions _in vitro_ and/or _in vivo_, the reactivity with nucleophiles of the π-allyl complexes and their possible oxidation to allylic alcohols (Jones and Knox, 1975) may discourage their use in MIA.

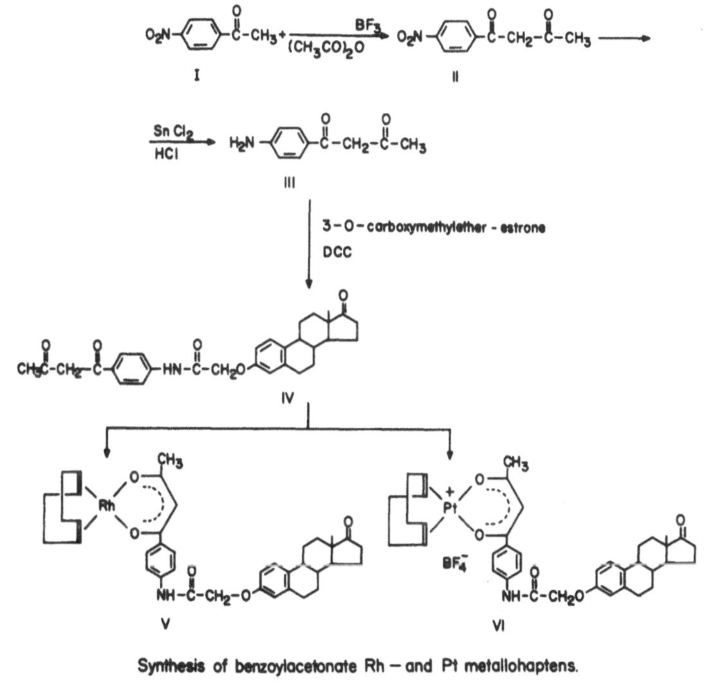

Synthesis of benzoylacetonate Rh – and Pt metallohaptens.

Fig. 5. Scheme for labeling estrone with benzoylacetonates of Rh and Pt.

One of the several systems we have investigated (Cais, 1979, 1980, 1983) for the direct introduction of metal atoms into the organic molecule has been the mercuration of the aromatic ring of estradiol and estriol (Cais and Josephi, 1977). We have prepared the 2-chloromercuri-, 4-chloromercuri- and 2,4-(bis-chloromercuri)derivatives of estradiol and estriol as shown in Figure 10.

The three chloromercuri-derivatives (Fig. 10, II-IV) can be easily characterized by their NMR spectrum, in particular that of the aromatic ring

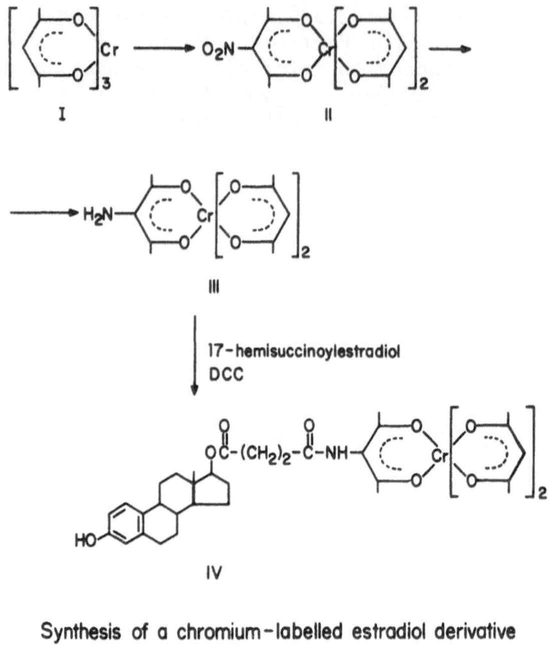

Synthesis of a chromium-labelled estradiol derivative

Fig. 6. The use of amino-substituted Cr-acetylacetonate
for labeling 17-β-hemisuccinoylestradiol.

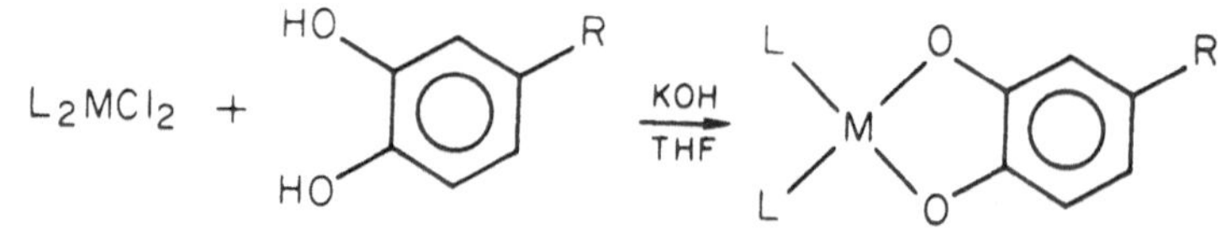

Fig. 7. General scheme for synthesis of o-catecholatometal complexes

L = Neutral ligand, e.g. PPh_3; M = Pd(II), Pt(II);
R = -COOH, $-CH_2-COOH$, $-(CH_2)_2COOH$; $-(CH_2)_2-NH_2$

protons, which very clearly indicates the position and degree of substitution: Incidentally, the availability of the pure chloromercuri-estrogens provides a convenient method for the synthesis of pure radioactive-iodo derivatives for use in radioimmunoassay. We have used these mercury-labeled estrogens for studying the effect of metal-labeling on the antibody-antigen and hormone-receptor reactions (vide infra).

EFFECT OF METAL-LABELING ON THE ANTIBODY-ANTIGEN AND RECEPTOR-HORMONE REACTIONS

It is well established (Abraham, 1977) that in general the labeling of antigens and/or antibodies with radio-isotope (as well as with enzyme and fluorescent) labels may introduce structural and/or sterical modifications

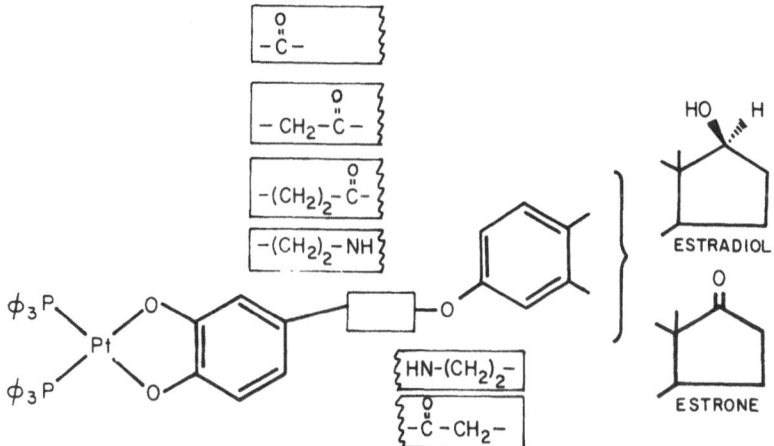

Fig. 8. Platinum (II)-o-catecholato-3-0-estrogen derivatives.

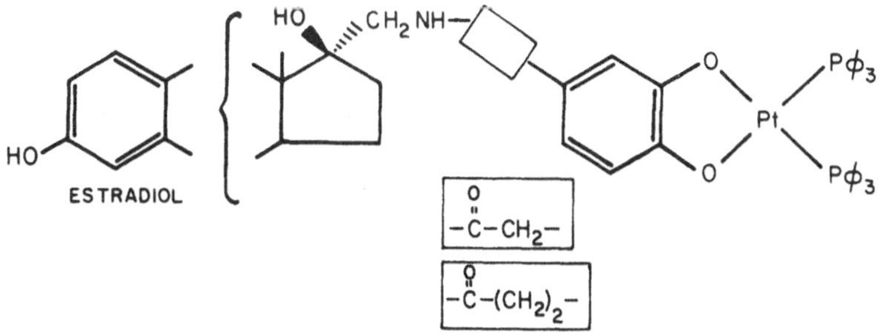

Fig. 9. Platinum (II)-o-catecholato-17α-substituted estradiol derivatives.

which can affect the recognition pattern and the binding constant in the antibody-antigen reaction. It was therefore important to investigate this factor also for metal-labeling in the MIA system. In this section we present some of our results of this investigation with anti-estrogens and anti-barbiturate antisera as well as with the estradiol-estrogen receptor system in tumor tissues.

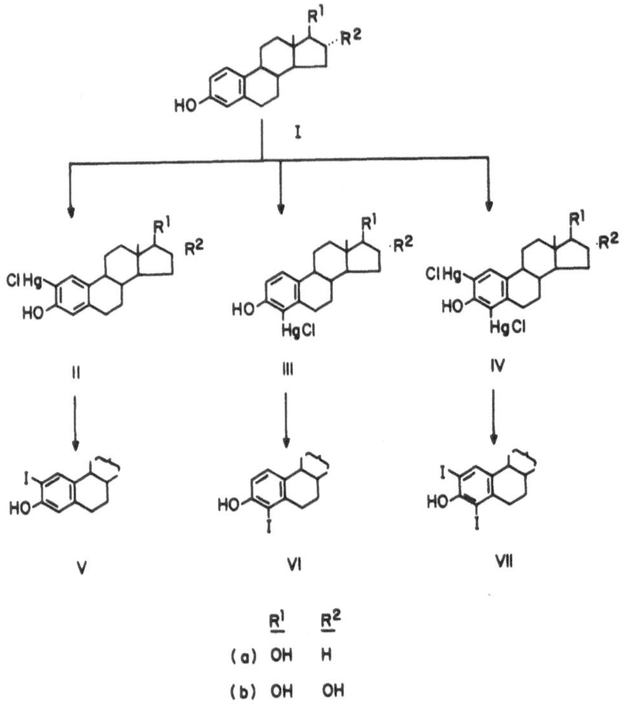

Synthesis of Chloromercuri — and iodoestrogens.

Fig. 10. Synthesis of chloromercuri- and iodoestrogens.

In our immunization protocols three groups of three rabbits each were immunized by multiple subcutaneous injections along the flank, nape and rump with an emulsion of complete Freund's adjuvant and 2 mg of the respective BSA conjugate. A booster injection (2 mg) was given one week later and then six additional boosters were given at regular six-week intervals.

Bleeding was carried out prior to immunization (for normal serum) and at weekly intervals subsequent to immunization. Following the second booster injection, the sera from each bleeding was worked up separately. Immuno-globulins were prepared by precipitating once with ammonium sulfate, dialysis against 0.0175 M phosphate buffer, pH 6.3, followed by purification on a DE-52 column. The purified IgG fraction from each bleeding was then dialyzed against 0.1 M phosphate buffer saline (PBS), pH 7.3 (with 0.1% sodium azide) and the dialyzed solution was distributed in 1 ml vials which were placed in storage at -20 °C. Detection of antibodies to the steroid-protein conjugate, the titer of the antisera and the binding constants of the antibodies with the respective steroid were carried out by both free radical immunoassay techniques as well as by RIA with either ^{3}H-(estrone and estradiol) or ^{125}I-(estriol) labeled haptens. These antisera were used in our experiments to determine the effect of metal labeling in our systems. Two typical examples will be presented here: (a) one case in which metal labeling is done on a functionalized hapten; (b) the second case in which the metal label is substituted directly into the molecule of the hapten.

(a) We metal-labeled a phenobarbitone derivative with iron-, manganese- and chromium complexes, to obtain the derivatives shown in Figure 11. The same phenobarbitone derivative was conjugated to BSA and anti-phenobarbitone anti-bodies were obtained by the method described above. This antiserum was then

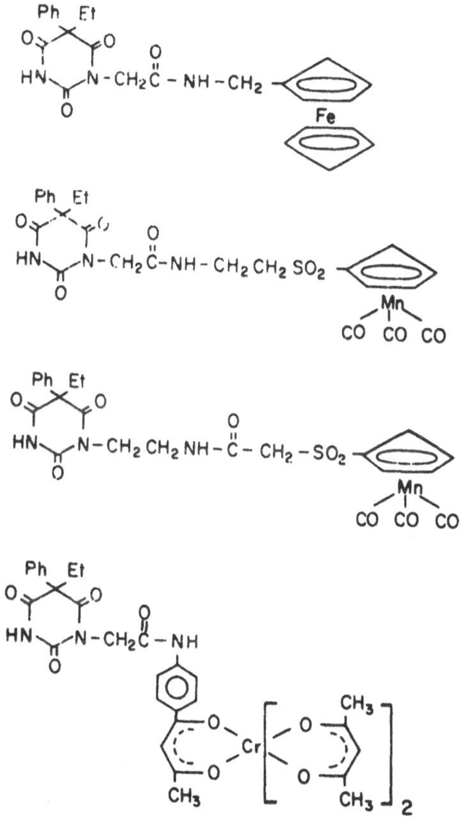

Fig. 11. Metal-labeled derivatives of phenobarbitone.

tested by competitive binding experiments with three metal-labeled pheno-barbitone derivatives in competition with a free-radical nitroxide-pheno-barbitone tracer. The results of the three separate experiments were plotted on the same graph as shown in Figure 12. As can be seen from the graph, all three metallo-phenobarbitone reagents compete equally well with the tracer for the phenobarbitone antibodies, irrespective of the kind of metal or the type of coordination ligands present in the metal complex.

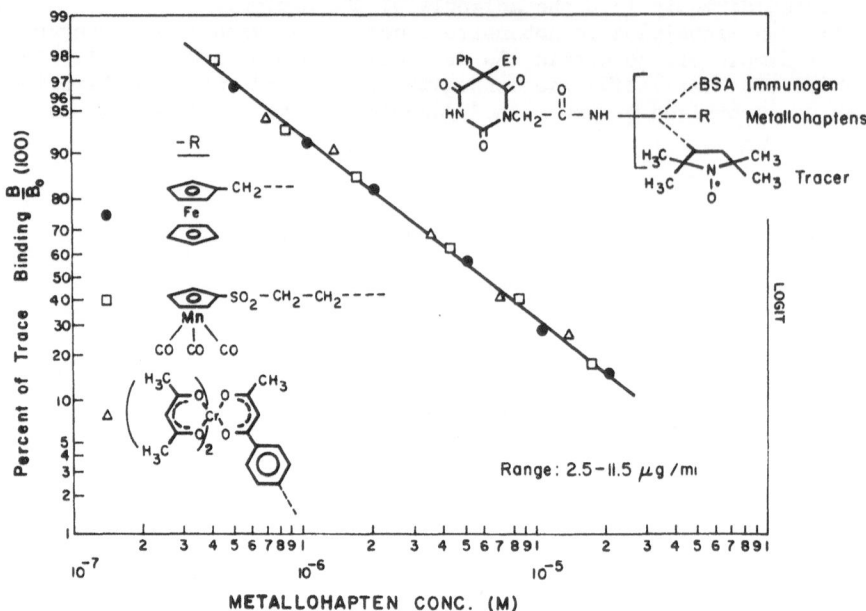

Fig. 12. Competitive binding of metallohaptens
with anti-phenobarbitone antibodies.

    This experiment also indicated to us that by judicious choice of
position for metal complex attachment to the hapten molecule in a particular
system one might have a great deal of latitude in the choice of label since
the type of metal will not affect the antibody-antigen recognition pattern.
(b) On the other hand, we tested 4-chloromercuri-estradiol (compound IV_a in
Fig. 10 in competition with a radioisotope labeled estradiol for the antibody
binding sites of antisera raised against several types of estradiol-BSA
conjugates. The results shown in Figure 13 indicate that introduction of a
bulky atom (atomic volume of Hg is about the same as that of the aromatic
ring A of estradiol) in a position which is probably an important immunogenic
determinant has a significant effect on the antibody-antigen recognition
pattern. However, it should be noted that the recognition is not totally
eliminated, the cross reactivity of the mercurated compound being about 5-10%
relative to the unsubstituted estradiol molecule.
    A similar experiment was carried out by RIA with antiestriol anti-serum
(produced with conjugate VIc in Fig. 1) where the iron-labeled and mercury-
labeled estriol derivatives were compared with unlabeled estriol in the
inhibition of binding the tracer ^{3}H-estriol. The results shown in Fig. 14
indicate that the position of labeling and/or the metal label may have some
effects. Additional experiments are in progress to investigate this problem.
However, we tentatively conclude at present that all other parameters being
equal, the metal-label would not adversely affect the antigenicity of a
hapten.
    As part of the program of metal-labeling proteins, and investigating the
effect of metal-labeling on antibodies, the following experiments were
carried out:

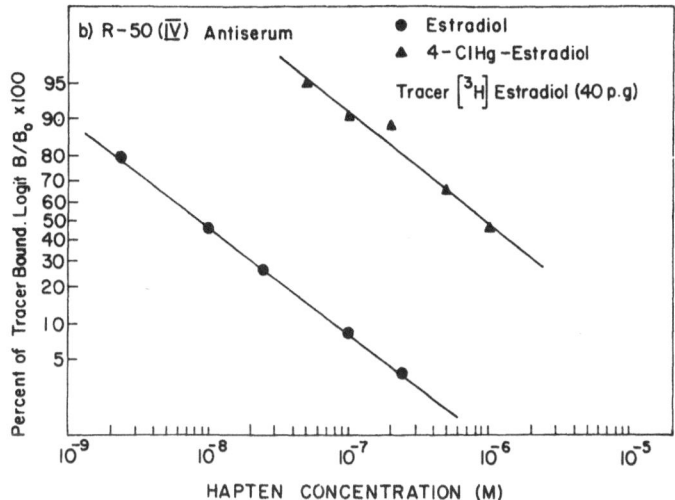

Fig. 13. Competitive binding experiment for 4-chloromercuri-estradiol
(with antibodies raised against 17-β-hemisuccinoyl-estradiol-BSA
conjugate).

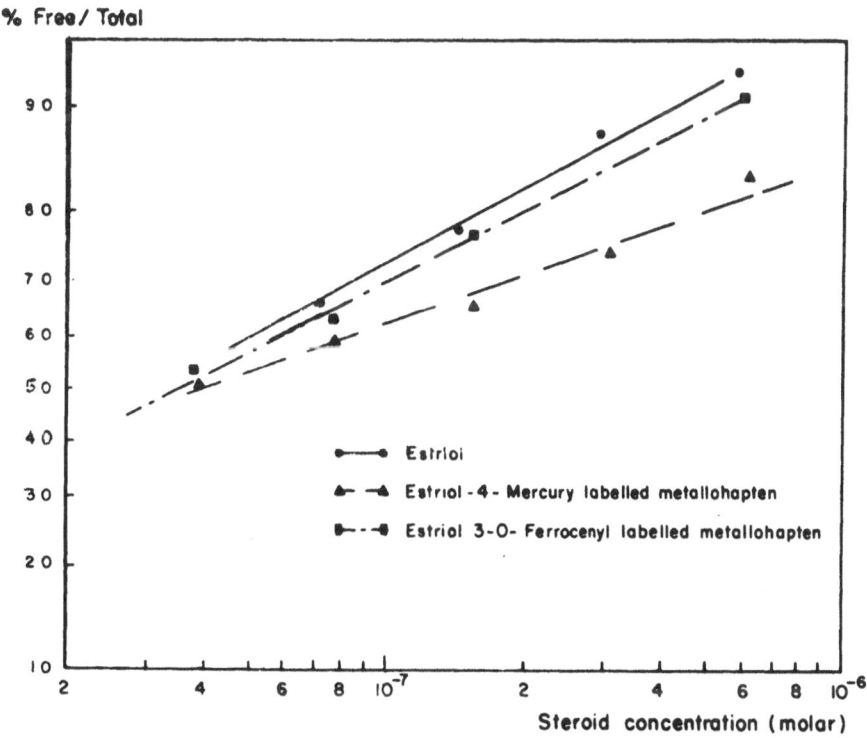

Fig. 14. Cross-reactivity of anti-estriol antiserum with iron-
and mercury-metallohaptens. [Separation with goat anti-
rabbit IgG; Tracer:3H-estriol (90,000 cpm); Concentrations
adjusted to range suitable for MIA experiments).

The N-hydroxysuccinimide active ester of carboxycobalticenium in
acetonitrile was added to a 10 ml solution of anti-bis-hemisuccinate estriol
antibodies (which had been isolated from antisera by ammonium sulfate
precipitation). The reaction mixture stirred at 4 ^{o}C for 48 hr and then
purified by column chromatography (DEAE cellulose) to collect the IgG
fraction (designated Preparation D, see Table 1). Atomic absorption
measurements indicated incorporation of 4-5 Co atoms/mole IgG.

In a control experiment, 5 ml of the same antisera preparation was
reacted with acetonitrile, without the cobalt complex, and worked up under
indentical conditions to those described above (Preparation C, see Table 1).

In another control experiment, the antibodies were purified by
precipitation three times with ammonium sulfate (Preparation A) and followed
by DEAE cellulose column chromatography (Preparation B).

The above four preparations, A-D were assayed for anti-estriol activity
by determining the ability of each preparation to bind 3H-estriol and
^{125}I-estriol tracers, using an anti-Rabbit IgG-sepharose conjugate for
separation of the bound from free.

The results summarized in Table 1 indicate that the anti-estriol
activity of the antibodies was not impaired either by the labeling procedure
or by the presence of the Co-labels in the IgG molecules.

The results described above indicated to us that the antibody-antigen
system provided a model which might be suitable for testing the feasibility
of our concepts for studying the carrier-target approach to obtain
specificity in antitumor chemotherapy. Particularly so, in view of the
recent advances made in investigating the use of monoclonal antibodies
against antigens of clinical interest (Monoclonal Antibodies and ..., 1981).

In parallel with the above studies, we have been carrying out work on a
research program to investigate the receptor-hormone system using the same
target-carrier concept to arrive at specificity of interaction.

A careful analysis of the problem and an extensive literature search led
us to choose the steroid hormone-receptor system as substrate for our
investigation. The presence of specific protein receptors for estradiol in
the hormone-dependent tissues of many species has been well established
(Gorski et al., 1968; Jensen et al., 1969; Jensen and De Sombre, 1973; Liao,
1975). Jensen et al., (1967) suggested that by measuring the uptake of
17β-estradiol by tumor tissue, it might be possible to predict the outcome of
endocrine ablation therapy in patients with breast carcinoma. This
suggestion triggered extensive studies of estrogen receptors in human breast
tumors in laboratories throughout the world (McGuire et al., 1975; King,
1979).

Table 1: Effect of Co-labeling on Antibody Activity
(see text for explanation)

Antibody Preparation	Dilution factor	% Bound (tracer)	
		^{3}H-Estriol	^{125}I-Estriol
A	1:4	30	28
B	1:4	43	30
C	1:4	44	40
D	1:2	62	58

Our studies with estrogens in the antibody-antigen model provided us with information which could be transferred to the estrogen receptors system. It appeared to us that we could use the antibody-antigen metalloimmunoassay concept as an in vitro screening analytical tool for selecting the most suitable metal-labeled steroids for in vitro and in vivo hormone receptors studies. Provided, of course, that some analogy could be found between the antibody-antigen recognition pattern with that in the protein receptor-hormone system. We carried out preliminary studies at the Sloan-Kettering Cancer Center in New York and tested human breast cancer cytosols with mercurated estradiol (Cais and Shimoni, 1980) in collaboration with Dr. Merry R. Sherman. Some of the results of this work are shown in Figures 15-18.

Linear density gradient studies with estrogen-receptor active human breast tumor cytosol showed that 4-chloromercuri-estradiol was recognized by the estrogen receptors in competition with [^{3}H]-estradiol, (Figures 15-17) even though there was a high degree of non-specific binding due to chemical reaction of the -HgCl moiety with -SH groups in the cytosol proteins. Furthermore, as shown in Figure 18, 4-chloromercuriestradiol was a cross-reactant (about 3%) to unsubstituted estradiol for human breast tumor cytosol estrogen receptors in competition with the tracer [^{3}H]-estradiol. It might be accidental that the degree of cross-reaction of 4-chloromercuriestradiol with cytosol estrogen receptor was of about the same degree as that found for

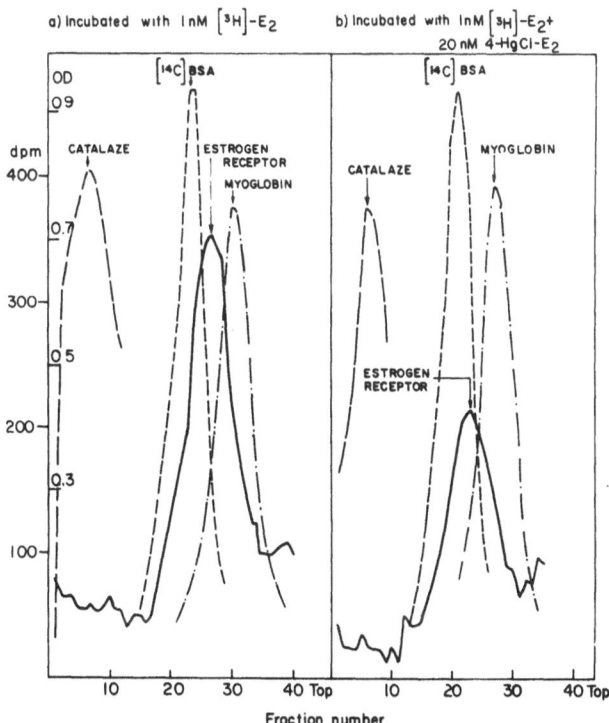

Fig. 15. Competitive binding of [3H]estradiol and 4-chloromercuriestradiol to estrogen receptors in human breast tumor cytosol determined by linear density gradient sedimentation pattern.

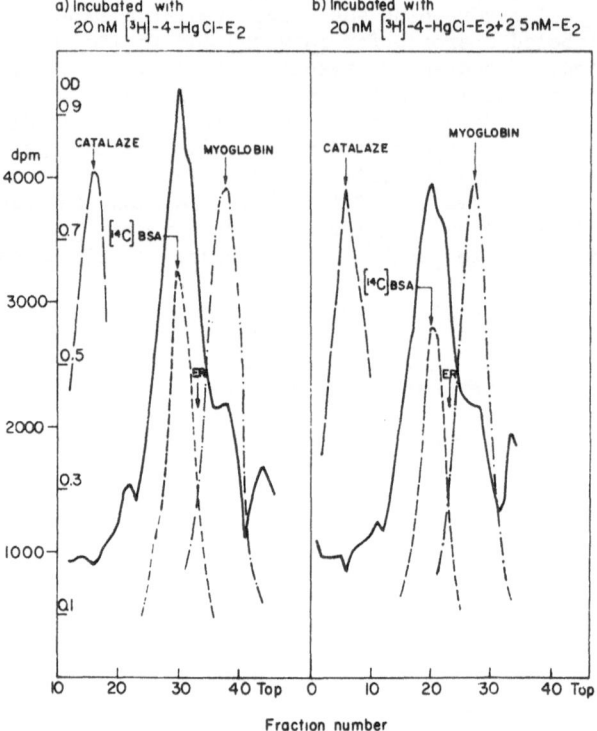

Fig. 16. Binding of 4-chloromercuri-[3H]-estradiol with estrogen receptors (ER) in human breast tumor cytosol.

the antiestradiol antibodies but at least the result indicated to us that there was some analogy and that the idea of using the antibody-antigen system as a screening method was not unfeasible.

In further experiments (Shani et al., 1982), the gamma-emitters 2-[203]Hg-estradiol and 4-[203]Hg-estradiol, steroids that mimic 17β-estradiol in its production of several biological responses and in its binding to soluble uterine receptor sites (Muldoon, 1969, 1971, 1980), were synthesized and injected into Fischer rats bearing transplanted mammary adenocarcinoma and into Sprague-Dawley rats bearing spontaneous mammary tumors. Four days after injection, tumor to blood ratios in the various groups were between 5 and 14, while the healthy mammary glands had a ratio of unity. Similar ratios were obtained for the uterus. When injected into male rats, 4-[203]Hg-estradiol accumulated slightly in the prostate and in the epidimis.

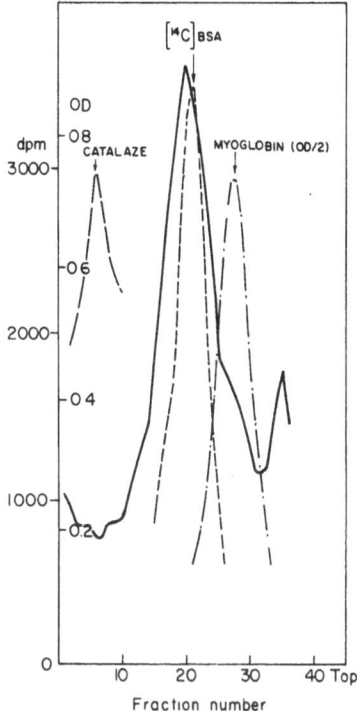

SEDIMENTATION PATTERN IN LINEAR GLYCEROL
DENSITY GRADIENT OF [³H]-4-HgCl-E₂ WITH
ER NON−ACTIVE HUMAN BREAST TUMOR CYTOSOL (Z-2)

Fig. 17. Non-specific binding of 4-chloromercuri-[3H]-estradiol to
human breast tumor cytosol non-active in estrogen receptors.

When minute doses of $4-^{203}Hg$-estradiol were injected into healthy male
rats and retention of the label was followed with a whole-body counter - the
residual drug decreased exponentially with time according to typical first
order kinetics. With pharmacological doses, the clearance exhibited a
biphasic pattern. Even though the mercurated estrogens might be unsuitable
because of high non specific binding, the above preliminary results indicated
to us the feasibility of using metallohormones as imaging agents for
delineating the distribution of estrogen receptors in reproductive tissues
and breast tumors by external gamma detectors.

In general, procedures for separating free from bound antigen exploit physicochemical or immunochemical differences between free and bound fractions. A variety of techniques have been devised and evaluated (Daughaday and Jacobs, 1971; Oddel et al., 1975; Collins et al., 1975; Thorell and Larson, 1978) and many of the practical and theoretical aspects of the problem have been reviewed. The main requirements of an ideal separation technique have been enunciated as follows. (i) It should

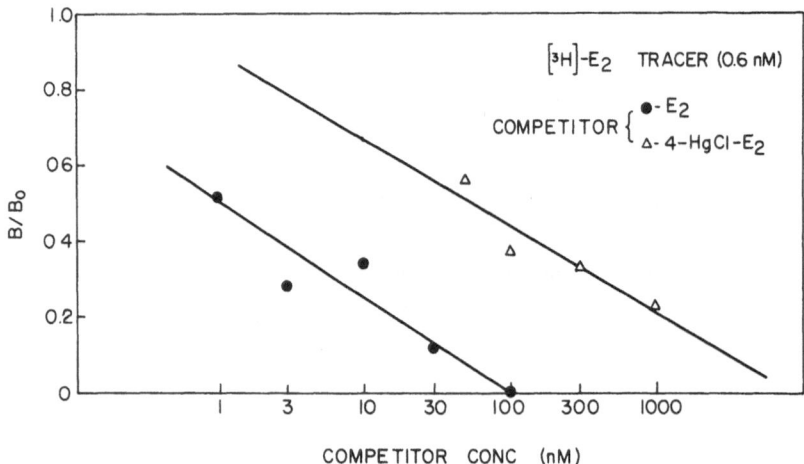

Fig. 18. Competitive binding of 4-chloromercuri-estradiol to estrogen receptors in human breast tumor cytosols.

completely separate bound and free fractions, with a wide margin for error in the conditions used for separation. (ii) It should not interfere with the primary antigen–antibody binding reaction. (iii) It should be simple, easy, and rapid to use. (iv) It should be inexpensive and use reagents and equipment that are readily available. (v) It should not be affected by plasma or serum. (vi) All manipulations should be performed in a single tube. (vii) It should be suitable for automation. (viii) It should be applicable to a wide range of antigens. (ix) The manipulative steps in radioimmunoassays should be designed so that they ensure maximum safety from radiation hazards resulting from handling the radioactive reaction system.

A critical review of the variety of techniques available and the extent to which each technique approaches all or any of the ideal requirements mentioned above is beyond the scope of this report.

The selection of any particular procedure is determined by consideration of many inter-related factors such as solubility of compound, characteristics of antiserum, fraction to be analyzed, degree of non-specific binding and type of label.

Many of the practical methods available require the addition, after the primary incubation step, of a precipitating or adsorbing reagent to be followed by incubation, centrifugation, and then decantation steps. For MIA systems, the required addition of a precipitating reagent denotes a highly critical and error-prone step since the reagent must be ultra-pure in as far as metal ion content is concerned. The analytical method in the MIA concept must determine the labeling-metal atom content at concentrations of 10^{-6} g/l or lower. Consequently, the introduction and mixing-in of any additional reagent into the protein-binding reaction system carries with it the danger of contamination with the same metal atom or ion being used for the labeling of the antigen.

A potentially useful solution to this problem emerged upon analysis of the separation steps in terms of mass transfer principles as applied to an extraction reagent rather than to a precipitating or an adsorbing reagent. As a result we were prompted to investigate the application of solvent extraction principles to the development of a separation technique suitable for the MIA system. After a number of initial experiments it transpired that the use of an extraction reagent for separating bound from free eliminated the need for centrifugation. Furthermore, it became clear that solvent extraction need not be limited to the MIA system and that this approach had interesting potential for use also with immunoassays in which markers other than metal atoms or ions were used for labeling the hapten (Cais and Shimoni, 1981a). In conjunction with the solvent extraction methodology we also developed a novel device which performed the segregation of two spontaneously separated immiscible liquid phases (into which the tracer had been partition-ed) without the need for decantation. Furthermore, we also developed (Cais and Shimoni, 1981b; Cais 1983) a device which allowed for mass transport through selective barriers to achieve a liquid-solid separation without the need for centrifugation and decantation, thus reducing significantly the manipulative steps in the assay with concomitant reduction of contamination opportunities. Both these methodologies were developed primarily with the needs of MIA in mind but so far most of the results available have come from applications to radioimmunoassays.

The next component to be considered in the description of the metallo-immunoassay concept is the choice of a suitable analytical method to determine the metal concentration in the "bound" and/or "free" phase of the analyte, following the separation step. Various methods are now available for metal analysis such as emission, absorption and fluorescence spectro-metry, microwave excitation emission spectrometry, anodic stripping volta-metry and other electrochemical methods, neutron activation, etc. In our own studies so far we have made use only of flameless atomic absorption spectrometry (AAS) employing an electrically heated graphite furnace for the drying, charring and atomization of the metal-containing sample under analysis. The choice of this method has been governed by several factors, such as relative simplicity of operation, general availability of instru-ments in clinical chemistry laboratories (over 30,000 atomic absorption spectrometers are in use world-wide), the requirement of small aliquots of analyte solution (20-30 l) and potential high sensitivity of metal atom detection (down to picogram levels).

The detection sensitivity of metal atoms in AAS instruments fitted with graphite furnace varies from metal to metal and it depends on a number of variables connected both with instrumental parameters and the matrix of the assay sample. Therefore it is always necessary to perform a "method development" operation in order to determine the optimal conditions for each assay system. Once this has been done, it is then possible to use the same instrument settings for the required analysis with that particular assay system without having to go again through the procedure of "method development".

There have been several reports on the application of electrochemical detection methods to MIA. Using a ferrocene-labeled morphine tracer, Weber and Purdy (1979) developed a system described as "voltammetric immunoassay", in which they monitored the oxidation of the "free" ferrocene-labeled tracer, without the need for separating the "bound" and "free" fractions. In the presence of antibody, the oxidation current due to the "free" ferrocene-labeled morphine was reduced. On the other hand, when codeine was added to the same concentration of antibody and tracer, the oxidation current of the "free" ferrocene-labeled morphine was increased, indicating the displacement by codeine of the ferrocene tracer from the antibody-tracer conjugate. No interference due to the reduction of oxygen was observed. However, since the work was done at +500 mV <u>vs</u> S.C.E., species other than the ferrocene-labeled tracer were prone to oxidation.

A similar approach has been reported by Heineman et al., (1979) who used a mercuri-labeled estriol tracer. These workers also claimed that separation of the "free" and "bound" fractions was not required since reduction of the antibody-bound tracer occurred at a more negative potential than the "free" mercurated tracer. One serious limitation seemed to be due to the fact that reduction of the mercuri-labeled estriol at −300 mV <u>vs</u> S.C.E. was prone to oxygen interference if the analytical samples were not carefully degassed.

A metalloimmunoassay system using indium (In^{+3}) as the labeling metal and anodic pulse stripping voltametry as the analytical detection method has been reported by Doyle et al., (1982). In this assay, human serum albumin was derivatized with diethylenetriaminepentaacetic acid (DTPA) and subsequently labeled with In^{+3} which forms a very strong coordination bond with DTPA. The assay described involved the need for separating the "bound" and "free" fractions, and then releasing the In^{+3} ion from the DTPA complex by lowering the pH.

It has recently been shown that ferrocene and its derivatives can act as a mediator to glucose oxidase (Cass et al., 1984) and certain flavoproteins (Cass et al., 1985). Based on these observations, an amperometric metallo-immunoassay system was designed (DiGleria et al., 1986) with ferrocene-labeled lidocaine (α-diethylamino-2,6-dimethyl acetanilide) as tracer. This compound was found to act as a mediator to glucose oxidase due to the ferrocene moiety. Furthermore it was shown that since antibodies to lidocaine also recognized the lidocaine-ferrocene tracer, the diffusion properties of the latter were greatly affected in the presence of the anti-lidocaine antibodies and was no longer able to act as a mediator to various oxido-reductases. The metalloimmunoassay configuration consisted (DiGleria et al., 1986) of a fixed amount of antibodies, the ferrocene-labeled drug as a tracer, enzyme and substrate; the only variable being the unknown concentration of the drug. In the absence of any drug the ferrocene-labeled tracer was bound to the antibodies and no ferrocene-bearing species was available to mediate to the flavoprotein. In the presence of the drug, competition occurred between the ferrocene-labeled tracer and the unlabeled drug for the antibody sites. The greater the concentration of the drug in the analyte sample, the greater the amount of the "free" ferrocene-labeled tracer and the greater the catalytic current measured. The results obtained so far by the Oxford group indicate that this type of configuration for MIA may turn out to provide the basis for a successful, practical and convenient non-radioisotopic immunoassay system which could become a useful addition to the diagnostics armamentarium.

A NOTE ON NOMENCLATURE

When we first published (Cais et al., 1977) our concept of using metal

complexes as labeling agents, we chose the term METALLOIMMUNOASSAY (MIA) for
the new system in accordance with established practice at that time of
designating the immunoassay according to the type of labeling agent. Thus we
had at the time RADIOIMMUNOASSAY (RIA) for radio-isotope labels, ENZYME
IMMUNOASSAY (EIA) for enzyme labels and FLUORESCENCE IMMUNOASSAY (FIA) for
fluorescent labels. We would like to suggest that this practice be continued
in the future and avoid designating the immunoassay system by the type of
analytical method or instrument used in the detection of the labeled species.
There is every reason to expect that in the area of non-radioisotopic immunoassays, more than one analytical method and/or type of instrumentation will
become available for any particular labeling system. We feel that adherence
to our suggestion may help avoid confusion in the future as this immuno-
diagnostic area will undoubtedly continue to grow and expand.

ACKNOWLEDGMENTS

The author wishes to express thanks and appreciation to all his coworkers, named in the list of references, whose enthusiasm and devoted labors
have contributed to the development of the MIA concept. I also wish to thank
Dr. Monika J. Green for information on the lidocaine work prior to publication. I thank Mrs. Charlotte Diament for her help in preparing the
manuscript.

REFERENCES

Abraham, G.E., 1977, "Handbook of Radioimmunoassay," Decker, New York.

Barton, H.R. and Patin, H., 1977, Palladium Chloride-π-allyl complexes from
    calciferol and related compounds, J. Chem. Soc. Chem. Commun. 15:799-
    800.

Beck, W.M., Calabrese, J.C. and Kottmair, N.D., 1979, Palladium(II) and
    platinum(II) complexes with nucleobases and nucleosides. Crystal
    structure of trans-bis(adeninato)bis(tri-n-butylphosphine)palladium
    (II), Inorg. Chem. 18:176-182.

Cais, M., Dani, S., Eden, Y., Gandolfi, O., Horn, M., Isaacs, E.E., Josephi,
    Y., Saar, E., Slovin, E., and Snarsky, L., 1977, Metalloimmunoassay,
    Nature (London) 270:534-535.

Cais, M., Slovin, E. and Snarsky, L., 1978, Metalloimmunoassay. II., Iron-
    metallohaptens from estrogen steroids, J. Organometal. Chem. 160:223-
    230.

Cais, M., 1979, Tests immunologiques à l'aide des complexes organométal-
    liques. Un nouveau concept, 1979, L'actualité chimique 7:14-26.

Cais, M. and Josephi, Y., 1977, unpublished results; also Josephi, Y., 1979,
    D.Sc. Thesis, Technion-Israel Inst. of Technology, Haifa, Israel.

Cais, M. and Adler, R., 1979, unpublished results; also Adler, R., 1980,
    M.Sc. Thesis, Technion-Israel Inst. of Technology, Haifa, Israel.

Cais, M., 1980, Specific binding assay method and reagent means, U.S. Patent
    No. 4205952.

Cais, M. and Shimoni, M., 1980, unpublished results; also Shimoni, M., 1981,
    D.Sc. Thesis, Technion-Israel Inst. of Technology, Haifa, Israel.

Cais, M. and Tirosh, N., 1981, Metalloimmunoassay. IV. Manganese-labelled
    metallohaptens via cymantrene derivatives, Bull. Soc. Chim. Belg.
    90:27-35.

Cais, M. and Shimoni, M., 1981a, A feasible solvent separation system for
    immunoassays, Ann. Clin. Biochem. 18:317-323.

Cais, M. and Shimoni, M., 1981b, A novel system for mass transport through
    selective barriers in non-centrifugation immunoassays, Ann. Clin.
    Biochem. 18:324-329.

Cais, M., 1983a, Metalloimmunoassay: principles and practice, Methods in
    Enzymology 92:445-458.

Cais, M., 1983b, Non-centrifugation immunoassays: novel systems, Methods in Enzymology 92:336-344.

Cais, M., Cwikel, D., Agmon, I., and Kapon, M., 1986, Studies towards the synthesis of potentially selective antitumor cis-platinum(II)-complexes, Abstracts, XXIVth Int. Conf. on Coordination Chem., Athens, Greece.

Cass, A.E.G., Davis, D., Francis, G.D., Hill, H.A.O., Aston, W.J., Higgins, I.J., Plotkin, E.V., Scott, L.D.L., and Turner, A.P.F., 1984, Ferrocene-mediated enzyme electrode for amperometric determination of glucose, Anal. Chem. 56:667-671.

Cass, A.E.G., Davis, G., Green, M.J., and Hill, H.A.O., 1985, Ferricinium as one electron acceptor for oxido-reductases, J. Electroanal. Chem. 190:117-127.

Chu, G.Y.H., Duncan, R.E. and Tobias, R.S., 1977, Heavy metal-nucleoside interactions. 10. Binding of cis-diammine platinum(II) to cytidine and uridine in aqueous solutions: necessary conditions for formation of platinum-uridine "blues," Inorg. Chem. 16:2625-2636.

Collins, W.P., Barnard, G.J.R. and Hennam, J.F., 1975, Factors affecting the choice of separation technique, in "Stereoid Immunoassays" (Cameron, E.H.D., Hillier, S.G. and Griffiths, K., eds.) pp. 223-228, Alpha Omega, Cardiff.

Daughaday, W.H. and Jacobs, L.S., 1971, in "Principles of Competitive Protein Binding Assays," (Odell, W.D. and Daughaday, W.H., eds.) pp. 303-316, J.B. Lippincott Co., Philadelphia.

Dehand, J. and Jordanov, J., 1977, Binding of cis-dichlorobis(1-methyl-imidazole-2-thiol)-palladium(II) and -platinum(II) to nucleosides: synthesis and hydrogen-1 and carbon-13 nuclear magnetic resonance studies, J. Chem. Soc. Dalton Trans. 1588-1593.

DiGleria, K., Hill, H.A.O., McNeil, J., and Green, M.J., 1986, Homogeneous ferrocene-mediated amperometric immunoassay, Anal. Chem. 58:1203-1205.

Doyle, M.J., Halsall, H.B. and Heineman, W.R., 1982, Heterogeneous immunoassay for serum proteins by differential pulse anodic stripping voltametry, Anal. Chem. 54:2318-2322.

Gandolfi, O., Dolcetti, G., Ghedini, M. and Cais, M., 1980, Metal catecholato complexes: a source for metallo-labelling antigens, Inorg. Chem. 19:1785-1791.

Gandolfi, O., Cais, M., Dolcetti, G., Ghedini, M., and Modiano, A., 1981, Metalloimmunoassay. V. Steroid-platinum(II)-o-catecholato complexes: a novel set of metallohaptens, Inorg. Chim. Acta 56:127-133.

Gorski, J., Toft, D., Shyamala, G., Smith, D., and Notides, A., 1968, Hormone receptors: studies on the interaction of estrogen with the uterus, Rec. Progr. Hormone Res. 24:54-80.

Hadjiliadis, N., Kourounakis, P. and Theophanides, T., 1973, Interactions of platinum(II) with adenosine and its acetyl derivatives, Inorg. Chim. Acta 7:226-230.

Hadjiliadis, N. and Theophanides,T., 1976, Platinum purine nucleosides. I. Interaction of K_2PtX_4 (X=Cl,Br) with adenosine, triacetyladenosine, adenosine-1-oxide and 9-methyladenine, Inorg. Chim. Acta 16:67-75.

Hadjiliadis, N. and Theophanides,T., 1976, Platinum purine nucleosides. II. Interaction of K_2PtX_4 (X=Cl,Br) with inosine and guanosine, Inorg. Chim. Acta 16:77-88.

Hadjiliadis, N. and Pneumatikakis, G., 1978, Complexes of palladium (II) with nucleosides. Preparation and properties of complexes of the type potassium trichloro(nucleoside)palladate(II), J. Chem. Soc. Dalton Trans. 1691-1695.

Harvie, I.J. and McQuillin, F.J., 1978, Steric mechanism of formation of the PdCl π-allyl derivative from $[6\beta-^2H]$cholest-4-ene, J. Chem. Soc. Chem. Commun. 747-748.

Heineman, W.R., Anderson, C.W. and Halsall, H.B., 1979, Immunoassay by differential pulse polarography, Science 204:865-866.

Henderson, K. and McQuillin, F.J., 1978, Mechanism of formation of 4-6η-3-oxo stereoid-PdCl complexes, J. Chem. Soc. Chem. Commun. 15-16.

Jensen, E.V., De Sombre, E.R. and Jungblut, P.W., 1967, in "Endogenous Factors Influencing Host-Tumor Balance" (Wissler, R.W., Dao, T.L. and Wood, S., Jr., eds.) University of Chicago Press, Chicago.

Jensen, E.V., Numata, M., Smith, S., Suzuki, T., and De Sombre, R.E., 1969, Estrogen receptor interaction in target tissues, in Action Horm. Symp., 20-39 (Foa P.P. ed.), Thomas, Springfield, Ill.

Jensen, E.V. and De Sombre, E.R., 1973, Estrogen-receptor interaction, Science 182:126-134.

Jones, D.N. and Knox, S.D., 1975, Regioselective and stereoselective oxidation of steroidal palladium π-allyl complexes to allylic alcohols, J. Chem. Soc. Chem. Commun. 166-167.

Katzenellenbogen, J.A., Heiman, D.F., Carlson, K.E., Payne, D.W., and Lloyd, J.E., 1974, in "Cytotoxic Estrogens in Hormone Receptive Tumors" (Raus, J., Martens, H. and Leclercq, G., eds.), Academic Press, New York.

King, R.T.B., 1979, "Steroid Receptor Assays in Human Breast Tumors," Alpha Omega, Cardiff.

Lange, R.C., Spencer, R.P. and Harder, H.C., 1972, Synthesis and distribution of a radio labeled antitumor agent: cis-diamminedichloroplatinum(II), J. Nucl. Med. 13:328-330.

Lange, R.C., Spencer, R.P. and Harder, H.C., 1973, Antitumor agents cis-diamminedichloroplatinum. Distribution studies and dose calculations for platinum-193m and platinum-195m, J. Nucl. Med. 14:191-195.

Leh, F.K.V. and Wolf, W., 1976, Platinum complexes: a new class of antineoplastic agents, J. Pharm. Sci. 65:315-328.

Liao, S., 1975, Cellular receptors and mechanisms of action of steroid hormones, Int. Rev. Cytol. 40:87-172.

Marzilli, L.G. and Kistenmacher, T.G., 1977, Stereoselectivity in the binding of transition metal-chelate complexes to nucleic acid consistuents: bonding and nonbonding effects, Acc. Chem. Res. 10:146-152.

McGuire, W.L., Carbone, P.P. and Volmer E.P., 1975, "Estrogen Receptors in Human Breast Cancer," Raven Press, New York.

Monoclonal Antibodies and Ultrasensitive Immunoassay in Human Diagnosis and Monitoring of Therapy, 1981, RIA 81 Symposium, Gardone Riviera, Italy.

Muldoon, T.G. and Warren, J.C., 1969, Characterization of steroid-binding sites by affinity labelling. II. Biological activity of 4-mercuri-17β-estradiol, J. Biol. Chem. 244:5430-5435.

Muldoon, T.G., 1971, Characterization of steroid-binding sites by affinity labeling. Further studies of the interaction between 4-mercuri-17β-estradiol and specific estrogen-binding proteins in the rat uterus, Biochemistry 10:3780-3784.

Muldoon, T.G., 1980, Molecular and functional anomalies in the mechanism of the estrogenic action of 4-merciou-17β-estradiol, J. Biol. Chem. 255:1358-1366.

Oddel, W.D., Silver, C. and Grover, P.K., 1975, Competitive protein binding assays: methods of separation of bound from free, "Steroid Immunoassays" (Cameron, E.H.D., Hillier, S.G. and Grover, P.K., eds.) pp. 207-222, Alpha Omega, Cardiff.

Pneumatikakis, G., Hadjiliadis, N. and Theophanides, T., 1977, Palladium reactions with guanosine. Definitive evidence of O_6N_6 chelation, Inorg. Chim. Acta 22:L1-L2.

Pneumatikakis, G., Hadjiliadis, N. and Theopanides, T., 1978, Complexes of inosine, cytidine and guanosine with palladium(II), Inorg. Chem. 17:915-922.

Ratcliffe, J.G., 1974, Separation techniques in saturation analysis, Br. Med. Bull. 30:32-37.

437

Rosenberg, B., VanCamp, L., Trosko, J.E., and Mansour, V.H., 1969, Platinum compounds: a new class of potent antitumor agents, Nature (London) 222:385-386.

Shani, J., Lieberman, L.M., Samuni, T., Schlesinger, M., and Cais, M., 1982, Tissue distribution of 2- and 4-[203 Hg]estradiol in mammary-tumor-bearing rats, Int. J. Nucl. Med. Biol. 9:251-255.

Thorell, J.I. and Larson, S.M., 1978, "Radioimmunoassay Techniques," C.V. Mosby, Saint Louis.

Trost, B.M. and Verhoeven, T.R., 1976, New synthetic reactions. Catalytic vs. stoichiometric allylic alkylation. Stereocontrolled approach to steroid side chain, J. Am. Chem. Soc. 98:630-632.

Trost, B.M. and Verhoeven, T.R., 1978, Stereocontrolled approach to steroid side chain via organometallic chemistry. Partial synthesis of 5α-cholestanone, J. Am. Chem. Soc. 100:3435-3443.

Weber, S.G. and Purdy, W.C., 1979, Homogeneous voltametric immunoassay: a preliminary study, Anal. Letters 12:1-9.

SPECIFIC PROTEINS ANALYSIS BY NEPHELOMETRY AND TURBIDIMETRY

Peter Horsewood

Department of Pathology
McMaster University
Hamilton, Ont.

INTRODUCTION

The phenomenal rise in the past decade of clinical chemistry methods has brought about increased demands for more rapid, precise and economic assays. In the assay of serum proteins for either diagnostic or medical care purposes, these demands have been met largely by the development of automated immunoassays employing modern instrumental hardware. Occupying a prominent position among the analytical methods are the related light scattering techniques of nephelometry and turbidimetry since they meet many of the requirements for fast, efficient assays. While these light scattering methods may not have the glamour of newer immunoassay methods, they still remain an essential workhorse for most large clinical laboratories.

The present report deals with nephelometric and turbidimetric analysis of clinically relevant proteins using specific antibodies. In principle, the reaction of any substance with a specific reagent that causes particle formation may be analyzed by these two techniques and indeed, the theory and principle of the methods were formulated long before the existence of antibodies were known. Excellent reviews dealing with the background, theory, and instrumentation of the methods have appeared and are exemplified by those of Deverill and Reeves (1980), Ritchie, (1978), Whicher and Perry (1984).

When a light beam is directed at a fluid containing suspended particles light is absorbed, scattered, reflected and transmitted. The quantification of the residual transmitted light is the basis for turbidimetry while the measurement of scattered light is employed in nephelometry. The underlying physics, reviewed by Kerker (1969), which applies to both nephelometry and turbidimetry, describes the scattering process resulting from the interaction of the incident electromagnetic radiation with electrons in the scattering particles. The early studies of light scattering were undertaken by Rayleigh (1871), and later extended by Mie (1908) and Debye (1915). The theories and concepts formulated by these investigators showed that the dispersion of light was dependent on the relationship of the incident wavelength to the size and shape of the particle. This wavelength to size and shape dependency determines whether light waves reradiated from the particles are synergic or destructive and this in turn determines the angular distribution of the scattered light. Particles smaller than the wavelength of the incident beam scatter light almost symmetrically (Fig. 1, Rayleigh scatter, envelope A) and

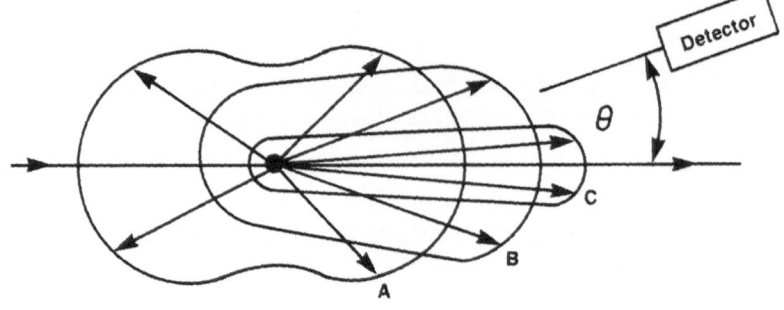

Fig. 1

A	Rayleigh	$d < 0.1\lambda$
B	Rayleigh-Debye	$d \approx \lambda$
C	Mie	$d > \lambda$

as the particle size increases an increasingly larger proportion of forward scatter occurs. (Fig. 1, Rayleigh-Debye and Mie scatter, envelopes B and C).

The majority of the soluble proteins are of such a size (1-100nm) and shape that they are small compared to the wavelengths of visible light and so scatter light in an almost symmetrical fashion. Antibodies and antigens form heterogenous complexes having a typical size range of 250-1400nm (Buffone et al., 1975a) which is close to the wavelengths used in light-scattering analyses. Thus, the blank or background for an antigen-antibody solution gives a near symmetrical Rayleigh scatter while the resulting complexes produce a predominantly forward scatter as described by Rayleigh-Debye and Mie theories. Nephelometric analyses capable of measuring at forward scatter angles accordingly have greater potential sensitivity.

The principles of antigen-antibody reactions are formulated generally in terms of noncovalent, intermolecular forces and how these relate to complexation with subsequent lattice formation and precipitation. These intermolecular forces are influenced by the medium in which they occur and consequently such variables as reagent concentration, ionic composition, ionic strength, pH, temperature and degree of hydration will all play a part in complex-precipitate formation. Detailed investigation of these variables have been carried out by several workers. Killingsworth and Savory (1973) and Van Munster et al., (1977), have shown that pH and temperature has a minimal effect on the reaction when studied over normal operating ranges. The influence of selected anions on the important initial stages of antigen-antibody interaction have been described by Anderson and Sternberg (1978), and Marrack and Richards (1971). The effects generally are in accord with the chaotropic nature and associated hydration spheres of such ions. Relative to chloride anions, low charge-density anions such as thiocyante, bromide and perchlorate promote macromolecular unfolding and discourage protein interactions, while higher charge-density anions such as fluoride, sulphate and phosphate promote interaction. There appears to be little difference in light scattering for the various cations Li, Na, K and Cs when used in salts with the same anion as shown by Marrack and Richards (1971). These latter authors along with Hawkins (1964), have also shown that antigen-antibody interactions increase with decreasing ionic strength although a possible exception has been reported by Killingsworth and Savory (1973).

440

It is likely that the observed effects of ions on complex formation is related, in part, to their charge density and in turn their degree of hydration. A more striking consequence of this effect is seen with non-ionic polymers. A marked enhancement of the antigen-antibody reaction is brought about by inclusion into buffer solutions those polymers which have large hydration spheres and are effective in concentrating the solute molecules into a smaller exclusion volume. Studies of the enhancing properties of various polymers, which have been reviewed by Hellsing (1978), have led to the routine use of polyethylene glycol 6000 (PEG 6000) in turbidimetry and nephelometry. This innovation increases sensitivity, enables end-point assays to be run in minutes instead of hours and increases peak rates in kinetic assays.

Probaby the best studied and most important variable in antibody-antigen complex formation is the concentration of reactants. Polyclonal antisera and soluble macromolecular proteins have multiple interaction sites and form lattice-like soluble complexes which further associate to form insoluble precipitates. Adding increasing amounts of antigen to antibody in the region of antibody excess, up to the equivalence point, results in an increasing number of scattering centres; and there appears to be only a small variation in the size of these complexes (Buffone et al., 1975; Marrack and Richards, 1971). In antigen excess particles continue to increase in number but decrease in size and scattering potential, until at the limit only complexes of a single antibody molecule bound to two antigen molecules exist. Overall, under conditions of limiting antigen, these processes result in light scattering being directly proportional to the antigen concentration and give a scatter curve similar to the classical precipitation curve of Heidelberger (1935), (Fig. 2a, 2b). Analyses are run, therefore, in antibody excess since light scatter from a sample in large antigen excess may not be differentiated from the response given by some other lower antigen concentration.

MATERIALS AND METHODS

Instrumental and Practical Considerations

Many factors, some of which have been discussed above, contribute to the

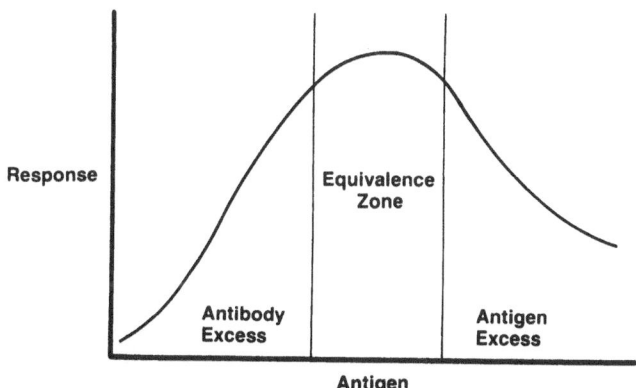

Fig. 2
a) Response = amount precipitate
b) Response = light scatter
c) Response = peak rate of scatter

light scattering process and their effects are largely predictable from an understanding of the physics of the phenomenon. Some of these factors such as particle size, shape and refractive index etc. are an inherent part of the antigen-antibody process and are only partially controllable. Other factors such as intensity and wavelength of incident light, concentration of particles, angle of detection etc. are much more controllable and can be manipulated favorably to optomize the technique. Thus, treatment and choice of reagents along with choice and design of instruments greatly affect the outcome of specific protein analysis using light scattering techniques.

Instrumentation Principles

Turbidimetric measurements are aimed at the quantification of forward transmitted light after it has passed through the suspension to be analysed (Fig.3). Since the majority of assays are conducted in dilute solutions the residual light, after losses due to scatter and absorption, represents a large fraction of the incident beam and consequently large amplification is unnecessary. The ability to detect the small modulation in signal between the incident and transmitted beam is fullfilled by most modern spectrophotometers. Many design features determine the overall performance of spectrophotometers with some, such as wavelength accuracy and band width, being of lesser importance to turbidimetric analysis. The critical feature, of being able to detect small signals above a large range of blank absorbances, requires low optical noise with good detector stability and these parameters are found in the majority of instruments (Spenser and Price, 1981).

Both the measured turbidity and the measured rate of formation increase with decreasing wavelength (Hawkins, 1964), therefore, instruments employing low wavelength light sources would appear to have an advantage. In practice the presence of proteins which absorb strongly below 300nm and of endogenous serum porphyrins which absorb between 400-425 nm limit usuable wavelengths to the 320-380 or 500-650 nm range. The use of tandem cuvettes and reference cell blanking has allowed turbidimetric measurements to be made over a wide range of visible and u.v. wavelengths (Jacobsen and Steensgaard, 1979).

Good turbidimetric measurements may be made on commonly available laboratory spectrophotometers which have stable light sources, high precision detectors and wavelength selection. Most automated analysers including the

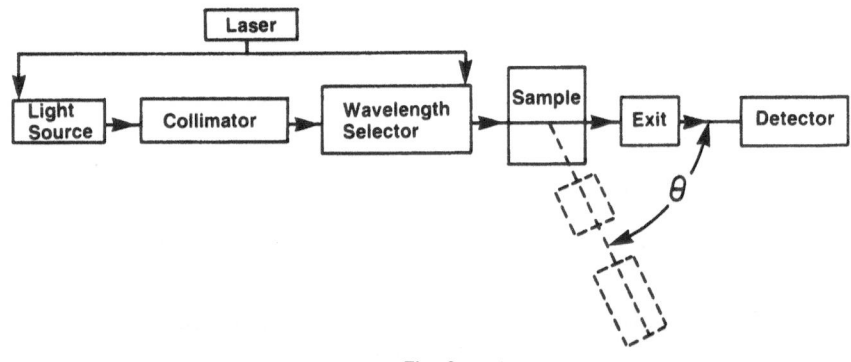

Fig. 3

a) Turbidimeter (spectrophotometer) θ = 0°; source is tungsten/deuterium lamp; wavelength selector is prism or grating; exit is exit slit and detector is photodiode.

b) Nephelometer 0° < θ ≤ 90°; source is tungsten lamp, discharge lamp or laser; wavelength selector (if necessary) is filter; exit is filter and light trap (if necessary) and detector is photomultiplier or photodiode.

centrifugal analysers which employ photometric detection also represent good instruments for turbidimetry and they have the added advantage of being able to perform kinetic assays.

Instruments for nephelometry must be capable of measuring the small amount of light scattered from a turbid solution and consequently high-intensity light sources and/or large amplification are necessary. In the absence of turbidity there is a blank scatter and sensitivity of the method is largely determined by the efforts to decrease this blank scatter and the ability of the detector system to discriminate increases over the background. Measurement of scattered light is either at right angles or some forward angle to the incident beam (Fig. 3).

As predicted by light scattering theory and discussed for turbidimetry, sensitivity increases with decreasing incident light wavelength. Whereas most turbidimetric measurements are made on spectrophotometers possessing visible and ultra violet light sources with variable wavelength selection, nephelometers usually operate at a fixed wavelength in the visible range. Selection of the required wavelength is by the use of optical filters and commonly used sources are tungsten lamps, mercury or xenon discharge lamps and one newer nephelometer employs a high power light emitting diode (LED) as a source. The spectral response characteristics of the photomultiplier or photodiode used for detection are matched to the chosen source and filter wavelength.

Of paramount importance in any nephelometric system is the control of stray light reaching the detector since the method depends critically on discriminating very small amounts of scattered light above an almost null or blank background level. Theoretically, a specific signal above any level of background signal should be measurable but in practice stray light is non-constant and gives rise to a small signal-to-noise ratio. Most worthwhile instruments, therefore, employ highly collimated, high-intensity incident beams and some have reference photodetectors to compensate for fluctuations in source intensity. Various other techniques such as the use of polarizing filters and the use of coated tubes have been employed to decrease stray light and increase the signal-to-noise ratio. Electronic devices have been incorporated into nephelometers to reduce the problem of stray light arising from the optical components and dust particles.

Lasers are an alternative light source which have many advantages and are employed in several commercial instruments. Unlike conventional light sources which use combinations of lenses and slits for collimation, lasers produce beams with very little divergence (typically 1 milliradian) and require no additional collimation. Laser light is highly stable and very instense further contributing towards sensitivity, although the latter advantage is limited somewhat by detector saturation. Because of the highly collimated nature of the laser light the forward transmitted beam can be very effectively masked allowing low forward angles of scatter detection. Disadvantages of lasers are their increased cost and availability in limited fixed wavelengths. The three commonly used commercial laser instruments all use a helium-neon laser which emits red light at 632.8nm. Lasers operating on lower wavelengths, for example, a helium-cadmium laser emiting light at 442 nm, would offer greater sensitivity but are more expensive and pose the potential problems of safety and contributing to background scatter from endogenous fluorophores in serum samples.

Endpoint versus Kinetic Instruments

A major drawback of the early light scattering assays was the long time required to attain the maximum signal but considerable improvement resulted from the introduction of enhancing polymers. A further decrease in assay

443

time, with concomitant increased sample throughput, was achieved by the introduction of kinetic analysis. Rate analysis can be carried out by measuring a difference in signal between two fixed time points, by monitoring the observed rate of change or by measuring the observed rate at a fixed time after initiation of turbidity. Analysis using two fixed time points is only meaningful if the reaction rate is constant or if a very small time interval between measurements is used. True kinetic measurements, in which there is a continuous differentiation of the signal, are accomplished readily using modern electronic processing devices. Many aspects of the rate method, particularly as it applies to nephelometry have been investigated and reported by Anderson and Sternberg (1978).

The kinetic approach offers important advantages such as, smaller effects due to instrument drift and freedom from non-specific side reactions which give a significant, cumulative signal over the longer times employed in end point assays but contribute little during rate measurements. At the initiation of reaction the intrinsic rate is zero and therefore no blank or background correction is needed as is the case with end-point methods. The rate of reaction, as followed by light scattering, increases with increasing antigen up to a certain concentration after which the rate decreases. Plots of peak rate versus antigen concentration therefore, give curves that resemble antigen precipitation and end-point scatter curves (Fig.2c). Because the high affinity antibodies contribute more to peak rate measurements than they do to the end-point signal, the end-point and kinetic "precipitation" curves are not necessarily the same or have the same equivalence points. Operating variables usually have to be more vigorously defined and controlled in kinetic methods since many of the instrument conditions are optimally fixed. Thus, mixing, temperature, buffer composition, antiserum and sample dilution are critical parameters that require careful consideration.

Basic instrumentation for end-point and rate light scattering measurements is similar with the latter requiring extra signal processing devices. The addition of a chart recorder for instance, adapts a spectrophotometer for kinetic turbidimetric analysis, albeit a crude conversion. Anderson and Sternberg (1978) claim that for kinetic nephelometry the use of a tungsten lamp with a 400-550 nm broadband filter is superior to a single wavelength source because the average particle scattering function over a broad wavelength range is relatively constant for the constantly changing size of the scattering species. Also, single wavelength light sources employing lasers potentially produce problems through polarization effects giving variable optimum scatter angles for the various sizes of particles that may be present during the critical early measurement stages of kinetic assays. Buffone (1975) has shown that in antibody excess, as is used for normal operating conditions, the particle size formed for different antigen concentrations is relatively constant. Therefore, the use of a wide band width light source may not be essential and in fact commercial rate nephelometers using a fixed wavelength laser source operate well.

Turbidimetric assays are readily carried out on centrifugal analysers which are ideally suited for kinetic measurements (Deverill, 1980; Austin and Maznicki, 1983). They are high throughput instruments employing high resolution spectrophotomers, microprocessors for data treatment and process samples and standards under the same conditions. An excellent text, edited by Price and Spencer (1980), deals with the use of centrifugal analysers in clinical chemistry and contains chapters specific for kinetic turbidimetric assays. Kinetic nephelometric assays have been run on modified centrifugal analysers (Buffone et al., 1975; Tiffany et al., 1974) and commercial machines are available. Reaction rate analysers also may be conveniently adapted for use in kinetic immunoturbidimetric assays and possess many of the advantages discussed above (van Munster et al., 1977, Malkus et al., 1978).

Commercial Instruments

Some instruments are specifically designed and marketed as turbidimeters but these are normally low specification devices not suitable for specific protein analysis. Turbidimetric measurements are made on laboratory photometers, spectrophotomers, kinetic analysers and centrifugal analysers and there are numerous manufacturers of these. A description of instruments is beyond the scope of this article.

Nephelometers are specialized machines specifically designed for light scattering measurements, although many fluorometers either can be adapted easily or are already equipped as dual purpose instruments. The available, commercial instruments range from committed, unsophisticated, manual, end-point types such as the Thorp micronephelometer to the versatile and automated, kinetic nephelometers using centrifugal analysers with light scattering detectors. Between these extremes are all types of nephelometers employing combinations from the various instrumental options of end-point versus kinetic analysis, manual versus automated sample handling and laser optics versus other light sources. These options have been used by Deverill and Reeves (1980) to classify light scattering instruments while Price et al., (1983) have grouped nephelometers into one of four optical systems. No attempt will be made here to give a comprehensive and descriptive list of commercial nephelometers but Table 1 gives brief details of some of the popular instruments currently available. A more detailed description of the commonly used nephelometers has been given by Whicher and Perry (1984). A measure of the utility of the nephelometric method may be judged by the fact that one manufacturer has developed a device, to convert an instrument from its original purpose, to light scattering detection. Thus, the Abbott TDx[R] fluorescence polarization analyser can be fitted with an accessory carousel, containing a light-emitting diode as an auxilliary light source, to allow nephelometric assays.

The choice of instrument for light scattering analysis is dictated largely by the particular application, the sample load, the versatility desired and the cost. Consequently, while centrifugal analysers and rate analysers require a large capital investment, such machines may be used for many purposes other than nephelometry or turbidimetry. Conversely, a reference centre laboratory with a large demand for specific protein analysis may well justify a committed, automated nephelometer having a high throughput. A further consideration in choosing an instrument is the cost and availability of operating reagents. Many of the manufacturers supply antisera and assay kits for their own particular machines but in most cases it is possible to develop good assays using reagents from other commercial sources or even those generated in-house, provided the necessary testing and quality control is ensured. The Beckmann Immunochemistry system (ICS) is a kinetic nephelometer and is available in manual or automated sampling handling models. The necessary information for characteristics of antisera and calibrator are encoded on machine readable cards which code for analyte identification, sample dilution, computation data, data for antigen excess detection and parameters for optimum instrument conditions. Although this control may seem restrictive the system can be used in a manual mode using the operator's own reagents but for routine use the suppliers' reagents and code cards are required.

Reagents

Specific protein analysis is normally carried out on serum or plasma samples but any body fluid containing sufficient levels of antigen may be employed, for example, cerebrospinal fluid, amniotic fluid, urine and seminal

TABLE 1   Commercial Light-Scattering Instruments

	Light Source (nm)	Kinetic (K) or End Point (E)	Automated (A) or Manual (M)	Throughput Samples/hr	Comments
Nephelometers					
Baker 410	Tungsten Lamp (610)	E + blank	M	*	Inexpensive and basic instrument
420	Laser (632.8)	K	A	25	Preprograms do an automatic antigen excess check by post addition of Ab.
Beckman ICS	Tungsten Lamp (400–550)	K	M or A	30–50	Difficult to use own reagents – does multiple assays on single sample
Behring	Laser (632.8)	E + blank	M or A	* 120–240	Normal operation is without PEG and blanking – uses large amounts antisera
Hyland	Laser (632.8)	E + blank	M or A	* 120–160	Basic model is similar to Behring – automated versions available
Technicon AIP	Mercury Arc (357)	E + blank	A	60–120	Segmented flow controls time. Allows Ag excess detection – Uses large amounts antisera
Centrifugal Analyser†	Zenon Arc (467 or 405)	K	A	100	Expensive outlay but versatile instrument. Can run other assays.
Turbidimeters					
Spectro-photometer	Tung/Deut. Lamp (vis – uv)	E + blank	M	*	Basic manual instrument with many other uses.
Centrifugal Analyser	Variable (vis – uv)	K	A	100–300	Expensive outlay – low reagent costs and versatile
Kinetic Analyser	Variable (vis – uv)	K	A	30–100	High outlay – may be batch or selective. Like cent. analyser is versatile

* Operator dependent – usually in the range of 30–60/hr
† E.g. I.L. Multistat III and Monarch

plasma. The presence of endogenous light scattering species in a sample will limit its use and this intrinsic turbidity, which determines the signal-to-noise ratio, is more critical in nephelometry. Turbidimetry uses spectrophotometers having high resolution detectors that can measure a small signal above the blank or background reading more precisely than nephelometers where both background and signal are normally amplified.

Endogenous scattering species found in serum include chylomicrons, lipoproteins, immune complexes and protein aggregates and measures may be taken to reduce these. Thus, lipids can be reduced by using fasting-serum samples, by centrifugation, by treatment with precipitating agents (Kallner, 1977; Hellsing and Enstrom, 1977) or by solubalization (Whicher and Blow, 1980). Treatment with protamine sulphate also reduced high plasma backgrounds possibly through removal of fibrinogen (Wood et al., 1978).

Avoidance of freezing-thawing or excessive heating of samples reduces protein aggregation. For most studies, providing the antigen of interest is present in sufficient quantity, it is adequate to decrease background scatter by using a high dilution of the sample, which for nephelometry generally means dilutions equal to or greater than 1:50. Reagent buffers should also be filtered (0.2u filter) to remove dust particles and other contaminating scattering species. The overall effect of these measures generally gives blank values which though small, nevertheless are sufficient to require their separate measurement for background corrections. As discussed previously such corrections are compensated for automatically in kinetic assays.

A novel method for compensating for blank antigen and antibody turbidity has been described by Jacobsen and Steensgaard (1979). These investigators employed tandem cuvettes, whereby antigen and antibody solutions are separated in the sample cuvette; a similar reference cuvette eliminates protein self-absorption and allows readings in the more sensitive ultra violet range. The sample cell is inverted to mix the reagents and the difference in turbidity recorded.

Antisera for use in nephelometry and turbidimetry should be clear, monospecific, high titred and have high avidity. The net light scattering produced is the sum of all the various antigen-antibody complexes formed in the mixture and the detector system has no discrimination for the specificities present, unlike gel based precipitation methods whereby minor rings and rockets can be differentiated and ignored. The requirements for monospecificity are, therefore, very stringent and all antisera should be carefully screened and when necessary absorbed. Because of the problems of endogeonous scattering species antisera should be of sufficient titer to allow a reasonable dilution which still retains a sufficient number of reactive antibody molecules. The avidity of the antiserum will determine the rate of complex formation and this is important both for end point and kinetic analysis but because of the shorter time interval employed it is probably more critical with kinetic assays. The combined requirements of titre and avidity will determine which antisera are useable for light scattering studies. They dictate the working range such that the chosen dilution still represents a large excess of antibody over the highest sample antigen concentration that may be analysed. While these ideal demands are met in many cases, certain pathological conditions occur where grossly elevated antigen levels exist and a single useable working range is not feasable, e.g., for myeloma immunoglobulins and C-reactive protein. Criteria for analyzing antisera for suitability in nephelometry has been advanced by Hudson et al (1981) and Cambiaso et al., (1974) and antisera designated as nephelometric grade is commercially available.

In the regional reference laboratories here at McMaster University Medical Centre, a large number of samples for immunoglobulin analysis are run each day and simultaneous determinations for IgG, IgA and IgM are carried out using a three channel flow system on a Technicon AIP set-up coupled with an electronic peak-picker and computer for data processing. Antisera requirments, which are large, are raised in-house in sheep and individual bleeds are screened for monospecificity using Ouchterlony and immunoelectrophoresis and absorbed if necessary. Antiserum samples are then tested on the system using the regular calibrators to ensure good performance and standard curves at a minimum 1:40 dilution. The high titred/high avidity bleeds are pooled to give large batches that are centrifuged (45 min. @ 125,000 g), aliquoted and stored @ -20° and this allows operation over a prolonged time using a single reagent.

The requirement for good standards or calibrators is common to all quantitative analytical methods but presents special problems in the case of serum proteins (Hjelm, 1981). Because of genetic variation, damage during venipuncture, subsequent sample purification and storage the protein in samples or standards may not be equally recognised by a particular antisera and this results in inaccuracy. The problem is further compounded by the differing specificity of an antisera against the immunizing protein to that protein in general and also by the different analytical methods used. Commercial reference serum are available from the various manufacturers of kits and reagents and preparations are also available from the World Health Organization. The quality of reference and control materials has been investigated by Shulman (1980), who noted a lack of accuracy between different lot numbers from the same manufacturer and he advocates standardization of subsequent working standards against a parent reference. Many of the aspects concerning reference materials for protein analysis have been dealt with by Ritchie (1978b). Until generally accepted international standards are available the problem of accuracy will remain but with good internal controls and standardization this should not detract from the clinical utility of the methods.

Assay Set Up

Assays may be set up totally from "scratch" or commercially available kits used, either specifically designated for a particulary instrument or for general use. In those cases where a kit is not available some initial effort is needed to establish a good working assay but often this is justified economically for those situations when routine assays of large numbers of samples are required. Such requirements are necessary of course for new developmental assays. The precise practicalities will be dependent upon the method (turbidimetric or nephelometric), upon the assay type (end point or kinetic) and upon the particular instrument used; but some generalities are relevant.

A good starting point is to choose a set of standards that encompass the range of interest, where this is known, and be of sufficient dilution so has to have low blank values. Standards are prepared from either primary or secondary reference sera and are available commercially. Dilutions of antisera, usually incorporating 4% PEG, are then prepared and a series of precipitin type curves generated with the standards. Volumes of reagents, and thus final amounts of reacting antigen and antibody, are dictated by particular instrument requirements or are readily chosen using common sense principles. From the precipitin curves generated, that which best conveys economy of reagents, usable range, necessary sensitivity and in the case of end-point assays, acceptable blank values, is chosen for further refinement.

Conditions, such as PEG concentration, dilutions, instrument settings, etc., are then optimised around the initial values to allow a good working assay to be carried out in a reasonable time. The effect of these changes on signal-to-noise ratio is an important consideration. Other factors in the optimisation process include the necessary dilution of samples and standards to achieve the desired sensitivity while maintaining acceptable blank values and being able to achieve a signal in a reasonable time. A balancing of all these factors may necessitate some moving backwards and forwards among all the variables to obtain the best conditions. A detailed account of the important variables and their optimisation for nephelometric assays has been published by Whicher & Blow (1980) and is recommended for anybody intending to set up a light scattering assay. The basic tenet is to aim for as high a dilution of sample or standards as possible using the highest dilution of antisera that will give a good standard curve with economy of reagents yet still maintain an antibody excess over the working range. In this way the initially chosen antibody/antigen ratio conditions may be optimised by a simple dilution process whereby the ratio is maintained along with the necessary sensitivity. Instrument parameters will have to be adjusted to maximise the signal for the given changes.

DISCUSSION

Nephelometry versus Turbidimetry

Much has been made of the relative values of the two techniques when choosing nephelometry or turbidimetry for specific protein analysis and while both methods find their particular uses some points on their various merits will be considered. Although initially the analysis of serum proteins by turbidimetry was a more used technique than nephelometry, the latter has enjoyed a more widespread use over the last fifteen years. This was mainly because of the availability of specific automated nephelometers and the lack of protocols and interest in turbidimetric methods. The situation has changed somewhat in the last few years with the increased awareness of the turbidimetric utility and the widespread availability of suitable instrumentation.

In choosing between the techniques the first consideration of many analysts is the sensitivity of the methods. While this is important it is by no means the only, or even the major, criteria on which a decision should be made. The sensitivity of an assay is governed by many variables and comparisons can be misleading. Method selection involves many considerations some of which are obvious, such as accuracy, precision, cost, flexibility, throughput and speed; others are not so obvious and include available personnel, training, space, safety and disposal problems, manufacturers warranties and repairs. The particular application may also dictate suitability, with sample size being a consideration for pediatric analysis for instance. In the majority of cases, the question resolves itself into deciding if the required assay can be conveniently and economically run with available reagents and equipment and only for some assays will sensitivity be a factor. With the greater availability of automated kinetic analyses and centrifugal analysers, this has meant that many assays now are being done on these machines by turbidimetry which were done previously by nephelometry. For clinical assays that are to be run on a routine basis a careful consideration of reagent cost, capital outlay and versatility of the apparatus must be undertaken. For research purposes the turbidimetric assay of human (and animal) proteins on the spectrophotometer, which is a readily available instrument, represents an economical and convenient method.

Sensitivity and Precision

Sensitivity, which defines the lowest, reliable amount of an analyte that can be distinguished from a blank will decide if a particular assay is clinically useful. For nephelometry and turbidimetry the detection limits are similar and they allow useful analysis of serum proteins occurring in the

449

low mg/litre range. A theoretical treatment for the nephelometric detection limit has been given by Anderson and Sternberg (1978) as about 400 ug/litre for a protein in the presence of large amounts of other proteins (eg. serum) and about one quarter of this value for a pure protein in aqueous solution. Beck and Kaiser (1982) have used a manual, kinetic nephelometric assay for IgE in serum which has a low standard of 24ug/litre and a manual, laser nephelometric assay for B_2-microglobulin in body fluids gave a detection limit of 40 ug/litre (O'Reilly et al., 1983). The detection limit is determined by many factors but the quality of the antisera and the initial analyte concentration are important, the latter factor determines the sample dilution which in turn determines the background and hence signal resolution. Fluids, like CSF and urine, which have low endogenous light scattering properties thus have low blank values and allow a more sensitive detection of specific proteins than possible in serum. Pretreatment of samples to reduce background values has been discussed previously and this results in better sensitivity but such treatment must not affect the analyte of interest. A comparison of sensitivities of nephelometry and turbidimetry using the same instrument has been done by Tiffany et al., (1980), and a similar study reported by Price et al. (1983), indicates that nephelometry had a limit of detection about one third lower than that of turbidimetry. These latter authors give detection limits for selected serum proteins for both turbidimetry and nephelometry using various instruments and these are in the range 1-100ng/cuvette. Compared to other analytical methods light scattering techniques have capable sensitivities about one tenth of Mancini and Laurell rocket assays but two to three orders of magnitude less than enzyme or radiolabelled immunoassays.

For any chosen method an acceptable, analytical error must be attained in order that useful diagnostic decisions can be made. Using presently available instruments and procedures, the imprecision achievable is comparable for nephelometry and turbidimetry. Intra and inter assay coefficients of variation generally are in the 2-5 and 4-10% range respectively, the lower limit figures being typical for automated, kinetic assays.

Antigen Excess Detection

The working range for optimized light scattering assays is in the region of antibody excess and a generally accepted rule is to employ a dynamic range whose upper limit represents about one half to two thirds of the antigen equivalence level. For many proteins reagents and conditions can be found that give a dynamic range which accomodates the normal clinical concentrations. Under certain circumstances, notably with myeloma immunoglobulins, haptoglobin and C-reactive protein (CRP), the concentrations are such that no suitable antisera is available to maintain an antibody excess over the whole pathologic range. A further problem with the monoclonal myeloma proteins is that the limited antigenic heterogeneity may selectively deplete reactive antibodies in the antiserum resulting in anomalous values and false interpretations.

Samples containing an excess of antigen result in reduced formation of scattering complexes with a concommitant decrease in signal which may be construed as a falsely low antigen level. Several methods have been used to overcome the problem and detect the presence of antigen excess. In end-point assays for immunoglobulins, which have large pathologic ranges, samples have been run routinely at two dilutions, thereby eliminating the possiblity of misinterpretation and give the added benefit of extra reliability from two readings (Walker and Gauldie, 1978). Monitoring for increased light scattering following the post-addition of either extra antibody or antigen to the final reaction mixture will demonstrate antigen excess or antibody excess respectively (Deaton et al., 1976). While these are frequently used procedures they are uneconomic in terms of both assay time and reagents, particularly when extra antibody is used. Instrumental methods for detecting conditions of antigen excess exist for both end-point assays employing the Technicon segmented flow system and for kinetic nephelometers. With the former instrument antigen is pulsed, along with air, into a continuous flow of antibody resulting in leading and trailing segments of low antigen concentration and central segments containing the full antigen concentration. If antigen excess conditions exist in the central segments, which normally correspond to a peak maximum, peak profiles have degrees of bifid deformity which correspond to the antigen excess (Larson et al., 1970). With kinetic analysis, parameters may be chosen such that the peak rate and/or time to achieve peak rate can be analysed to differentiate conditions of antigen excess (Anderson and Sternberg, 1978). Such electronic surveillance would seem to be an optimal solution, but in practice, antigen excess normally is detected for rate nephelometry using the post-addition method.

While the problem of antigen excess is by far the most common, occasional problems occur because of antibody excess. Conditions of antibody excess manifest themselves in a similar way as antigen excess, with decreasing signal resulting from an increasing antibody concentration. These conditions are met normally only at the very low end of the antigen range and for most assays occur below the lower clinical range and therefore present no problem.

Present and Future Developments

Currently, immunonephelometry represents the most widely used technique for specific protein analysis, being more convenient, faster and precise than immunodiffusion based methods. A recent trend has been towards the use of turbidimetry over nephelometry, in particular by those laboratories that either already possess or have justification for acquiring kinetic or centrifugal analysers. Many proteins can be analysed on a "stat" basis using the newer "random access" analysers fitted with turbidimetric detection or the newer Technicon capsule chemistry systems with nephelometric detection. Both light scattering methods are restricted to analysis of those proteins that occur in sufficient concentration such that after suitable dilution to give acceptable backgrounds they are still within the lower detectable limit. A list of some common proteins that have been analysed by the techniques is given in Table 2.

To broaden the potential application of the methods, various modifications of the techniques have been undertaken. These include the analysis of drugs and other haptens using an inhibition modification and the use of particle-enhanced turbidity. Both of these methods have been commercially adopted and are described elsewhere in this text. Further refinements in the detection of scattered light would allow nephelometry to come into its own and fulfill the promise of being a highly sensitive technique for protein analysis; without such improvements the technique may become a poor second cousin to turbidimetry for routine clinical use. Several potential developments to increase the sensitivity of detection of scattered light have been reported. Thus, investigations by Benedek and co-workers have shown the utility of using quasi-elastic laser-light scattering spectroscopy (Cohen, 1975; von Schulthess, 1976) and anisotropic light scattering (von Schulthess, 1980) for immunoassay, and similarly, Uzgiris (1981) has used a laser Doppler-spectrometry method for probing immune reactions. To date, these methods rely on particle enhancement to achieve the high sensitivity reported. Lower wavelength sources along with various other optical and detection modifications to improve signal-to-noise ratio would increase sensitivity and such improvements are to be expected in newer generation instruments.

Table 2

Plasma Proteins - Determined by Nephelometry/Turbidimetry

Albumin	Fibronectin
α_1-Acid Glycoprotein (orosomucoid)	Haptoglobin
α_1-Anti-trypsin	Hemopexin
α_2-Macroglobulin	Immune Complexes
Anti-DNA antibodies	Immunoglobulin A
Anti-Thrombin III	Immunoglobulin E
Apolipoprotein AI	Immunoglobulin G (& subclasses)
Apolipoprotein AII	Immunoglobulin M
Apolipoprotein BI	Immunoglobulin light chains
B_2-Microglobulin	Pre-albumin (Transthyretin)
C1-Inactivator	Rheumatroid Factor
C1-Esterase Inhibitor	Serum Amyloid P Component (SAP)
C3	Serum Retirol Binding Protein
C4	Somatotropin
Ceruloplasmin	Thyroxine Binding Globulin (TBG)
C-Reactive Protein	Transferrin

The requirement for high affinity antibodies for nephelometry and turbidimetry has been fulfilled by specific immunization protocols and careful screening and testing of antisera. Such requirements result in limited supplies of reagents and necessary checks on assay specificity, sensitivity and accuracy when changing antisera. Monoclonal antibodies have been touted as a solution to many of the problems associated with polyclonal antisera, being theoretically capable of providing a limitless supply of ideally selected reagent. The use of these reagents in precipitation based methods would appear to be limted, however, since lattice theory would predict complex formation to be restricted to those antigens possessing repeat determinants. The immunoglobulin molecule which is symmetrical and therefore minimally antigenically divalent, has been quantitated by radial immunodiffusion using monoclonal antibody (Lowe et al., 1982). For turbidimetric determination it was found that a minimum of two monoclonal antibodies, directed against spatially different determinants, was required (Steensgaard, et al., 1980). A recent study (Marcovina et al., 1985) on the use of monoclonal antibodies for the determination of apolipoprotein B by radial immunodiffusion found a minimum of two antibodies was required for precipitation. A similar study here at McMaster University Medical Centre using monoclonal antibodies against apolipoprotein B with nephelometric detection failed to find any combination that gave a significant scatter signal (Gauldie, J. - personal communication). With suitable selection techniques for high affinity antibodies, the future for monoclonal antibody based nephelometric and turbidimetric protein assays seems bright. Hapten inhibition and particle enhanced light scattering assays are also ideal techniques for monoclonal antibody exploitation and being multiple site assays require only reagents of a single specificity.

SUMMARY

Specific protein analyses are an essential part of the clinical chemistry laboratory and light scattering assays employing turbidimetric or nephelometric detection play a major role in those analyses. The advent of reliable, automated equipment has allowed these techniques to attain a key position in the quantification of many, clincally important proteins found in plasma and other body fluids. Some of the qualitative information from gel-based immunoassays may not be available from nephelometry or turbidimetry but this is more than compensated for by the high sample throughput

achievable with automation of the latter methods. To date, nephelometry has
been the more widely used of the two light scattering techniques but with the
more widespread availability of kinetic and centrifugal analysers an increase
in turbidimetric methods is to be expected. The limits of detection and
imprecision of the methods are similar but in a few cases the extra
sensitivity achievable with nephelometry may make this method either
desirable or necessary.

A continuing, important role for light scattering assays is forseen and
future developments, both in terms of reagents and instrumentation, will
strengthen this role and may extend the techniques into new areas which are
presently the domain of the more sensitive immunoassay methods.

REFERENCES

Anderson, R.J., and Sternberg, J.C., 1978, A rate nephelometer for
    immunoprecipitation measurement of specific serum proteins, in:
    "Automated Immunoanalysis, Part 2," R.F. Ritchie, ed., Marcel Dekker,
    New York, 409-470.
Austin, G.E., and Maznicki, E., 1983, Automated turbidimetric assay of serum
    apolipoprotein A1 using the COBAS-BIO centrifugal analyser, Clin.
    Biochem., 16: 338-340.
Beck, O.E., and Kaiser, P.E., 1982, Rate nephelometry of human IgE in serum,
    Clin. Chem., 28: 1349-1351.
Buffone, G.J., Savory, J. and Hermans, J., 1975a, Evaluation of kinetic light
    scattering as an approach to the measurement of specific proteins with
    the centrifugal analyser. II. Theoretical considerations, Clin.
    Chem., 21: 1735-1746.
Buffone, G.J., Heintges, M.G., Savory, J., and Killingsworth, L.M., 1975b,
    The kinetic nephelometric measurement of IgG and albumin in serum and
    CSF using a laser modified aminco rotochem centrifugal fast analyser,
    Clin. Chem., 21: 943A.
Cambiaso, C.L., Masson, P.L., Vaerman, J.P., and Heremans, J.F., 1974,
    Automated nephelometric immunoassay (ANIA) I. Importance of antibody
    affinity, J. Immunol. Methods, 5: 153-163.
Cohen, R.J., and Benedek, G.B., 1975, Immunoassay by light scattering
    spectroscopy, Immunochem., 12: 349-351.
Deaton, C.D., Maxwell, K.W., Smith, R.S., and Creveling, R.L., 1976, Use of
    laser nephelometry in the measurement of serum proteins, Clin. Chem.,
    22: 1465-1471.
Debye, P., 1915, Zerstreuung von Rontgenstrahen, Ann. Physik., 46: 809.

Deverill, I., and Reeves, W.G., 1980, Light scattering and absorption -
    developments in immunology, J. Immunol. Methods, 38: 191-204.
Deverill, I., 1980, Kinetic measurement of the immunoprecipitation reaction
    using the centrifugal analyser, in: "Centrifugal Analysers in Clinical
    Chemistry", C.P. Price and K. Spencer, eds., Praeger, Eastbourne,
    109-124.
Hawkins, J.D., 1964, Some studies on the precipitin reaction using a
    turbidimetric method, Immunol., 7: 229-238.
Heidelberger, M., and Kendell, F.E., 1935, A quantitative theory of the
    precipitin reaction III. The reaction between crystalline, egg
    albumin and its homologous antibody, J. Exp. Med., 62: 697-720.
Hellsing, K., 1978, Enhancing effects of nonionic polymers on immunochemical
    reactions, in: "Automated Immunoanalysis, Part I", R.F. Ritchie, ed.,
    Marcel Dekker, New York, 67-112.
Hellsing, K., and Enstrom, H., 1977, Pre-treatment of serum samples for
    immunonephelometric analysis by precipitation by polyethylene glycol,
    Scand. J. Clin. Lab. Invest., 37: 529-536.

Hjelm, M., 1981, Components in a model for the production of reference materials with special reference to immunoassays, in: "Immunoassays for the 80s", A. Voller, A. Bartlett and D. Bidwell, eds., University Park Press, Baltimore, 185–192.

Hudson, G.A., Ritchie, R.F., and Haddow, J., 1981, Method for testing antiserum titer and avidity in nephelometric systems, Clin. Chem., 27: 1838–1844.

Jacobsen, C., and Steensgaard, J., 1979, Measurements of precipitation reactions by difference turbidimetry: A new method, Immunology, 36: 293–298.

Kallner, A., 1977, Removal of background interference in nephelometric determination of serum proteins, Clin. Chem. Acta, 80: 293–297.

Kerker, M., 1969, "The Scattering of Light and Other Electromagnetic Radiation", Academic Press, New York.

Killingsworth, L.M., and Savory, J., 1973, Nephelometric studies of the precipitin reaction: A model system for specific protein measurements, Clin. Chem., 19: 403–407

Lowe, J., Brid., P., Hardie, D., Jefferis, R., and Ling, N., 1982, Monoclonal antibodies (McAbs) to determinants on human gamma chains: Properties of antibodies showing subclass restriction or subclass specificity, Immunol., 47: 329–336.

Larson, C., Orenstein, P., and Ritchie, R.F., 1970, An automated method for quantitation of proteins in body fluids, in: "Advances in Automated Analysis, Vol 1", Thurman Associates, Miami, 101–104.

Mie, G., 1908, Beitrage zur optik truber medien, speziell kolloidaler metallosungen, Ann. Physik., 25: 377.

Marcovina, S., France, D., Phillips, R.A., and Mao, S.J.T., 1985, Monoclonal antibodies can precipitate low-density lipiprotein. I. Characterization and use in determining apolipoprotein B., Clin. Chem., 31: 1654–1658.

Marrack, J.R., and Richards, C.B., 1971, Light-scattering studies of the formation of aggregates in mixtures of antigen and antibody, Immunology, 20: 1019–1040.

Malkus, H., Buschbaum, P., and Castro, A., 1978, An automated turbidimetric rate method for immunoglobulin assays; Clin. Chem. Acta, 88: 523–530.

O'Reilly, D., Whicher, J., and Sanderson, A.R., 1983, A simple, precise and sensitive nephelometric assay for B2-microglobulin in body fluids, J. Immunol. Methods, 57: 265–273.

Price, C.P. And Spencer, K., eds., 1980, "Centrifugal Analysers in Clinical Chemistry", Praeger, Eastbourne.

Price, C.P., Spencer, K., and Whicher, J., 1983, Light-scattering immunoassay of specific proteins: A review, Ann. Clin. Biochem., 20: 1–14.

Lord Rayleigh, 1871, On the scattering of light by small particles, Phil. Mag., 41: 447.

Ritchie, R.F., ed., 1978a, "Automated Immunoanalysis, Parts 1 and 2", Marcel Dekker, New York.

Ritchie, R.F., 1978b, Reference materials for plasma protein analysis, in: "Automated immunoanalysis, Part 1", R.F. Ritchie, ed., Marcel Dekker, New York, 159–180.

Shulman, G., 1980, Quality of commercially available controls in laser immunonephelometry, Ann. Clin. Biochem., 17: 178–182.

Spencer, K., and Price, C.P., 1981, Clinical chemistry instrumentation and light scatter measurement, U.V. Spectro. Group Bull., 8: 38–50.

Steensgaard, J., Jacobsen, C., Lowe, J.A., Hardie, D., Ling, N.R., and Jefferis, R., 1980, The development of difference turbidimetric analysis for monoclonal antibodies to human IgG, Mol. Immunol. 17: 1315–1318.

Tiffany, T.O., Manning, G.B., Hills, L.P., Huey, E.E., and Frankart, M.L., 1980, The application of fluorescence measurements to the centrifugal analyser, in: "Centrifugal Analysers in Clinical Chemistry", C.P. Price and K. Spenser, eds., Praeger, Eastbourne, 329–361.

454

Tiffany, T.O., Parella, J.M., Johnson, W.F., and Burtis, C.A., 1974, Specific protein analysis by light-scatter measurement with a miniature centrifugal fast analyser, Clin. Chem., 20: 1055-1061.

Uzgiris, E., 1981, Probing immune reactions by laser light scattering spectroscopy, in: "Methods in Enzymology, Vol. 74", Langone, J.J., and van Vunakis, H., eds., Academic Press, New York, 177-198.

van Munster, P.J.J., Hollen, G.E.J.M., Samwel-Mantingh, M. and Holtman-van Meures, M., 1977, A turbidimetric immunoassay (TIA) with automated individual blank compensation, Clin. Chim. Acta, 76: 377-388.

von Schulthess, G.K., Cohen, R.J., and Benedek, G.B., 1976, Laser light scattering spectroscopic immunoassay in the agglutination-inhibition mode for human chorionic gonadotropin (hCG) and human luteinizing hormone (hLH), Immunochem., 13: 963-966.

von Schulthess, G.K., Giglio, M., Cannell, D.S., and Benedek, G.B., 1980, detection of agglutination reactions using anisotropic light scattering: An immunoassay of high sensitivity, Mol. Immunol., 17: 81-92.

Walker, W.H.C., and Gauldie, J., 1978, Automated determination of immunoglobulins, in: "Automated Immunoanalysis, Part 1", R.F. Ritchie, ed., Marcel Dekker, New York, 203-226.

Whicher, J.T., and Blow, C., 1980, Formulation of optical conditions for an immunonephelometric assay, Ann. Clin. Biochem., 17: 170-177.

Whicher, J.T., and Perry, D.E., 1984, Nephelometric methods, in: "Practical Immunoassay", W.R. Butt, ed., Marcel Dekker, New York, 117-177.

Wood, P.J., Cockett, D., and Mason, P., 1978, A rapid and inexpensive laser nephelometric assay for plasma pregnancy specific B_1-glycoprotein levels, Clin. Chem. Acta, 90: 87-91.

NEPHELOMETRIC INHIBITION IMMUNOASSAY FOR SMALL MOLECULES

Chan S. Oh and James C. Sternberg

Applied Research Department
Diagnostic Systems Group
Beckman Instruments, Inc.
Brea, California

INTRODUCTION

The use of nephelometry, and especially rate nephelometry (Tiffany, Parella, Johnson and Burtis, 1974), for the measurement of specific serum proteins has become very well established. In nephelometric methods, the marked increases in light scattering which occur as a consequence of the immunochemical reaction between polyvalent antigens (such as proteins) and their antibodies is utilized. Thus, while solutions of protein molecules scatter only a small to moderate amount of visible light, their reactions with their specific antisera lead to the formation of cross-linked three dimensional immuno complexes large enough to scatter a significantly increased amount of light, and, eventually, to precipitate. In rate nephelometry, the rate of increase of light scattering during the course of this reaction is utilized to provide a measure of the protein concentration.

While small molecules (haptens) are not generally antigenic, an antibody response specific to the hapten can usually be obtained for conjugates of the hapten with macromolecules (especially proteins). Because of its small size, each hapten molecule will normally be able to accommodate only one antibody binding site, so that the reaction of a hapten with its antiserum cannot lead to cross-linking and immunoprecipitation. A conjugate consisting of several molecules of the hapten covalently attached to a protein molecule, however, is immunochemically polyvalent, and the addition of an antiserum directed towards the hapten can lead to immunoprecipitation of such a conjugate. Under these conditions, a three dimensional immunocomplex can form, and, when the resulting particles reach a certain critical size, turbidity can be measured. Alternatively, the intensity and rate of increase of intensity of light scattered at a particular angle can be measured.

At a fixed antiserum titer, increasing amounts of the hapten-protein conjugate produce increased rate nephelometric signals. With further increases in the hapten-protein conjugate, the signals eventually reach a maximum value, and then decrease. This dependence of the rate response on the concentration of the conjugate is similar to the well-known "Heidelberger" immunoprecipitation curve (Heidelberger and Kendall, 1935). At a fixed concentration of the hapten-protein conjugate, increasing amounts of antiserum lead to a monotonic increase of rate signals. The upper limit of the rate response would be reached at the antibody concentration of the antiserum. However, in practice, neat antiserum can not be used, because other serum components tend to precipitate "nonspecifically". If purified antibody is used, an even higher rate signal can be obtained. The upper limit of the rate response is ultimately determined by the solubility of the purified IgG in the reaction buffer.

When a sample solution containing a small hapten molecule to be analyzed (the analyte) is brought into contact with the antiserum, no nephelometric signal is obtained, because the monovalency of the hapten, it cannot cross-link with other molecules. If a multivalent hapten-protein conjugate is then introduced, cross- linking of the conjugates by antibody can occur, leading to a nephelometric signal. The nephelometric signal obtained upon the subsequent addition of the conjugate is dependent upon the concentration of the analyte in the sample introduced. Increasing concentrations of the hapten occupy increasing numbers of antibody sites, diminishing their availability for cross- linking of the conjugate. The presence of the analyte thus results in a concentration-dependent <u>inhibition</u> of the signal obtained when the conjugate is added, and such methods are called "nephelometric inhibition immunoassay" (NIIA) methods (Riccom, Masson, Vaerman and Hereman, 1972). Since no signal would occur with the analyte only, in the absence of the conjugate, the conjugate is the substance which enables an analyte concentration-dependent response to develop; the multivalent hapten-protein conjugate is therefore called the "developer antigen".

CURRENT STATE OF THE ART

Commercial nephelometric inhibition assay kits are available for the measurement of several therapeutic drugs. An automated rate nephelometer, capable of performing the calibration and calculations required for these measurements is also commercially available (ICS, Immunochemistry System, Beckman Instruments, Brea, CA).

While NIIA methods provide very convenient and rapid means for the measurement of therapeutic drugs, the existing commercial assay kits have certain limitations. Thus, under inadequately controlled operating conditions, frequent calibration may be necessary and/or poor precision can result. One of the major causes of these problems has been found to be the temperature sensitivity of the rate signal.

The sensitivity of NIIA methods has thus far been limited to about 10^{-7} molar serum concentrations (concentrations of analyte in the final measuring cuvet, however, may be as low as 10^{-9} molar). Most of this sensitivity limitation results from scatter by other components of the serum sample, which must therefore be significantly diluted before injection into the assay buffer.

In order to better understand the true potential and the limitations of rate nephelometric inhibition immunoassay, a study of the immunochemical and physico-chemical principles underlying these methods was undertaken.

PRINCIPLES OF RATE NEPHELOMETRIC INHIBITION IMMUNOASSAY

As it was pointed out by one of the authors (Sternberg, 1977), the nephelometric rate signal observed upon adding the developer antigen results from a complex sequence of events. In the primary immunochemical reaction step, the antibody first forms an immuno-complex with the developer antigen. Since the antibody is divalent and the developer antigen is multivalent, a secondary step will occur, in which the primary complexes form oligomers through additional immunochemical bonding between them. The primary and oligomeric immuno-complexes are probably formed

458

quite rapidly (within a few seconds) under the conditions normally employed for rate nephelometry.

The molecular weight (or diameter) of the oligomeric complex probably reaches a few million Daltons, which is far too small to cause significant light-scattering. However, as the population (or concentration) of this oligomeric immuno-complex reaches a certain critical value, aggregation of these complexes, based on a combination of charge and hydrophobic interactions, begins to take place. When the size of the resulting aggregates reaches sixty to a hundred million Daltons, light in the visible wavelength region is scattered (Sittampalam and Wilson, 1984). This tertiary process (aggregation) is the relatively slow step. It is affected by several factors, including pH, ionic strength, concentration of polyethylene glycol (PEG) or other additives, and temperature, as well as such intrinsic properties of the immunoreagents themselves as hydrophobicity and electrical charge distribution.

In carrying out the entire competitive binding inhibition reaction for measurement of the analyte, the free hapten is permitted to bind first to the antibody, and the developer antigen is then added. Since the tertiary aggregation reactions are much slower than the primary and secondary immunoreactions, there will be ample time during the aggregation process for some displacement of the free hapten from the anti-hapten antibody sites by the hapten conjugate. The extent of inhibition will therefore be dependent upon the relative affinities of the antibody for the free and the conjugated hapten.

The temperature dependence of the rate signal is especially pronounced when there are marked differences in the antibody affinity constants for the free and conjugated hapten. Since the anti-hapten antibody has normally been generated against a synthetic antigen consisting of the same or another conjugate of the hapten to a protein, it is quite possible that the affinity of the antibody for the conjugate may be greater than that for the free hapten. In general, it would be expected that the binding of the antibody to the entity displaying the weaker of these two affinities would be more affected by temperature. An increase in temperature would therefore tend to decrease the extent of inhibition if the affinity of the antibody for the conjugated hapten exceeded the affinity for the free hapten; the extent of inhibition would be increased with increasing temperature if the affinity of the antibody for the free hapten exceeded that for the conjugated hapten.

Since fewer immunochemical primary and secondary events would be required for the reaction to reach the stage where significant aggregation could occur, it would be expected that the use of a developer antigen having a greater molecular weight should provide a greater rate nephelometric signal.

The studies reported here were carried out in attempts to better understand the factors that control responses in nephelometric inhibition immunoassay.

EXPERIMENTAL INVESTIGATIONS OF NIIA

In these studies, theophylline was selected as the model system. In order to minimize any effects of the cross reactivity of the antibody towards various drug analogs and metabolites, a monoclonal antibody highly specific to theophylline (with a discrimination ratio of greater than 100:1 towards caffeine) was utilized.

Theophylline was first modified to provide a carboxyl terminated pendant group. The carboxy group-derivatized theophylline was then activated with dicyclohexylcarbodiimide (DCCI). This activated theophylline derivative will readily react with the amino groups of any selected protein.

Theophylline conjugates to BSA, IgG, apoferritin, and thyroglobulin were prepared and studied. These theophylline-conjugated proteins were diluted to approximately equivalent protein concentration (based on the optical density at 280 nm) and tested in the nephelometer (Beckman Immunochemistry System, ICS) using the monoclonal anti-theophylline antibody. At a given antibody titer, the apoferritin-theophylline conjugate was found to give the highest rate signal of the conjugates studied.

For each of these conjugates, the highest rate signal was that found at the peak of its Heidelberger curve (i.e., the curve of peak rate signal vs. conjugate concentration). Contrary to our expectations, the thyroglobulin-theophylline conjugate gave no immunoprecipitation at any concentration of the antibody used, although the theophylline moiety of the thyroglobulin-theophylline conjugate was shown to inhibit the immunoprecipitation of the antibody by the apoferritin-theophylline conjugate. Thus the naive assumption that a developer antigen of higher molecular weight would necessarily lead to a higher rate signal is not correct. Apparently the distribution of the theophylline moieties on the surface of the protein has a more significant effect on the ability of the antibody to bind. In particular, as we will see later, the antibody would have difficulty in binding if the surface configuration of the conjugate led to appreciable steric hindrance, or if the rather hydrophobic theophylline moieties tended to be sequestered into hydrophobic regions on the carrier protein surface.

Avidin-biotin interaction as a model for nephelometric inhibition

The interaction of avidin with biotin has many features in common with antibody-hapten interactions, and provides a well-defined model system for the study of the inhibition process. Thus, the bioaffinity precipitation of avidin with biotinylated protein gives a rate nephelometric response analogous to that obtained for the precipitation of antibody with developer antigen. Each molecule of avidin has four biotin binding sites with very high binding constants (about 10^{15} liters moles^{-1}). The four biotin binding sites of avidin occur in two pairs, at opposite ends of the molecule. The binding sites within each pair are so close together that it may not be possible for the two closely located biotin binding sites within the pair to simultaneously bind two bulky biotin derivatives. When appropriately substituted biotins are employed, therefore, the avidin behaves as only bivalent with respect to biotin, and such location of the functional sites leads to a linear profile, with the groups bound to the two biotins pointed in opposite directions.

We have biotinylated BSA, IgG (human, rabbit, mouse, goat), apoferritin and thyroglobulin, using biotin-NHS (biotin-N-hydroxysuccinimide) or biotin-X-NHS (biotin-6-aminocaproic N-hydroxy-succinimide). In each case, avidin was found to cross-link the biotinylated proteins to provide a very high rate signal, inducing precipitation with a few micrograms of the biotinylated protein. In each of the reactions of avidin with biotinylated proteins, the nephelometric signal developed so rapidly that it was necessary to override the time gating used by the instrument to eliminate the effects of the injection transient--otherwise, much of the signal would be missed.

The observed rate was found to be independent of the molecular weight of the protein. Thus, despite the failure of the thyroglobulin-theophylline conjugate to precipitate with anti-theophylline antibody, the biotinylated thyroglobulin precipitates with avidin just as well as the biotinylated BSA or apoferritin! It appears that the high affinity constant of avidin for biotin, and the excellent cross-linking capability of avidin resulting from its linear binding configuration, outweigh other factors in bringing about this precipitation.

At a fixed amount of avidin, increasing amounts of the biotinylated protein gave increasing rate signals, which reached a maximum value and then decreased rapidly with further increases of the biotinylated protein (see Fig. 1A). In comparison with the behaviour of anti-theophylline with protein-theophylline conjugates, the slopes of the ascent and descent of the rate signal due to the changes in concentration of the biotinylated protein were very much steeper. This unique behavior also appears to be due to the highly effective cross-linking resulting from the linear binding configuration of the avidin.

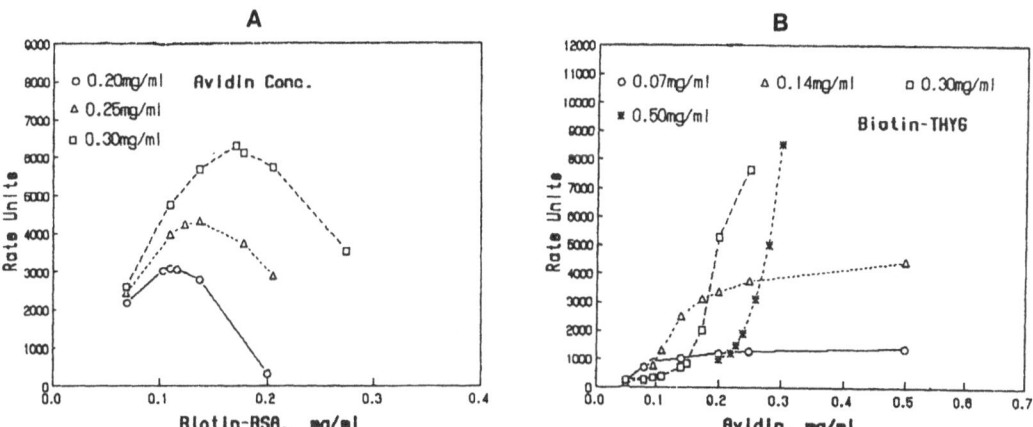

Fig. 1A.   Effect of concentration of biotin-bovine serum albumin (biotin-BSA) conjugate on rate nephelometric response at a series of concentrations of avidin.

Fig. 1B.   Effect of concentration of avidin on rate nephelometric response at a series of concentrations of biotin-thyroglobulin (biotin-THYG) conjugate.

(All rate nephelometric measurements were carried out in a Beckman ICS using a manual mode M33 card and with 42 microliters of each of the two solutions injected into 600 microliters of ICS buffer.)

With a given amount of biotinylated protein, however, the addition of increasing amounts of avidin produced increasing rate signals, which were found to reach a maximum value, beyond which higher amounts of avidin did not cause either further increase or a decrease. The height of this maximum rate plateau value was found to increase with increasing amounts of the biotinylated protein (Fig. 1B). Excess avidin does not cause the expected decrease in rate response, apparently because the oligomeric complex formed when avidin saturates the biotin sites on each biotinylated protein molecule is already large enough to aggregate rapidly, even without further cross-linking.

Inhibition behavior was also readily demonstrated in the avidin biotinylated protein system. Thus, the addition of either free biotin or biotinylated drug molecules inhibited the rate nephelometric response obtained in the reaction of avidin with the biotinylated proteins.

Improved NIIA for haptens using avidin-biotin coupling

The use of the unique properties of the avidin-biotin interaction provides a powerful additional tool for exploring the factors influencing the signals obtained in nephelometric inhibition immunoassays for haptens. The above studies of the rate nephelometric behavior of avidin with biotinylated proteins have led to improved methods for the rate NIIA for small molecules using specific monoclonal antibodies. We have proposed and studied several alternative uses of the avidin-biotin system to provide increased rate signals and improved performance over conventional NIIA.

In attempting to reduce the amount of the rather costly monoclonal antibody required for the measurement of theophylline using the multivalent apoferritin-theophylline conjugate as the developer antigen, it was recognized that the developer antigen could bind much of the antibody in a "non-productive" manner, in which the two binding Fab legs of the antibody may be bound to two adjacent theophylline moieties on the same apoferritin molecule. In this kind of binding, the antibody molecule cannot cross-link to form the oligomeric antigen-antibody complexes; this results in inefficient use of the expensive monoclonal antibody. This is quite probable in view of the "hinge region flexibility" of antibody molecules (Oi, Vuong, Hardy, Reidler, Dangl, Herzenberg, Stryer, 1984). In contrast, avidin lacks such a flexibility, so that its two effective binding sites are less likely to link to biotins on the same biotinylated protein molecule. This causes it to be highly effective in cross-linking biotinylated proteins.

An additional problem was also identified in the use of apoferritin as the core protein for the preparation of the developer antigen. Apoferritin preparations often contain various amounts of dimers, oligomers and polymers. The distribution of these species was analyzed using high performance liquid chromatography (HPLC) with a TSK 3000 type size exclusion column, before and after the apoferritin underwent reaction with a theophylline derivative which had been activated with a water soluble carbodiimide. The distribution of apoferritin species with different molecular weights was found to be similar in both cases. The most pronounced chromatographic peak is due to the monomer. The next large peak has a molecular weight corresponding to the dimer, and the remaining peaks are associated with the varying degrees of oligomers and polymers.

A batch of developer antigen preparation was separated in a large size exclusion chromatography column (A1.5M of BioRad). Fractions corresponding to monomer, dimer and polymer were separately collected, concentrated, and reconstituted in such a manner that each pool contained an equal protein concentration. These monomeric, dimeric, and oligomeric developer antigens were found to have different nephelometric rate signals in their reaction with the same antibody solution. Dimeric developer antigen gave the highest rate signal, the monomeric compound gave some rate signal, and the polymeric compound gave virtually no signal. This heterogeneity of the developer antigen can create problems in the manufacture, quality control and shelf-life of such a reagent.

To overcome these problems, we have proposed the use of simple bidentate molecules, composed of the hapten of interest linked to biotin, as a chemically well-defined replacement for the developer antigen (patent pending). To prepare one of these bidentate molecules, a reactive theophylline derivative was covalently attached to biotin, instead of to a protein, so that the theophylline moiety could bind to its antibody and the biotin moiety could bind to avidin. Since the antibody has two and avidin has two to four binding sites, it should be possible to form a three dimensional polymer or oligomer containing the antibody, avidin and theophylline-biotin conjugate. For a stable oligomer to form, the distance between the theophylline and biotin moieties in the binary conjugate should be such that the antibody molecules do not sterically interfere with the binding of the avidin molecules, and the avidin molecules do not sterically interfere with the binding of the antibody molecules to the conjugate.

The preparation of biotin-theophylline conjugates

To evaluate the requirements for such an analytical system, it was necessary to synthesize a series of bidentate biotin-theophylline

462

conjugates of differing connecting chain length. A series of amine-terminated theophylline derivatives was first synthesized. 8-bromo-theophylline was refluxed in excess of a diamine ($NH_2-(CH_2)_N-NH_2$) under a nitrogen atmosphere for a period of two to seventy-two hours, depending upon the amine used. The end of the reaction was determined by a thin layer chromatographic (TLC) analysis of the reaction mixture, using glass TLC plates coated with silica gel, and ultra-violet indicator. The eluant used contained 20% methanol and 4% ammonium hydroxide in chloroform. Most of the impurities were found to have high RF values, while the monoamine showed a higher RF than the starting diamine (which usually stayed at the origin). The product (I, below) is UV-indicator positive and ninhydrin positive.

The product was purified by means of column chromatography on silica gel. The reaction mixture was evaporated to a small volume under vacuum. The concentrated reaction mixture was mixed with a small quantity of silica gel and dried on a hot plate. The mixture was carefully loaded on top of a silica gel column using chloroform as the starting eluant. The column was eluted with solvent containing varying amounts of methanol in chloroform. When the gradient composition reached 20% methanol in chloroform, the column was washed with the 20% methanol/chloroform, until the eluant contained no UV-indicator-positive material. The column was then eluted with a mixture containing 20% methanol and 4% ammonia in chloroform. The fractions containing the pure monoamine derivative of theophylline were pooled and evaporated to dryness in a rotary evaporator. White-yellowish crystalline solids were used for the next reaction without further purification. Theophylline monoamine derivatives have a molar absorptivity of 1.2×10^4 at 295 nm in methanol.

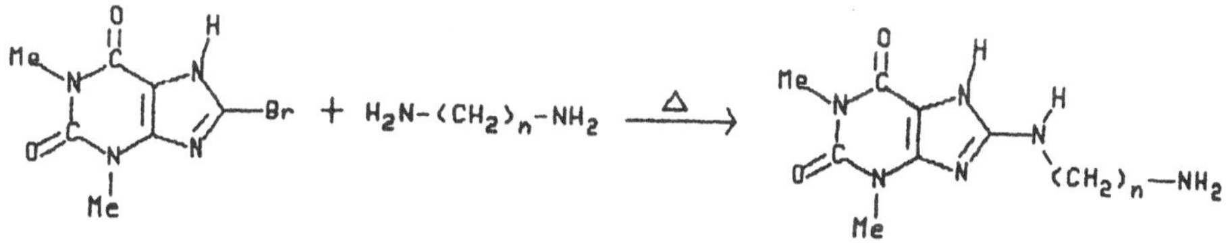

I (n=2-8)

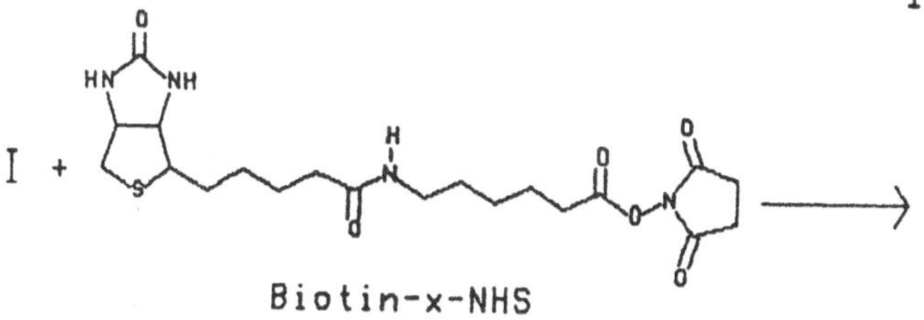

Biotin-x-NHS

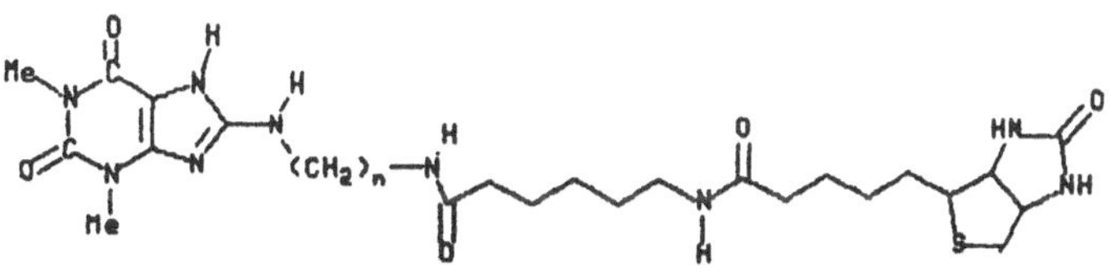

II (n=2-8)

Biotin-theophylline conjugates (II, above) were then prepared by means of the following procedure. The theophylline monoamine was dissolved in dry dimethylformamide (DMF) and then mixed with an equimolar quantity of the N-hydroxysuccinimide ester of caproamidobiotin (biotin-X-

NHS). The mixture was stirred at room temperature. The desired products usually separate out of DMF as white, flocculent solids. The product was collected on a filter paper and then was purified (using either a preparative TLC or a column chromatography method) to show only a single spot in a TLC test.

The products (II, above) displayed UV-absorption spectra identical to those of the starting amino-theophyllines. The compounds on TLC plates were shown to be biotin-positive with dimethylaminocinnamaldehyde spray (obtained from SIGMA, D-5388). The theophylline and biotin contents of the conjugates were determined by UV-absorption, NIIA, and hydroxyazobenzoic acid (Green, 1965) methods, and found to be in a one-to-one molar ratio.

Nephelometric studies on the bidentate conjugate

In order to evaluate the theophylline-biotin conjugates, the relative quantities of the conjugate, antibody and avidin should be in the proper stoichiometric proportions. Three-component optimization was carried out to provide the maximum rate units for each conjugate.

The various theophylline-biotin (T-B or B-T) conjugates synthesized and nephelometrically evaluated are listed in Table 1 for structures of the forms shown as II (above) and III (below). The approximate distance from the theophylline ring carbon to the alicyclic ring carbon in biotin is shown in the right hand column.

It is seen in Fig. 2 (see below) that the lowest member of the homolog (II, N=2), having only 16 atoms between the theophylline ring carbon and the alicyclic ring carbon of biotin, gave no precipitation. The next higher member (II, N=3) began to produce scatter response. The higher homologs produced higher and higher rate units until a plateau was reached for the bidentate conjugate (II, N=8). These increased rate and scatter signals can be attributed to the "apparent binding avidity" increase of theophylline to antibody and biotin to avidin as the steric hindrance becomes relieved through the use of a longer spacer. This "apparent avidity" difference must arise from "steric hindrance" between the two binding proteins, because the avidities of the antibody and avidin have not been changed.

Table 1. Spacing Between Theophylline and Biotin in a Series of Synthesized Bidentate Conjugates

Bidentate Conjugate		Number of Atoms in the Spacer	Approximate Spacing (A^O)
Form	Number of CH_2 Groups		
II	N=2	16	21.0
II	N=3	17	22.2
II	N=4	18	23.5
II	N=5	19	24.8
II	N=6	20	26.0
II	N=7	21	27.3
II	N=8	22	28.5
III	N=5	16	21.0
III	N=6	18	22.2

The data obtained for these compounds is shown in Fig. 2.

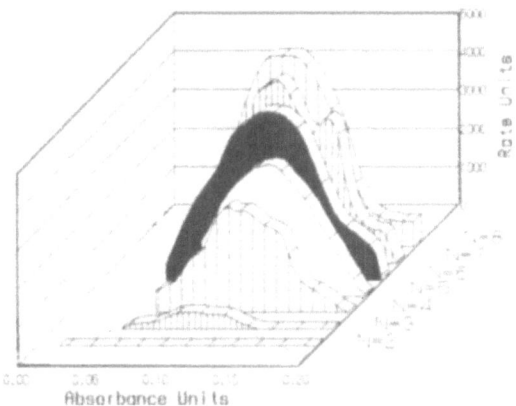

Fig. 2. Rate nephelometric responses for the reaction of avidin with
various bidentate theophylline-biotin (THE-BIOTIN) conjugates.
Individual curves are shown for each of the series of conjugates
from (II, N=2) to (II, N=8) (see Table 1). Each curve shows the
rate response for that conjugate as a function of the
concentration of conjugate (expressed in terms of its
absorbance), using monoclonal anti-theophylline ascites fluid
diluted 1/13.3 in ICS*Diluent. Avidin was present at 0.13 mg/ml
in ICS*Diluent. The units on the vertical axis are ICS rate
units, obtained using the Manual Mode card M33. For rate
signals exceeding 2000 units, a lower gain card was used, and
the results were calculated to correspond to the M33 gain
setting.

Green, Konieczny, Toms and Valentine, 1971, reported that bis-biotin
having a spacer arm 27.3 A^O long failed to form a linear polymer with
avidin, while higher homologs, with longer spacer arms, formed polymers.
It is interesting to note that our lowest member of the bidentate T-B
conjugate to give a rate signal has a spacer arm 22.2 A^O long. Although
the method utilized by Green and our rate nephelometric method are quite
different, these data suggest that the depth of the pocket
of theophilline binding site in the antibody is shallower than
that of avidin for biotin.

Another series of T-B conjugates (III, below) was prepared from
theophylline-8-butyric acid. This series differed from the first series
in having a methylene group adjacent to the theophylline moiety in place
of the amino group located there in the first series.

The first conjugate of the second series was prepared by coupling
biotin with theophylline butyric acid hexanediamine monoamide. The number
of atoms from the "ring carbon" of theophylline to the alicyclic ring
carbon of biotin was 17 atoms. This conjugate gave immunoprecipitation in
the presence of the antibody plus avidin. The usual "self-inhibition"
curve was obtained when the conjugate concentration was varied against the
fixed concentrations of antibody and avidin. The maximum rate obtained
was 1100 rate units on the M33 Manual Mode card. Conjugate (II, N=3), in
which the spacer group contains 17 atoms, gave immunoprecipitation with
antibody and avidin. The rate and scatter signals were comparable to
those obtained with conjugate (III, N=6), also containing 17 spacer atoms.

However, the bidendate conjugate (III, N=5) having 16 spacer atoms, failed to give any precipitation, while the bidendate conjugate (II, N=3), also with 16 spacer atoms gave no immuno-precipitation signal, as described above. Thus the minimum length of the spacer (between the theophylline ring carbon and the biotin alicyclic ring carbon) required to give precipitation is approximately 22.2 A^{O}.

III

Inhibition by theophylline and its derivatives

Inhibition by free theophylline. The inhibition by free theophylline of the rate nephelometric responses in the interactions of a series of bidentate theophylline-biotin conjugates with avidin and monoclonal anti-theophylline is shown in Fig. 3.

Inhibition by theophylline derivatives. Somewhat greater inhibition was observed when free theophylline was replaced by either of two of its derivatives, theophylline-8-aminobutylamine, and theophylline-8-butanoic acid, as shown in Fig. 4.

It appears, therefore, that the antibody recognized the amino and acid derivatives of theophylline better than it recognized free theophylline. Theophylline-8-butanoic acid had been used in the preparation of the immunogen for the monoclonal antibody production. Thus it should not be surprising to observe the better affinity exhibited by theophylline-8-butanoic acid. It is, however, more difficult to account for the superior inhibitory power of the amino-theophylline derivative. Attachment of nitrogen atom at the 8-position of theophylline molecule apparently increases the electron density of the theophylline ring, causing a bathochromic shift of max of theophylline from 270 nm to 295 nm. The electron pair of the nitrogen donates electrons to the ring. In spite of this electronic effect, however, the antibody recognized substitution of either carbon (8-butanoic acid) or nitrogen (8-aminobutylamino) at the 8-carbon of theophylline with approximately equal efficiency. It appears, therefore, that the anti-theophylline antibody must have recognized the shape of the pendant group along with theophylline structure.

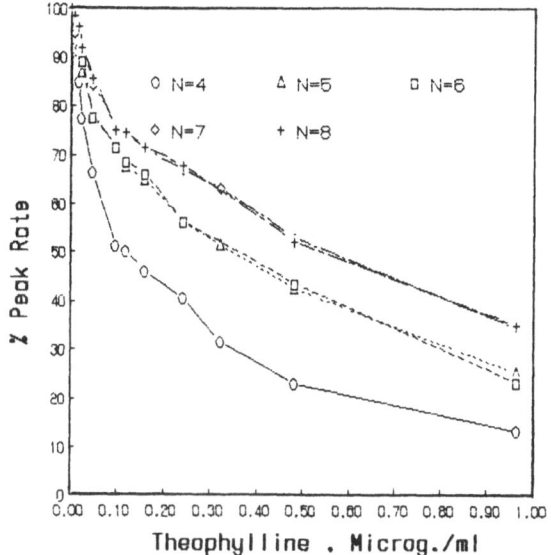

Fig. 3. Nephelometric inhibition responses as a function of free theophylline concentration in the reaction of avidin and monoclonal anti-theophylline with each of a series of theophylline-biotin conjugates having different spacer lengths. Monoclonal antibody was used at a dilution of 1:13.3 and avidin at 0.13 mg/mL. In each case, the conjugate concentration used was that giving the highest rate signal at that avidin concentration (see Fig. 2). Measurements were made in the manual mode on the ICS using an M33 card.

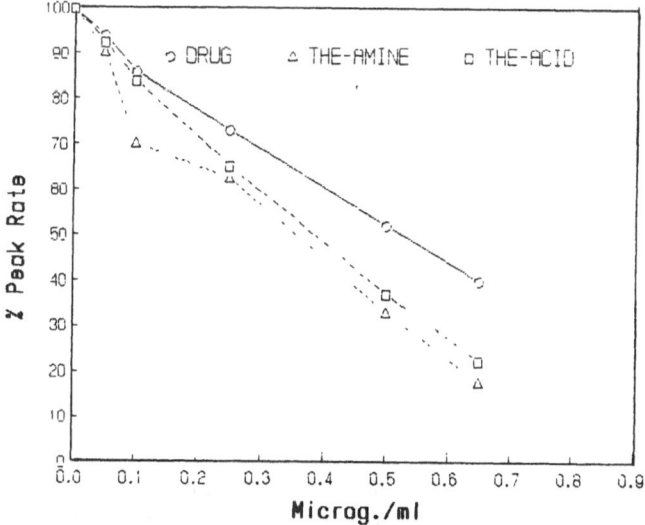

Fig. 4. Comparison of the inhibitions of the rate nephelometric interaction of conjugate (II, N=4) with avidin and monoclonal anti-theophylline by free theophylline, theophylline-8-butylamine (THE-AMINE), and theophylline-8-butanoic acid (THE-ACID). Conditions are same as in Fig. 3.

The effect of the order of reagent addition upon the nephelometric rate signal was also studied.   It was anticipated that the rate signal would be dependent upon the order of reagent addition.   Theophylline-biotin conjugate was first allowed to react with antibody; then the immunoprecipitation was triggered by addition of avidin. Because of the high affinity of avidin for biotin, an enhanced rate was anticipated. For comparison, the conjugate was allowed to react with avidin first, and then the immunoprecipitation was triggered by addition of the antibody.

```
                    ┌──────────┐          ┌──────────────┐
      T—B  +        Y            ───────→   B—T   T—B
                                              \Y/

                                               │ + Avidin
                                               ▼

              ⎛                                          ⎞
          A┤B—T   T—B—A┤B—T                               Y
              ⎝   \Y/                        \Y/          ⎠ n


                                         +Antibody         ▲
      T—B  +  A ───────→A (B—T)₄ ─────────────────────────│
                                                           ╪
                                                           ▼

          ⎛                T   T—B                         ⎞
          ⎜                 \Y/                            ⎟
          ⎜                  B                             ⎟
          ⎜                  │                             ⎟
         ─┤B—T   T—B—A—B—T   T—B├─
          ⎜   \Y/         │  \Y/                           ⎟
          ⎜               B                                ⎟
          ⎜               │                                ⎟
          ⎜               T                                ⎟
          ⎝                \Y/                             ⎠ m
```

The observed rate signals with these two different orders of addition were found to be almost identical, within experimental error.   Thus these rate signals are apparently not affected by the difference in the binding constant of avidin-biotin vs. antibody-theophylline.   This is probably because the forward rates of binding of avidin to the available biotin and antibody to the available theophylline are both fast, so that the primary and secondary immuno reaction steps are completed virtually immediately after mixing the reagents.   The PEG in the ICS buffer then more gradually coagulates the "oligomers" to give rise to the observed scatter increases and precipitation.

As we have seen, however, the scatter rate signal is dependent upon the length of the spacer chain between the theophylline and biotin in the conjugate. Conjugates with shorter spacer chain lengths produce smaller scatter signals. These shorter conjugates would be expected to cause greater steric hindrance between the avidin on one end and the antibody on the other, leading to less stable immuno bond formation and resulting in oligomers of smaller average molecular weight (see Fig. 2), which aggregate less readily.

It is likely that when a stoichiometric amount of avidin is brought into contact with B-T conjugate, tetra-substituted $Av(B-T)_4$ can be formed. Subsequent addition of the antibody to $Av(B-T)_4$, however, may not give rise to $Av((B-T)_2Ab)_4$, because of steric hindrance. According to Green, Konieczny, Toms and Valentine,(1971), two adjacent biotin binding sites of avidin are located on one face of the avidin molecule only about 15 A^o apart. The avidin face is 55 A^o x 55 A^o. The depth of the biotin binding site is about 23 A^o. The end-on view of an Fab fragment of antibody shows a surface area of about 35 A^o x 35 A^o (Valentine and Green, 1967), and the depth of an antibody binding site is estimated to be less than 15 A^o. It is easy to visualize that the above $Av(B-T)_4$ would be able to accommodate the binding of only two Fab groups.

The longer theophylline-biotin conjugates. Further studies were carried out on the conjugates (II, N=5) and (II, N=8). As discussed previously, the amount of the final triggering reagent added was varied to maximize the rate signal. At two antibody concentrations (1:12 and 1:18 dilutions), varying amounts of avidin were added to the antibody solution. Avidin concentrations were varied from 0.10 mg/ml to 0.26 mg/ml. The antibody-avidin mixture was introduced into the ICS vial, and then the immunoreaction was triggered by the addition of the T-B conjugate.

In Fig. 5A, the observed rate signal is shown as a function of the concentration of the conjugate (expressed in absorbance units). The rate is seen to increase as the avidin concentration increases and the step increments in the rate signal become smaller on further increase of avidin; a maximum of the rate signal is eventually reached at an avidin concentration of about 0.20 mg/ml. The concentration of the conjugate required to provide the maximum rate signal at each avidin concentration moves towards higher values as the avidin concentration is increased. This is the behaviour expected for a three-component precipitation reaction.

In Fig. 5B, the antibody titer has been increased by changing the dilution from 1:18 (in Fig. 5A) to 1:12. At the higher antibody titer, much higher rate signals are observed. The rate signal under these conditions was found to increase monotonically with the avidin concentration, and did not reach a plateau; the maximum observed value was obtained at the highest avidin concentration studied, 0.26 mg/ml.

Under the conditions of Figure 5A (antibody dilution of 1:18), antibody is in excess for avidin concentrations below 0.20 mg/ml, while, at higher avidin concentrations, the antibody is the limiting reagent, as can be seen from the increase and subsequent drop of the curve maxima. Under the conditions of Fig. 5B, with the higher antibody titer of 1:12, the antibody-limited region was not reached within the range of avidin concentrations used. For each avidin level, there is an optimum concentration of the bidentate which gives the maximum rate for that avidin level; the bidentate level for each avidin level was found to be independent of the antibody concentration. The stoichiometry for maximum rate signal can thus be seen to be determined by avidin and the conjugate, independent of the antibody concentration. For the maximum rate signal, then, the ratio between avidin and conjugate plays the predominant role, regardless of the antibody to conjugate ratio. This suggests that the rate signal is primarily controlled by the avidin-to-biotin ratio.

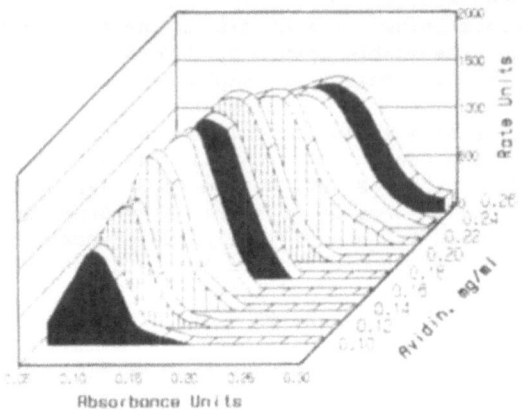

A

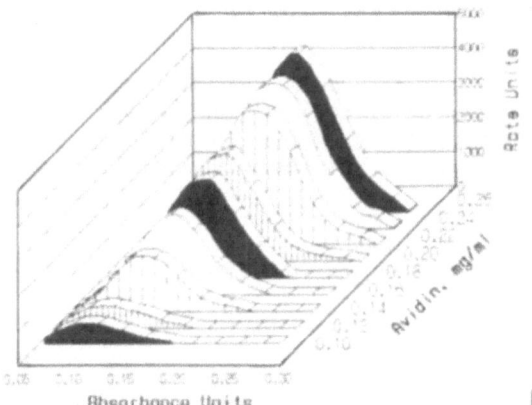

B

Fig. 5A. Rate nephelometric signals for the interaction of
avidin, monoclonal anti-theophylline, and the theophylline-biotin
conjugate (II, N=5) as functions of conjugate and avidin
concentrations, using an antibody dilution of 1:18.
Other conditions as in Fig. 3.

Fig. 5B. Rate nephelometric signals for the interaction of
avidin, monoclonal anti-theophylline, and the theophylline-biotin
conjugate (II, N=5) as functions of conjugate and avidin
concentrations, using an antibody dilution of 1:12.
Other conditions as in Fig. 3.

When the theophylline molecule was connected to biotin through a
longer spacer moiety, using conjugate (II, N=8), the data shown in Figures
6A and 6B were obtained. A comparison of these results with those shown
in Fig. 5 for T-B conjugate (II, N=5) shows that lower avidin and·conju-
gate concentrations give higher rate signals with the longer spacer than
with the shorter spacer. This enhanced rate signal can be attributed

470

solely to the reduction in steric hindrance resulting from the extension of the spacer chain length by three methylene groups.

Increasing the avidin concentration from 0.10 mg/ml to 0.12 mg/ml, further increased the rate signal, and the maximum rate signal moved to a higher conjugate concentration. With a further increase in the concentration of avidin to 0.14 mg/ml, however, the rate maximum split into two peaks. The two peaks became more pronounced at 0.16 mg/ml avidin, and continued to develop into a multi-peak pattern at higher avidin concentrations.

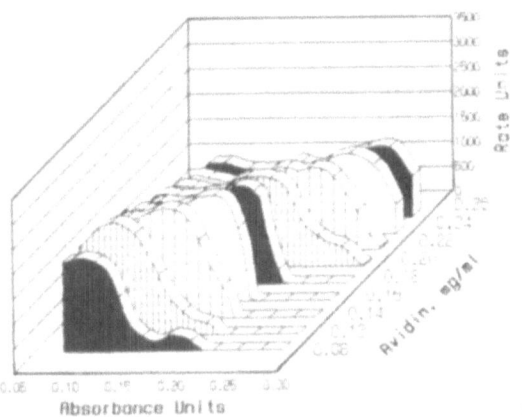

A

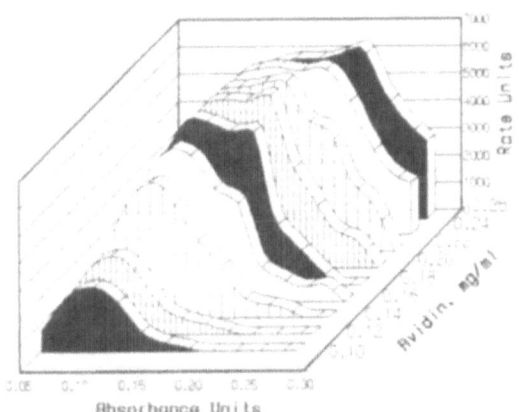

B

Fig. 6A. Rate nephelometric signals for the interaction of avidin, monoclonal anti-theophylline, and the theophylline-biotin conjugate (II, N=8) as functions of conjugate and avidin concentrations, using an antibody dilution of 1:18. Other conditions as in Fig. 3.

Fig. 6B. Rate nephelometric signals for the interaction of avidin, monoclonal anti-theophylline, and the theophylline-biotin conjugate (II, N=8) as functions of conjugate and avidin concentrations, using an antibody dilution of 1:12. Other conditions as in Fig. 3.

This may be explained by the fact that the longer spacer between the theophylline and biotin permits the conjugate to bend in such a way as to markedly diminish steric interference between two antibody molecules attaching to theophyllines coupled to biotins on the same end of the avidin molecule. This allows an effective valence change of avidin with respect to conjugate when antibody is bound. The first maximum is reached when the molar ratio of antibody to avidin to conjugate is 2:1:2; the second maximum is reached when the ratio is 2:3/2:3, and the third maximum is reached when the ratio is 2:2:4, etc.

At higher antibody concentration (Fig. 6B), the multipeak patterns coalesce into a large dome-shaped curve, for which the rate signals are relatively insensitive to the changes of concentration of avidin or conjugate. This complex agglutination behavior remains to be explained.

<u>The optimization of conditions for nephelometric inhibition.</u>
Using the bidentate T-B conjugate (II, N=5), various combinations of concentrations of the conjugate, avidin and antibody were selected for inhibition studies. The percent inhibition under three different sets of conditions is shown as a function of the concentration of theophylline-amine in Fig. 7.

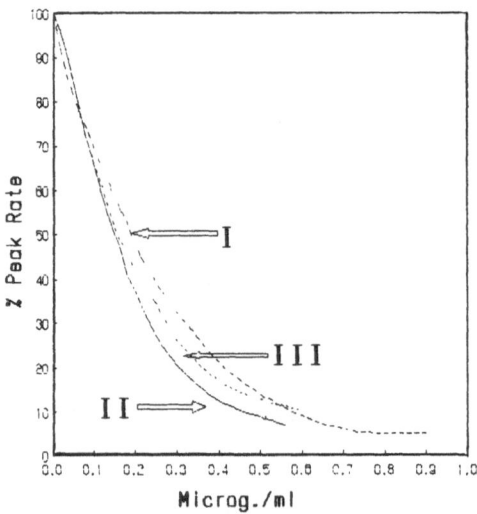

Fig. 7. Effects of selected operating ranges on the nephelometric
        inhibition by theophylline-amine for conjugate (II, N=5). Curve
        I: avidin, 0.12 mg/ml; absorbance of conjugate, 0.125. Curve II:
        avidin, 0.26 mg/ml; absorbance of conjugate, 0.204. Curve
        III: avidin, 0.20 mg/ml; absorbance of conjugate, 0.118.
        The antibody dilution was 1/18 in all three cases.

Curve I shows the inhibition behavior using a composition at a relatively low avidin concentration and slightly on the conjugate excess side occurring at the beginning of the descending part of the curve of rate signal vs. conjugate concentration (see Fig. 5A). Curve II shows the inhibition behavior obtained for a composition characterized by a relatively high avidin concentration, near the the beginning of the descending part of curve (Fig. 5A). Greater inhibition is obtained with the composition corresponding to curve II than with that corresponding to curve I. A third ternary composition, representing the ascending part of the curve obtained at an avidin concentration of 0.20 mg/ml in Fig. 5A, was observed to have an inhibition intermediate between those found in curve I and curve II.

472

The inhibition behavior obtained using the conjugate (II, N=8), which has a longer spacer between the theophylline and biotin, was also studied, using an avidin concentration of 0.20 mg/ml, because of the well-developed two-peak pattern in the dependence of rate signal on concentration of conjugate observed under these conditions (see Fig. 6A). Ternary compositions at the descending parts of the first and second peaks of the curve of Fig. 6A were selected for comparison. The results are shown in Fig. 8. It is seen that lower concentrations of bidendate conjugate exhibited greater inhibition by theophylline amine.

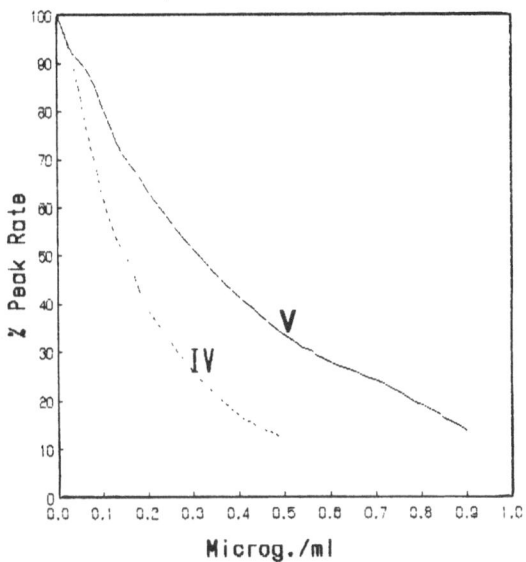

Fig. 8. Effects of selected operating ranges on the nephelometric inhibition by theophylline-amine for conjugate (II, N=8). Curve IV: avidin, 0.20 mg/ml; absorbance of conjugate, 0.118. Curve V: avidin, 0.20 mg/ml; absorbance of conjugate, 0.196. The antibody was at a dilution of 1/18 in both cases.

A comparison of curves IV and V in Fig. 8 with curves B and C in Fig. 9 shows that the concentration of the bidentate conjugate has a far more pronounced effect on inhibition by free theophylline than on inhibition by theophylline amine at a fixed avidin concentration. In both inhibition studies, the higher concentration of the bidentate conjugate led to poorer inhibition, since the free inhibitor was statistically less able to compete with the bidentate.

Each avidin molecule can bind four bidentate conjugate molecules to form the species avidin-$(B-T)_4$. When a short spacer conjugate is used, steric considerations permit only two antibodies (one at each end of the avidin molecule) can bind to the theophylline moieties in this species, forming a linear complex. With the longer spacer conjugates, however, each avidin-$(B-T)_4$ can bind four antibody molecules, and thus becames more analogous to a highly labeled apoferritin-theophylline "developer antigen," which exhibits poor "leverage" (i.e., a shallow dependence of inhibition on concentration) in conventional NIIA. This effect is especially pronounced when free theophylline is the inhibitor, because the free drug molecule competes even less favorably with the conjugate for antibody binding.

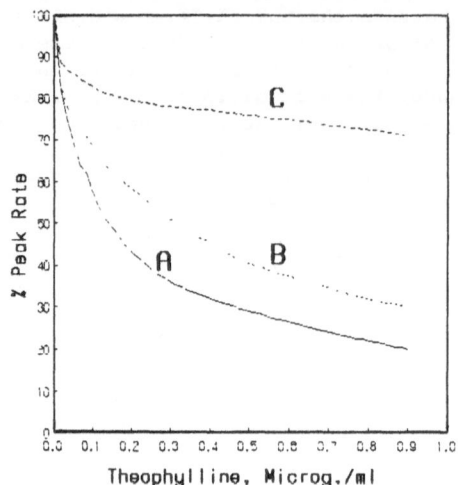

Fig. 9. Effects of selected operating ranges on the nephelometric
         inhibition by free theophylline for conjugate (II, N=8).
         Curve A: avidin, 0.20 mg/ml; absorbance of conjugate, 0.073.
         Curve B: avidin, 0.20 mg/ml; absorbance of conjugate, 0.118.
         Curve C: avidin, 0.20 mg/ml; absorbance of conjugate, 0.196.
         The antibody was at a dilution of 1/18 in both cases.

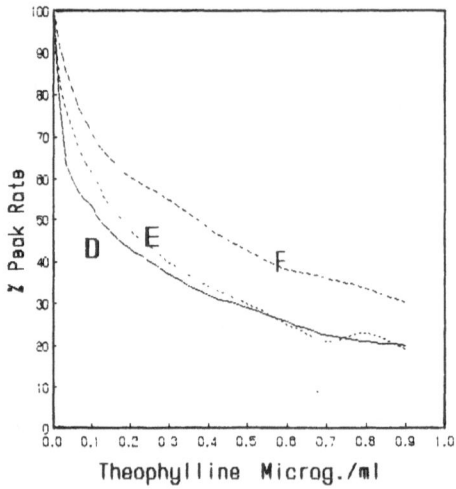

Fig. 10. Effects of selected operating ranges on the nephelometric
          inhibition by theophylline-amine for conjugate (II, N=8). Curve D:
          avidin, 0.26 mg/ml; absorbance of conjugate, 0.082. Curve E:
          avidin, 0.26 mg/ml; absorbance of conjugate, 0.118. Curve F:
          avidin, 0.26 mg/ml; absorbance of conjugate, 0.082. The antibody
          dilution was 1/18 for curves D and E, and 1/12 for curve F.

     In Fig. 10, a higher avidin concentration (0.26 mg/ml) was used.
Curve D represents an inhibition study using a low bidentate
concentration of 0.082 A. Increasing the bidentate concentration to
0.116 A did not noticeably affect the inhibition, as shown in curve E.
Although curve D shows better inhibition at low drug concentration, the

difference between curves D and E diminishes at higher drug
concentrations. An increase of the antibody titer from a 1/18 dilution
(curve D) to a 1/12 dilution (curve F) led to a markedly poorer
inhibition response. This effect of excess antibody has been observed
frequently in conventional NIIA.

SUMMARY

     Rate nephelometry provides a very rapid and convenient, and
moderately sensitive, method for the measurement of many proteins of
clinical and research interest. It has been extended, through the
application of nephelometric inhibition methods, to the measurement of
several haptens. The nephelometric inhibition methods for haptens, based
on the use of polyhaptenylated developer antigens, have met with only
limited success thus far. Critical parameters peculiar to nephelometric
inhibition include the relative affinities of the antibodies
for the hapten of interest and for the developer antigen, the tendency
for flexible antibody molecules to form non-bridging immunochemical bonds
to the developer antigen, and the ratio of concentrations of the antibody
to the developer antigen.
     The use of biotinylated hapten molecules in place of the
conventional developer antigen, and the special properties of avidin-
biotin binding, provide a means for developing a new class of
nephelometric inhibition methods. The conjoint binding of avidin to one
end, and antibody to the other end, of a bidentate biotin-hapten
conjugate requires a sufficient spacing between the bidentate components
to reduce steric hindrance between the two binding proteins.
     A spacing of 22 - 27 A° between the biotin and the hapten is
sufficiently long to permit the two proteins to bind simultaneously to
opposite ends of the bidentate conjugate, yet sufficiently short that
steric hindrance allows only one bidentate to bind to each end of the
avidin molecule. This range of bidentate spacings leads to the formation
of essentially linear repeating complexes of the form:

     ...(antibody)(bidentate conjugate)(avidin)(bidentate conjugate)....

With a smaller spacing, the avidin and antibody cannot simultaneously
bind to the same bidentate, while if the spacing is too great, each
avidin can bind three or four bidentate molecules, and a single antibody
molecule can have both of its binding sites tied to the hapten moieties
of bidentates on the same avidin molecule. This partial ring closure
diminishes the average size of the aggregates formed, results in a
decreased nephelometric signal, and ties up antibody molecules in a non-
functional manner.

     The linear configuration is ideal for inhibition assays, since in an
inhibition assay, the binding of a single hapten molecule to an antibody
binding site effectively terminates the chain. This leads to an improved
sensitivity and a greater economy of antibody than can be obtained with
conventional nephelometric inhibition methods using polyhaptenylated
developer antigens.
     The antibody titer to be used is determined by the concentration range
desired for the hapten. The maximum uninhibited signal obtained in the
absence of sample is then determined by the ratio of biotin-containing
bidentate conjugate to avidin. When the linear aggregates are formed, a
concentration of hapten equal to the molar concentration of antibody is
sufficient to cause complete inhibition of the nephelometric signal, if
the antibody has a sufficiently high binding constant. The inhibition
response obtained has been found to be 10- to 50-fold more sensitive than
has been obtained using polyhaptenylated developer antigens.

REFERENCES

Green, N. M., 1965. A Spectrophotometric Assay for Avidin and Biotin Based on Binding of Dyes by Avidin. Biochemistry, 94: 23C.

Green, N. M., Konieczny, L., Toms, E. J., and Valentine R.C., 1971, The Use of Bifunctional Biotinyl Compounds to Determine the Arrangement of Subunits in Avidin, Biochemistry, 125: 781-791.

Heidelberger, M. and Kendall, F. E., 1935. A Quantitive Study and a Theory of the Reaction Mechanism., J. Experimental Medicine, 61: 563-591.

Oi, V. T., Vuong, T. M., Hardy, R., Reidler, J., Dangl, J. Herzenberg, L. A. and Stryer, L., 1984. Correlation Between Segmental Flexibility and Effector Function of Antibodies, Nature, 307: 136-140.

Riccom, H., Masson, P. L., Vaerman, J. P. and Heremans, J. F., 1972. An Automated Nephelometric Inhibition Immunoassay (NIIA) for Haptens, 9-11, in J. D. Hamm (ed)., Automated Immunoprecipitin Reactions. Technicon Instruments, Terrytown, N.Y.

Sittampalam, G., and Wilson, G. S., 1984. Experimental Observations of Transient Light Scattering Complexes Formed During Immunoprecipitin Reactions, Anal. Chem.. 56: 2170-76.

Sternberg, J. C., 1977. A Rate Nephelometer for Measuring Specific Proteins by Immunoprecipitin Reactions. Clin. Chem. 23: 1456-64.

Sternberg, J. C., 1986. Rate Nephelometry, in "Manual of Clinical Laboratory Immunology" Rose, N. R. Freidman, H., and Fahey, J. L., eds., American Society for Microbiology, 33-39.

Tiffany, T. Q., Parella, J. M., Johnson, 1974, W. F. and Burtis, C. A. Specific Protein Analysis by Light Scatter Measurement with Miniature Centrifugal Fast Analyzer. Clin. Chem., 20: 1055.

Valentine, R. C., and Green, N. M., 1967. Electron Microscopy of an Antibody-Hapten Complex. J. of Molec. Biol. 27 615.

CONTRIBUTORS

AMIR-ZALTSMAN, Y.
    Departments of Hormone Research, and of Biophysics, The
    Weizmann Institute of Science, Rehovot, 76100, Israel
ARWIN, HANS
    Laboratory of Applied Physics, Department of Physics
    and Measurement Technology, Linkoping Institute of
    Technology, S-581 83 Linkoping, Sweden
AUSHER, Y.
    Department of Hormone Research, Weizmann Institute of
    Science, Rehovot, Israel
BAILEY, M.P.
    Biochemistry Department, Royal Sussex County Hospital,
    Eastern Road, Brighton BN2 5BE, Sussex UK
BARNARD, G.J.R.
    Dept. of Obs./Gyn., King's Coll. School of Medicine
    and Dent., London U.K.
BAYER, E.A.
    Department of Biophysics, The Weizmann Institute of
    Science, Rehovot, 76100, Israel
BERKE, CARL M.
    Hygeia Sciences, 330 Nevada Street, Newton, MA 02160
BLUESTEIN, BARRY
    Biomedical Research Laboratories, Ciba Corning
    Diagnostics Corp., Cambridge, MA, USA
CAIS, MICHAEL
    Department of Chemistry, Technion - Israel Institute of
    Technology, Haifa 32000, Israel
CALDINI, ANNA L.
    Endocrinology Unit, Department of Clinical Physio-
    pathology, University of Florence, Italy
DAHNE, CLAUS
    Biomedical Group, Battelle Institute, Geneva,
    Switzerland
DE BOEYER, J.
    Akademisch Ziekenhuis, Vrouwenkliniek/Poli III,
    Gent, Belgium
FRICKE, HARALD
    Klinische Laboratorien, Klinik fur Innere Medizin,
    Medizinische Universitat zu Lubeck, D-2400 Lubeck,
    Federal Republic of Germany
GAYER, B.
    Departments of Hormone Research, and of Biophysics, The
    Weizmann Institute of Science, Rehovot, 76100, Israel
GILAD, S.
    Dept. of Hormone Research, Weizmann Institute of
    Science, Rehovot, Israel

GOULD, B.J.
    Department of Biochemistry, University of Surrey,
    Guildford, Surrey GU2 5XH, U.K.
GUESDON, JEAN-LUC
    Laboratoire des Sondes Froides, Institut Pasteur,
    Paris, France
HASHIDA, SEIICHI
    Department of Biochemistry, Medical College of Miyazaki
    Kiyotake, Miyazaki 889-16, Japan
HORSEWOOD, PETER
    Department of Pathology, McMaster University, Hamilton,
    Ont.
ISHIKAWA, EIJI
    Department of Biochemistry, Medical College of Miyazaki
    Kiyotake, Miyazaki 889-16, Japan
ISHIMORI, YOSHIO
    Laboratory of Biological Products, The Institute of
    Medical Science, The University of Tokyo, Minato-Ku,
    Tokyo 108
KHANNA, PYARE L.
    Syva, 900 Arastradero, Palo Alto, California 94303
KIM, J.B.
    College of Animal Husbandary, Kon-Kish University,
    Seoul, Korea
KOHEN, F.
    Department of Hormone Research, The Weizmann Institute
    of Science, Rehovot, 76100, Israel
KOHNO, TAKEYUKI
    Department of Biochemistry, Medical College of Miyazaki
    Kiyotake, Miyazaki 889-16, Japan
KRONICK, MEL N.
    Applied Biosystems, 850 Lincoln Centre Drive, Foster
    City, CA 94404 (U.S.A.)
LITCHFIELD, W.J.
    E.I. du Pont de Nemours & Company, Inc., Biomedical
    Products Department, Glasgow Research Laboratory,
    Wilmington, DE 19898
LOVGREN, TIMO
    Wallac Oy, P.O. Box 10, SF-20101 Turku, Finland
LUNDSTROM, INGEMAR
    Laboratory of Applied Physics, Department of Physics
    and Measurement Technology, Linkoping Institute of
    Technology, S-581 83 Linkoping, Sweden
MANSELL, RICHARD L.
    Biology Department, University of South Florida, Tampa,
    Florida 33620
MARKS, VINCENT
    Department of Biochemistry, University of Surrey,
    Guildford, Surrey GU2 5XH, U.K.
MATTIASSON, BO
    Department of Biotechnology, Chemical Center, P.O.Box
    124, S-221 00 Lund, Sweden
McGOWN, LINDA B.
    Department of Chemistry, Oklahoma State University,
    Stillwater, Oklahoma
McINTOSH,CECILIA A.
    Biology Department, University of South Florida, Tampa,
    Florida 33620
MESSERI, GIANNI
    Endrocrinology Unit, Department of Clinical Physio-
    pathology, University of Florence, Italy

NGO, THAT T.
    Department of Developmental and Cell Biology,
    University of California, Irvine, CA   92717
NIELSEN, K.H.
    Agriculture Canada, Animal Diseases Research Institute,
    Nepean, P.O. Box 11300, Station H, Nepean, Ontario,
    Canada K2H 8p9
OH, C.S.
    Applied Research Department, Diagnostic System group,
    Bechman Instruments Inc, Brea, California
ORLANDO, CLAUDIO
    Endrocrinology Unit, Department of Clinical Physio-
    pathology, University of Florence, Italy
PAZZAGLI, MARIO
    Endocrinology Unit, Department of Clinical Physio-
    pathology, University of Florece, Italy
PORSTMANN, BARBEL
    Institute of Pathological and Clinical Biochemistry and
    Institute of Medical Immunology, Faculty of Medicine
    (Charite), Humboldt University, Berlin, GDR
PORSTMANN, TOMAS
    Institute of Pathological and Clinical Biochemistry and
    Institute of Medical Immunology, Faculty of Medicine
    (Charite), Humboldt University, Berlin, GDR
RILEY, C.
    Biochemistry Department, Royal Sussex County Hospital,
    Eastern Road, Brighton BN2 5BE, Sussex UK
ROCKS, B.F.
    Biochemistry Department, Royal Sussex County Hospital,
    Eastern Road, Brighton BN2 5BE, Sussex UK
·SLOVACEK, RUDOLF
    Biomedical Research Laboratories, Ciba Corning
    Diagnostics Corp. Cambridge, M.A., U.S.A.
SOINI, ERKKI
    Wallac Oy, P.O. Box 10, SF-20101 Turku, Finland
STERNBERG, J.C.
    Applied Research Department, Diagnostic Symtem Group
    Bechman Instruments, Inc, Brea, California
STRASBURGER, CHRISTIAN J.
    Klinische Laboratorien, Klinik fur Innere Medizin,
    Medizinische Universitat zu Lubeck, D-2400 Lubeck,
    Federal Republic of Germany
SUTHERLAND, RANALD
    Biomedical Group, Battelle Institute, Geneva,
    Switzerland
TANAKA, KOICHIRO
    Department of Biochemistry, Medical College of Miyazaki
    Kiyotake, Miyazaki 889-16, Japan
UMEDA, MAMORU
    Laboratory of Biological Products, The Institute of
    Medical Science, The University of Tokyo, Minato-ku,
    Tokyo 108
VISTNES, A.I.
    Department of Physics, University of Oslo, Blindern,
    0316 Oslo 3, Norway
WELIN, STEFAN
    Laboratory of Applied Physics, Department of Physics
    and Measurement Technology, Linkoping Institute of
    Technology, S-581 83 Linkoping, Sweden

WILCHEK, M.
        Department of Biophysics, The Weizmann Institute of
        Science, Rehovot, 76100, Israel
WOOD, WILLIAM G.
        Klinische Laboratories, Klinik fur Innere Medizin,
        Medizinische Universitat zu Lubeck, D-2400 Lubeck,
        Federal Republic of Germany
WRIGHT, P.F.
        Agriculture Canada, Animal Diseases Research Institute,
        Nepean, P.O. Box 11300, Station H, Nepean, Ontario,
        Canada K2H 8p9
YASUDA, TATSUJI
        Laboratory of Biological Products, The Institute of
        Medical Science, The University of Tokyo, Minato-ku,
        Tokyo 108, Japan

Lipoprotein, 375
Liposome, 16, 361, 363, 366, 368, 377, 382, 389, 391, 394, 396, 397
  antibody-bearing, 392
  antigen-bearing, 391
  fluorescent dye entrapped, 389
  multilamellar, 361, 368, 392
  preparation, 391
  sensitized, 361
  small unilammellar, 368
Liposome immune lysis assay, 389
Lipoxygenease, 151
Long-lifetime label, 188
  pyrene, 188
  rare-earch chelate, 189
  erythrosine, 189

LPS, 133, (see also lipopoly-saccharide)
Luciferase, 33
Luminescence immunoassay, 257
  kinase label, 257
  aryl hydrazide label, 257
Luminol, 261, 293
  detection limit, 293
Lymphadenopathy, 173
Lymphocyte, 173
Lysozyme, 17-19

Maize prolamin, 155
Malate dehydrogenase, 17, 19
  NADP, 150
Malate dehydrogenase, 21, 81
Maleimide, 27, 33, 35, 43
  measurement, 37
  stability, 35
Maleimide-alkaline phosphatase, 44
Maleimide-B-D-galactosidase, 45
Maleimide-peroxidase, 43, 45, 46
m-Maleimidobenzoyl-N-hydroxy-succinimide, 36, 260
Manganese-labeled barbiturate, 417, 420
Manganese-labeled estrogen, 417, 420
Marker, 372
  ATP, 372
  dichromate, 372
  enzyme, 372
  fluorescent dye, 372
  glucose, 372
  ion, 372
  Rb+, 372
  spin label, 372
Matrix effect, 6
Matrix material, 304
MBS, 260, (see also m-

maleimidobenzoyl-N-hydroxysuccinimide ester)
MBTH, 61, 63, 71, 73, (see also 3-methylbenzothia-zolinone hydrazone)
MCS, 260, (see also succinimidyl-4-(N-maleimidomethyl)-cyclohexane-1-carboxylate)
Meldola blue, 78, 79, 81
Membrane immunoassay, 361, 362, 363, 364, 368, 370, 372, 374, 376, 384
Meningitides, 363, 364, 370, 371, 375, 381
  antibody, 363
  meningococci, 364
  polysaccharide, 363, 371, 375
Mercaptosuccinylated avidin, 45
Mercaptosuccinylated IgG, 43, 44
Metal-labeled phenobarbitone, 425, 426
Metal-labeling, 416, 417, 422, 423
Metalloantibody, 417
Metalloantigen, 416
Metallohapten, 416, 425, 426, 427
  phenobarbitone, 425, 426
Metalloimmunoassay, 415, 416, 434
  amperometric, 434
Methotrexate, 15
o-Methoxyphenol, 70, (see also guaiacol)
3-Methyl-2-benzothiazolinone hydrazone, 61, 63, 71
Methyl-D-mannoside, 89
4-Methylumbelliferone, 32
4-Methylumbelliferyl-B-D-galactoside, 32, 99
4-Methylumbelliferylphosphate, 32, 89, 92
MIA, 361, 367, 369, 370, 371, 372, 373, 375, 376, 377, 378, 381, 382, 383, 416, (see also membrane immunoassay an metallo-immunoassay)
Micelles, 250, 251
Microperoxidase, 258, 263, 272, 279, 287, 288, 289
MLL, 368, (see also multi-lammelar liposome and liposome)
MMT, 80, (see also thiazolyl blue)
Monoclonal antibody, 7, 9, 88,

Wave guide, 334, 339
  fibre optics, 339, 354

Zein, 150, 155
Zymogen, 16
  activation, 16